W9-CSP-377

ACS SYMPOSIUM SERIES 315

Evaluation of Pesticides in Ground Water

Willa Y. Garner, EDITOR
U.S. Environmental Protection Agency

Richard C. Honeycutt, EDITOR
CIBA-GEIGY Corporation

Herbert N. Nigg, EDITOR
Citrus Research and Education Center
University of Florida

Developed from a symposium sponsored by
the Division of Pesticide Chemistry
at the 189th Meeting
of the American Chemical Society,
Miami Beach, Florida,
April 28–May 3, 1985

American Chemical Society, Washington, DC 1986

53305826

SEP/AE CHEM

Library of Congress Cataloging-in-Publication Data

Evaluation of pesticides in ground water.
(ACS symposium series; 315)

Includes indexes.

1. Water, Underground—Pollution—Congresses. 2. Agricultural pollution—Environmental aspects—Congresses. 3. Pesticides—Environmental aspects—Congresses.

I. Garner, Willa Y., 1936- . II. Honeycutt, Richard C., 1945- . III. Nigg, Herbert N., 1941- . IV. American Chemical Society. Division of Pesticide Chemistry. V. American Chemical Society. Meeting (189th: 1985: Miami Beach, Fla.) VI. Series.

TD426.E96 1986 628.1'6842 86-14153
ISBN 0-8412-0979-0

Copyright © 1986

American Chemical Society

All Rights Reserved. The appearance of the code at the bottom of the first page of each chapter in this volume indicates the copyright owner's consent that reprographic copies of the chapter may be made for personal or internal use or for the personal or internal use of specific clients. This consent is given on the condition, however, that the copier pay the stated per copy fee through the Copyright Clearance Center, Inc., 27 Congress Street, Salem, MA 01970, for copying beyond that permitted by Sections 107 or 108 of the U.S. Copyright Law. This consent does not extend to copying or transmission by any means—graphic or electronic—for any other purpose, such as for general distribution, for advertising or promotional purposes, for creating a new collective work, for resale, or for information storage and retrieval systems. The copying fee for each chapter is indicated in the code at the bottom of the first page of the chapter.

The citation of trade names and/or names of manufacturers in this publication is not to be construed as an endorsement or as approval by ACS of the commercial products or services referenced herein; nor should the mere reference herein to any drawing, specification, chemical process, or other data be regarded as a license or as a conveyance of any right or permission, to the holder, reader, or any other person or corporation, to manufacture, reproduce, use, or sell any patented invention or copyrighted work that may in any way be related thereto. Registered names, trademarks, etc., used in this publication, even without specific indication thereof, are not to be considered unprotected by law.

PRINTED IN THE UNITED STATES OF AMERICA

S.P. 4/10/86

TD426
E961
1986
CHEM

ACS Symposium Series

M. Joan Comstock, *Series Editor*

Advisory Board

Harvey W. Blanch
University of California—Berkeley

Alan Elzerman
Clemson University

John W. Finley
Nabisco Brands, Inc.

Marye Anne Fox
The University of Texas—Austin

Martin L. Gorbaty
Exxon Research and Engineering Co.

Roland F. Hirsch
U.S. Department of Energy

Rudolph J. Marcus
Consultant, Computers & Chemistry Research

Vincent D. McGinniss
Battelle Columbus Laboratories

Donald E. Moreland
USDA, Agricultural Research Service

W. H. Norton
J. T. Baker Chemical Company

James C. Randall
Exxon Chemical Company

W. D. Shults
Oak Ridge National Laboratory

Geoffrey K. Smith
Rohm & Haas Co.

Charles S. Tuesday
General Motors Research Laboratory

Douglas B. Walters
National Institute of Environmental Health

C. Grant Willson
IBM Research Department

FOREWORD

The ACS SYMPOSIUM SERIES was founded in 1974 to provide a medium for publishing symposia quickly in book form. The format of the Series parallels that of the continuing ADVANCES IN CHEMISTRY SERIES except that, in order to save time, the papers are not typeset but are reproduced as they are submitted by the authors in camera-ready form. Papers are reviewed under the supervision of the Editors with the assistance of the Series Advisory Board and are selected to maintain the integrity of the symposia; however, verbatim reproductions of previously published papers are not accepted. Both reviews and reports of research are acceptable, because symposia may embrace both types of presentation.

CONTENTS

REGULATORY ASPECTS

PREFACE

THE INTENSE PUBLIC INTEREST IN CHEMICAL CONTAMINATION of our nation's ground water has been well documented by the media. Pesticide contamination of ground water through agricultural use is a unique situation in comparison to contamination by point source leaching from dumps, landfills, and spills. Consequently, data bases for agricultural pesticide contamination are small, and published research and textbooks are limited.

This volume explores all aspects of the factors that impinge on ground water contamination from agriculturally applied pesticides. Although only pesticide contamination is dealt with in this volume, the principles examined apply to all xenobiotics and natural substances that can reach our ground waters.

As organizers of the symposium and editors of this volume, we thank the contributors, whose generous time and combined expertise have made this book a valuable reference for those working in the ground water area. We also thank our symposium session chairpersons, Jeff Wagenet (Cornell University), Sam Creeger (U.S. Environmental Protection Agency), and Chris Wilkinson (Cornell University), for their special contributions that made the symposium from which this book was developed a success.

WILLA Y. GARNER
U.S. Environmental Protection Agency
Washington, DC 20460

RICHARD C. HONEYCUTT
CIBA-GEIGY Corporation
Greensboro, NC 27409

HERBERT N. NIGG
Citrus Research and Education Center
700 Exp. Station Road
Lake Alfred, FL 33850

October 1985

PHYSICAL AND CHEMICAL PARAMETERS

1

Processes and Factors Affecting Transport of Pesticides to Ground Water

H. H. Cheng[1] and W. C. Koskinen[2]

[1]Department of Agronomy and Soils, Washington State University, Pullman, WA 99164
[2]Southern Weed Science Laboratory, Agricultural Research Service, U.S. Department of Agriculture, Stoneville, MS 38776

The process of transporting pesticides down through the soil horizons and the vadose zone into the ground water is affected by a number of other processes taking place in the soil profile. The sorption process can retard or retain the chemicals from moving with leaching water. The commonly used pesticides, mostly being organic chemicals, can be degraded partially or completely to inorganic end products by chemical, photochemical, or biochemical means. The degradation process reduces or eliminates the presence of pesticides in the environment. Pesticides can also be transported to the atmosphere by volatilization or to surface water by runoff from soil, or be removed from soil by plant uptake. Whether circumstances for transport of pesticides to groundwater exist will depend upon a combination of factors including the nature of the pesticide chemical, the properties and conditions of the soil, and climatic and environmental variables. Realistic assessment of the potential for transport of pesticides to groundwater must include simultaneous evaluation of all the processes and factors that may impact the transport process.

The topic of this symposium is both timely and of special concern to many sectors of the public as it combines two contrasting subject matters. Groundwater, being an essential natural resource, has been assumed to be pristine in quality and must not be violated by contamination. After all, 86% of the water in the United States is stored in aquifers; over 50% of the U. S. population and 95% of rural U. S. use groundwater as drinking water. On the other hand, any mention of pesticides often connotes substances that are toxic, hazardous, and even life-threatening. A tremendous amount of pesticides is produced and used in the U. S. each year. In 1984, 1.1 billion pounds was synthesized; even this amount was down from the 1.5 billion pounds produced in 1975 (1).

0097-6156/86/0315-0002$06.00/0
© 1986 American Chemical Society

Some 68% of the pesticides was used for agricultural purposes, 17% by industries and commercial concerns, 8% by homes and gardens, and 7% by governmental uses. Over 60% of the pesticides produced are herbicides, the others being mostly insecticides and fungicides. There is increasing evidence for the presence of a variety of pesticides in groundwater at certain locations, such as in shallow, unprotected aquifers or in karst formations. A report by the U. S. Environmental Protection Agency (EPA) in 1984 (2) listed some 12 pesticides that have been detected in drinking water wells in Florida, New York, California, Hawaii, and other locations. Recently, the EPA announced its intention to include 60 pesticides in its drinking water survey (3), more than half of which are herbicides. The seriousness of the potential contamination of groundwater by pesticides certainly cannot be minimized.

Although the concerns by EPA for groundwater quality are to be lauded, the planned program on groundwater monitoring, particularly the drinking water wells survey, is of limited value. The monitoring program could at best alert the people of existing hazard. It would not be useful in assessing the magnitude of the potential danger nor in deducing the sources of contamination. An effective management program must not only include a groundwater monitoring effort but also support a planned program of research that can isolate and identify the potential causes of contamination by pesticides and devise ways to minimize or eliminate the sources of contamination. The heart of such a program must be the search for a basic understanding of the processes and factors that affect the transport of pesticides to groundwater.

The Setting

It is not always obvious that groundwater in most cases is a renewable resource. Few of the aquifers that supply drinking water are of ancient origin. Most of them are dynamic in nature, and are recharged repeatedly by water moving slowly or rapidly downward from the surface of the earth through the soil profile to a level that accumulates the water. In the process, dissolved chemicals including pesticides can also be carried downward into the soil profile and eventually reach the groundwater. Thus, the concern for contamination of groundwater is legitimate and justified. A number of papers of this symposium are dealing with the specific process of transport of pesticides to the groundwater. The objective of this paper is to set the stage for this symposium by relating the transport process to other processes occurring simultaneously in soil that affect the transport process. We will attempt to view the transport process in perspective of the fate of pesticides in the total environment, and to share some cautions needed in assessing these processes.

The natural environment can be viewed as consisting of several environmental zones (Figure 1), extending from the atmosphere and the above-ground crop zone into the soil from the surface down, through the root zone, the unsaturated soil or vadose zone below, eventually down to the saturated zone where ground water is situated. For a pesticide to contaminate groundwater, the chemical must first reach the soil, either directly by application or indirectly such as by drift. Foliar-applied chemicals are subject to photodecomposition, plant absorption, dripping to soil, or falling on soil when plants

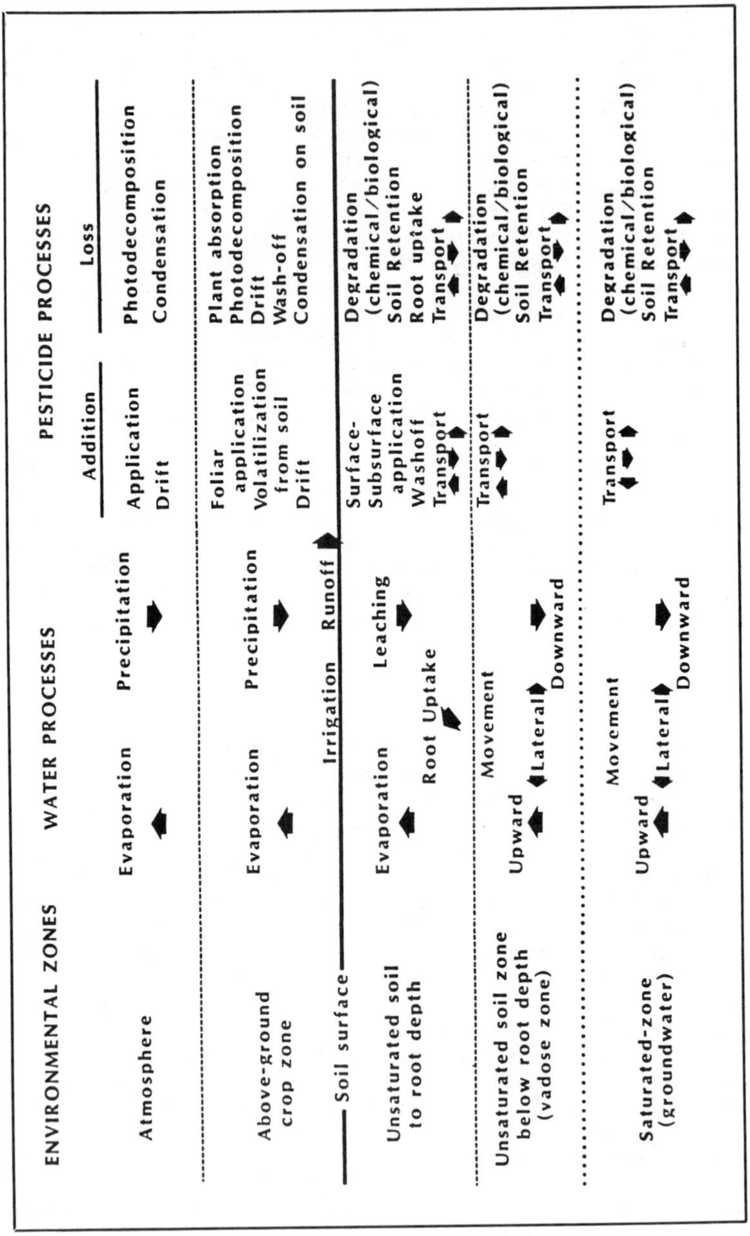

Figure 1. Processes affecting water movement and pesticide transport in various environmental zones.

die. Once in the soil, the chemical must be transported, usually by water, through the various environmental zones down to the groundwater. Factors that affect water infiltration and movement within a zone or from one zone to another will also affect the location of the chemical. In addition to moving downward, water can also be transported upward to the soil surface and evaporated into the atmosphere, taken up by plants and transpired, or discharged over soil surface into surface water bodies. Pesticides can be transported by water in similar manners. Before these transport processes can be properly characterized at any particular setting, all the processes and factors that impact pesticide transport must be characterized.

Two major types of processes that can affect the amount of pesticides present and available for transport through the soil profile are retention and transformation. The retention processes do not affect the total amount of pesticide present in the soil but can decrease or eliminate the amount available for transport. On the other hand, the transformation processes actually reduce or totally eliminate the amount of pesticide present and available for transport.

Retention Processes

The literature abounds with references on the retention of pesticides in soils (e.g., 4,5). The term 'retention' is used here in an all encompassing sense, but it is most frequently equated with adsorption or simply sorption. In a strict sense, adsorption is a reversible process involving nonspecific attraction of a chemical to the soil particle surface and retention of the chemical on the surface for a longer or shorter period of time depending on the affinity of the chemical to the surface. However, whether a chemical is actually sorbed to a particular surface is often not confirmed by the technique used to characterize retention. Few of the techniques generally used for retention characterization can differentiate the mechanisms involved in attracting the chemical to the soil surface. For instance, the commonly used batch equilibration method merely determines the decrease or disappearance of a chemical from solution when soil is added to the solution, under the assumption that what does not remain in the solution would be adsorbed. This method does not provide any information on the mechanism of adsorption, or the strength of adsorption, or whether the reduction in solution concentration was related to adsorption at all. Studies have shown that this method could lead to erroneous estimation of adsorption if precaution was not taken in eliminating or accounting for degradation of the chemical during the adsorption-desorption processes (6,7).

A number of quick-test techniques have been used widely to estimate the extent of sorption of pesticides to soils, and these estimates are often used in pesticide transport models. The most commonly used technique is to determine the ratio of distribution of a chemical, often at one concentration, between the solution and soil solid phases (K_d) or simply the distribution between water and octanol phases (K_{ow}). The use of K_d as an index of adsorption assumes that the distribution ratio is constant over a range of concentrations of the chemical in the soil. In other words, the amount of chemical adsorbed increases linearly with that remaining in the solution. The linear relationship may be valid over a narrow range of concentrations. However, a plotting of K_d values taken from the literature vs. the equilibrium concentration of pesticide in solution readily demonstrates the disparity of the K_d values (8). Most literature values indicate that pesticide adsorption data obtained by the batch equilibration method can be better fitted to a curvilinear equation, such as the Freundlich equation. The wide disparities in the Freundlich adsorption constants (K_f) reported in the literature should be critically evaluated, as the true significance of the data obtained by different studies can be a consequence of the parameters used in the testing procedure. Koskinen and Cheng (9) have shown that variations in experimental parameters could result in variations of up to several fold. As a result, mobility of the pesticide 2,4,5-T (2,4,5-trichlorophenoxyacetic acid) in the Palouse silt loam soil, for instance, could be classified as low or high depending on the experimental parameters used.

Many studies have shown that pesticide adsorption can be correlated to the soil organic matter contents, but not with soil mineral or clay contents. Thus, adsorption constants have often been expressed in terms of the soil organic matter or organic carbon contents (K_{oc}) (4). Such correlations have often been shown to be statistically significant (8). Significant correlations between sorption (K_{oc}) and estimates of sorption using water octanol partition (K_{ow}) (10), reverse-phase high-performance liquid chromatographic retention (R_t) (11), and water solubility (S) of the chemical (12) have led to wide use of these parameters to estimate K_{oc}. These estimates may be adequate as a first approximation, especially for hydrophobic, nonionic chemicals. However, a rough correlation between soil sorption of a chemical and its K_{ow}, R_t, or S does not automatically impart any theoretical meaning to the empirical measurement. Such factors as soil solution composition and temperature can affect the amount of pesticides adsorbed but cannot be accounted for by these indirect methods. Thus, any measurement which does not take into consideration the role of soil or the environmental variables in the adsorption process can at best be simply an estimation.

Certain cautions should also be mentioned so that the meaning of these correlations is not extended beyond what the data warrant. A significant correlation between sorption and soil organic carbon contents does not imply that only one mechanism of sorption is involved or that all pesticides interact with all components of soil organic matter by the same mechanism. Furthermore, the lack of correlation between pesticide sorption and soil mineral or clay contents should not be taken to mean that adsorption on soil minerals is not important. Adsorption of organic chemicals on soil mineral surfaces is a well-established fact (13,14). The lack of correlation may only imply that soil mineral content by weight is not a good index for the extent of mineral surface available for adsorption. The significance of adsorption on mineral surfaces should not be ignored, especially in materials low in organic matter, such as the geological materials in the vadose zone underneath the surface soil.

The complexity of the adsorption process should be understood by those interested in assessing the impact of sorption on the transport of chemicals to the groundwater. Mechanisms or forces involved in adsorption can range from van der Waals-London forces, hydrogen bonding, ligand and ion exchange, charge transfer, ion-dipole and dipole-dipole forces, hydrophobic bonding, to chemisorption (5). Any simplistic representation of adsorption such as depicting the process as partitioning into an organic phase is to ignore the true nature of the processes involved as pointed out by Minglegrin and Gerstl (15). Even characterization of sorption based solely on solute and solvent factors may not be adequate. The composition of the aqueous phase can affect the pesticide-soil-solution equilibrium of certain pesticides. For example, soluble organic matter and metal ions in soil solution can complex or bind small organic molecules and stabilize them in the aqueous phase (16). Koskinen and Cheng (9) found that leaching soil decreased the amount of soluble organics in soil solution and increased adsorption of 2,4,5-T in the soil solid phase.

Currently a number of laboratories are devoting considerable effort in developing methodology to characterize the various bonds involved in adsorption, many of which are not readily reversible by simple equilibration with water (e.g., 17). Thus retention cannot be viewed as processes that merely retard transport of pesticides in water, but can bind the chemicals irreversibly to soil surfaces (the so-called 'bound residue') and totally remove the chemicals from transport. Such process as chemisorption may not be easily distinguished with chemical transformation of the compound in the soil media (18).

Transformation Processes

The term 'transformation' is used here to encompass all changes in the chemical structure or composition of the pesticide compound. The chemical structure may be modified by such reactions as oxidation, reduction, hydrolysis, substitution and removal of functional groups, complexation with metal ions, polymerization, and others. The structure may also be broken down into fragments of the original molecule and eventually into inorganic endproducts, such as H_2O, CO_2, halide, ammonium, phosphate, and other salts. The term degradation should only be associated with the breakdown process, although it is commonly used to describe other transformation processes. One should be aware that while most transformation processes modify the structure to detoxify the chemical, other reactions may lead to more toxic products. A notable example is the formation of 3,3'4,4'-tetrachloroazobenzene by condensation of 3,4-dichloroaniline which is a degradation product of many anilide herbicides (19). Only by degradation can a pesticide be totally eliminated from the environment.

Pesticides can be transformed by chemical, photochemical, and biochemical means. Soil can provide the conditions or serve as the catalyst or component for chemical reactions. Chemical reactions are mediated by such soil properties as pH or catalyzed by soil minerals (20). Photolysis of a chemical can result directly from absorbing radiation or indirectly by reaction with another chemical which is activated by absorbed radiation. However, the predominant means of transformation is microbial or enzymatic. Mechanisms of these reactions have been extensively reviewed and summarized (21-23).

Attention should be given to experimental methods both for characterizing the process and for assessing the kinetics of degradation. Characterization of degradation under field conditions is often complicated by other processes such as retention and transport acting simultaneously on the chemical (24). Studies on persistence, dissipation, disappearance, or loss of effectiveness of pesticides are usually not a direct measure of pesticide degradation. The influence of all processes must be sorted out before that attributable to degradation can be evaluated.

The kinetics of pesticide degradation is affected by (a) the quantity and availability of the pesticides, (b) presence of microorganisms or enzyme systems capable of degrading the pesticide, and (c) activity level of the microorganisms as affected by the nutrients available to sustain the microbial population; by environmental conditions such as temperature, moisture, oxygen supply, aeration; and by various soil parameters. Many models for estimating the fate of pesticides in the environment assume that degradation can be simply expressed as a first order reaction with respect to pesticide concentration, although the inadequacy of this approach has been pointed out repeatedly (e.g., 25-27). Expressions for biodegradation rates should include considerations of the substrate concentration as well as the activities of microorganisms or enzyme systems present, such as the Monod equation or the analogous Michaelis-Menten equation for enzyme kinetics. Lewis et al. (27) have called attention to the need to have a multiphasic approach in expressing the kinetics of pesticide degradation, depending on the concentration of the chemical in the environment. A number of studies from Alexander's laboratory (28) have shown that the kinetics of degradation calculated from the rates normally used for such studies may not be applicable for pesticides at extremely low concentrations. The same could also be said for very high concentrations of pesticides (e.g. 29).

In addition to the amount of pesticide present, the degradation rate could be affected by the availability of the chemical for degradation. Ogram et al. (30) have recently presented evidence suggesting that only the 2,4-D (2,4-dichloro phenoxyacetic acid) in soil solution, but not that adsorbed on soil colloids, could be degraded by soil microbes both in soil solution and sorbed on soil colloids. Other considerations should also be given to the nature and quantity of soil microbial biomass present in relation to nutrient availability (26,28,31) and the adaptability of microbes, either by natural selection or by genetic manipulation, to attack and utilize the pesticide chemical (32).

Most of our knowledge on pesticide degradation has been accumulated from studies with surface soils and under laboratory incubation conditions. However, it is difficult to predict the behavior of pesticides under field conditions from data obtained under a controlled laboratory condition. The degradation in soil containing plants may be entirely different. Root exudates and decaying root fragments can provide energy and nutrients for microbial growth (33) and lead to an accelerated mineralization of pesticides in the rhizosphere (34,35). The presence of plants will also affect soil water potential, which in turn affects soil microbial activities and the degradation processes. Furthermore, indications are that organic chemicals do degrade in the vadose zone and in groundwater (e.g., 36-38), but the mechanisms and kinetics of degradation are mostly unknown. More attention is needed to better characterize the mechanisms and kinetics of pesticide degradation in the vadose zone.

Transport Processes

Although the downward transport of pesticides by water is of ultimate concern in evaluating the potential of groundwater contamination, other modes of pesticide transport should also be taken into consideration. These processes include the upward transport by water to soil surface, evaporation or volatilization from soil surface, transport by water in surface runoff or by soil particles in erosion, and uptake by plants. The volatilization process involves two stages: the upward movement to soil surface and the escape from the soil surface. Volatilization is a function of the vapor pressure of the pesticide and is affected by pesticide concentration, soil-water content, adsorptivity of the soil, diffusion rate in soil, temperature, and air movement (39-41). A model accounting for most of the processes affecting volatilization has been tested by Jury et al. (42). Although volatilization losses are usually most rapid immediately following pesticide application, the continued slow loss over an extended period in a water-deficient environment such as in the arid west can also be significant.

Pesticide transport by surface runoff and soil erosion is a function of time lag between rainfall and application; the chemical nature and persistence of the pesticide; the hydrological, soil, and vegetative characteristics of the field; and the method and target of application (43). Wauchope (44) found that unless severe rainfall occurred shortly after pesticide application, total losses for the majority of pesticides due to runoff were less than 0.5% of the amount applied in most cases, although single-event losses from small plots or watersheds can be much greater.

In assessing the fate of pesticides in the environment, the process of plant uptake and its consequences have often been ignored, even though its importance is readily recognized in any study on the efficacy of the chemical (45,46). Plants not only degrade pesticides and enhance their degradation, but they can also participate in pesticide transport. If the pesticide is not degraded after being taken up by plants, the pesticide could be passed through the food chain when the plants are harvested and consumed, or could be recycled back to soil if the plant parts fall back on the ground and are not removed.

Systems Approach

A number of modeling approaches attempting to depict the process of downward transport of pesticides to the groundwater have been published (47) or presented at this symposium. A conceptual framework for any such considerations will include not only the retention, transformation, and transport processes involved, but also the factors affecting all the processes as inputs to the model, before the outcome of all processes acting simultaneously on the pesticides can be predicted (Figure 2). Some of the major factors can be broadly divided as follows:

I. Pesticide factors:

Chemical properties: structure, solubility, volatility
Application methods: formulation, rate, mode
Degradation patterns: pathways, metabolite formation

II. Soil factors:

Soil properties: type and amount of organic matter, clay, and amorphous materials, pH, structure, permeability
Soil conditions: moisture, aeration, nutrient status, microbial activity, heterogeneity, depth to water table
Land forms: topography, slope length and steepness, drainage

III. Plant factors:

Species characteristics
Stage of growth
Root system and rhizosphere

IV. Environmental factors:

Temperature
Precipitation
Air movement
Radiation

This list is by no means an exhaustive one, but it reveals the multifarious interactions of the factors involved that affect all the processes. What we have attempted in this presentation was to point out some of the difficulties and pitfalls one should be aware of in any attempt to model pesticide transport as well as other factors affecting the fate of pesticide in the environment. While modeling can be an important tool for estimation of pesticide movement and fate in the environment, the current lack of knowledge of the mechanisms and interactions of factors and processes affecting pesticide behavior in the environment has led to assumptions and simplifications in the systems to be modeled. Errors either in estimation simplifications or inherent in the assumptions are difficult to quantify. Moreover, errors associated with inputs for each factor or process in the model can be compounded by errors in subsequent interactions. Thus predictive values obtained from many current models must all be accepted with caution if they are to be used for assessment purposes.

As a final item of food for thought, we would like to mention that one sometimes has the impression that pesticide movement into groundwater is an inevitability, especially if the chemical moves beyond the root zone into the subsoil where microbial activities are much lower and degradation would be lessened. We are reminded of the soil genesis process in that a great deal of soluble organic matter has been leached from the surface soil down into the subsoil over centuries and even millennia of soil development. The question is why have we not seen a massive amount of natural organic matter in

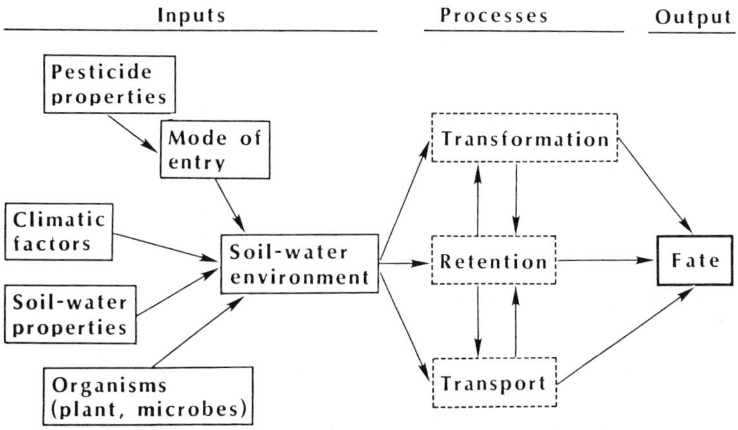

Figure 2. Interactions of the factors and processes affecting the fate of pesticides in the soil environment.

the groundwater? Perhaps we have not yet attained nearly the level of understanding of all the processes involved in the retention, transformation, and transport of chemicals in the soil as we should have. The challenge is still in front of us.

Acknowledgments

Joint contribution from Department of Agronomy and Soils, Washington State University, Pullman, and Southern Weed Science Laboratory, USDA-ARS, Stoneville, MS. Scientific Paper No. 7215. College of Agriculture and Home Economics Research Center, Washington State University, Pullman, Project 1858.

Literature Cited

1. Storck, W. J. Chem. Engin. News 1984, 62(15), 35-59.
2. Cohen, S. Z.; Creeger, S. M.; Carsel, R. F.; Enfield, C. G. In "Treatment and Disposal of Pesticide Wastes"; Kruager, R. F.; Seiber, J. N., Eds.; ACS SYMPOSIUM SERIES No. 259, American Chemical Society: Washington, D. C., 1984; pp. 297-325.
3. Pesticide & Toxic Chemical News January 23, 1985.
4. Hamaker, J. W.; Thompson, J. M. In "Organic Chemicals in the Soil Environment"; Goring, C. A. I.; Hamaker, J. W., Eds.; Marcel Dekker: New York, 1972; Vol. 1, pp. 49-143.
5. Calvert, R. In "Interactions Between Herbicides and the Soil"; Hance, R. J., Ed.; Academic Press: New York, 1980; pp. 1-30.
6. Koskinen, W. C.; O'Connor, G. A.; Cheng, H. H. Soil Sci. Soc. Am.J. 1979, 43, 871-4.
7. Koskinen, W. C. Weed Sci. 1984, 32, 273-8.
8. Koskinen, W. C. Weed Sci. Soc. Am. Mtg. Abstr. 1985, No. 264.
9. Koskinen, W. C.; Cheng, H. H. J. Environ. Qual. 1983, 12, 325-30.
10. Briggs, G. G. Proc. 7th British Insect. Fung. Conf. 1973, pp. 83-6.
11. Swann, R. L.; Laskowski, D. A.; McCall, P. J.; Vander Kuy, K.; Dishburger, H. J. Residue Rev. 1983, 85, 17-28.
12. Dishburger, H. J. Bull. Environ. Contam. Toxicol. 1980, 24, 190-5.
13. Greenland, D. J. Soils Fert. 1965, 28, 415-25 & 521-32.
14. Green, R. E. In "Pesticides in Soil and Water"; Guenzi, W. D., Ed.; Soil Science Society of America: Madison, WI, 1974; pp. 3-37.
15. Minglegrin, M.; Gerstl, Z. J. Environ. Qual. 1983, 12, 1-11.
16. Hayes, M. H. B.; Stacey, M.; Thompson, J. M. In "Isotopes and Radiation in Soil Organic Matter Studies". International Atomic Energy Agency, Vienna, 1968; pp. 75-90.
17. Harper, S. S.; Cheng, H. H. Agron. Abstr. 1984, p. 177.
18. Stevenson, F. J. In "Bound and Conjugated Pesticide Residues"; Kaufman, D. D. et al., Eds.; ACS SYMPOSIUM SERIES No. 29, American Chemical Society: Washington, D. C., 1976; pp. 180-207.
19. Bartha, R. J. Agric. Food Chem. 1968, 16, 602-4; 1971, 19, 394-5.
20. Wang, T. S. C.; Wang, M. C.; Ferng, Y. L.; Huang, P. M. Soil Sci. 1983, 135, 350-60.
21. Kearney, P. C.; Kaufman, D. D. "Herbicides: Chemistry, Degradation, and Mode of Action" Vol. 1 & 2; Marcel Dekker: New York, 1976.

22. Alexander, M. Science 1981, 211, 132-8.
23. Miller, G. C.; Zepp, R. G. Residue Rev. 1983, 85, 89-110.
24. Cheng, H. H.; Lehmann, R. G. In "Symposium on Assessment of Methodology for Field Evaluation of Herbicide Behavior in Soils". Weed Sci. 1985, (in press).
25. Hamaker, J. W. In "Organic Chemicals in the Soil Environment"; Goring, C. A. I.; Hamaker, J. W., Eds.; Marcel Dekker: New York, 1972; pp. 253-340.
26. Howard, P. H.; Banerjee, S. Environ. Toxicol. Chem. 1984, 3, 551-62.
27. Lewis, D. L.; Holm, W. H.; Hodson, R. E. Environ. Toxicol. Chem. 1984, 3, 563-74.
28. Alexander, M. Environ. Sci. Technol. 1985, 18, 106-11.
29. Cheng, H. H.; Majka, J. T.; Mangin, R.-M.; Farrow, F. O. ACS 185th Nat. Mtg., Div. Pesticide Chemistry Abstr. 1983, No. 11.
30. Ogram, A. V.; Jessup, R. E.; Ou, L. T.; Rao, P. S. C. Appl. Environ. Microbiol. 1985, 49, 582-7.
31. Frehse, H.; Anderson, J. P. E. In "Pesticide Chemistry"; Mimamoto, J., Ed.,; Pergamon Press: New York,1983; pp. 23-32.
32. Ghosal, D.; You, I.-S.; Chatterjee, D. K.; Chakrabarty, A. M. Science 1985, 228, 135-42.
33. Rovira, A. D. Pestic. Sci. 1973, 4, 361-6.
34. Cheng, H. H.; Führ, F.; Mittelstaedt, W.; In "Environmental Quality and Safety Suppl. Vol. III: Pesticides"; Coulston, F., Korte, F.; Eds.; G. Thieme: Stuttgart, 1975; pp. 271-6.
35. Seibert, K.; Führ, F.; Mittelstaedt, W. Landw. Forsch. 1982, 35, 5-13.
36. Bouchard, D. C.; Lavy, T. L.; Marx, D. B. Weed Sci. 1982, 30, 623-32.
37. Barcelona, M. J.; Naymik, T. G. Environ. Sci. Technol. 1984, 18, 257-61.
38. McCarty, P. L.; Rittman, B. E.; Bouwer, E. J. In "Groundwater Pollution Microbiology"; Britton, G.; Gerba, C. P., Eds.; Wiley: New York, N.Y., 1984; pp. 90-115.
39. Spencer, W. F.; Farmer, W. J.; Cliath, M. M. Residue Rev. 1973, 49, 1-47.
40. Guenzi, W. D.; Beard, W. F. In "Pesticides in Soil and Water"; Guenzi, W. D., Ed.; Soil Science Society of America, Madison, WI, 1974; pp. 107-22.
41. Hance, R. J. In "Interactions Between Herbicides and Soil"; Hance, R. J., Ed.; Academic Press: New York, 1980; pp. 59-81.
42. Jury, W. A.; Spencer, W. F.; Farmer, W. J. J. Environ. Qual. 1984, 13, 573-9 & 580-6.
43. Bailey, G. W.; Swank, A. R., Jr.; Nicholson, H. P. J. Environ. Qual. 1984, 3, 95-102.
44. Wauchope, R. D. J. Environ. Qual. 1978, 7, 459-72.
45. Schmidt, R. R.; Pestemer, W. In "Interactions Between Herbicides and Soil"; Hance, R. J., Ed.; Academic Press: New York, 1980, pp. 179-201.
46. Appleby, A. P. In "Symposium on Assessment of Methodology for Field Evaluation of Herbicide Behavior in Soils". Weed Sci. 1985, (in press).
47. Wagenet, R. J.; Rao, P.S.C. In "Symposium on Assessment of Methodology for Field Evaluation of Herbicide Behavior in Soils". Weed Sci. 1985, (in press).

RECEIVED April 7, 1986

2

Soil Characteristics Affecting Pesticide Movement into Ground Water

Charles S. Helling and Timothy J. Gish

Agricultural Research Service, U.S. Department of Agriculture, Beltsville, MD 20705

Processes that modify convective transport of pesticides through soil and into groundwater include adsorption/desorption, degradation, volatilization, runoff, and plant uptake. These processes, in turn, are affected by soil characteristics, climate, pesticide properties, and agricultural practices. A screening model based on the convection-dispersion equation (assuming 1st-order degradation) was used to rank several soil properties that may affect atrazine leaching. Transport was most retarded by low hydraulic conductivity and high soil organic matter content; increased bulk density attenuated leaching to a lesser extent. A literature survey, with emphasis on atrazine, aldicarb, and DBCP (pesticides that have leached to groundwater), tended to confirm that sandy soils (with high hydraulic conductivity and low organic matter) were usually associated with leaching. Restricted drainage has led to lateral subsurface movement or occurrence of residues in perched groundwater. At the other extreme, karst topography allowed rapid recharge and high probability of pesticide leaching.

Groundwater is estimated to supply 40-50% of U.S. drinking water needs, and constitutes at least part of the water source for 75% of American cities (1). About 95% of the rural population depends on groundwater for their drinking water. Reliance on groundwater for domestic consumption and agricultural uses becomes increasingly important in the more arid Western states.

Groundwater contamination in the U.S. was reviewed by Pye et al. in 1983 (1). The most common sources of such contamination included human and animal wastes, industrial wastes, petroleum products, landfill leachate, and (along coastal regions) saltwater intrusion. Contamination from the agricultural use of pesticides was apparently far less common. Nevertheless, by 1984, Cohen et al. (2) reported that

This chapter not subject to U.S. copyright.
Published 1986, American Chemical Society

12 pesticides were found in groundwaters of 18 states as a consequence of such use. Typical positive residue levels varied widely, although most commonly they ranged from ca. 1-100 ppb.

The extent of pesticide leaching depends on a combination of factors that include the physicochemical characteristics of the pesticide (or its degradation products), soil properties, climate, and agronomic factors such as the timing, rate, and method of pesticide application, the use of irrigation, and the influence of crop cover. Whether actual contamination of groundwater occurs is also influenced by the depth to groundwater and the permeability of overlying soil.

In this paper, we will first focus on soil factors that affect pesticide leaching through the root zone and into the subsoil. The general convection-dispersion equation with first-order degradation will be used to characterize the effects of various soil properties on atrazine movement. In the remainder of the paper, we will discuss laboratory and field experiments that focus on adsorption and leaching. Generalizations from these studies will be compared with soil properties at sites of known groundwater contamination. Finally, the results from the transport model analysis will be used in conjunction with the literature review to propose a hierarchical ranking of properties affecting leaching.

Background: Soil Properties

Soils can be characterized in many ways, depending, for example, on whether primary concern relates to agronomic applications, engineering utility, or soil genesis. From the standpoint of predicting pesticide transport, a series of physical, chemical, and biological properties---some transient---could be considered. For convenience, we have listed many such parameters in outline form within Table I. Numerous references describe them and their analysis (3-7).

In addition to the classification of properties in Table I, the site can be characterized further as in Table II. Surface or subsurface drainage systems, if present, would be important additional descriptions.

Background: Other Related Properties

Metereological conditions affect transport of water and solutes since, in the absence of irrigation, they determine how much water reaches the soil surface, what the intensity and frequency of that precipitation is, and how much water is recycled from the soil via evapotranspiration losses. Temperature influences the rate of pesticide degradation and the rates of water and pesticide volatilization. Properties of the pesticide strongly affect the tendency to leach and degrade, but are addressed elsewhere in this symposium. Agricultural factors such as the manner and timing of pesticide application and the cropping tillage practice have potential impact on the ultimate fate of a chemical. These various nonsoil factors are listed in Table III.

Table I. Classification of Some Soil Properties

A. Physical composition
   1. Soil texture (% Sand, silt, clay; gravel)
   2. Soil organic matter content (% OM)

B. Chemical composition
   1. Clay mineralogy
   2. Organic matter type

C. Physical properties
   1. Bulk density
   2. Field moisture capacity
   3. Hydraulic conductivity
   4. Pore size distribution, macropores; tendency to crack on drying

D. Chemical properties
   1. pH
   2. Cation-exchange capacity (CEC); anion-exchange capacity
   3. % Base saturation
   4. Redox potential, Eh

E. Transient soil properties
   1. Soil moisture content (volumetric)
   2. Soil temperature

F. Biological/biochemical properties
   1. Number and type of microorganisms
   2. Activity of specific enzymes

Table II. Classification of Some Macro Soil Properties

A. Surface
   1. Relief
   2. Slope

B. Subsurface
   1. Profile changes (type, depth, and areal homogeneity)
   2. Restricting layers
   3. Depth to groundwater (perched and unconfined aquifer)

Table III. Classification of Nonsoil Factors Potentially Affecting Transport to Groundwater

A.	Climate
	1. Rainfall (temporal distribution, intensity)
	2. Temperature
	3. Evapotranspiration
B.	Pesticide properties
	1. Soil adsorption coefficient (K)
	2. Water solubility
	3. Octanol:water partition coefficient (K_{ow})
	4. Ionization constant (pK_a, pK_b)
	5. Chemical and biological stability (persistence in soils)
	6. Volatility
C.	Pesticide application
	1. Formulation
	2. Method of application (foliar, soil surface, soil incorporation)
	3. Rate
	4. Timing
	5. History of pesticide use (accelerated degradation; buildup)
D.	Agricultural practices
	1. Cropland
	a) Conventional tillage
	b) Conservation tillage
	c) Irrigation
	2. Noncropland
	a) Fallow
	b) Rangeland, forests, etc.
	3. Soil amendments

Processes Affecting Leaching

The distribution of pesticides throughout the soil profile, as a function of time, represents the integration of processes such as mass flow, diffusion, adsorption/desorption, degradation, volatilization, runoff, and plant uptake (the latter, mainly as it affects water movement in the root zone). These have been the subject of many reviews (8-12), and therefore only limited attention will be given to the subject in the following sections.

Pesticide Transport Model

Models for describing pesticide transport on a field scale generally fall into one of two categories, deterministic or stochastic. Deterministic models seek to account for pesticide leaching by describing the mechanisms governing volatilization, adsorption, degradation, convection and diffusion, while at the same time accounting for the physical and chemical characteristics of the soil medium. Due to the complex nature of such an interaction, it is often necessary to make assumptions which are at best a first approximation of what occurs under field conditions.

Stochastic models assume that although pesticide movement on a small homogeneous scale obeys certain physical laws, the random component associated with those laws in a heterogeneous system will override their deterministic behavior. Consequently, transport is ascertained by evaluating transfer function models or by evaluating the probability distribution of some transport process (13). The probability density function (PDF) for a specific transport process, on a field site, can be combined with deterministic theory to account for spatial heterogeneity (14-18). However, a solely stochastic model for screening pesticides may be of little value for several reasons. First, the PDF for a specific soil property would require the analysis of numerous soil samples. Second, a purely stochastic model lacks the ability of predicting the location of potentially hazardous areas within a field. Third, if large variations in soil or soil water properties are present, they may affect the relative behavior of most pesticides in a similar manner.

Our theoretical development of a screening model will focus on the deterministic approach since it may still be applied in a semi-stochastic manner. This model is conducive to analyzing the effect of a specific soil property while holding others constant. Our theoretical consideration of pesticide transport begins (Equation 1) with an expression of the rate at which water moves through soil, where J_w is the water flux (volume of water flowing through a cross section of area per time), ψ is the matric potential, ∇H is the hydraulic gradient, and $K(\psi)$ is the hydraulic conductivity.

$$J_w = -K(\psi)\nabla H \quad (1)$$

Frequently termed the Buckingham-Darcy equation, Equation 1 may be used to describe pesticide movement through a soil volume by employing mass balance equations for both water and pesticide

$$\frac{\partial \theta}{\partial t} + \nabla J_w = 0 \qquad \text{mass balance for water} \qquad (2a)$$

$$\frac{\partial C_r}{\partial t} + \nabla J_s = \phi \qquad \text{mass balance for pesticide} \qquad (2b)$$

where C_r is the volume-averaged pesticide concentration, J_s is the solute flux, θ is the volumetric water content, and ϕ is a reaction term describing the stability or the rate of plant uptake for a particular pesticide.

The solute flux of the pesticide consists of two terms. The first term corresponds to the convective or bulk transport of the pesticide with the moving soil solution; the second, a diffusion-dispersion term, accounts for the random thermal motion of the pesticide molecules (19) as well as any hydrodynamic dispersion that may occur due to variations in the pore water velocity (20). The mathematical representation of J_s is

$$J_s = C_r J_w - D \nabla C_r \qquad (3)$$

where D is the diffusion-dispersion coefficient. Combining Equations 2a, 2b, and 3, a solute transport equation in one dimension can be written where

$$R \frac{\partial C_r}{\partial t} = D \frac{\partial^2 C_r}{\partial x^2} - V \frac{\partial C_r}{\partial x} - \mu C_r \qquad (4a)$$

where

$$R = 1 + \frac{\rho_b K_d}{\theta} \qquad (4b)$$

and

$$S = K C_r \qquad (4c)$$

Here, ρ_b is the soil bulk density, μ is the first-order degradation coefficient, K_d is the distribution coefficient for the soil/water phases, V is the average pore water velocity ($V = J_w/\theta$), x is the soil depth, S is the adsorbed concentration per unit of mass, and R is a dimensionless variable.

The assumption of a linear adsorption isotherm, $K_d = K$ in Equation 4b, may be valid under low solution concentrations (21). However, Rao and Davidson (22) showed that this assumption could produce errors within a factor of 2 or 3. Additionally, K can be estimated from soil OM content (22). Consequently, the solution of Equation 4a, subjected to appropriate initial and boundary conditions, can be used to perform a sensitivity analysis on the effect of ρ_b, OM and J_w on pesticide movement.

To solve Equation 4a, boundary conditions are imposed that describe the initial soil conditions with respect to the pesticide and the method of chemical application. Initially, there may be some finite concentration in the soil due to the previous year(s) of pesticide application. This residual concentration will be denoted in the solution as C_i (see Appendix I). If the pesticide is applied as a one-time application (per growing season), a pulse boundary condition at the surface and a flux bottom boundary condition are well suited (23). The solution of Equation 4a with these restrictions is given by van Genuchten and Alves (24).

Although the deterministic model presented assumes steady state conditions, laboratory studies have shown that solute transport under transient flow conditions may be approximated by assuming an equivalent uniform water flux and water content (25-6).

Model Assumptions

The model presented can be used for screening nonvolatile pesticides under various field conditions. The assumptions behind its derivation should be clearly stated, as we attempt to do for the following attenuation and transport mechanisms.

Adsorption. The solution used to evaluate the pesticide transport equation, Equation 4a, assumes a linear adsorption isotherm that is constant with depth. However, linearity may not be the case for some pesticides and the adsorption coefficient will almost never be constant with depth. The rationale for using a linear model is initially based on the Freundlich isotherm

$$S = KC^{1/n} \qquad (5)$$

where K is the Freundlich constant, and 1/n is an exponent that generally ranges between ca. 0.5 and 1.2. If 1/n is assumed to be 1, the resulting equation is linear, i.e., $K = K_d$ in Equation 4c. Although the validity is still under discussion, Karickhoff (21) considered that for the low solution concentrations typically associated with pesticides, the linear model is appropriate. Rao and Davidson (22) showed that the assumption of linearity could produce errors within a factor of 2 or 3. The value of K is critical since it indicates the proportion of pesticide in the mobile water phase. K has often been used to predict the extent of leaching by assuming only convective movement and adsorption in a retardation factor, R, as in Equation 4b (14, 27-8).

Since the adsorption coefficient is critical to the theoretical development, caution should be exercised in using a particular K value for a particular soil and pesticide. The most common method used to measure adsorption is by the batch equilibrium technique, in which soil samples are equilibrated with a series of pesticide concentrations. However, the equilibrium time is critical, and may not represent adsorption under field conditions where the pesticide is moving in the solution phase. Consequently, flow equilibrium methods have also been developed (29).

The adsorption K can also be estimated from the soil OM content. Organic matter has been shown to be a primary site for adsorption (unless the pesticide is permanently charged). As a result, the adsorption coefficient may be approximated by the equation

$$K = f_{oc} K_{oc} \tag{6}$$

where K_{oc} is the coefficient of linear adsorption normalized on organic carbon and f_{oc} is the fractional content of organic carbon (22). K_{oc} can often be obtained from previously published values or estimated from the octanol:water partition coefficient (30). One advantage of employing Equation 6 is that it allows one to model the effect of organic matter on transport. Additionally, the use of Equation 6 has often resulted in reducing the coefficient of variation associated with K (22, 31-2). Thus, Equation 6 will be employed in solving Equation 4a.

Diffusion-Dispersion. A diffusion-dispersion coefficient was used in transport Equation 4. Depending upon the water velocity, either diffusion or dispersion would be the dominating mechanism. For relatively mobile chemicals, the variability in D may be linked directly to the water velocity (33-4). Consequently, D would be dominated by hydrodynamic dispersion with $D = \varepsilon V$, where dispersivity ε ranges from 0.1-4 cm (35). On the other hand, diffusion may be the dominant mechanism controlling the magnitude of D, especially if water movement is slow. In a field setting, the time between precipitation events will be much greater than the duration of precipitation events, allowing more time for diffusion than dispersion. This being the case, the diffusion coefficient can be estimated by using the Millington and Quirk tortuosity model (36)

$$D = (\theta^{10/3}/\phi^2)\ D_{water} \tag{7}$$

where ϕ is the porosity of the bulk soil and D_{water} is the diffusion coefficient in water; for most pesticides, D_{water} can be approximated as $4.3 \times 10^{-5}\ m^2\ day^{-1}$.

Convection. Convection is the bulk transport of pesticide with the moving soil solution. Consequently, the water velocity is the major mechanism governing convective transport. Numerous studies have shown that a stochastic representation of the water velocity does a better job of describing chemical transport (14-8). However, for a screening mode, an accurate depiction of the water velocity is not essential since velocity variations would affect the relative behavior of most chemicals in a similar manner once the partition coefficient between the liquid and adsorbed phases has been established. Predictions of the average water velocity (expected or mean value) have been made for field experiments by monitoring meteorological events and subtracting estimates of the evapotranspiration from the water inputs (33).

Degradation. Pesticide degradation is a complex phenomenon since the process may be purely chemical or dependent upon the presence of microorganisms. The assumption of first-order kinetics may be misleading since the degradation rate will depend upon temperature as well as the particular phases in which the pesticide resides. Thus, a specific pesticide may degrade by different mechanisms at different rates. Assuming isothermal conditions, the volumetric water content will dictate the overall rate of degradation (37). The water content affects both aeration as well as the fraction of pesticide undergoing degradation in the solution phase. In addition, the soil pH will also affect the rate of degradation for some pesticides (38).

Assuming a first-order degradation process, the degradation rate may be measured according to

$$\mu = \ln (C_0/C)\ t^{-1} \tag{8a}$$

or estimated from published values of $t_{1/2}$, the half-life of a particular pesticide, as in Equation 8b

$$\mu = \ln (1/2)\ t_{1/2}{}^{-1} = 0.693\ t_{1/2}{}^{-1} \tag{8b}$$

Pesticide Transport Simulations

The relative importance of ρ_b, OM, and J_w on pesticide movement was accomplished by conducting a series of computer simulations and subjecting the results to a sensitivity analysis. The sensitivity analysis compares the peak concentrations in the liquid phase, since this phase will be the major vehicle for pesticide transport to groundwater. The atrazine application rate used in the simulations was equivalent to 2.8 kg/ha of active ingredient. The diffusion-dispersion coefficient was assumed to be constant, 1 $cm^2\ day^{-1}$. The ranges chosen for ρ_b, OM, J_w, and θ_v in the simulations were 0.8-1.45 g m^{-3}, 1-5%, 1-4 cm day^{-1}, and 0.15-0.35 m^3 water/m^3 soil, respectively. These ranges correspond to typical field values. Since a family of curves was generated, only a few representative curves will be shown, Figure 1a-c. So that visual comparisons can be made between different simulations, the atrazine profile corresponding to $\rho_b = 0.8$, $\theta_v = 0.15$, and $J_w = 1$ cm day^{-1} was used in Figure 1a-c. All simulations assume that 20 days have transpired since atrazine application and that $\mu = 0.0098\ day^{-1}$ (27, 39).

In Figure 1a the effect of an increased bulk density on the atrazine concentration profiles is shown, while holding θ_v, J_w, and OM constant. The soil bulk density of 0.8 g cm^{-3} corresponds to a very light soil and(or) a soil that has been recently plowed, while $\rho_b = 1.45$ g cm^{-3} represents a soil that has a natural high density or a soil that has been compacted by farm implements. As the soil density increases, the maximum pesticide concentration in the liquid phase decreases. Additionally, the maximum pesticide concentration will occur closer to the soil surface as ρ_b increases. Since ρ_b increases with depth for most agricultural soils, Figure 1a indicates that pesticide movement would be more retarded as it moves through the soil profile, all other factors held constant. Since the coefficients of variation for $_b$ are generally between 5-10%, assuming a constant density could introduce errors within a factor of 2.

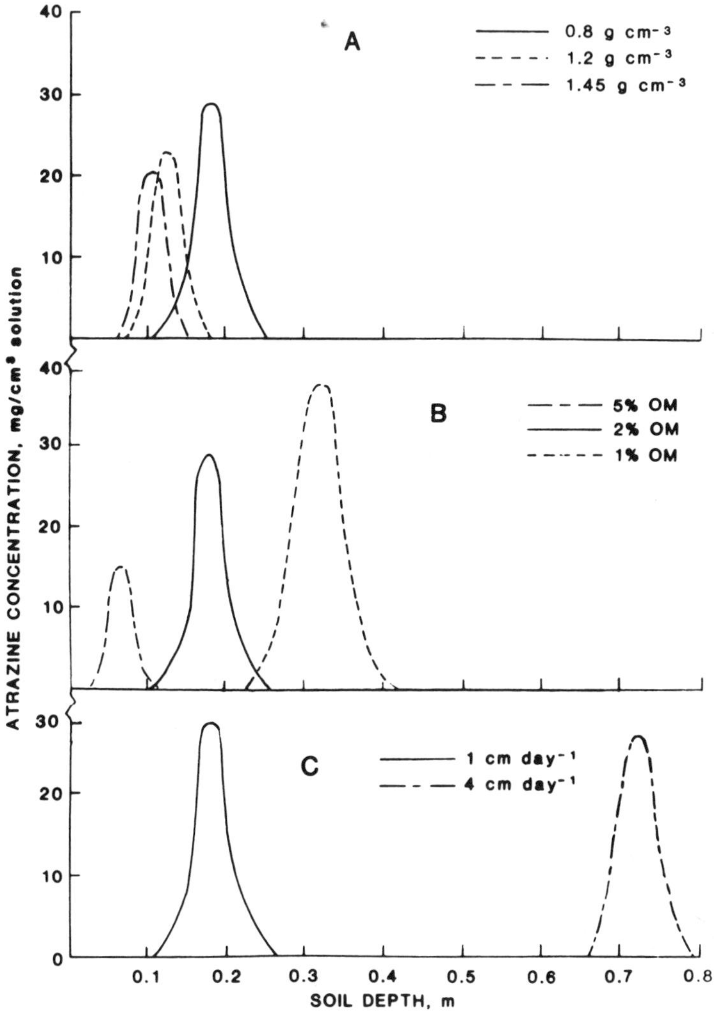

Figure 1. Simulated effect of soil properties on atrazine leaching: (a) bulk density [ρ_b]; (b) organic matter [OM] content; and (c) hydraulic conductivity [$K(\psi)$]. Other symbols used include θ_v (volumetric water content), V (average pore water velocity), and $\partial H/\partial z$ (hydraulic gradient). Assumptions are: Fig. 1a--2% OM and J_w = 1 cm day^{-1}; Fig. 1b--ρ_b = 0.8 and J_w = 1 cm day^{-1}; Fig. 1c--2% OM and ρ_b = 0.8.

The effect of organic matter content on atrazine movement is shown in Figure 1b. Like the trend for ρ_b, the maximum pesticide concentration decreases as OM increases. Although the simulations assume a constant OM content with depth, most of the peak concentrations were in the top 0.3 m where the organic matter contents are generally large. Since the organic matter content generally decreases with depth, Figure 1b suggests that pesticide movement will be more retarded near the soil surface. The relative change in peak concentrations between Figures 1a and 1b indicates that the typical values of OM govern atrazine movement to a greater extent than does ρ_b. Although the coefficients of variation range between 20-50%, a predetermined mean value should yield results within a factor of 3.

Kinetics may play an important role in pesticide adsorption (28, 39). Rapid transport of pesticide in large soil pores (rocks, void root channels, worm holes, etc.) could take place, thus exposing the pesticide to only a fraction of the adsorption sites. Since the organic matter content decreases with depth, the occurrence of macropore flow could result in pesticide movement beyond the surface layers where most of the OM resides.

The effect of the hydraulic conductivity on pesticide transport can be simulated by assuming unit gradient ($\nabla H = 1$); thus, $J_w = K(\psi)$. As a result, J_w can be adjusted in such a way that pores of different radii are simulated. For comparison, two hypothetical soils are simulated in Figure 1c. The pore radii are 0.023 and 0.033 cm, which correspond to hydraulic conductivities of 1 and 4 cm day^{-1}. Rawls et al. (40) evaluated data from 1,320 soils and found that mean saturated conductivities ranged from 1.44 cm day^{-1} in clayey soils to 504 cm day^{-1} in sandy textured soils. Although the conductivities simulated in Figure 1c differ by only a factor of 4, they reflect the greatest difference in atrazine peak concentrations simulated within Figure 1a-c. The peak concentration corresponding to 4 cm day^{-1} is shallower than one would expect under actual field conditions since the model is assuming a constant OM content with depth. As a result, the hydraulic conductivity is the most important soil factor governing pesticide contamination of groundwater.

Summary of Model Simulations. Using a one-dimensional convection-dispersion transport model with first-order degradation, the relative importance of ρ_b, OM, and J_w was evaluated. The hydraulic conductivity was the most important soil parameter, followed by OM and ρ_b. Both the organic matter content and ρ_b can be estimated, resulting in solutions that are within a factor of 2 or 3. However, pesticide movement is very sensitive to the magnitude of the hydraulic conductivity. Soils with hydraulic conductivities within a factor of 4 can yield dramatically different concentration profiles. Additionally, soils containing soil cracks or large pores may be prone to groundwater contamination.

Spatial Variability

To this point, several assumptions have been built into the theoretical development. The soil medium is being characterized as a homogeneous volume allowing the one-dimensional solute transport equation

(4a) to be used. Field soils are not homogeneous, but consist of horizons, or horizontal layers of soil material where each layer differs with respect to organic and inorganic composition. Additionally, textural deposits may be randomly located below the soil surface. These textural deposits or lenses also vary in their composition. The lenses may be composed of fine particles forming a dense layer, or possibly a very coarse sandy lens. The dense lens, as is intuitive, can restrict downward water movement, thereby producing a perched water table. However, the retardation of flow by a sand lens is not so obvious. Soil water will not enter a buried sand layer until the finer textured soil overlaying the sand lens is essentially saturated, thus retarding vertical transport. This mechanism was suggested for the lateral movemtent, for ca. 15 m, of solutes eluting from a landfill; it is depicted in Figure 2 based on Gerhardt's study (41). If or when the finer textured soil is saturated at the lens interface, the soil water will flow through preferential pores, a process termed channeling.

A hierarchy of soil parameters was established in the previous section. If the coefficients of variation are high, on the order of 100%, several hundred soil samples would need to be taken to characterize that soil property. Consequently, the variability of the hydraulic conductivity dictates to what extent we can accurately predict pesticide transport. Even within the same soil texture 42-4), coefficients of variation for the hydraulic conductivity were found to be well in excess of 100%.

Laboratory and Field Leaching Tests

Pesticide registration requirements have long included leaching tests; the reports submitted in registration petitions are confidential, however, so such information often remains unpublished. Despite that, a vast data bank exists in the literature on movement of pesticides in soils, although no recent, comprehensive bibliography on leaching is---to our knowledge---available. Methods of conducting soil leaching tests for toxic organic chemicals have been reviewed (45).

Textural differences within and among soils were shown (indirectly, via the hydraulic conductivity) to influence pesticide leaching markedly in one transport model. We also discussed how they contribute to spatial variability, especially within a profile. In lab and field experiments, textural differences among surface soils have often been associated with leaching. Thus, chemicals are intuitively expected to move deeper into coarse-textured sandy soils than in medium-textured silt loams or fine-textured clay soils. This was illustrated in a laboratory study of metribuzin leaching in 16 soils (46); clay and sand contents were highly correlated (negatively and positively, respectively) with movement. A still better predictor was the 0.33-bar (0.033 MPa) soil moisture content (previously considered to be field moisture capacity, FMC; FMC is now generally accepted as 0.01 MPa). This property is largely determined by soil texture and organic matter content, so the FMC relationship is expected. Organic matter content and pH were less well correlated with leaching.

In an earlier investigation, Helling (47) studied the influence of soil properties on leaching by using 12 pesticides in 14 soils. With this soil thin-layer chromatography (soil TLC) method, FMC gave the best prediction of mobility for seven nonionic pesticides, whereas soil pH was best correlated (positively) to movement of four acidic chemicals. Clay and organic matter contents were negatively correlated with leaching of the nonionic pesticides.

Figure 3 illustrates the effect of soil differences on the herbicide propham's movement in soil TLC plates [the summary data were previously published (48)]. Soil organic matter content increases markedly going from Norfolk through Celeryville soil and is thought to be the dominant retarding factor. Based on studies with chemical homologue chlorpropham (47), the best predictor of relative mobility would likely have been a combination of FMC (negative) plus water flux (positive). The inherent risk of leaching in Norfolk sandy loam is clearly greater than in muck soil.

Leaching of the nonfumigant nematicide oxamyl was recently studied in unsaturated columns of three soils (49). Figure 4, adapted from this research, shows that oxamyl is highly mobile in the Arredondo sand, some of the solute moving with the water front (movement is here expressed in R_f units, i.e., distance eluted relative to that of the water front, by analogy with soil TLC presentation). Both Arrendondo and Cecil plots are notably asymmetric, which the authors ascribe to nonequilibrium adsorption. It is apparent from Figure 4 that leaching in Webster silty clay loam, with 38% clay and 4% organic carbon content (ca. 6.8% OM), is much retarded.

It is commonly recognized that an inverse relationship exists between adsorption of pesticides and their tendency to leach in soils (11, 50-1). Adsorption is usually incorporated into leaching models as the adsorption constant normalized on the fractional content of soil organic carbon (see Equation 6), i.e., K_{oc}, since much research has correlated the adsorption of neutral organic compounds with soil organic matter content. As indicated earlier, parameters that more directly indicate the soil water-holding capacity and the size and range of conducting pores may actually correlate even better with transport of pesticides.

In a previously cited laboratory study (47), soil pH was positively correlated with movement of acidic pesticides. Figure 5 is presented as a comparison in which soil pH differences may have differentially affected the mobility of two herbicides, simazine and terbacil. Based on the work of Hogue et al. (52), who did not suggest this hypothesis, Figure 5 shows the expected migration of residues away from the soil surface as water input increases from 20 to 80 cm, simulating seasonal application of irrigation water to the orchard soils. Terbacil was relatively mobile and simazine, less so. However, whereas, terbacil's mobility was significantly greater in Rutland sandy loam than in Penticton loam, simazine behaved similarly in both soils. The loam had much higher clay and organic matter contents, so diminished terbacil leaching in that soil is expected. We suggest that simazine movement in Rutland sandy loam is less than otherwise anticipated because at this soil's low pH (4.6), more of the simazine is protonated, and therefore more strongly adsorbed, than in Penticton loam (pH 7.5).

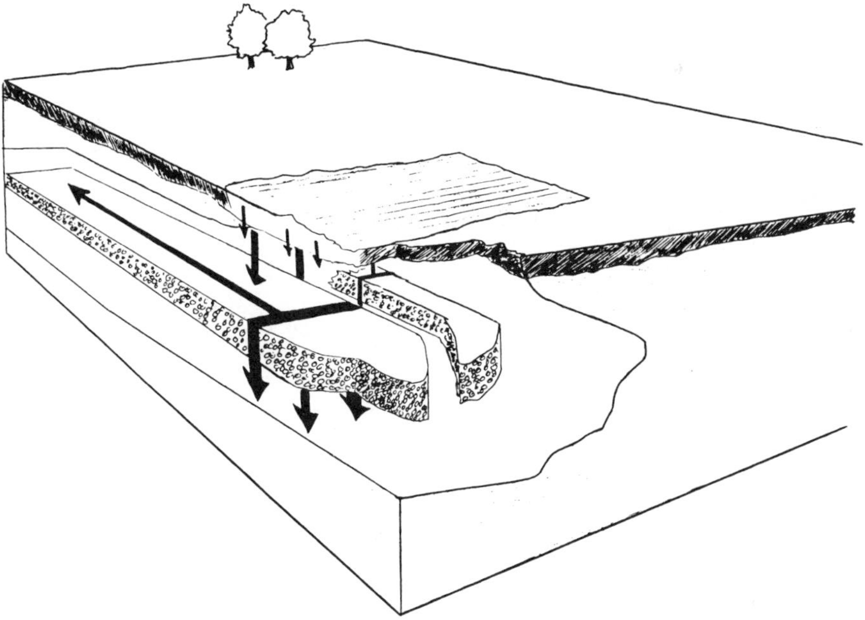

Figure 2. Lateral subsurface movement of water and solutes caused by coarse-textured soil strata (after Gerhardt, 41).

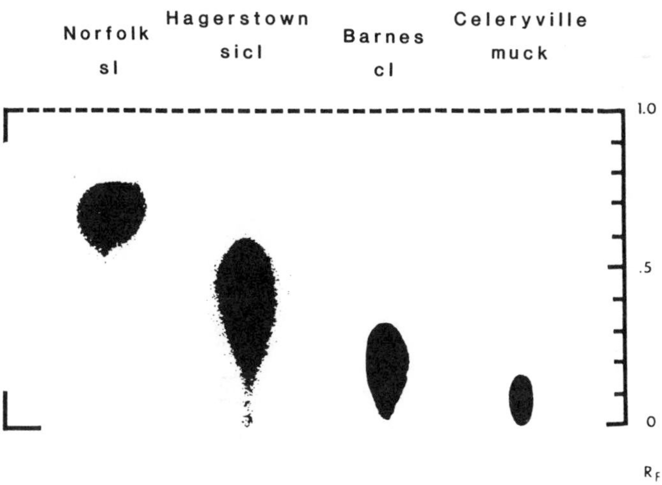

Figure 3. Effect of four soils on leaching of propham, using soil TLC plates (after Helling, 48).

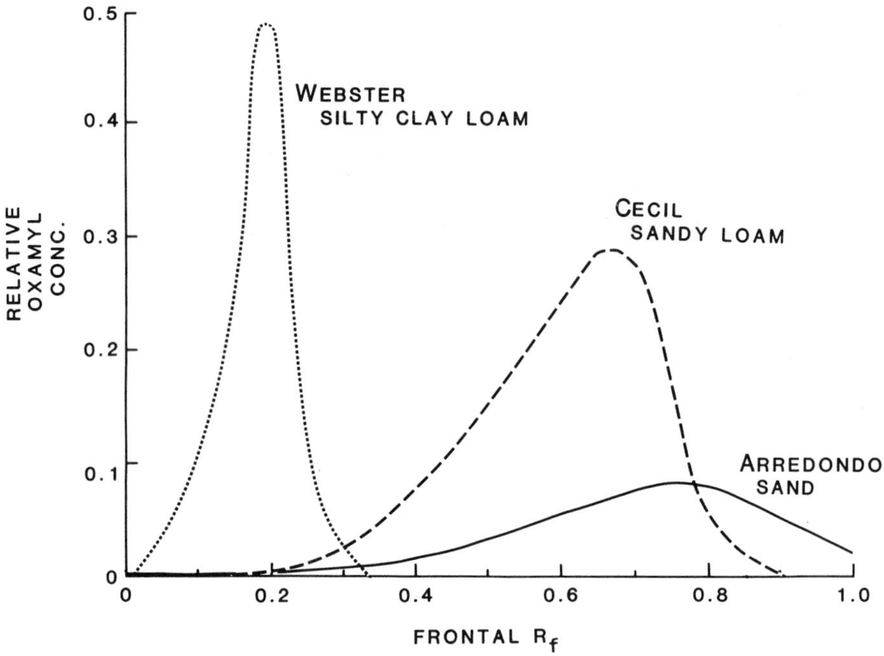

Figure 4. Effect of three soils on leaching of oxamyl, using unsaturated soil columns. Adapted from reference 49.

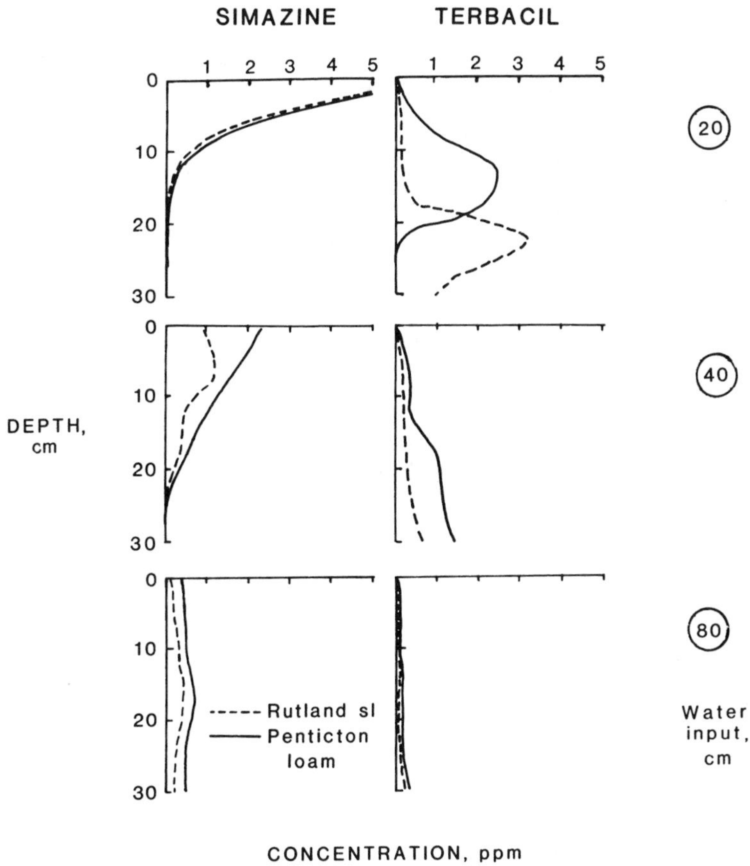

Figure 5. Relative leaching of simazine and terbacil, showing effects of added water, soil texture, and (possibly) pH. Adapted from reference 52.

Groundwater Contamination

Of the 700-800 U.S.-registered pesticidal active ingredients, only a relatively small number (2) have leached to groundwater. Unfortunately, some of these chemicals are or were among the major pesticides used in American agriculture (2). Three of these will be discussed in the following section, with special attention to soil characteristics that may have influenced movement.

Atrazine. This widely used herbicide was detected in the mid-1970s in all sampled Iowa municipal water supplies in which that water was derived from shallow wells in the alluvial plains of rivers contaminated from atrazine runoff (53). Levels ranged up to 483 ppb.

An extensive series of studies in Nebraska (54-8) established the common occurrence of sub-phytotoxic (58) levels of atrazine in groundwater. The sites were all located in intensively irrigated counties adjacent to the Platte River. Where depth to groundwater ranged from 1-7 m (55), soils tended to be somewhat poorly drained, grading to sand and gravel at ca. 0.25-1 m; deeper groundwater sites had well-drained silty soils that graded to mixed sand and gravel. Nearly all atrazine levels were <1 ppb and residues tended to be highest in shallow wells. Because of very high permeability of subsoil in the river bottomlands and the direction of groundwater flow (away from the river, ca. 3 m/day), atrazine levels in shallow groundwater matched those in the Platte River by September (55), contrasting with the attenuation seen when soils are predominantly finer textured (54). Secondary aquifers, i.e., groundwaters isolated by clayey silt layers, contained only trace amounts of atrazine (55).

The peak atrazine concentrations in Nebraska were observed down-gradient from irrigated fields, at the end of the irrigation season (55); residue levels diminished with lateral distance (58). There was some areal and vertical association with nitrate concentration, but no correlation could be drawn with dissolved organic carbon (of which atrazine would constitute <<1%).

Atrazine-contaminated water seeped from an irrigation tailwater pit (54). Because residues would be highest early in the growing season, seepage in late spring-early summer would increase atrazine residues in groundwater, whereas relative dilution of the already-contaminated groundwater would occur during the balance of the year. In this study (54), well water beneath somewhat poorly drained surface soils had less atrazine (average, 0.2 ppb) than did wells under moderate- and well-drained (1.5 ppb) sites. The thickness of the vadose zone in the three classes was <1.5, 2.4-3.6, and 3.0-7.8 m, respectively. Poorly drained soils generally had higher OM contents in the surface horizons. This factor, typically associated with both greater adsorption and microbial activity, may have limited atrazine leaching by the direct effect of retardation and the indirect effect of longer residence time---leading to more complete degradation.

Although atrazine was regularly detected in Nebraska groundwater, 11 other pesticides were monitored, but not found (56). These included herbicides 2,4-D, EPTC, and silvex, and insecticides DDT, DDE, dieldrin, endrin, heptachlor, heptachlor epoxide, lindane, and methoxychlor. In this study, alachlor was found in 2 of 14 samples, 0.02 and 0.07 ppb; similarly low residues of alachlor were found in 2 of 33 samples in another Nebraska study (55).

Atrazine leaching was far more limited in conventionally-tilled silty clay loam and clay loam soils in Pennsylvania (59). These Typic Hapludults were deep, well-drained soils of moderate permeability and high available water capacity. Suction lysimeter measurements showed atrazine leaching to 122 cm, but generally less movement occurred, in part because dissipation was fairly rapid and the high evapotranspiration rate in summer minimized downward flow of water. The greatest leaching occurred at the highest herbicide rates. Typical soil solution concentrations of atrazine were 10-40 ppb.

Atrazine was shown to leach more readily in irrigated, permeable Nebraska soils (54-8), but to move much less readily below the plow layer in finer textured, nonirrigated Pennsylvania soils (59). A different geological environment, the karst-carbonate aquifers of northeast Iowa, have been recognized as being potentially fragile with respect to groundwater quality. A recent report (60) documented leaching of atrazine, alachlor, cyanazine, and metribuzin into groundwater there.

Four geological regions were defined in the study area (60), as described in Table IV. The area characterized by considerable depth to carbonate bedrock had no contaminated well water. In addition to their thickness, the overlying soil deposits were said to be of low permeability, increasing the potential for surface lateral transport of water and solutes. In the other three regions, 67% of sampled wells contained pesticide residues, usually atrazine, during at least part of the year. The highest concentrations, in mid-summer and early fall, corresponded to leaching after the annual application of herbicide.

Table IV. Aquifer Characteristics and Herbicide Contamination of Groundwater in Floyd and Mitchell Counties, Iowa (after 60)

Geological region	Depth to bedrock (m)	Special features	Pesticide residues	
			Samples: positive/ total	Chemicals*
Deep Bedrock	>15		0/12	
Shallow Bedrock	<15	Few sinkholes	17/29	At
Karst	Very shallow	Numerous open sinkholes; rolling terrain	12/23	At, Al, M
Incipient Karst	1-3	Numerous incipient sinkholes; flat terrain	7/20	At, Al, Cy, M

*At = atrazine; Al = alachlor; Cy = cyanazine; M = metribuzin.

Significant features of northeastern Iowa and southeastern Minnesota include the presence of open and incipient sinkholes as well as large areas where bedrock is exposed or within 1-3 m of the surface. Sinkholes allow very rapid recharge from surface water and thus direct introduction of solutes such as pesticides. The Incipient Sinkhole regions contained hundreds of small, soil-filled depressions; pesticide infiltration must, therefore, occur through the soil mantle. Water movement through this soil is rapid. Within the Incipient Karst region, one well showed consistently high residues of alachlor, cyanazine, and metribuzin. Analyses showed high herbicide concentration in the soil profile and a very shallow (3-5 m) water table; probable high herbicide application rates were thought to contribute to the contamination problem. Libra et al. (60) found that residues in tile drain water were a good indicator of the quality of infiltrating water.

Aldicarb. The insecticide/nematicide aldicarb and its mobile degradation products ("aldicarb residues") reportedly have been found in the groundwaters of 13 states (2). In general, aldicarb residues were highest for wells sampled nearest the field application sites. The influence of soil differences on the pattern of aldicarb movement is well illustrated in a recent study (61-2). Potato fields in northeastern Florida were located on sandy soil underlain at 1-2 m by an impervious clay layer (61). Residues leached downward, then laterally, emerging into drainage ditches at concentrations of ca. 1-7 ppb (highest was 190 ppb 49 days after application). Thus, under restricted percolation, the surficial water was contaminated whereas the aquifer beneath this clay layer apparently was not.

Investigations in citrus fields elsewhere in Florida (62) were done in two types of sites: (a) shallow, poorly drained "flatwoods" soils, and (b) deep, well-drained "ridge" soils. Both had sandy soils. At site (a), aldicarb residues leached only to 1.25 m in 60 days, presumably a response to restricted drainage, higher FMC, and slightly higher adsorption in the surface 30 cm (OM content was greatest in the flatwoods soil). Leaching in the ridge soils was more extensive, with residues detected to ca. 3 m by 45 days. Subsoils [i.e., below 30-45 cm in (a) and 60-90 cm in (b)] showed no calculated retardation of leaching due to adsorption.

Soil properties were undoubtedly a major factor in the widespread occurrence of aldicarb residues in Long Island, New York groundwater. Soils there are sandy and permeable. According to a report (their reference #39) cited by Cohen et al. (2), "at least several percent of the aldicarb applied to certain Long Island fields" has leached to groundwater; generally, pesticide residues reaching groundwater seem to represent <<1% of the agronomic dose. Even under maximum assumed rates of degradation and plant uptake, modeling of Pacenka and Porter (63) predicted a substantial amount of aldicarb to leach below the root zone. Such movement is accelerated by a cold, wet spring, due to the combination of slower breakdown and greater net water infiltration.

The possibility of aldicarb leaching is recognized by the manufacturer's (64) warning on the product label that "...a combination of permeable and acidic soil conditions, moderate to heavy irrigation and/or rainfall, use of 20 or more pounds per acre, and soil temperature below 50°F at application time, tend to reduce degradation and promote movement of residues to groundwater." Aldicarb's use is prohibited in Suffolk County, Long Island, New York.

DBCP. During monitoring of California wells in 1979, the nematicide DBCP was detected in 59 of 119 wells tested (65). Residues in the San Joaquin Valley exceeded 10 ppb in 11 cases. Among the reasons for finding DBCP contamination include its long history of use in California (1960-1977), high required application rates, direct injection into soil or incorporation via irrigation water, and moderate mobility in laboratory or field leaching tests (66-8). DBCP is chemically hydrolyzed only very slowly (69); reliable estimates of half-life from microbial degradation seem to be lacking, perhaps because of difficulty in distinguishing breakdown from loss by volatilization. One of the major reasons for deep leaching of DBCP in California is probably that soils in the treated area are typically very sandy, with low OM content and high percolation (70).

At about the same time that DBCP was discovered in California water, it was also found in 26 of 93 Arizona well samples (71), though various short-circuiting mechanisms were then considered as alternative routes besides normal percolation through the soil profile. DBCP was absent in 282 samples (mostly groundwater) from Florida, Georgia, and South Carolina (72). A slightly smaller survey in South Carolina showed DBCP to occur in ca. 37% of surface waters from a high-use area; >1 ppb DBCP was found in 10% of well waters (73). It was not detected in areas where little or no application had been made. A followup study suggested that the groundwater contamination was due to a single point source (74). In general, the monitoring studies in the southeastern U.S. seemed to indicate that DBCP had not become a serious contamination problem there.

Continued monitoring in California showed that DBCP was present in soils of the San Joaquin Valley at depths to 15 m (75), but more recent cores collected 5 years following termination of DBCP's use in vineyards did not contain the nematicide (70). In the second case, sampling was to groundwater (7.5 and 10 m). Both soils were highly permeable, and the shallower groundwater had contained DBCP, so leaching may have contributed significantly to the chemical's loss. A California groundwater survey in 1982 showed no pesticides (i.e., DBCP, EDB, carbofuran, or simazine) in the Santa Maria or Salinas Valley groundwater basins (76). DBCP was present in 6 of 23 wells in the Upper Santa Ana basin (0.1-8 ppb), and in 21 of 166 wells in the San Joaquin basin (0.1-10 ppb). The authors were unable to correlate well characteristics with DBCP concentration in water. Simazine was also found in 5 wells, carbofuran in 1 well, and EDB in 2 wells.

A recent field experiment was reported (77) in which DBCP was chisel-injected into fallow plots, in Georgia, at normal and 3X normal rates. Under accelerated irrigation, traces of DBCP had leached to at least 12 m by 8 months, though at normal rates of pesticide and water, leaching below 4 m in 6 months probably had not occurred. The pesticide was present in perched groundwater and there was evidence, due to the existence of restricting clay layers, that lateral movement into control plot subsoil was important.

Summary

The major emphasis of this discussion centered on the importance of surface and subsurface soil characteristics in influencing deep pesticide leaching. Some factors, such as the depth to groundwater and the amount of incipient rainfall or irrigation, are clearly important factors that do affect the probability of pesticide residues reaching groundwater. Similarly, properties of the pesticide itself (especially its inherent mobility and chemical/biological stability) correlate closely with pollution potential, but their evaluation is outside the scope of this review.

We attempted to select, from many soil properties, those characteristics best linked to pesticide transport. Two approaches were used: (a) sensitivity analyses of several properties through the use of a one-dimensional convection-dispersion transport model and (b) qualitative review of laboratory and field studies. The model was thus intended to screen soils with respect to their potential for permitting leaching. A moderately mobile, nonvolatile pesticide (atrazine) was tested in the deterministic model.

The hierarchy of soil characteristics affecting leaching, taking into consideration their typical field variation, was bulk density < soil organic matter < hydraulic conductivity. Lab and field results support particularly the last two factors, although hydraulic conductivity is usually only inferred based on soil texture (i.e., more leaching in coarse-textured, permeable soils). The importance of macropore flow was recognized as a potentially important route for pesticide migration deep into the vadose zone. It is not easily quantified, however, and so does not seem to be amenable for consideration in screening models. On a local scale, soil surface and subsurface special characteristics should be incorporated--at least qualitatively--into prediction of pesticide movement.

APPENDIX

$$C_T(x,t) = \begin{cases} C_i A(x,t) + C_o(B(x,t) & o < t \leqslant t_o \\ C_i A(x,t) + C_o B(x,t) - C_o B(x,t-t_o) & t > t_o \end{cases}$$

where

$$A(x,t) = e^{(-\mu t/R)} \left\{ 1 - \frac{1}{2} \operatorname{erfc}\left[\frac{Rx-vt}{2(DRt)^{1/2}}\right] - \left(\frac{v^2 t}{\pi DR}\right)^{1/2} e^{\left[\frac{(Rx-vt)^2}{-4DRt}\right]} \right.$$

$$\left. + \frac{1}{2}\left(1 + \frac{vx}{D} + \frac{v^2 t}{DR}\right) e^{\left(\frac{vx}{D}\right)} \operatorname{erfc}\left[\frac{Rx+vt}{2(DRt)^{1/2}}\right] \right\}$$

$$B(x,t) = \frac{V}{v+u}\, e^{\left[\frac{(v-u)x}{2D}\right]} \operatorname{erfc}\left[\frac{Rx - ut}{2(DRT)^{1/2}}\right] + \frac{v}{v-u}\, e^{\left[\frac{(v+u)x}{2D}\right]} \operatorname{erfc}\left[\frac{Rx - ut}{2(DRT)^{1/2}}\right]$$

$$+ \frac{v^2}{2\mu D}\, e^{\left[\frac{vx}{D} - \frac{\mu t}{R}\right]} \operatorname{erfc}\left[\frac{Rx + vt}{2(DRt)^{1/2}}\right]$$

and

$$u = v\left(1 + \frac{4\mu D}{v^2}\right)^{1/2}$$

Literature Cited

1. Pye, V. I.; Patrick, R.; Quarles, J. "Groundwater Contamination in the United States"; Univ. Pennsylvania Press: Philadelphia, 1983; p. 38.
2. Cohen, S. Z.; Creeger, S. M.; Carsel, R. F.; Enfield, C. G. In "Treatment and Disposal of Pesticide Wastes"; Krueger, R. F.; Seiber, J. N., Eds.; ACS SYMPOSIUM SERIES No. 259, American Chemical Society: Washington, D.C., 1984; pp. 297-325.
3. Black, C. A., Ed. "Methods of Soil Analysis"; Agronomy 9, American Society of Agronomy: Madison, Wisc., 1965.
4. Brady, N. C. "The Nature and Properties of Soils", 9th Edit.; Macmillan: New York, 1984.
5. Gieseking, J. E., Ed. "Soil Components"; Springer-Verlag: New York, 1975.
6. Jackson, M. L. "Soil Chemical Analysis"; Prentice-Hall: Englewood Cliffs, N.J., 1958.
7. Revut, I. B.; Rode, A. A., Eds. "Experimental Methods of Studying Soil Structure"; Amerind Publ. Co. Pvt. Ltd.: New Delhi, 1981.
8. Goring, C. A. I.; Hamaker, J. W., Eds. "Organic Chemicals in the Soil Environment"; Marcel Dekker: New York, 1972.
9. Graham-Bryce, I. J. In "The Chemistry of Soil Processes"; Greenland, D. J.; Hayes, M. H. B., Eds.; John Wiley: New York, 1981; Chap. 12.
10. Helling, C. S.; Kearney, P. C.; Alexander, M. Adv. Agron. 1971, 23, 147-240.
11. Hartley, G. S.; Graham-Bryce, I. J. "Physical Principles of Pesticide Behavior"; Academic Press: New York, 1980; Vol. 1.
12. Scheunert, I.; Klein, W. In "Appraisal of Tests to Predict The Environmental Behaviour of Chemicals"; Sheehan, P.; Korte, F.; Klein, W.; Bourdeau, P., Eds.; John Wiley: New York, 1985; Chap. 5.

13. Jury, W. A. Water Resour. Res. 1982, 18, 363-8.
14. Parker, J. C.; van Genuchten, M. Th. "Determining Transport Parameters from Laboratory and Field Tracer Experiments"; Virginia Agric. Exp. Sta. Bull. 84-3; 1984.
15. Gelhar, L. W.; Gutjahr, A. L.; Neff, R. I. Water Resour. Res. 1979, 15, 1387-97.
16. Gelhar, L. W.; Axness, C. L. "Stochastic Analysis of Macrodispersion in Three-Dimensionally Heterogeneous Aquifers"; Hydrology Res. Center, New Mexico Instit. Mining Technol.: Las Cruces, NM; Rep. H-8; 1981.
17. Bakr, A. A.; Gelhar, L. W.; Gutjahr, A. L.; MacMillian, J. R. Water Resour. Res. 1978, 14, 263-71.
18. Simmons, C. S. Water Resour. Res. 1982, 18, 1193-1214.
19. Nye, P. H. Adv. Agron. 1979, 31, 228-72.
20. Saffman, P. G. J. Fluid Mech., Part B 1959, 6, 321-49.
21. Karickhoff, S. W. Chemosphere 1981, 10, 833-46.
22. Rao, P. S. C.; Davidson, J. M. In "Environmental Impact of Nonpoint Source Pollution," Overcash, M. R., Ed.; Ann Arbor Sci. Publ.: Ann Arbor, Mich.; 1980.
23. Parker, J. C.; van Genuchten, M. Th. Water Resour. Res. 1984, 20, 866-72.
24. van Genuchten, M. Th.; Alves, W. J. "Analytical Solutions of the One-Dimensional Convective-Dispersive Solute Transport Equation"; U.S. Dept. Agric.: Washington, D.C.; U.S. Dept. Agric. Tech. Bull. No. 1661; 1982.
25. Wierenga, P. J. Soil Sci. Soc. Am. J. 1977, 41, 1050-5. De Smedt, F.; Wierenga, P. J. Soil Sci. Soc. Am. J. 1978, 42, 7-10.
26. De Smedt, F.; Wierenga, P. J. Soil Sci. Soc. Am. J. 1978, 42, 7-10.
27. Jury, W. R.; Spencer, W. F.; Farmer, W. J. J. Environ. Qual. 1984, 13, 573-9.
28. van Genuchten, M. Th.; Wierenga, P. J. Soil Sci. Soc. Am. J. 1976, 48, 437-80.
29. Davidson, J. M.; Rieck, C. E.; Santlemann, P. W. Soil Sci. Soc. Am. Proc. 1968, 32, 629-33.
30. Karickhoff, S. W.; Brown, D. S.; Scott, T. A. Water Res. 1977, 13, 231-48.
31. Osgerby, J. M. In "Sorption and Transport Processes in Soils"; S.C.I. Monogr. No. 37; Soc. Chem. Industry: London, 1970; pp. 63-78.
32. Hamaker, J. W.; Thompson, J. M. In "Organic Chemicals in the Soil Environment"; Goring, C. A. I.; Hamaker, J. W., Eds.; Marcel Dekker: New York, 1972; pp. 51-143.
33. Gish, T. J.; Parker, J. C.; Coffman, C. B., unpublished data.
34. Amoozegar-Fard, A.; Nielsen, D. R.; Warrick, A. W. Soil Sci. Soc. Am. J. 1982, 46, 3-9.
35. Gish, T. J.; Jury, W. A. Trans. Am. Soc. Agric. Eng. 1983, 26, 440-44, 451.
36. Millington, R. J.; Quirk, J. M. Trans. Faraday Soc. 1961, 57, 1200-7.
37. Walker, A. J. Environ. Qual. 1974, 3, 396-401.
38. Hance, R. J. Pestic. Sci. 1979, 10, 83-6.
39. Gish, T. J.; Helling, C. S.; Parker, J. C.; Zhuang, W.; Isensee, A. R.; Kearney, P. C., unpublished data.

40. Rawls, W. J.; Brankensick, D. L.; Saxton, K. E. Trans. Am. Soc. Agric. Eng. 1982, 25, 1316-20, 1324.
41. Gerhardt, R. A. Ground Water Monit. Rev. 1984, 4, 56-65.
42. Nielsen, D. R.; Biggar, J. W.; Erh, K. T. Hilgardia 1973, 42, 215-59.
43. Libardi, P. L.; Reichardt, K.; Nielsen, D. R.; Biggar, J. W. Soil Sci. Soc. Am. J. 1980, 44, 3-6.
44. Willardson, L. S.; Hourot, R. C. J. Irrig. Drain. Div. Am. Civil Eng. 1965, 91, 1-9.
45. Helling, C. S.; Dragun, J. In "Test Protocols for Environmental Fate and Movement of Toxicants"; Assoc. Offic. Anal. Chemists: Arlington, Va., 1981; pp. 43-88.
46. Savage, K. E. Weed Sci. 1976, 525-8.
47. Helling, C. S. Soil Sci. Soc. Am. Proc. 1971, 35, 743-8.
48. Helling, C. S. Soil Sci. Soc. Am. Proc. 1971, 35, 737-43.
49. Bilkert, J. N.; Rao, P. S. C. J. Environ. Sci. Health 1985, B20, 1-26.
50. Helling, C. S. Residue Rev. 1970, 32, 175-210.
51. Letey, J.; Farmer, W. J. In "Pesticides in Soil and Water"; Guenzi, W. D., Ed.; Soil Sci. Soc. Am.: Madison, Wisc., 1974; Chap. 4.
52. Hogue, E. J.; Khan, S. U.; Gaunce, A. Can. J. Soil Sci. 1981, 61, 401-7.
53. Richard, J. J.; Junk, G. A.; Avery, M. J.; Nehring, N. L.; Fritz, J. S.; Svec, H. J. Pestic. Monit. J. 1975, 9, 117-23.
54. Spalding, R. F.; Exner, M. E.; Sullivan, J. J.; Lyon, P. A. J. Environ. Qual. 1979, 8, 374-83.
55. Junk, G. A.; Spalding, R. F.; Richard, J. J. J. Environ. Qual. 1980, 9, 479-83.
56. Spalding, R. F.; Junk, G. A.; Richard, J. J. Pestic. Monit. J. 1980, 14, 70-3.

Wehtje, G.; Leavitt, J. R. C.; Spalding, R. F.; Mielke, L. N.; Schepers, J. S. Sci. Total Environ. 1981, 21, 47-51.

58. Wehtje, G. R.; Spalding, R. F.; Burnside, O. C.; Lowry, S. R.; Leavitt, J. R. C. Weed Sci. 1983, 31, 610-18.
59. Hall, J. K.; Hartwig, N. L. J. Environ. Qual. 1978, 7, 63-8.
60. Libra, R. D.; Hallberg, G. R.; Ressmeyer, G. G.; Hoyer, B. E. "I. Groundwater Quality and Hydrogeology of Devonian-Carbonate Aquifers in Floyd and Mitchell Counties, Iowa"; Iowa Geological Survey: Iowa City, Iowa. Iowa Geolog. Survey Open File Rep. 84-2; 1984.
61. Weingartner, D. P.; Fong, G. In "Aldicarb Research: Task Force Report. September 1, 1983"; Inst. Food Agric. Sci. and College of Engineering, Univ. Florida: Gainesville, Fla.; pp. 23-5.
62. Rao. P. S. C.; Hornsby, A. G.; Nkedi-Kizza, P.; Ou, L. T.; Edvardsson, S.; Bilkert, J. N.; Woodburn, K.; Jessup, R. E.; Wheeler, W. B.; Rocca, J. In "Aldicarb Research: Task Force Report. September 1, 1983"; Inst. Food Agric. Sci. and College of Engineering, Univ. Florida: Gainesville, Fla.; pp. 7-21.
63. Pacenka, S.; Porter, ; "Preliminary Regional Assessment of the Environmental Fate of the Potato Pesticide, Aldicarb, Eastern Long Island, New York", Cornell Univ.: Ithaca, New York; 1981.
64. Union Carbide Agricultural Products. "1985 Chemical Guide"; Union Carbide Agric. Products Co., Inc.: Research Triangle Park, N.C.; 1985; pp. 212, 224.

65. Peoples, S. A.; Maddy, K. T.; Cusik, W.; Jackson, T.; Cooper, C.; Frederickson, A. S. Bull. Environ. Contam. Toxicol. 1980, 24, 611-18.
66. Johnson, D. E.; Lear, B. Soil Sci. 1968, 105, 31-5.
67. Wilson, J. T.; Enfield, C. G.; Dunlap, W. J.; Cosby, R. L.; Foster, D. A.; Baskin, L. B. J. Environ. Qual. 1981, 10, 501-6.
68. Youngson, C. R.; Goring, C. A. I.; Noveroske, R. L. Down to Earth 1967, Summer, 27-32.
69. Burlinson, N. E.; Lee, L. A.; Rosenblatt, D. H. Environ Sci. Technol. 1982, 16, 627-32.
70. Zalkin, F.; Wilkerson, M.; Oshima, R. J. "Pesticide Movement to Groundwater. Vol. II. Pesticide Contamination in the Soil Profile at DBCP, EDB, Simazine and Carbofuran Application Sites"; Calif. Dept. Food and Agric.: Sacramento, Calif.; Final Rep., EPA Grant #E009155-79; 1984.
71. Filleman, T.; Nemecek, E. A. Proc. Deep Percolation Symp. 1980, Arizona Dept. Water Resour. Rep. No. 1, 85-97.
72. Mason, R. E.; McFadden, D. D.; Iannacchione, V. G.; McGrath, D. S. "Survey of DBCP Distribution in Ground Water Supplies and Surface Water Ponds"; Research Triangle Inst.: Research Triangle Park, N.C.; EPA Final Rep. RTI/1864/05-08F; 1981.
73. Carter, G. E., Jr.; Riley, M. B. Pestic. Monit. J. 1981, 15, 139-42.
74. Carter, G. E., Jr.; Ligon, J. T.; Riley, M. B. Water, Air, Soil Pollut. 1984, 2, 201-8.
75. Nelson, S. J.; Isklander, M.; Volz, M.; Khalifa, S.; Haberman, R. Sci. Total Environ. 1981, 21, 35-40.
76. Weaver, D. J.; Sava, R. J.; Zalkin, F.; Oshima, R. J. "Pesticide Movement to Groundwater. Vol. I. Survey of Groundwater Basins for DBCP, EDB, Simazine, and Carbofuran"; Calif. Dept. Food and Agric.: Sacramento, Calif.; Final Rep., EPA Grant #E009155-79; 1983.
77. Helling, C. S.; Wehunt, E. J.; Feldmesser, J.; Kearney, P. C. Abstr., 187th ACS Natl. Mtg. 1984, PEST 10.

RECEIVED April 7, 1986

3

Determining Uncertainty in Physical Parameter Measurements by Monte Carlo Simulation

David W. Coy[1], Gregory A. Kew[2], Michael E. Mullins[1], and Phillip V. Piserchia[1]

[1] Research Triangle Institute, Research Triangle Park, NC 27709
[2] Office of Health and Environmental Assessment (RD-689), U.S. Environmental Protection Agency, Washington, DC 20460

A statistical approach, often called Monte Carlo simulation, has been used to examine propagation of error and to better characterize the uncertainty associated with measurement of several parameters important in predicting environmental transport of chemicals. These parameters are vapor pressure, water solubility, octanol-water partition coefficient, and "volatilization from water" (based on the ratio of laboratory-measured volatilization rate constant to oxygen reaeration rate constant for a specific system). Column chromatographic and high pressure liquid chromatographic (HPLC) methods are replacing more traditional equilibrium methods (e.g. shake flask, isoteniscope) for measuring the first three parameters. The newer methods tend to underpredict aqueous solubility and vapor pressure and overpredict octanol-water partition coefficient, although deviations for both the equilibrium and dynamic systems are similar. Measurement error proves not to be normally distributed, with differing bias for each parameter. For "volatilization from water", determinations of the ratio of rate constants for compounds whose Henry's Law constant equals or exceeds 1,000 torr/mole/liter typically report 95% percent confidence limits equal to 5 to 10 percent of the ratio. Analysis of a regression approach often used to determine the ratio suggests underestimation of both the ratio and its variance. Monte Carlo simulation did not confirm underestimation of the ratio but suggests variances may be underestimated by a factor of 2.3. Using this statistical approach in other cases might allow an investigator to choose levels of a parameter (e.g. a drinking water standard) knowing the uncertainty associated with the choice, or the converse.

0097-6156/86/0315-0039$06.50/0
© 1986 American Chemical Society

The purpose of this study is to illustrate a statistical approach for determining overall uncertainty in a parameter calculated from more than one input, when distribution of error for the inputs can be estimated. The approach incorporates this information into a concise statement of overall uncertainty. In particular, it permits determining the probability of exceeding a specific level. The results then are compared with error estimates for specific experimental procedures in the literature.

This approach seems applicable to broader investigations than are explored here, such as predicting environmental transport and fate of chemicals where properties of various compounds are used in models. Even more generally, it might be applied to exposure assessment where, usually, too few data are available to characterize distributions of exposure to particular chemicals.

General approaches for expressing uncertainty most evident in the literature may be viewed as various forms of "propagation of error." Variants range from "sensitivity analysis" where the effects of individual input variables are examined by holding all other variables at midrange while the one under study is varied from minimum to maximum, to the use of fixed formulae derived for various mathematical relationships in specifying absolute maxima and minima for outputs. None of these methods can supply a probability of occurrence for a given value within the range of possible error.

What appears to be yet another variant, a statistical approach often called Monte Carlo simulation, recently has been demonstrated to apply by Walentowicz and Falco if the distribution of values for each input variable is known or can be approximated (1). Thereafter, this information can be manipulated to yield a distribution of outputs expressing probability of occurrence for any given value. Also, a recent study by Whitmore (2) develops the concepts of Walentowicz and Falco and offers conclusions on the form of final limiting distributions (e.g., simulated outputs for product functions approach log normal) if specific mathematical conditions are met. The precise degree to which the conditions must be fulfilled for the technique to yield a useful result in the present application have not yet been fully explored.

While the examples considered here are experimental determination of (1) water solubility, (2) octanol-water partition coefficient, (3) vapor pressure, and (4) volatilization from water, the resulting distributions presumably then could become inputs to larger problems such as determining media-specific environmental concentration, in turn expressed as a well-characterized distribution.

Simulation of Method Error

The process used for determining a distribution of aqueous solubility values involves assuming a "true" solubility value equal to 1.00 arbitrary units, assigning an error distribution for each recognized error source, and calculating the effect on true value of the several sources of error. After assigning the "true" value, random numbers generated by computer program were used to pick input values within the solubility range of concern. The corresponding error was then calculated for the source first affecting the solubility determination. The resulting discrete error was added to or subtracted from the true value and the resultant used as input to the distribution for the second-occurring error source. Contributions from sources operating simultaneously, rather than sequentially, were calculated and summed simultaneously. This process is repeated to include all recognized sources and results in a single solubility value corresponding to the single random number originally chosen.

The process of choosing a random number and calculation of cascading error is repeated 10,000 times, generating a distribution of solubility values relative to the "true" value which can be characterized mathematically, plotted, and used to furnish the likelihood or probability of any particular solubility value in the range being exceeded. Essentially the same steps were followed in simulations for other parameters investigated, even for those on vapor pressure, where effects of temperature variation are compound-specific.

The calculation procedures used are available in software of the Statistical Analysis System (SAS), version 82-3. Central processing unit time per output distribution on a main frame computer (IBM Model 3280) was around 20 seconds and cost per output distribution was less than $6.

Distributions generated by the process described above may not be easy to label as "normal" or "skewed log-normal", for example, but in general (for Pearson distributions) they are completely characterized by specifying the first four statistical moments, i.e. mean, variance, skew, and kurtosis (3). Consequently, rather than supplying less quantitative labels, these moments are provided for all output distributions which follow. Comparisons with experimental results in the text are limited to discussion of the mean and the square root of variance or standard deviation, because few experiments (and none located in this study) are repeated often enough to generate data sets allowing meaningful calculation of third and fourth moments.

"Beta functions" have been used to represent several sources of error. The beta function is extremely flexible in the sense that it can assume a great variety of shapes depending on values chosen for exponents, as seen in Figure 1, and seems particularly useful in representing negative bias.

Experimental Methods for Solubility Determination

Solubility data for organic compounds in water are obtained readily for highly soluble materials, but for sparingly soluble compounds measurements can be difficult due to potential losses to air or to the container walls, long equilibration times, need for extreme purity in starting materials, and possibly unanticipated minor reactions in solution. As a result, errors of an order of magnitude for some compounds have been reported by MacKay et al. (4).

For slightly soluble compounds, two methods dominate the current literature, the shake-flask method, and the generator column method. In the shake-flask (or stir-flask) method, a known amount of solute exceeding the anticipated solubility limit is added to a measured water sample in a sealed container. This mixture is shaken or stirred thoroughly and then allowed to sit for eight hours to several days to ensure that solute-solvent equilibrium is attained. The mixture is then filtered or centrifuged to remove suspended particles, and the solution concentration is determined using customary chromatographic or spectrascopic techniques. Changing range within a method may affect error.

The column generator (and related HPLC) method of determining aqueous solubility has become the predominant method in the literature over the past 5 years. In this method, a column is packed with an inert solid support which provides a high surface area for the solute to insure quick equilibration between it and the aqueous phase. After the support is impregnated with the solute of interest and excess solute has been displaced, the column is brought to the desired temperature and the aqueous concentration is determined as a function of flow rate. The equilibrium concentration corresponds to the highest eluting flow rate where solute concentration remains flow-independent.

Sources of Error in Solubility Determination

The shake-flask method for determining solubility has four major sources of determinate error associated with it. In the first category are losses from adsorption of the solute on the flask walls or to evaporation. This quantity is somewhat dependent on the vapor pressure of the solute in question, but if the quantity of solution used is fairly large (i.e., 250 ml), these losses may represent a maximum of 5 percent of the total solute. A beta distribution was chosen for this error source. The presence of undissolved solute suspended in the water phase may cause a sample to register anomalously high solubility. Centrifuging or filtering reduces this contribution, although the problem may persist for more neutrally bouyant solutes. This error is represented by an exponential distribution with the very low standard deviation expected if the technique described by Karickhoff and Brown (5) is properly followed (±2 percent with 95-percent confidence).

The third source arises from the extraction efficiency for the removal of solute from water, a technique common to most methods. Peters (6), has shown this to introduce another negative bias within 8 percent; therefore a beta distribution with a 95-percent confidence level of -8 percent was selected. Finally, the accuracy of the detection method was modeled by a normal distribution with a 95-percent confidence level of ±2 percent as recommended by Mallon and Harrison (7).

Note that the error introduced by each step is based upon the exit concentration of solute from the previous step and, therefore, is multiplicative in nature. In line with the conclusions of Whitmore (2), the resulting output distribution might then be expected to approximate log-normality. However, the results of the computer simulation of error for aqueous solubility by the shake-flask method shown in Figure 2 clearly exhibit a negative bias. This bias was introduced by potential losses and by extraction efficiency, although the standard deviation of 3.9 percent compares quite favorably to that cited by several researchers. The mean simulated value underestimates the true value by slightly over 2 percent.

The column method also has several steps which may result in a negative bias on measured aqueous solubility. The lack of water-solute equilibrium in the generator column itself may produce an outlet concentration lower than the "true" solubility according to Stolzenburg and Andren (8). This result also has been modeled with a beta distribution (95-percent confidence level of -3 percent).

The efficiencies for adsorption and extraction steps again are estimated to be similar to those suggested by Peters (6), therefore, the same beta distributions as before are employed. However, note that for the column generator, two-step organic removal typically is used. Finally, one may ascribe the same detection limits as those obtained for the shake-flask method.

The column method results (Figure 3) exhibit a mean simulated value 7 percent under the true value, or 5 percent less than that obtained simulating the shake-flask method. The difference may be attributed to the additional extraction step and solute-equilibration step in the column method. The tendency of this method to underpredict reaffirms the need for calibration of the column system via shake-flask standards prior to use, although the predicted standard deviation of 4 percent is quite good.

Methods and Error in Octanol-Water Partition Coefficient Determinations

The octanol-water-partition coefficient (K_{ow}) is the most frequently cited measure of environmental partitioning behavior. Partition coefficients traditionally have been determined by some variation on the shake-flask method, however, over the past several years high-pressure liquid chromatography has been shown to measure octanol-water partition coefficients accurately over a wide range of values with greater ease.

Since the column/HPLC and shake-flask methods are also the main source of solubility values, the evaluation of error for K_{ow} is very similar. Loss of solute to walls or the atmosphere are still of concern, as are the extraction efficiencies for solute removal from water. Equilibration of each of the three phases involved (instead of two for solubility) is of concern.

The shake-flask method for K_{ow} differs from the solubility determination only in the measurement of solute concentration in the octanol phase. Since this typically involves standard chromatographic or spectroscopic techniques, error again was assumed to be normally distributed about the true concentration with a 95-percent confidence level of ± 2 percent.

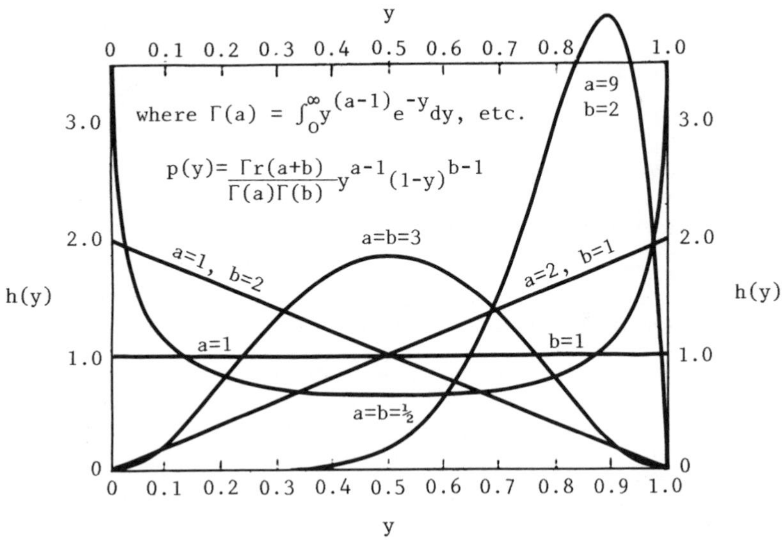

Figure 1. Beta density functions. Reproduced with permission from Ref. 20. Copyright 1970, Houghton Mifflin Co..

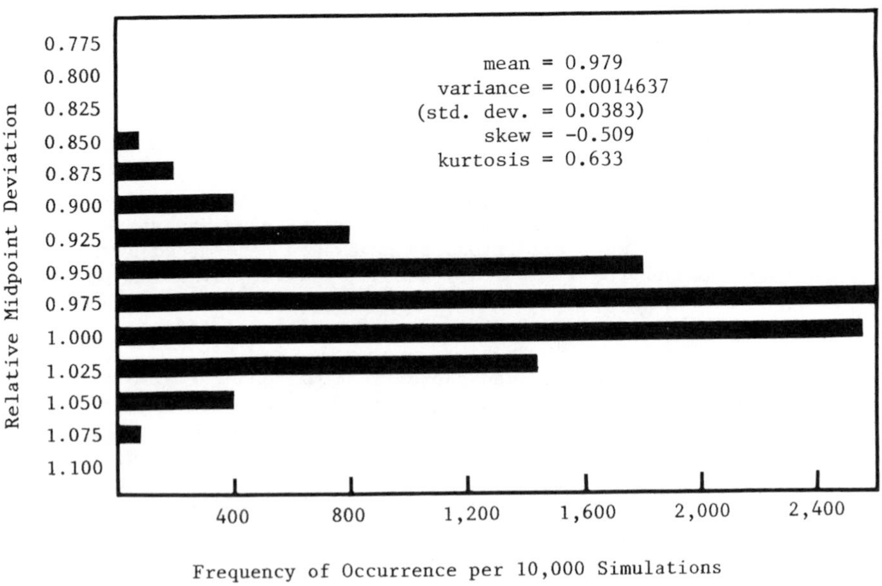

Figure 2. Distribution of simulated aqueous solubility from the shake-flask method (true value = 1.00).

Examination of Figure 4 illustrates the tendency of this technique to slightly overpredict K_{ow}. The mean value observed exceeds the true value by 2.3 percent. The standard deviation of 10 percent is somewhat in excess of the 6.1 percent given by Unger et al. (9) and the 8 percent given by Bowman and Sans (10).

The column method for determining K_{ow} involves the equilibration of a solute from a stationary phase (in this case a simulated octanol phase) into an aqueous phase (or methanol-water phase). The solute is then extracted from water and analyzed using an appropriate detector.

Comparison of the column method K_{ow} results with the results given by the column method for solubility in Figure 3 shows the shape of the distributions to be virtually identical with standard deviations being 4.0 percent for solubility and 4.2 percent for K_{ow}. This corresponds to the similarities noted by DeVoe et al., (11). The column method, however, differs from the shake-flask by 9.2 percent in the case of K_{ow} (versus 5 percent for solubility). Woodburn et al. (12) quote a difference of 10 percent between techniques. As indicated by comparing standard deviations, the column method seems as reproducible as the shake-flask method, but accurate calibration is even more necessary to prevent underprediction than was the case for column method for solubility.

Methods for Vapor Pressure Determination

Methods used to determine the vapor pressure of pure compounds are usually divided into two groups based on pressure ranges to which they apply: (1) between 1.0 and 760 torr, and (2) below 1.0 torr.

Manometric Methods. Direct manometry methods such as the isoteniscope method are most often used for the pressure range of 1.0 to 760 torr. The method for the isoteniscope is described in detail in ASTM Method D-2879-70 (13). Figure 5 shows a schematic of the isoteniscope. The sample bulb and short leg of the manometer are filled with the liquid of interest and isoteniscope is attached to a vacuum source. Dissolved gases are removed from the liquid by reducing the system pressure to approximately 1 torr and carefully warming the liquid to its boiling point. The resulting vapor is used to force a small amount of the liquid into the manometer section which traps some vapor between the manometer and bulb. The entire device is placed in a constant temperature bath until equilibrium between liquid and vapor is achieved. Nitrogen is added to the side of the manometer opposite to the sample bulb, until both liquid legs in the manometer section are equal. At that point, the nitrogen pressure is determined and recorded. A McLeod gauge usually is used between 15 and 760 torr. This procedure typically is replicated for temperature increments of 25°C until a system pressure of 760 torr is obtained.

The data obtained in this manner are plotted in terms of the Antoine equation, which relates the natural logarithm of pressure to the reciprocal of absolute temperature:

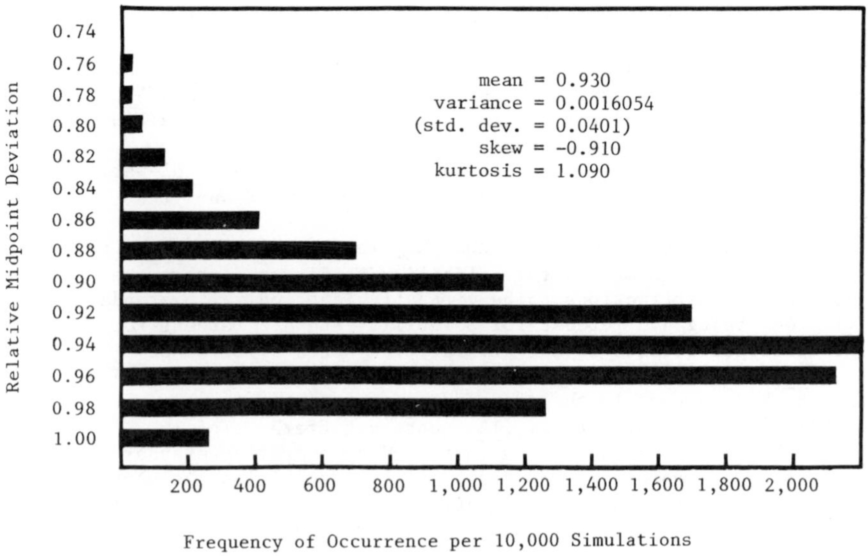

Figure 3. Distribution of simulated aqueous solubility from the column method (true value = 1.00).

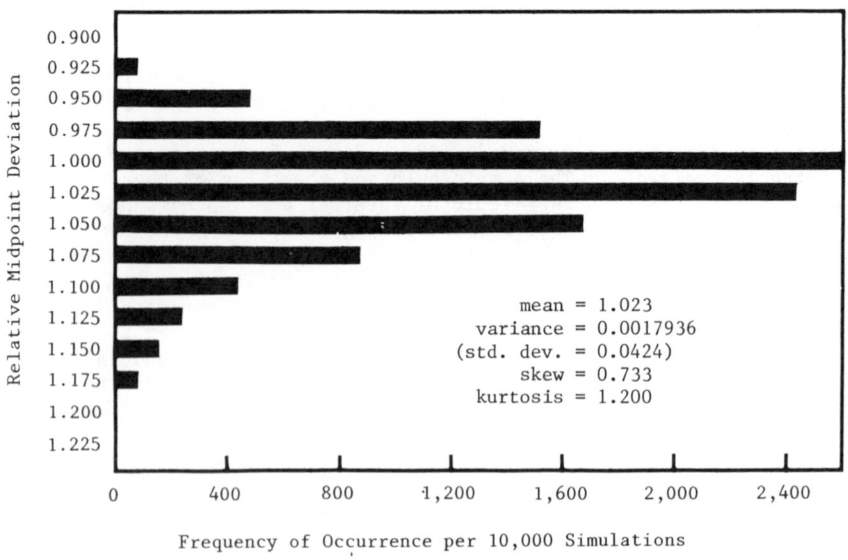

Figure 4. Distribution of simulated octanol-water partition coefficient from the shake-flask method (true value = 1.00).

$$\ln P_{vapor} = A + \frac{B}{(T + C)} \quad (1)$$

where A, B, and C are constants for a given compound. (For this study, rounded "true values" were assumed for vapor pressure, as noted in the text and figures which follow, rather than inserting specific values of A, B, and C for each compound.)

The isoteniscope measurement contains three major sources of error. Two of these sources are independent of pressure. The mercury manometer accuracy of ± 1 torr is set as a standard normal distribution about the "true" pressure. The question of sample purity is addressed in the ASTM methodology, with a maximum upward bias of 1 torr on the vapor pressure assured by using this technique. A log-normal distribution with a mean to 0.1 torr is chosen as the representative model based upon suggestions by Osborn and Scott (14). Since these errors are simultaneous in nature and independent of sample pressure, they are taken to be additive.

To determine the overall effect on pressure error of normally distributed variation in temperature, the Antoine equation must be employed. Thus, the pressure error is a function of the compound of interest. To focus on typical situations, three compounds representative of the range of application of the isoteniscope were used. These were: benzene with a relatively high vapor pressure ("true value" = 95 torr at 25°C); toluene for a medium vapor pressure ("true value" = 28 torr at 25°C); and chlorobenzene for a low vapor pressure ("true value" = 12 torr at 25°C).

The results of isoteniscope simulations show a slight over-prediction for all three compounds with a skewed distribution in each case (Figures 6, 7, and 8). Predictably, the lower the vapor pressure, the larger the relative standard deviation. On the other hand, the absolute standard deviation is relatively unchanged over the range of vapor pressures examined, varying between about 0.72 and 0.77 torr for the extreme cases. The maximum skewedness of the distribution is found for toluene, the medium-vapor pressure case. The standard deviation of 2.7 percent found for toluene corresponds quite well to the 3 percent error estimate made by MacKay et al. (4). The 6 percent standard deviation for the chlorobenzene case shows it to be at the lower end of the range recommended for reliable measurement. Figure 9 summarizes the isoteniscope simulations, indicating a small tendency of the method to overpredict in the overall pressure range for the three compounds considered.

Gas Saturation Method. For vapor pressures less than 1 torr, a gas saturation method is most often employed. The effusion method or Knudsen method also has been used in this pressure range, but it has proved quite imprecise (15).

The gas saturation method relies upon measurement of vapor loss rates from surfaces for a known volume of gas passed over the surface. While there are several variations on the technique, including ones using HPLC, the one described by Spencer et al. (15) is typical. Dry nitrogen is passed through a quartz sand

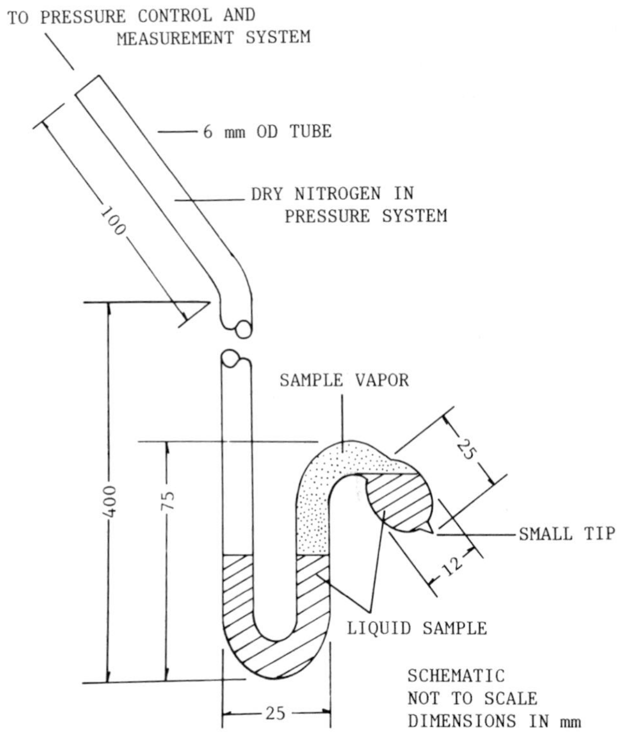

Figure 5. Schematic of an isoteniscope. Reproduced with permission from Ref. 13. Copyright 1974, American Society for Testing and Materials.

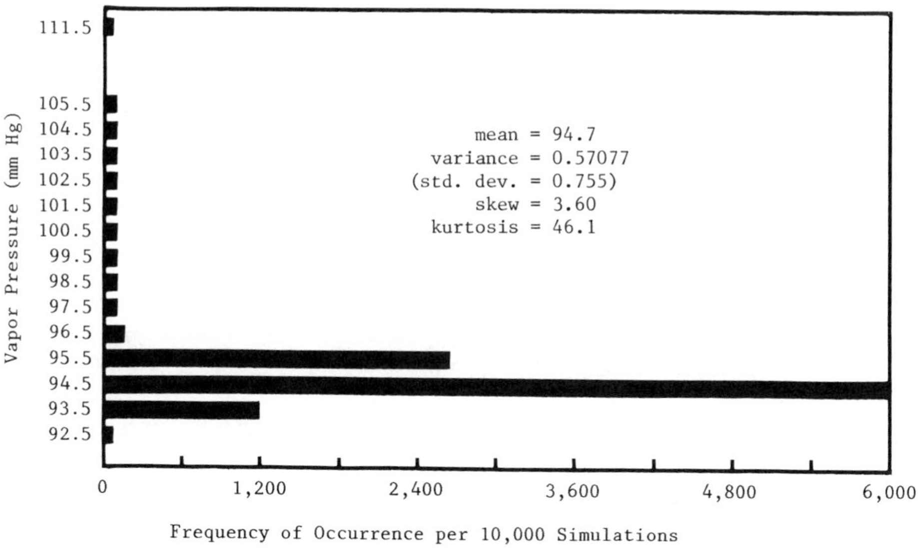

Figure 6. Distribution of simulated vapor pressure for benzene using the isoteniscope ("true value" = 95.0 mm Hg).

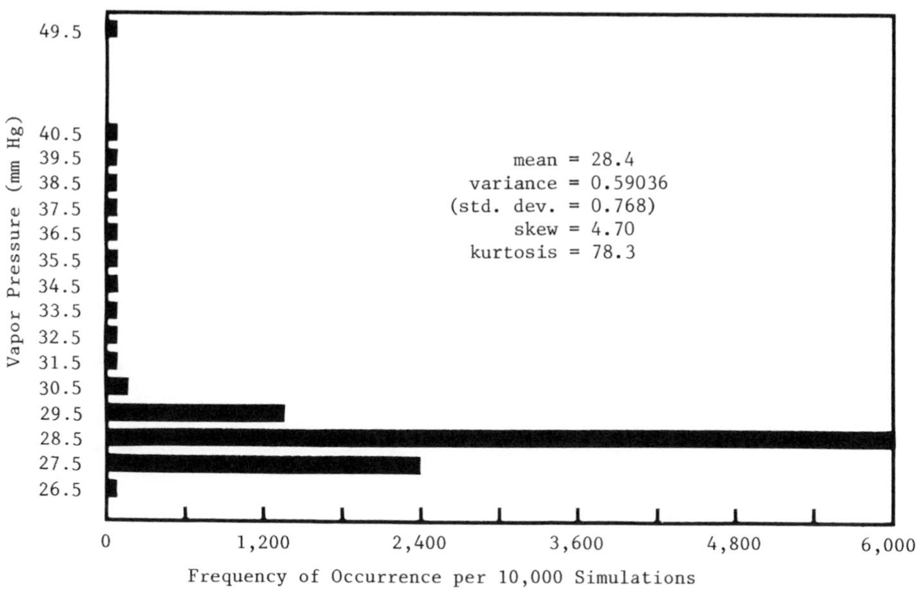

Figure 7. Distribution of simulated vapor pressure for toluene using the isoteniscope ("true value" = 28.0 mm Hg).

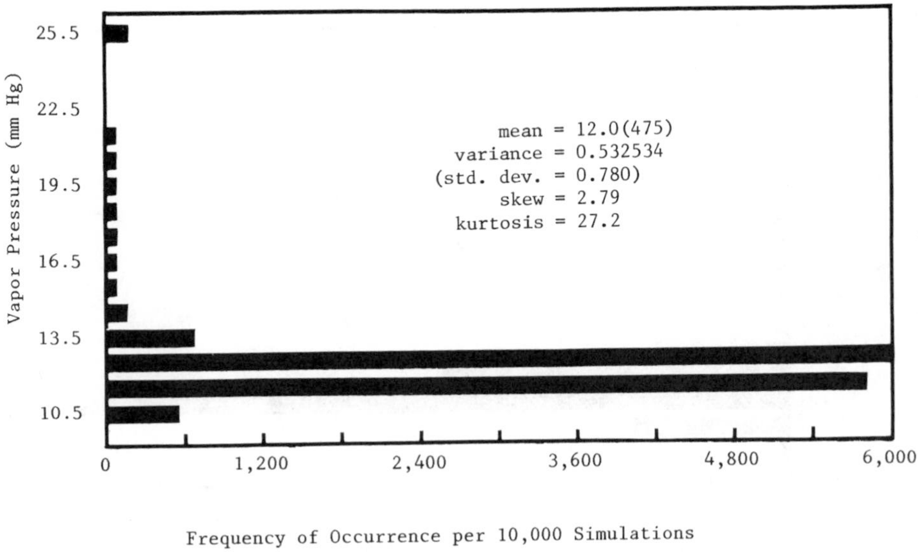

Figure 8. Distribution of simulated vapor pressure for chlorobenzene using the isoteniscope ("true value" = 12.0 mm Hg).

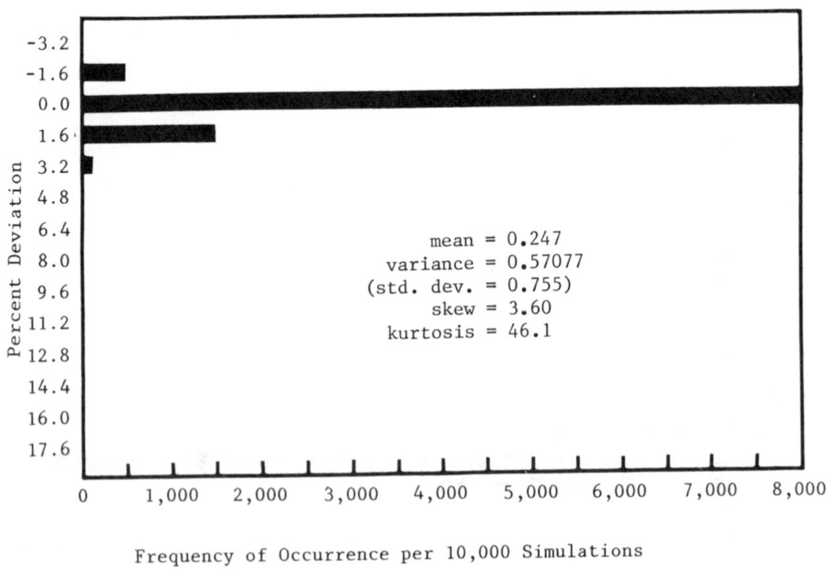

Figure 9. Distribution of relative error for the three vapor pressure simulations (true value = 0.00).

(or other support material) that has been saturated previously with the compound of interest. The vapor that is evolved from this column is trapped on a appropriate sorbent material, which is subsequently extracted and concentrated with a hexane/acetone solution. The concentration of the sample compound in the extract may then be determined using gas chromatography or other detection methods. The vapor phase concentration at equilibrium conditions may be calculated based upon the volume of carrier gas that is used to collect a measured amount of solute in the extract. This procedure is repeated for 10°C temperature increments, and the data are plotted as ln P_{vapor} versus reciprocal absolute temperature (Antoine equation).

The gas saturation method obviously suffers from errors of a different nature than the isoteniscope method. Vapor lost to surfaces within the measurement system and low vapor-trapping efficiency may lead to significant errors for low vapor pressure compounds. Spencer et al. (15) found vapor capture efficiencies as low as 90 percent, while Wasik, et al. (16) arrived at a 95-percent confidence level of ±9 percent. Thomas and Sieber (17) reportedly obtained an error of ±1 to 3 percent using a column-gas saturator. Spencer noted, however, that for some very low vapor pressure compounds, up to an order of magnitude variation has been reported.

For the gas saturation method, the error sources and their approximate values are very similar to previous column techniques. [At 95% confidence, temperature control = ±0.05°K (normal), lack of vapor-liquid equilibrium = -5% (beta), trapping efficiency = -8% (beta), extraction efficiency = -8% (beta), gas chromatographic measurement = ±2% (normal).] Consequently, the error distribution for all three column methods should be approximately the same. This means a general underprediction of vapor pressure as shown in Figure 3.

Effusion Method. The effusion or Knudsen cell method is much less reliable than the gas saturation method. The measurement is based upon weighing a pure sample before and after diffusion of vapor from the sample through an apperture of known cross section into a high vacuum system. Weight loss per unit time then is used to calculate vapor pressure. Unfortunately, the weight loss is the small difference between two relatively large numbers and also proves to be highly dependent on effusion times used, temperature control, and sample purity. Although effusion method data have been reported for materials with vapor pressures as low as 5 x 10^{-6} torr, DePablo (18) reported up to ±0.3 torr error at measurements of 1 torr (or 30-percent error with 95-percent confidence). Consequently, detailed statistical analysis was not considered worthwhile.

Volatilization from Water

Volatilization from water, the transport of a chemical in solution in a water body to the atmosphere, is believed to be the principal aquatic fate of low molecular weight, nonpolar compounds that are not otherwise rapidly degraded or transformed (19). The estimation of compound volatilization from a natural water body can be made on the basis of laboratory-measured rate constants, with extrapolation to estimate the activity in the natural environment typically performed in subsequent calculations. For compounds of differing levels of volatility, the rate constants that are required for extrapolation differ. Selection of the appropriate rate constant may be made using results of screening tests.

Test protocols are available for high, medium, and low volatility compounds. However, only the high volatility compound protocol has undergone more than preliminary validation studies. The categorization of compounds into high, medium, and low volatility classes is accomplished by performance of a screening test protocol on the selected compound. The principal reference for the test protocol description and background information is a report prepared by SRI International under contract to the U.S. Environmental Protection Agency by Mill, et al. (19).

The recommended error estimation methods for high volatility compounds from the reference are reviewed in this section of the report after describing the test protocol procedures. A Monte Carlo simulation of experimental results was then performed followed by statistical analyses to determine the distribution and amount of error associated with estimation of volatilization from water for a high volatility compound.

Description of Procedures -- Screening Test. Because the experimental protocols are of recent origin and the reference is not readily available to many readers, they are described here in some detail. The initial step in the SRI procedure is to classify the compound as low, medium, or high volatility and to identfy those compounds with a volatilization half-life of less than 3 months. The following physical property data are necessary to perform the screening test: (1) melting point, (2) Henry's Law constant, or (3) water solubility and vapor pressure at 20°C, and (4) heat of fusion, if the chemical is solid at 20°C and the vapor pressure data used to calculate the Henry's Law constant are for a liquid. The compounds are classified by calculating the Henry's law constant, H_c. The following classification scheme applies:

High volatility: H_c >1,000 torr/mole/liter
Intermediate volatility: 10 <H_c <1,000 torr/mole/liter
Low volatility: H_c <10 torr/mole/liter

The detailed test protocols are recommended for those compounds falling in the class of high or intermediate volatility. For low volatility compounds, the detailed test protocol is recommended if the screening study suggests that other processes, such as photolysis and biodegradation, are slow.

High Volatility Compounds Protocol. The estimated rate constant of volatilization in the natural environment, $(K_v^c)_{env}$, is based on the laboratory-measured ratio of volatilization rate constant to the oxygen reaeration rate constant $(K_v^c/K_v^o)_{lab}$ by the equation:

$$(K_v^c)_{env} = (K_v^c/K_v^o)_{lab}\,(K_v^o)_{env} \qquad (2)$$

The environmental oxygen reaeration rate constants $(K_v^o)_{env}$ are available for representative bodies of water in Mill, et al. (19). The detailed test protocol provides the procedures for estimating $(K_v^c/K_v^o)_{lab}$.

Measurements are performed in a 2-liter beaker with two tubing connections at the bottom, a constant-speed stirring motor with propeller to stir the solutions, and a recirculating pump that draws solution out of the beaker past a dissolved oxygen analyzer membrane and returns it to the beaker. The initial measurments determine the relationship between K_v^o and stirring rate in the beaker without any of the compound present in the distilled water. After establishing the stirring rate to K_v^o relationship, stock solution on the test compound is added to the distilled water, and the determination of $(K_v^c/K_v^o)_{lab}$ is begun.

Each experimental run is performed at a constant stirring speed. Oxygen content in the solution is monitored and the content recorded periodically. Aliquots of the solution are withdrawn from the beaker periodically, and the concentration of compound in the aliquot is measured. For each experimental run at a stirring speed, at least 10 measurements of concentration of compound must be made, and at least 15 observations of oxygen concentration must be made before the run is completed. Additional experimental runs are required so that data are obtained at no less than six stirring rates, producing a wide range of K_v^o values.

The volatilization rate of the chemical compound is given by the equation,

$$R_v^c = \frac{d[C]}{dt} = -K_v^c\,[C]_t \qquad (3)$$

where K_v^c is the volatilization rate constant and $[C]_t$ is the concentration of the chemical compound in solution at time t. The oxygen reaeration rate of the solution is given by the equation,

$$R_v^o = \frac{d[O_2]}{dt} = -K_v^o\,([O_2]_s - [O_2]_t) \qquad (4)$$

where K_v^o is the oxygen reaeration rate constant, $[O_2]_s$ is the dissolved oxygen concentration of the solution at saturation, and $[O_2]_t$ is the dissolved oxygen concentration in solution at time t.

Error Analysis -- High Volatility Compounds. The test protocol requires measurements of both dissolved oxygen and chemical compound as a function of time. Equations 2 and 3, when integrated, show the linear relationship between ln[C] and time t for the chemical compound, and ln $([O_2]_s - [O_2])$ and time t for dissolved oxygen content. The respective slopes for each line are K_v^c and K_v^o, the rate constants. Potential sources of error in this protocol are the individual dissolved oxygen and chemical compound concentration measurements.

The estimated volatilization rate constant K_v^c, is related to the measured concentration and time by the following regression equation:

$$K_v^c = \left[\frac{n\Sigma t \ln C - \Sigma t \ \Sigma \ln C}{n\Sigma t^2 - (\Sigma t)^2}\right] \quad (5)$$

where n is the number of concentration measurements. At time t = 0, the intercept C_o can be calculated from the following equation:

$$\ln C_o = 1/n \ (\Sigma \ln C - K_v^c \ \Sigma t) \quad (6)$$

No information presented in the reference document suggests typical values for errors associated with the concentration measurements, or the type of statistical distribution that might represent these errors. The document does suggest that the preferred measurement of error is a 95-percent confidence limit based on the t-statistic, implying that the error distribution is normal in appearance. The equation presented to compute the 95-percent confidence limit was incorrect in the reference. The following equation is the correct form:

$$CL\ (95\%) = t_{n-2,0.05} \left[\frac{\Sigma(\ln C)^2 - \ln C_o \Sigma \ln C - K_v^c \Sigma(\ln C)(t)}{(n-2)\ [\Sigma t^2 - (\Sigma t)^2/n]}\right]^{1/2} \quad (7)$$

where $t_{n-2,0.05}$ is the two-tailed t-statistic for n-2 degrees of freedom. The confidence interval on K_v^c is as follows:

$$K_v^c - CL < K_v^c < K_v^c + CL$$

The estimated oxygen reaeration rate constant is related to the dissolved oxygen concentration measurements and time by the following regression equations:

$$K_v^o = \left[\frac{n\Sigma(t)\,[\ln(C_s - C)] - \Sigma t\,\Sigma \ln(C_s - C)}{n\Sigma t^2 - (\Sigma t)^2} \right] \quad (8)$$

where C_s and C are substituted for $[O_2]_s$ and $[O_2]_t$ in Equation 3. Similarity between Equations 4 and 7 is readily apparent. The equation for the intercept C_o is:

$$\ln(C_s - C_o) = 1/n\,[\Sigma \ln(C_s - C) - K_v^o\,\Sigma t] \quad (9)$$

also similar to Equation 5. As in the case of $K_v{}^c$, there is no information in the reference text regarding typical error values for measurements of dissolved oxygen concentration or the type of statistical distribution representative of those errors. An equation is presented to calculate a 95-percent confidence limit using the t-statistic, thus making the assumption of near normally distributed errors apparent. The text listed an incorrect form of the equation; the correct form is as follows:

$$CL(95\%) = t_{n-2,0.05}$$

$$\times \left[\frac{\Sigma[\ln(C_s-C)]^2 - \ln(C_s-C_o)\Sigma\ln(C_s-C) - K_v^o\Sigma(t)[\ln(C_s-C)}{(n-2)\,[\Sigma t^2 - (\Sigma t)^2/n]} \right]^{1/2} \quad (10)$$

As in the case of Equation 6, t is the two-tailed t-statistic with n-2 degrees of freedom.

Given at lease six values of $K_v{}^c$ and $K_v{}^o$ obtained over a range of stirring rates, the ratio of $(K_v{}^c/K_v{}^o)$ can be obtained by plotting the data points and fitting a linear least squares line through the data points forced through the origin. The equation for this regression line is given as:

$$(K_v^c/K_v^o) = \left[\frac{\Sigma(K_v^o)_j(K_v^c)_j}{\Sigma(K_v^o)_j^2} \right] \quad (11)$$

where the subscript j refers to the jth measurement (stirring rate) of $K_v{}^c$ and $K_v{}^o$. The quotient $(K_v{}^c/K_v{}^o)$ is also assumed to be t-distributed in the reference. The equation for the 95-percent confidence level is given as:

$$\mathrm{CL}\ (95\%) = t_{n-1,\ 0.05} \left[\frac{\Sigma(K_v^c)_j^2 - (K_v^o/K_v^o)\Sigma(K_v^o)_j\ (K_v^c)_j}{(n-1)\ \Sigma(K_v^o)_j^2} \right]^{1/2} \tag{12}$$

The reference indicates that the 95-percent confidence limit on the average of (K_v^c/K_v^o) for high volatility compounds is about 5 to 10 percent of the magnitude of (K_v^c/K_v^o).

Simulation of Experimental Results and Error Estimation. An analysis of implied normality of error associated with measurements of K_v^c and K_v^o is not possible because of the lack of experimental data. However, even if the normality of error in measuring K_v^c and K_v^o is assumed, it is not ncessarily true that the estimated ratio of the two rate constants, (K_v^c/K_v^o), will have normally distributed errors as is assumed in the reference text. Further, the use of the linear regression approach to determine (K_v^c/K_v^o) may tend to underestimate (K_v^c/K_v^o) because the errors in determining K_v^o are taken into account. Likewise, the variance of (K_v^c/K_v^o) may be underestimated for the same reason.

To examine the reasonableness of the assumed near normality for (K_v^c/K_v^o) and the potential for the underestimation of the slope and its variance, Monte Carlo simulations were performed. For the simulations, a true mean value of (K_v^c/K_v^o) was selected as 0.71. The selection was based on an actually measured value of (K_v^c/K_v^o) for 1,1-dichloroethane over a range of K_v values of 0.3 to 12.0 hr^{-1} (19). Using this ratio and six assumed values of K_v^o corresponding to varied stirring rates) within the recommended experimental range of 1 to 15 hr^{-1}, a set of six corresponding K_v^c were calculated. These mean values are shown in Table I.

For the first simulation, it was assumed that the relative accuracy of determining K_v^o and K_v^c was such that 95 percent of the measured values would fall within the range of ± 5 percent of their true mean values, listed in Table I. It was also assumed that the measured value errors were t-distributed, thus allowing computation of standard deviations also shown in Table I. For the second simulation, it was assumed that 95 percent of the measured K_v^o and K_v^c would fall within the range of ± 10 percent using the same mean K_v^o and K_v^c as assumed for the first simulation. The resultant standard deviations are also shown in Table I.

After completing the computation of the mean K_v^o and K_v^c and standard deviations, a random number generator was used to develop 1,000 normally distributed values of each K_v^o and K_v^c, having the listed standard deviations. Verification of the normality of the randomly generated number sets was done by a Chi-square test on each number set.

The next steps were to compute the ratio of (K_v^c/K_v^o)

Table I. Simulation Mean and Standard Deviations[a]

Run No.		Mean Value	Simulation 1 Standard Deviation	Simulation 2 Standard Deviation
K_V^o	1	2.5	0.0486	0.0972
	2	4.0	0.0778	0.1556
	3	6.0	0.1167	0.2334
	4	8.3	0.1615	0.3229
	5	10.6	0.2062	0.4123
	6	13.1	0.2548	0.5095
K_V^c	1	1.78	0.0345	0.0690
	2	2.84	0.0552	0.1105
	3	4.26	0.0828	0.1657
	4	5.89	0.1146	0.2292
	5	7.53	0.1464	0.2927
	6	9.30	0.1809	0.3618

[a]A total of 1,000 values were used in each simulation.

corresponding to each of the six values of K_V^o (six stirring speeds) and to compute (K_V^c/K_V^o) from Equation 11 for each of the 1,000 trials. The standard deviation corresponding to each of the 1,000 estimated (K_V^c/K_V^o) were computed using the bracketed portion of Equation 12. The results of these computations are summarized in Table II for both simulations.

Table II. Simulation Test Results

Parameter	Simulation 1	Simulation 2
K_V^c/K_V^o mean	0.7098	0.7107
Low value	0.6779	0.6380
High value	0.7472	0.7766
Standard deviation of mean value	0.0104	0.0208
Average standard deviation of individual slopes	0.00693	0.01359
Other properties of distribution		
Coefficient of skewness	0.9551	0.03318
Kurtosis	-94.8748	4.6605
Chi-square sum	11.44	2.02

The estimated average values of (K_v^c/K_v^o) were in good agreement with the true value, 0.71 (0.7098 and 0.7107 for the ± 5 percent and ± 10 percent simulations, respectively). Chi-square tests performed on both distributions of (K_v^c/K_v^o) did not reject the hypotheses of normality at the 0.05 level of significance. Therefore, for the two simulations performed, the assumption of normality for the error distribution associated with (K_v^c/K_v^o) seems reasonable.

The question of whether the variance of (K_v^c/K_v^o) is underestimated by using the regression procedure of Equation 11 was studied in the following way. For each of the 1,000 estimated slopes, (K_v^c/K_v^o) the 95-percent confidence interval was calculated using Equation 12. If Equation 12 is a satisfactory estimator for the 95-percent confidence interval, then about 95 percent of the computed confidence intervals should contain the true slope, 0.71. When these calculations were performed on the simulation generated distributions of (K_v^c/K_v^o), only about 86 percent of the supposed 95 percent confidence intervals in each simulation contained the true slope, 0.71. Because the simulation generated estimates of average (K_v^c/K_v^o) were so near the true value of 0.71, it is presumed that the principal reason fewer confidence intervals contained 0.71 than expected is that the variance is underestimated by Equation 12.

As a potential means for correcting for the underestimated variance, the following computations were made. In computing the 95-percent confidence interval, the standard deviation of the individual (K_v^c/K_v^o) in the simulations were multiplied by S_{ga}/S_{iavg} where S_{ga} is the overall standard deviation of 1,000 (K_v^c/K_v^o) in each simulation and S_{iavg} is the average of the standard deviations computed for each of the 1,000 individual (K_v^c/K_v^o). When this correction was applied to computed confidence intervals in both simulations, about 96 percent of the 95-percent confidence intervals contained the true slope of 0.71. When applied to the standard deviation of individual (K_v^c/K_v^o) this correction factor had a value of about 1.5 for both simulations. This suggests that the bracketed portion of Equation 12 may underestimate the variance of (K_v^c/K_v^o) by as much as a factor of 2.3 (i.e., standard deviation by a factor of 1.5).

Conclusions

Column and high performance liquid chromatography (HPLC) methods for measurement of solubility, octanol-water partition coefficient, and vapor pressure which are replacing the older equilibrium methods tend to underestimate aqueous solubility and vapor pressure and tend to overestimate the octanol-water partition coefficient. The standard deviation for both the equilibrium and dynamic systems are similar, but calibration between systems is necessary to insure that they agree. The range of errors for both types of measurement as mentioned in the literature are well within the range predicted by the computer-simulated error distributions generated in this report. The measurement error distributions that were determined are not normally distributed and show a definite bias for each measured parameter.

The precise number of simulations needed to provide adequate "resolution" for the output distributions was not explored as part of the current study but seems likely to be a simple function of the number of subdividions for the range of output parameter being examined and also should depend, not so simply, on the number and resolution of input distributions, as well as on the probability requirements of the output (i.e., less resolution needed to assure a discrete value, "a", will be exceeded in only 1 out of 10 cases than is needed for only 1 out of 100 cases).

For volatilization from water, the measurement method error associated with determination of the ratio of rate constants, K_v^c/K_v^o, is typically reported in terms of 95-percent confidence limits on the average value. For high volatility compounds, the 95-percent confidence limits are reported to be 5 to 10 percent of the magnitude of (K_v^c/K_v^o). The regression approach typically used to determine (K_v^c/K_v^o) and the specification of confidence limits based on the t-distribution imply that the error distributions are assumed to be normal. However, even if the errors of K_v^c and K_v^o estimates are each normally distributed, the error distribution associated with the quotient of the two is not necessarily normally distributed. The regression approach assumed may produce an estimated value of (K_v^c/K_v^o) lower than the true value and underestimate the variance. Monte Carlo simulation of the regression approach did not confirm underestimation of (K_v^c/K_v^o); however, the variances appeared to be underestimated by a factor of 2.3.

Acknowledgments

The authors wish to express their gratitude to Dr. James W. Falco for the initial suggestion that a stochastic approach might be applicable to uncertainty considerations and also thank Mr. Richard Walentowicz for helpful discussion of Reference 1.

Literature Cited

1. Walentowicz, R.; Falco, J.W. "Stochastic Processes Applied to Risk Analysis of TCDD Contaminated Soil: A Case Study", Unpublished Report, 1984; Exposure Assessment Group, ORD, US EPA, Washington, DC 20460.
2. Whitmore, R.W. "Methodology for Characterization of Uncertainty in Exposure Assessments", 1985; EPA 600/8-85/009, NTIS No. PB85-240445, p. 26.
3. Kendall, M.G.; Stuart, A. "The Advanced Theory of Statistics; Volume 1, Distribution Theory"; Charles Griffin and Co. Ltd.: London, 1958; pp. 152-153.
4. MacKay, D.; Mascarenhas, R; Shiu, W.Y.; Chemosphere, 1980, 9, 257-264.
5. Karickhoff, S.W.; Brown, D.S. "Determination of Octanol-Water-Distribution Coefficients, Water Solubilities, and Sediment-Water-Partition Coefficients for Hydrophobic

Organic Pollutants", 1979; EPA-600/4-79-032, NTIS No. PB 80-103591, U.S. Environmental Protection Agency, Athens, GA.
6. Peters, R.L. Anal. Chem., 1982, 54, 1913-1914.
7. Mallon, B.J.; Harrison, F.L. Bull. Environ. Contam. Toxicol., 1984, 32(3), 316-323.
8. Stolzenburg, T.R.; Andren, A.W. Analytica Chimica Acta 1983, 151, 271-274.
9. Unger, S.H.; Cook, J.R.; Hollenberg, J. J. Pharm. Sci. 1978, 67(10), 1364-1367.
10. Bowman, B.T.; Sans, W.T. Env. Sci. Health, 1983, 18(6), 667-683.
11. DeVoe, H.; M. Miller; Wasik, S. J. Rsch. Natl. Bur. Stds., 1981, 86(4), 361-366.
12. Woodburn, K.B.; Doucette, W.; Andren, A. Environ. Sci. Technol. 1984, 18(6), 457-459.
13. American Society for Testing and Materials (ASTM), ASTM Method D-2879-70. ASTM Standards, Part 24, 1974, Philadelphia, PA.
14. Osborn, A.G.; Scott, D.W. J. Chem. Thermodynamics, 1980, 12, 429-438.
15. Spencer, W.F.; Shoup, T.; Cliath, M. J. Agric. Food Chem. 1979, 27(2), 273-278.
16. Wasik, S.; Tewari, Y; Miller, M. J. Rsch. Natl. Bur. Stds. 1982, 87(4), 311-315.
17. Thomas, T.C.; Sieber, A. Bul. Environ. Contam. Toxicol. 1974, 12, 17-21.
18. DePablo, R.S. J. Phys. D.: Appl. Phys., 13, 313-319.
19. Mill, T.; Mabey, W.R.; Bomberger, D.C. "Laboratory Protocols for Evaluating the Fate of Organic Chemicals in Air and Water." 1982, EPA-600/382-022, NTIS No. PB 3-150888, U.S. Environmental Protection Agency, Athens, GA.
20. Johnson, N.L.; Kotz, S. "Distributions in Statistics: Continuous Univariate Distributions-2"; Houghton Mifflin Co.: Boston, 1970; p. 44.

RECEIVED April 7, 1986

4

Quantifying Pesticide Adsorption and Degradation during Transport through Soil to Ground Water

W. Z. Zhong[1], A. T. Lemley[1], and R. J. Wagenet[2]

[1]College of Human Ecology, Cornell University, Ithaca, NY 14853
[2]Department of Agronomy, Cornell University, Ithaca, NY 14853

We describe a rapid experimental method and a basis for analyzing the results that will allow quantitative determination of pesticide adsorption and degradation during displacement through soil. A soil column methodology employing commercially available equipment developed for high-pressure liquid chromatography studies was used in the experiments. The movement and transformation of aldicarb, aldicarb sulfoxide and aldicarb sulfone were studied with this approach. Analytical solutions to the convection-dispersion equation that included description of linear adsorption and first-order transformation were used to interpret the data. Experiments were conducted in two soils, at two flow velocities, in sterile and non-sterile conditions, and at two initial influent concentrations to demonstrate the usefulness of the method, and to better define the adsorption and microbial and chemical conversion of aldicarb. Data from these experiments are presented and discussed in the context of this method and other studies on aldicarb. The method appears useful as a generalizeable protocol for the study of any pesticide in soil.

Pesticide leaching from the root zone and through the soil profile into groundwater is a serious concern. Contamination of groundwater by such chemicals may render that supply hazardous for human consumption. Additionally, passage of the chemical beyond the root zone prevents further impact upon the target pest. The residence time of applied pesticides within the root zone should therefore be maximized so that the dual objectives of high agricultural productivity and low environmental hazard from leaching can be met.

Pesticide and soil properties determine the mobility and degradation of applied chemicals. The interaction of the organic molecules with soil solids varies according to chemical structure, organic matter and clay content, soil pH, and in some cases concentration. Degradation rates are influenced by pH, substrate concentration,

0097–6156/86/0315–0061$06.00/0
© 1986 American Chemical Society

temperature, soil microbiological populations, and soil water content. Although it is possible to measure the relative effect of these variables in incubation-type laboratory studies, it is not obvious how to extrapolate the results of such experiments to field conditions. A relatively rapid method is needed for studying both mobility and degradation under conditions that more accurately represent the field regime. Soil column techniques provide the experimental framework for such a method. Although soil columns do not completely represent the field conditions, they do approximate transient flowing conditions better than incubation-type experiments. They may therefore be more useful in inferring the field condition. Analytic solutions of transport equations that consider adsorption and degradation provide theoretical models that can be used to interpret the results quantitatively. This paper presents an experimental and theoretical methodology that allows relatively rapid assessment of pesticide fate in soil, and illustrates the application of the method to aldicarb, a pesticide of both environmental and agricultural interest.

Theoretical Methods

Aldicarb (A) is both chemically and biologically oxidized to aldicarb sulfoxide (A-SO), which is then further oxidized by similar processes to aldicarb sulfone ($A\text{-}SO_2$). These compounds are simultaneously subject to other degradative chemical processes dominated by hydrolysis. These reactions have been successfully described by first-order kinetics (1), and can be generally summarized as:

$$\begin{array}{ccccc} & k_1^* & & k_2^* & \\ A & \rightarrow & A\text{-}SO & \rightarrow & A\text{-}SO_2 \\ \downarrow k_1 & & \downarrow k_2 & & \downarrow k_3 \\ A\text{-oxime} & & A\text{-}SO\text{ oxime} & & A\text{-}SO_2\text{ oxime} \end{array} \qquad (1)$$

where k_1, k_2 and k_3 represent hydrolysis, and k_1^* and k_2^* indicate oxidation. Expressed in first-order kinetic terms, assuming that oxidation and hydrolysis occur in solution, and with reference to a soil system in which only part of the total volume is occupied by water, the following can be formulated:

$$\theta \frac{d(A)}{dt} = -\theta\,(k_1 + k_1^*)(A) \qquad (2)$$

$$\theta \frac{d(A\text{-}SO)}{dt} = \theta k_1^*\,(A) - \theta(k_2 + k_2^*)(A\text{-}SO) \qquad (3)$$

$$\theta \frac{d(A\text{-}SO_2)}{dt} = \theta k_2^*(A\text{-}SO) - \theta k_3(A\text{-}SO_2) \qquad (4)$$

where θ = volumetric water content (cm^3cm^{-3}), the parentheses indicate concentration, and the transformation of chemical is defined with respect to the unit volume of the system. The assumption that degradation occurs in solution is a good one in the case of aldicarb and its metabolites since there is little adsorption.

These equations predict that the concentration of aldicarb is monotonically decreasing, while the sulfone and sulfoxide forms change in concentration depending on the value of the rate coefficients.

The description of these sequential and simultaneous reactions during transport through a soil system is accomplished using miscible displacement theory, which can be formulated for an interacting, degrading solute (e.g. an organic molecule) as (2):

$$\frac{\partial(\rho s)}{\partial t} + \frac{\partial(\theta c)}{\partial t} = \frac{\partial}{\partial z}\left[\theta D\ (\theta,v)\ \frac{\partial c}{\partial z} - v\theta c\right] \pm \phi \tag{5}$$

where c = solute concentration in the liquid phase ($\mu g\ ml^{-1}$), s = solute concentration in the sorbed phase ($\mu g\ g^{-1}$), ρ = soil bulk density ($g\ cm^{-3}$), θ = volume water content (cm^3cm^{-3}), $D(\theta,\ v)$ = apparent diffusion coefficient ($cm^2\ hr^{-1}$), v = pore water velocity, defined as q/θ, where q = volumetric water flux ($cm\ hr^{-1}$), and z and t are distance (cm) and time (hr), respectively. The formation or loss of the solute during displacement is conceptually represented by $\phi(\mu g\ ml^{-1}\ hr^{-1})$. Solution of Equation 5 for appropriate initial and boundary conditions will provide a method of predicting c(z,t) that can be used to fit measured effluent concentrations to quantify distribution of the chemical between sorbed and solution phases and the degradation rate. Equation 5 is general in form, and although developed below in more detail for aldicarb can be used as the starting point for description of the movement of any pesticide through soil. In this way, the theoretical methods used in this paper represent a broadly applicable approach to study of pesticide fate in soil.

In the case of laboratory soil column studies, uniformly packed columns with constant water contents and flow velocities are used (steady-state conditions). Equation 5 reduces for such cases to:

$$\rho K_d \frac{\partial c}{\partial t} + \theta \frac{\partial c}{\partial t} = \theta D \frac{\partial^2 c}{\partial z^2} - \theta v \frac{\partial c}{\partial t} \pm \phi \tag{6}$$

where it has been assumed that solution and sorbed concentrations are linearly related by $s=K_d c$, with K_d representing an instantaneous and reversible adsorption/desorption reaction. The values of D and v in Equation 6 are represented as constants when describing steady state water flow conditions, as they depend upon q and θ, which are themselves constant for a particular steady state experiment. Equation 6 can be simplified to (2):

$$\frac{\partial c}{\partial t} = D' \frac{\partial^2 c}{\partial z^2} - v' \frac{\partial c}{\partial z} \pm \phi' \tag{7}$$

where $D' = (D)/(1+\rho K_d/\theta)$, $v'=(v)/(1+\rho K_d/\theta)$, and $\phi'=(\phi/\theta)/(1+\rho K_d/\theta)$. When a non-interacting, non-degrading solute such as chloride is to be described, then K_d=0 and ϕ=0, and Equation 7 reduces to the classical dispersion-convection equation, which has often been used to describe the movement of such solutes as chloride or tritium in soil. The solution of Equation 7 subject to appropriate initial and boundary conditions can then be used to estimate the value of D by measuring q and θ (to give v) and curve fitting predicted c(z,t) to measured effluent chloride concentrations. If an interacting solute subject to degradation is to be described, then ϕ can be more explicitly defined, such as by first-order kinetics as in Equation 2-4, and used as the basis for determining K_{di} and k_i. A number of experimental studies (2,3) have used such curve-fitting techniques to estimate K_d and k in the presence of constant values of D and v. These techniques utilize the distinct and different influence of K_d, D, v and k upon the shape of the effluent concentration curve to determine the unique combination of the variables that describes the processes acting upon the solute during displacement.

Recognizing the above facts, the following equations based on Equation 7 and using Equation 2-4 to represent ϕ can be developed to describe the simultaneous transport and transformation of A, A-SO and $A\text{-}SO_2$:

$$(1+R_1)\frac{\partial c_1}{\partial t} = D\frac{\partial^2 c_1}{\partial z^2} - v\frac{\partial c_1}{\partial z} - (k_1+k_1^*)\ c_1 \qquad (8)$$

$$(1+R_2)\frac{\partial c_2}{\partial t} = D\frac{\partial^2 c_2}{\partial z^2} - v\frac{\partial c_2}{\partial z} + k_1^*\ c_1 - (k_2 + k_2^*)\ c_2 \qquad (9)$$

$$(1+R_3)\frac{\partial c_3}{\partial t} = D\frac{\partial^2 c_3}{\partial z^2} - v\frac{\partial c_3}{\partial z} + k_2^*\ c_2 - k_3\ c_3 \qquad (10)$$

where R_i is defined by $R_i = \rho K_{di}/\theta$. Equations 8-10 describe the transport and transformation in solution of A, A-SO, and $A\text{-}SO_2$, respectively, with the subscripts used to distinguish each species. In this study, Equations 8-10 were solved for the conditions of a soil column initially free of all aldicarb species, to which is applied a pulse of A containing no A-SO or $A\text{-}SO_2$. That is:

$c_1=c_2=c_3=0$	$z\geq0$	$t=0$	(11a)
$c_1=c_1^\circ$	$z=0$	$0<t\leq t_1$	(11b)
$c_1=c_2=c_3=0$	$z=0$	$t\geq t_1$	(11c)
$c_1=c_2=c_3=0$	$z\to\infty$	$t\geq0$	(11d)

The solution to a very similar set of equations with R_3 = 0 has been presented (4) and used in a study of the transformation of urea to NH_4^+ and NO_3^- during leaching (5). Equations 8-10 have been solved by similar techniques and the solutions used to provide estimates of c (z,t) as a function of D, v, c_1°, θ, ρ, K_{di} and k_i. It is therefore necessary to measure or estimate D, v, θ and ρ so that K_{di}

and k_i can be determined by curve-fitting the measured effluent concentrations of the aldicarb species.

Experimental Methods

The experimental apparatus (Figure 1) consists of a continuous delivery infinite volume syringe pump delivering influent solution to a soil column from which effluent is collected in distinct increments in a fraction collector. The column was adapted from equipment readily available for high pressure liquid chromatography (HPLC) work, and consists of a Beckman glass column 25 mm diameter and 250 mm long made of precision bore borosilicate glass. A bed support used on both ends consists of a woven FEP teflon diffusion mesh in contact with a porous teflon filter disk of pore size 30-60 µm. The diffusion mesh causes the injected liquid to spread radially and evenly before entering the filter disk on the inlet side, and works in reverse on the outlet side to prevent zone spreading. This column is surrounded by a glass water jacket with inlet and outlet connections to a controlled temperature water bath that is used to maintain a constant soil column temperature of 25°±0.5°C in all experiments. Influent solution is delivered by a Sage Model 220 syringe pump with two independently controlled syringes designed to continuously deliver a uniform and pulse-free flow. Effluent is collected in a Buchler LC-100 fraction collector.

Two soils, a Palmyra sandy loam from central New York state and a Riverhead sandy loam from eastern Long Island, New York were used (Table I). Soil was air-dried at room temperature and passed through a 2 mm sieve. A subsample of each soil was sterilized with dry heat (120°C for 5 hours) for use in selected experiments. Each column was filled to an average bulk density (with experimental error given in Table II) and was wetted from below by applying a slightly positive pressure. This resulted in 90% saturation for the Palmyra soil and 85% saturation for the Riverhead soil. Steady state flow at a predetermined water flux was established using 0.01 N $CaSO_4$, at which time the column was placed in a horizontal position and the experiment initiated. The influent was abruptly changed to a 0.01 N $CaSO_4$ solution containing 100 µg/ml chloride as KCl and 10 µg/ml A, A-SO or A-SO_2, depending upon the experiment. After application of approximately 100 ml of this solution, the unamended $CaSO_4$ solution was again continuously applied to displace the chloride species through the column.

The collected effluent was analyzed for Cl^-, A, A-SO, and A-SO_2. Chloride was determined using a Buchler-Cotlove chloridometer. Aldicarb, A-SO and A-SO_2 were extracted with methylene chloride and analyzed using a Hewlett-Packard model 5880A gas chromatograph with a nitrogen-phosphorus detector and a fused silica capillary column coated with methyl silicone. Further details of the aldicarb analytical techniques are reported elsewhere (6).

Since these studies were performed to elucidate rates of degradation by oxidation, (either chemically or biologically mediated) it

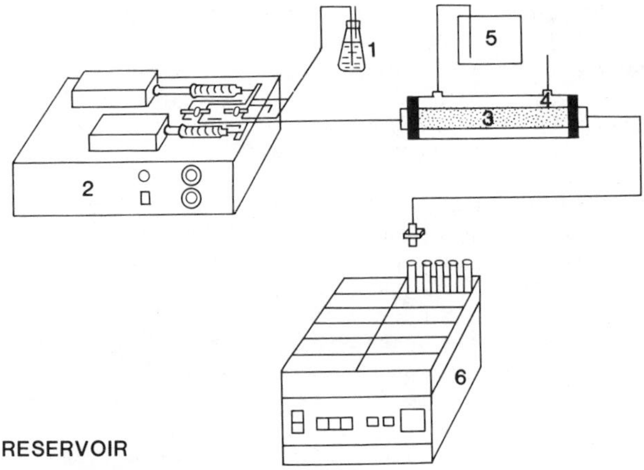

Figure 1. Experimental apparatus for laboratory pesticide leaching studies.

Table I. Physical and Chemical Soil Characteristics

Property	Palmyra	Riverhead
Organic Matter (%)	2.0	4.0
pH (1:1)	4.9	4.5
Cation Exch. Capacity (me/100g)	8.2	15.0
Exch. H^+ (me/100g)	7	16
Extractable Elements (µg/g)		
P	12.5	39
K	115	105
Mg	85	40
Ca	400	300
Particle Size (%)		
Sand	53	75
Silt	37	21
Clay	10	4

was important to establish that the column was being operated under aerobic and not anaerobic conditions. The soil columns were not completely saturated; some oxygen was available in the soil. In addition, a simple calculation demonstrates that there is approximately ten times more dissolved oxygen (8-10 mg/L) in the water than is needed to convert all of the applied aldicarb to aldicarb sulfone.

Thirteen experiments (Table II) were conducted on both soils in sterilized and unsterilized conditions, at two pore water velocities, and with several combinations of influent pesticide concentration. Chloride was applied in all experiments, and effluent data were analyzed using the solution to Equations 7, 8, 9 or 10 as appropriate.

Results And Discussion

The apparent diffusion coefficient, D, was determined for the particular leaching conditions of each of the thirteen experiments. This was accomplished using the measured chloride breakthrough (effluent concentration) curve and the analytical solution to Equation 7 with $K_d=\phi=0$. Examples of the observed and calculated chloride concentrations (determined by adjustment of D until a best fit was obtained) are presented for three different experiments (Experiments 7, 8, 11) in Figures 2-4. Values of D and the pore water velocity (v) determined for each experiment are presented in Table III. The value of D increased for cases with large v, and was different between soils for any particular v. This is consistent with the basic relationship be-

Table II. Experimental Designation

Experiment	Soil	Sterilized	Water Flux	ρ (±0.01) gcm^{-3}	θ cm^3cm^{-3}	C_1 μgml^{-1}	C_A^o μgml^{-1}	C_{A-SO}^o μgml^{-1}	$C_{A-SO_2}^o$ μgml^{-1}
1	Palmyra	yes	high	1.52	0.39	98.1	10.5	-	-
2	Palmyra	yes	low	1.47	0.39	99.9	9.02	-	-
3	Palmyra	yes	high	1.49	0.41	102	-	9.97	-
4	Palmyra	yes	low	1.49	0.40	94.0	-	9.88	-
5	Palmyra	yes	high	1.50	0.39	95.5	-	-	10.1
6	Palmyra	yes	low	1.49	0.40	97.0	-	-	10.0
7	Palmyra	no	high	1.49	0.40	98.6	10.9	-	-
8	Palmyra	no	low	1.51	0.38	99.8	10.5	-	-
9	Palmyra	no	low	1.47	0.41	99.5	98.6	-	-
10	Riverhead	no	high	1.27	0.42	99.7	10.1	-	-
11	Riverhead	no	low	1.29	0.44	103	10.8	-	-
12	Riverhead	no	low	1.25	0.42	105	-	11.6	-
13	Riverhead	no	low	1.27	0.44	102	-	-	10.1

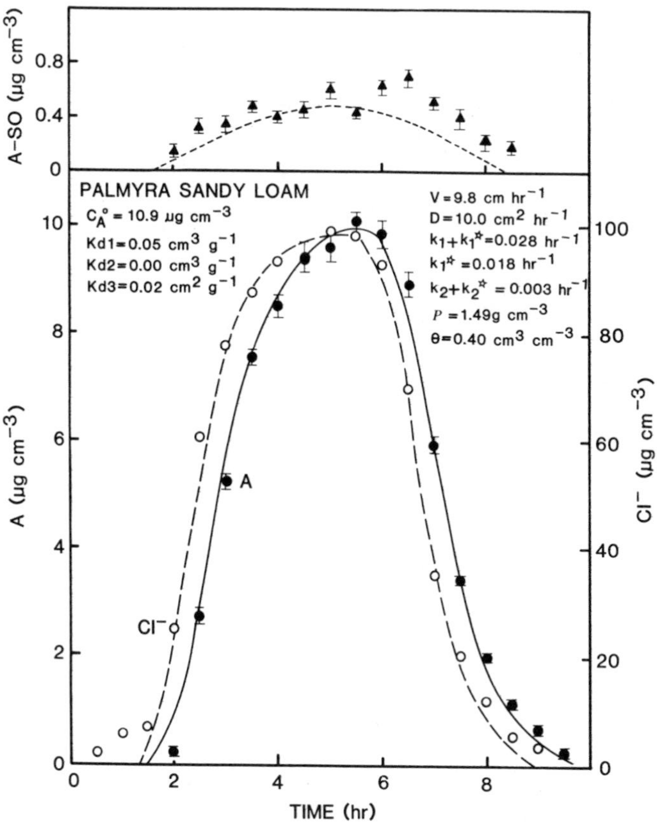

Figure 2. Measured and calculated effluent concentrations of chloride (Cl^-), aldicarb (A) and aldicarb sulfoxide (A-SO) from experiment 7.

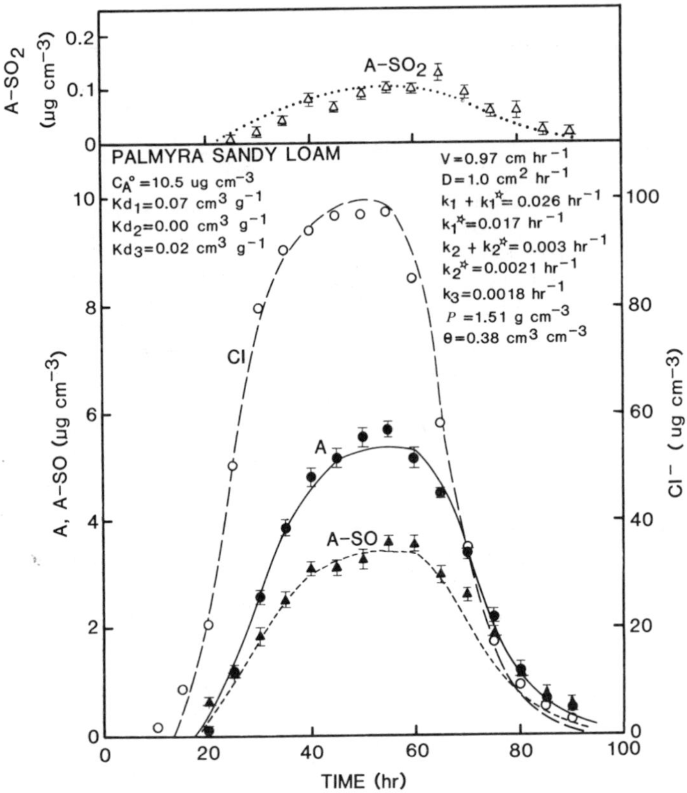

Figure 3. Measured and calculated effluent concentrations of chloride (Cl^-), aldicarb (A), aldicarb sulfoxide (A-SO) and aldicarb sulfone ($A\text{-}SO_2$) from experiment 8.

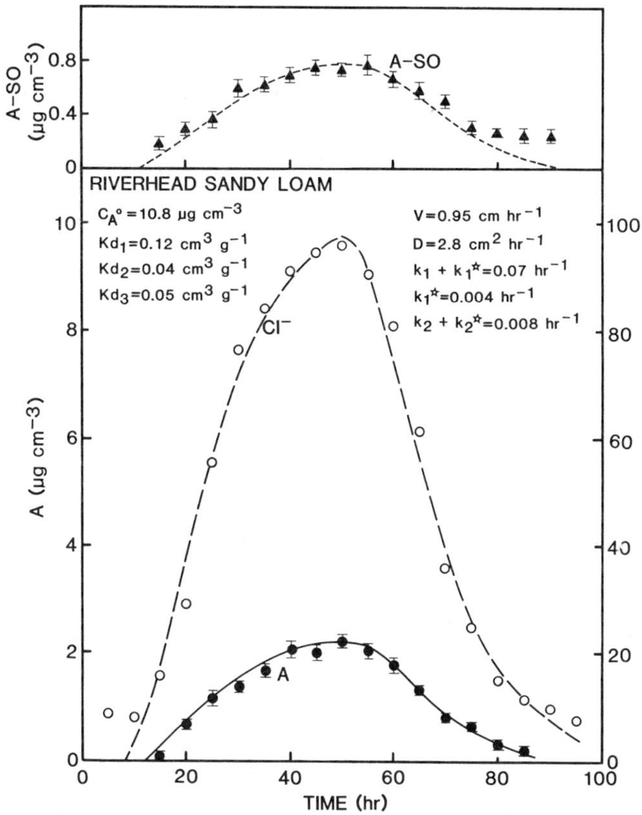

Figure 4. Measured and calculated effluent concentrations of chloride (Cl^-), aldicarb (A), and aldicarb sulfoxide (A-SO) from experiment 11.

Table III. Data for Study of the Degradation of Aldicarb

Experiment	V	D	K_{d1}	K_{d2}	K_{d3}	$k_1+k_1^*$	k_1^*	k_1	$k_2+k_2^*$	k_2^*	k_2	k_3
	cm hr^{-1}	cm^2hr^{-1}	⟵ ml g^{-1}		⟶	⟵			hr^{-1}			⟶
1	9.80	10.	0.05	-	-	0.015	-	-	-	-	-	-
2	0.93	1.0	0.07	0	0.02	0.013	0.004	0.009	0.002	0.0014	0.0006	0.0018
3	9.20	10.	-	0	-				(98% recovery)			
4	1.06	1.0	-	0	-	-	-	-	0.002	0.0014	0.0006	-
5	9.70	10.	-	-	0.02				(99% recovery)			
6	1.10	1.0	-	-	0.02	-	-	-	-	-	-	0.0018
7	9.80	10.	0.05	0	-	0.028	0.018	0.010	0.003	-	-	-
8	0.97	1.0	0.07	0	0.02	0.026	0.017	0.009	0.003	0.0021	0.0009	0.0018
9	0.97	1.0	0.06	0	0.02	0.024	0.015	0.009	0.003	0.0023	0.0007	0.0018
10	9.40	25.	0.15	0.04	-	0.060	0.007	0.053	0.008	-	-	-
11	0.95	2.8	0.12	0.04	-	0.070	0.004	0.066	0.008	-	-	-
12	0.97	2.8	-	0.04	-	-	-	-	0.008	0.0005	0.0075	0.006
13	0.95	2.8	-	-	0.05	-	-	-	-	-	-	0.006

tween water flow and dispersion and the dependence of the latter upon the geometry of the porous media (7). The value of D determined in this manner was used with v in the solution to Equations 8-10; K_{di} and k_i were then determined by curve-fitting.

Experiments 1-2, 3-4, and 5-6 represent paired cases of investigation of A, A-SO and A-SO_2 in Palmyra soil. Sterilized soil was used in these cases to isolate chemical oxidation and hydrolysis from biological conversion processes, and to determine whether the curve-fitted values of K_{di} and k_i depend on water velocity. The only difference within a pair of experiments was in the flow velocity (v). Between pairs, the difference was the aldicarb metabolite applied. The results (Table III) demonstrate that essentially identical values of K_{d1} were determined in each case, indicating that for this chemical the adsorption is independent of flow velocity. The value of K_{d1}, representing aldicarb adsorption by Palmyra soil, was 0.05-0.07, indicating very slight chemical-soil interaction. This is consistent with other studies (8, 9, 10) conducted in medium to coarse textured soils. No interaction (K_{d2}=0) of aldicarb sulfoxide with the soil was detected, and the interaction of aldicarb sulfone, K_{d3}=0.02, was even less than measured for aldicarb and also independent of flow velocity.

In both Experiments 1 and 2, the measured total rate of aldicarb degradation (k_1+k_1*) was approximately equal, ranging from 0.015 to 0.013 hr^{-1}. The rate coefficient for aldicarb degradation (k_1+k_1*) could not be separated into individual components of hydrolysis and oxidation using the results from the high velocity experiment (Experiment 1), as the residence time within the column was too short for analytically detectable levels of A-SO to be produced and appear in the effluent. However, in the same experiment conducted at slower velocity (Experiment 2), the production of A-SO was high enough that curve-fitting of effluent concentration data of A-SO and A provided the opportunity to separate the rate of hydrolysis, k_1=0.009 hr^{-1}, from the rate of oxidation, k_1*=0.004 hr^{-1}. Similarly, experiments 4 and 6, conducted at slow flow velocities (but with A-SO and A-SO_2 as the starting material, respectively) provided the opportunity to separate sulfoxide hydrolysis (k_2=.0006 hr^{-1}) from sulfoxide oxidation (k_2*=0.0014 hr^{-1}), with a total sulfoxide transformation rate (k_2+k_2*) of 0.002 hr^{-1} measured in both experiments. This is exemplified by Figures 2 and 3, which present results for an unsterilized case (discussed below) of fast and slow flow velocity. The two high velocity cases in which A-SO and A-SO_2 were applied (Experiments 3 and 5) resulted in such rapid leaching that no degradation could be measured, indicating that use of a high velocity may not be a suitable technique for study of weakly interacting chemicals. Similarly, chemicals with a long half-life could not be easily studied by this method even with a low flow velocity.

Transformation of aldicarb and its derivatives was also studied in three experiments in unsterilized Palymra soil (Experiments 7-9) to determine the biological component of each transformation process. The value of k_1 in unsterilized soil ranged from 0.009-0.010 hr^{-1}, compared to 0.009 hr^{-1} in the sterilized soil of experiment 2. This result implies that aldicarb hydrolysis is primarily a chemical process. The value of k_1* was approximately four times the value in unsterilized soil (0.015-0.018 hr^{-1}) than in sterilized soil (0.004

hr^{-1}). This indicates that aldicarb oxidation (k_1*) is equally mediated by biological and chemical processes. Similarly, hydrolysis of A-SO (k_2) was nearly identical in sterilized (0.0006 hr^{-1}) and unsterilized (0.0007 -0.0009 hr^{-1}) cases, but the rate of A-SO oxidation (k_2*) was approximately 50% greater (average 0.0022 hr^{-1} compared to 0.0014 hr^{-1}) in unsterilized soil. The degradation of A-SO_2(k_3) was not influenced by sterilization, indicating that only chemical processes were responsible. Such processes have been shown by other work (11) to be mainly hydrolysis. K_d values for each of the three species were unaffected by sterilization.

The usefulness of the technique developed here was further demonstrated by employing a second soil for limited comparison purposes (Experiments 10-13). The objective was not a comprehensive study of aldicarb, but the execution of several experiments identical to those accomplished on Palmyra soil. The second soil, Riverhead sandy loam, was coarser textured, slightly lower in pH and higher in organic matter (Table I). Agreement between theory and experiment was acceptable in all cases, as illustrated in Figures 4-5. The value of D at the high flow velocity in the Riverhead soil was 2.5 times greater than in Palmyra soil, but was well-described by the solution to Equation 7 (Figure 4). At the lower velocity, D was 2.8 cm^2/hr compared to 1.0 cm^2/hr in the Palmyra, but was equally well fitted by the solution to Equation 7.

Degradation and adsorption were measured only in unsterilized Riverhead soil, with the latter already shown to be independent of sterilization. The value of K_d for aldicarb in the Riverhead soil was 0.12-0.15, approximately double the value of 0.05-0.07 measured in the Palmyra soil. The value of K_{d2}, describing A-SO interaction with soil, was determined to be 0.04 in the Riverhead soil, indicating some slight interaction of the sulfoxide with the soil. The value of K_{d3} was nearly the same as K_{d2}, but was double the value of K_{d3} (0.02) found in the Palmyra soil. Clearly, the sorption of aldicarb and its derivatives increased in the Riverhead soil due to the presence of higher organic matter content and can be distinguished for the two soils by the experimental and theoretical methods used here.

The value of the distribution coefficient, K_{di}, varies with several soil properties, most notably the organic carbon content. This has led to the definition of a normalized sorption coefficient, K_{oc}, which is calculated as (12):

$$K_{oc} = K_{di}/f_{oc}$$

where f_{oc} is the fraction of organic carbon. Assuming that organic matter is 58% carbon, K_{oc} values for aldicarb and its metabolites can be calculated for both soils. The values are quite low compared with values for other organic chemicals in soil (12), and are (Table IV) nearly identical for the two soils. Such calculations indicate that even on a normalized basis, almost no aldicarb adsorption occurs in these two predominantly mineral soils.

Table IV. Values of K_d and K_{oc} for Aldicarb Species in Palmyra and Riverhead Soils

Species	Parameter	Palmyra	Riverhead
Aldicarb	K_d	0.05-0.07	0.12-0.15
	K_{oc}	4.3 -6.0	5.2 -6.5
Aldicarb Sulfoxide	K_d	0	0.04
	K_{oc}	0	1.7
Aldicarb Sulfone	K_d	0.02	0.05
	K_{oc}	1.7	2.2

Oxidation of aldicarb, found in Palmyra soil to be about double the rate of hydrolysis under unsterilized conditions, was in the Riverhead soil much less important than hydrolysis. In the latter case, the hydrolysis rate (k_1), was found to be 8-16 times greater than the oxidation rate (k_1*), with the total degradation rate (0.06-0.07 hr^{-1}) 2-3 times greater in the higher organic matter soil. Similar behavior of A-SO_2 was measured, with the degradation rate approximately three times greater in the Riverhead soil. It is expected that acid catalyzed hydrolysis might be enhanced by sorption to soil, but from previous work (13), the acid catalyzed hydrolysis rate for aldicarb sulfone in solution is 10^4 slower than the base hydrolysis rate. Acid catalyzed hydrolysis of aldicarb under flow conditions on a reactive ion exchange resin, although enhanced, is still 10^2 slower than nucleophilic cleavage (14). Since base hydrolysis predominates, this type of enhancement most probably does not account for the faster hydrolysis rate in the Riverhead soil. The reason for these differences is not clear, but could be studied by further application of the methods developed here, perhaps combined with study of the sorption properties of the soil or conditions in the column.

The values of the rate coefficients determined by the methods of this study compare well with those reported elsewhere (Table V). Exact agreement would not be expected because of different soils. We did limit comparisons to studies done at the same temperature. Contrary to other experimental designs, the methods used here require only 4-6 days per experiment. This makes the column design potentially useful as a screening method for studying the reactions of pesticides in a wide range of soils. Additionally, the method can be applied in a rigorous manner to study adsorption and degradation under dynamic conditions of water and solute movement, rather than in comparatively static batch-type experiments. An analytical method of curve-fitting (15) could be developed to precisely determine values of K_{di} and k_i, with methods developed by van Genuchten and Alves (16) demonstrating a useful approach. Such methods will become more important as more complicated relationships are used to represent partitioning between sorbed and solution concentrations, or as more subtle environmental effects upon degradation are studied.

Table V. Degradation Rates for Aldicarb and its Metabolites Measured in Other Studies at 25°C

Soil	Sterile	$k_1+k_1^*$ (hr^{-1})	$k_2+k_2^*$ (hr^{-1})	k_3 (hr^{-1})
Plow Layer[1]	yes	0.012	-	-
	no	0.029	-	-
Clay Loam[2] 4.5% o.m., pH = 7.2, 3% clay	no	-	0.0021 k_2^* = 0.0008 k_2 = 0.0013	0.0021
Greenhouse Soil[2] 16.5% o.m., pH = 6.0, 17% clay	no	-	0.016 k_2^* = 0.0011 k_2 = 0.0005	0.0014
Clay[3] pH = 8.0	no	0.0033	0.0015	0.0013
Silty Clay Loam[3] pH = 8.0	no	0.0046	0.00083	0.0006
Fine Sand[3] pH = 6.3	no	0.0021	0.0004	0.0013

[1]Reference (17); [2]Reference (18); [3]Reference (10)

Analytical solutions to transport equations that simultaneously consider adsorption and transformation, when used in concert with appropriate soil column techniques, offer substantial promise for study of the environmental fate of organic chemicals. Such methods can be widely used since a number of such solutions already exist in the scientific literature. It remains only to design research programs in which those who understand such equations are working with those who can design and execute complementary chemical studies.

Acknowledgments

The authors gratefully acknowledge the financial support of USGS under the Cooperative State Water Resources Research Program and of the United States Department of Agriculture.

Literature Cited

1. Lemley, A.T.; Zhong, W.Z. J. Agric. Food Chem., 1984, 32, 714-719.
2. Wagenet, R.J. Principles of salt movement in soils. In: Chemical Mobility and Reactivity in Soil Systems, Nelson, D.W.; Elrick, D.E.; Tanji, K.K., Ed. SSSA Special Publ. No. 11, Amer. Soc. of Agronomy, 1983, 123-140, (Ch. 9).
3. van Genuchten, M.Th.; Clearly, R.W. Movement of solutes in soil: Computer simulated and laboratory results. In: Soil Chemistry: B. Physico-chemical models, Bolt, G.H., Ed. Elsevier Scientific Publ. Co. Amsterdam, 1979, 349-386.
4. Wagenet, R.J.; Biggar, J.W.; Nielsen, D.R. Soil Sci. Soc. Amer. J., 1976, 41, 896-902.
5. Wagenet, R.J.; Nielsen, D.R.; Biggar, J.W. Analytical solutions of miscible displacement equations describing the sequential microbiological transformations of urea, ammonium and nitrate. Water Sci. and Eng. Papers No. 6001, Dept. of LAWR, University of California, Davis, CA.
6. Zhong, W.Z.; Lemley, A.T.; Spalik, J. Journal of Chromat., 1984, 299, 269-274.
7. Kirda, C.; Nielsen, D.R.; Biggar, J.W. Soil Sci. Soc. Amer. Proc., 1973, 37, 339-345.
8. Bromilow, R.H. Ann. Appl. Biol., 1973, 75, 473-479.
9. Bromilow, R.H.; Baker, R.J.; Freeman, M.A.H.; Gorog, K. Pestic. Sci., 1980, 11, 371-378.
10. Hough, A.; Thomason, I.J.; Farmer, W.J. J. Nematology, 1975, 7, 214-221.
11. Leistra, M. J. Environ. Sci. Health, 1978, B13(4), 343-360.
12. Rao, P.S.C.; Davidson, J.M. Estimation of pesticide retention and transformation parameters required in nonpoint source pollution models. In: Environmental impact of nonpoint source pollution, Overcash, M.R.; Davidson, J.M., Ed. Ann Arbor Science Publ., Inc., Ann Arbor, MI, 1980, 23-67.
13. Lemley, A.T.; Zhong, W.Z. J. Environ. Sci. Health, 1983, B18(2), 189-206.
14. Lemley, A.T.; Zhong, W.Z.; Janauer, G.E.; Rossi, R. In "Treatment and Disposal of Pesticide Wastes," Krueger, R.F.; Seiber, J.N.; Eds.; ACS SYMPOSIUM SERIES No. 259, American Chemical Society: Washington, DC, 1984; pp. 245-259.
15. Parker, J.C.; van Genuchten, M.Th. Virginia Agr. Exp. Station Bull., 1984, 84-3, 96p.
16. van Genuchten, M.Th.; Alves, W.J. U.S. Dept. of Agriculture Technical Bull., 1982, No. 1661, 151p.
17. Romine, R.R.; Hansen, J.L.; Jones, R.L. Aldicarb oxidation mechanism studies, Union Carbide Agricultural Products Company, Inc., Preliminary Draft, 1984.
18. Smelt, J.H.; Leistra, M.; Houx, N.W.H.; Dekker, A. Pestic. Sci., 1978, 9, 279-300.

RECEIVED April 7, 1986

5

Geohydrology of a Field Site

Study of Pesticide Migration in the Unsaturated and Saturated Zones of Dougherty Plain, Southwest Georgia

Sandra C. Cooper

Water Resources Division, U.S. Geological Survey, Doraville, GA 30360

A 10-acre area of a peanut field in southeastern Lee County, Georgia, was selected to investigate the migration of the pesticide aldicarb in the unsaturated and saturated zones, and to assess the potential for degradation of the ground-water resource. The hydrogeologic framework was determined from sample cuttings, a continuous core, and geophysical logs obtained from test wells drilled in the field, and from bimonthly and continuously recorded water-level measurements. The study site is underlain, in descending order, by the undifferentiated residuum, which forms the water-table aquifer, and by the Ocala Limestone, the Clinchfield Sand, and the Lisbon Formation of Eocene age, which form the Upper Floridan aquifer. Geohydrologic data indicate that in the area of the study site, the Upper Floridan aquifer consists of an upper and a lower permeable zone; the upper permeable zone is hydraulically connected to the water-table aquifer. Precipitation entering the residuum generally moves downward and recharges the Upper Floridan aquifer. Water recharging the Upper Floridan moves laterally downgradient to points of natural discharge and to pumping centers. Because of the hydraulic connection between the residuum and the Upper Floridan aquifer, the potential exists for pesticides to migrate into the ground-water system.

Agriculture is a major industry in the Southeastern United States and particularly in Georgia. Increased agricultural productivity in Georgia has been achieved through the use of large-scale irrigation and multicropping practices, and the application of fertilizers and

This chapter not subject to U.S. copyright.
Published 1986, American Chemical Society

pesticides. Although farming techniques have changed considerably over the years, one of the most significant changes has been the increased use of chemicals in controlling insects, weeds, and plant diseases. Early efforts in the development of chemical pesticides were concentrated on producing stable pesticides that were relatively persistent in soils. Although the persistence of these chemicals was advantageous in controlling pests, it became disadvantageous when the pesticides migrated into other parts of the environment posing risks to nontarget species. As a result, many of these chemicals were replaced by less persistent and less strongly sorbed, but generally more toxic, organophosphates and carbamates (1).

In 1965, a new class of carbamate insecticide known as oxime carbamate was developed by Union Carbide. The technical name assigned to the new soil-applied, systemic carbamate insecticide was aldicarb (2, 3).

Aldicarb [2-methyl-2-(methylthio) propionaldehyde O-(methylcarbamoyl) oxime] is a noncorrosive, nonflammable insecticide incorporated into the seeded furrow during planting to control several species of insects, mites, and nematodes. Aldicarb is extremely toxic to mammals, and its oxidation produces two toxic metabolites, aldicarb sulfoxide and aldicarb sulfone. Both aldicarb and its toxic derivatives are very soluble in water.

Recent incidents of chemical contamination of ground-water reservoirs have led to increased concern about the potential for pesticides, such as aldicarb, to leach through the soil profile into the ground water (4, 5, 6, 7, 8, 9, 10, 11). To develop needed information on the potential problem, the U.S. Geological Survey and the U.S. Environmental Protection Agency initiated a 5-year field study in a highly productive agricultural area of southwestern Georgia to investigate pesticide migration.

The objectives of the study are twofold. The first is to evaluate the long-term consequences of continued use of the pesticide aldicarb on the quality of the ground-water resource by field measuring the migration and degradation of aldicarb through the unsaturated and saturated zones. The second objective is to develop a data base to test mathematical models for use in evaluating the potential for degradation of the ground-water resource owing to increased applications of pesticides. This report describes work completed during the first year of the study. Background data collected on the geology, soils, and hydrology of the study site during the first year were used to characterize and define the hydrogeologic environment of the unsaturated and saturated zones.

Description of Study Area

The field site selected for the study of pesticide migration is in a major agricultural area within the Dougherty Plain topographic division of the Coastal Plain physiographic province (12). The 10-acre study site is in southeastern Lee County, Ga., approximately 10 mi northeast of the city of Albany, and 1.3 mi east of Georgia Highway 91 (Figure 1).

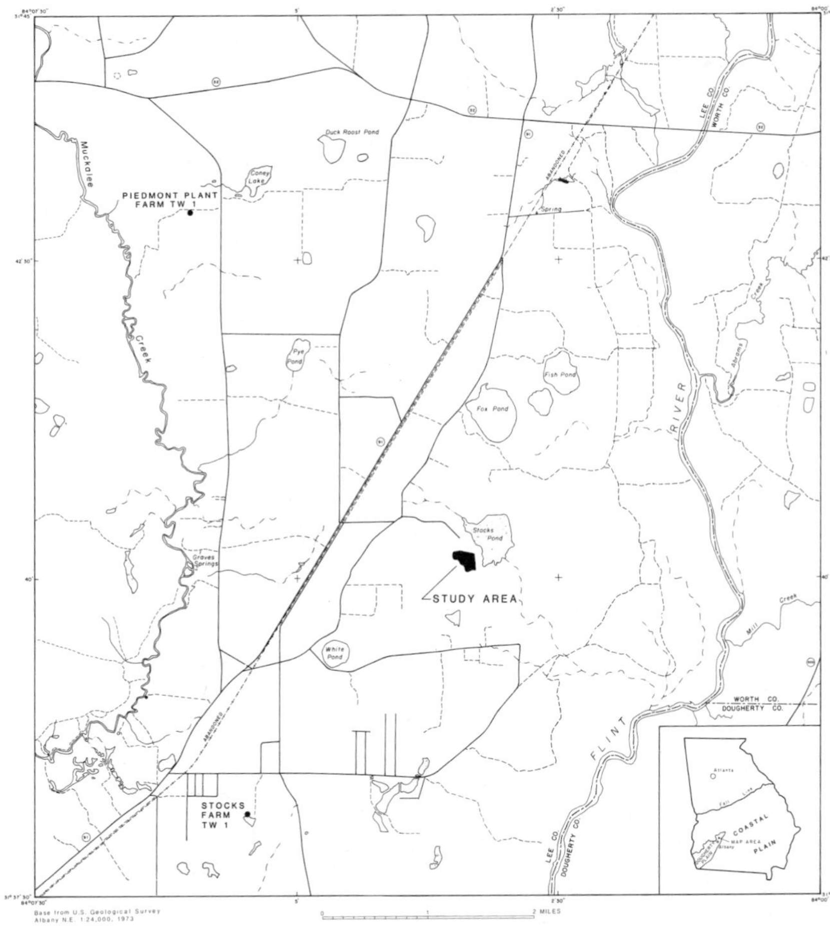

Figure 1. Location of the study site.

The study site was selected so that it would (1) have multi-layered soils, (2) have relatively flat topography (to minimize runoff), (3) be of manageable size (less than 20 acres), (4) have a shallow water table, (5) be isolated from domestic wells, (6) be within close proximity of a ground-water divide, (7) be available for a 5-year study, with owner cooperation, and (8) be in a major agricultural area.

The study site lies in a well-drained plain having a gently undulating land surface (13, 14). The maximum relief of the area surrounding the study site does not exceed 80 ft. The site has a relief slightly greater than 3 ft (Figure 2).

The site is underlain at shallow depths by the Ocala Limestone. Dissolution of the limestone and collapse of solution cavities has produced numerous shallow, saucer-shaped sinks or depressions in the area surrounding the study site (Figure 1). Most sinks are nearly circular and flat bottomed, and have gently sloping sides. Sinks vary in size from small, shallow depressions to large sinks covering several acres (15). Many of the larger sinks are filled with water throughout the year and form ponds or lakes, whereas others are wet only in winter and spring (16). Stocks Pond, about 500 ft east-northeast of the site (Figure 1), is an example of a large sink that is filled with water throughout the year. The north side of the site is bordered by a drainage ditch that is 6 ft wide and 8 ft deep. The drainage ditch is filled with water only during winter and spring when precipitation increases.

The study site lies entirely within the Flint River drainage basin. The Flint River flows into the Chattahoochee River, which discharges to the Gulf of Mexico. The Flint River lies east of the site and has downcut about 15 ft into the Ocala Limestone. As a result, the Flint River flows in a well-defined channel bordered by a relatively narrow flood plain.

West of the study site lies Muckalee Creek. It flows into Kinchafoonee Creek, which discharges to the Flint River just north of the city of Albany. Muckalee Creek has downcut about 10 to 15 ft into the limestone to produce a steep-sided, narrow channel. There is a distinct absence of small tributaries in this area because most of the drainage is subterranean (15, 17). Ground water that migrates into fractures in the limestone commonly discharges as springs along the banks of streams.

Soils in the study site and surrounding area are generally level to gently sloping. Soils in low-lying areas and depressions are moderately well drained to very poorly drained. These soils are associated with areas that have a seasonally high water table and, therefore, generally are flooded each year. Soils on broad ridges are well drained, and, because the water table lies several feet below land surface, flooding is not a problem (16).

Field Methods

Background data describing the hydrogeologic environment of the site were collected during 1983. The field work concentrated on defining soil properties, geologic characteristics of the residuum and of the Ocala Limestone, and existing ground-water conditions at the site.

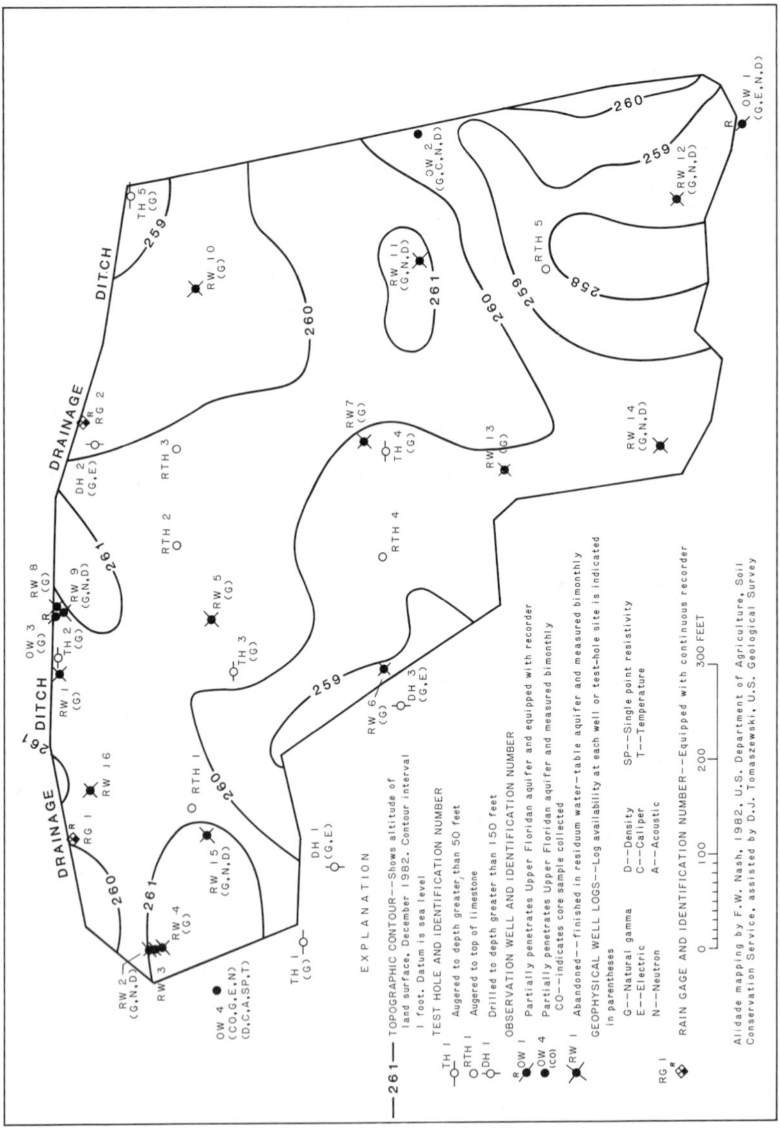

Figure 2. Location of test holes and observation wells, and topography of the study site.

A soil survey of the study site was conducted by the U.S. Soil Conservation Service to determine the number of soil series present. The soil survey provided a basis for determining the location and selecting the optimum number of monitoring sites.

Geologic information was gathered from 13 test holes and 21 observation wells drilled in the study site (Figure 2). Five of the 13 test holes were augered to the top of limestone; five were augered into the limestone; and three were drilled deeper than 200 ft. Geologic evaluation of the residuum and the Ocala Limestone was based on analyses of sample cuttings, a continuous core, field notes, and borehole geophysical logs.

Hydrologic information was gathered from 16 temporary wells cased in the residuum with slotted PVC pipe and four permanent wells cased through the residuum and open to the Ocala Limestone. Bimonthly water-level measurements were made and hydrographs plotted for 14 observation wells. Two observation wells (OW1 and OW3) open to the Ocala Limestone were equipped with continuous water-level recorders, and daily mean hydrographs were plotted for both wells. Precipitation data were collected continuously from two rain gages equipped with Fisher-Porter digital recorders (Figure 2).

Geology

Sedimentary rocks underlying the study site range in age from Upper Cretaceous to late Eocene. A thin surficial deposit of unconsolidated, post-Eocene residuum overlies the bedrock. The geologic units pertinent to the study are, in ascending order, the Lisbon Formation, the Clinchfield Sand, the Ocala Limestone, and the undifferentiated residuum (Figure 3).

Lisbon Formation. The Lisbon Formation of middle Eocene age is about 72 ft thick in the study area. The basal part of the Lisbon Formation consists of about 37 ft of greenish-gray, glauconitic, argillaceous sand interbedded with fossiliferous, sandy limestone. The basal sequence is succeeded by about 35 ft of sandy, glauconitic limestone that interfingers with thin sand stringers. This description of the basal sequence agrees with the stratigraphic description of the Lisbon Formation as given by Herrick (18) for the well drilled at Dixie Pines Plantation in northwestern Lee County, Ga., about 19 mi northwest of the site. The section contains abundant shell fragments, and the degree of cementation of the quartz sand is variable. The presence of the thin sand stringers with the limestone suggests that the quartz sand was originally deposited in a shallow marine environment. Minor transgressive pulses of the sea temporarily suspended clastic sedimentation in the shallow marine environment, thus allowing the accumulation of carbonate. The varying extent of calcareous cementation of the quartz grains is a secondary depositional feature attributed to the circulation of ground water after deposition.

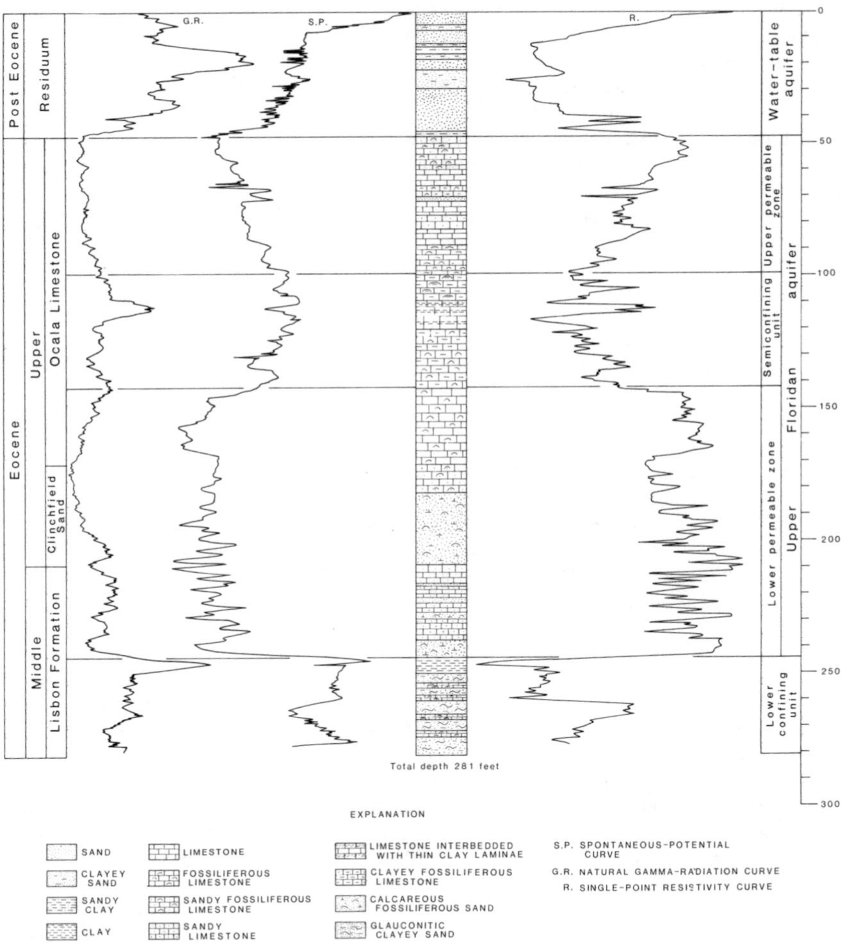

Figure 3. Stratigraphic section and borehole geophysical logs for test hole DH3.

Clinchfield Sand. The Lisbon Formation is unconformably overlain by about 27 ft of interfingering unconsolidated quartz sand and quartz sand containing calcareous cement. The medium-grained, well-sorted quartz sand is stratigraphically equivalent to the basal upper Eocene sand deposit that Herrick (19) described as the Clinchfield Sand.

The unconsolidated sand contains abundant shell material and echinoid fragments, and forms individual beds that have a maximum thickness of about 5 ft. The cemented sand contains abundant shell molds and forms beds up to 10 ft thick. The abundance of quartz, shell material, and echinoid fragments in the section indicates that it was deposited in a beach environment.

The sand grades upward into an 11-foot-thick zone of fossiliferous, sandy limestone interbedded with thin lenses of unconsolidated quartz sand. This part of the sequence was deposited in a shallow marine environment (19).

Ocala Limestone. The Clinchfield Sand is overlain by the Ocala Limestone of late Eocene age, which consists of about 124 ft of white, porous, calcitized, fossiliferous limestone. The Ocala Limestone was deposited in a marginal marine, coastal environment as a result of a major transgression of the sea that was punctuated by several minor regressions. The lithologic character of the Ocala Limestone beneath the study field is described in the stratigraphic section for test well DH3 (Figure 3), which penetrates the Ocala Limestone, the Clinchfield Sand, and the Lisbon Formation, and the location of DH3 is shown on Figure 2.

The basal part of the Ocala Limestone consists of about 29 ft of white, coarsely crystalline, fossiliferous limestone. This sequence is overlain by about 43 ft of dense argillaceous, fossiliferous limestone interbedded with thin clay laminae and about 5 to 8 ft of soft, chalky limestone. The void spaces within the argillaceous limestone have been partially filled with clay, and the unit has low permeability.

The upper part of the Ocala Limestone consists of about 52 ft of porous, coarsely crystalline limestone containing fine detrital quartz and abundant shell fragments. The fossiliferous limestone includes thin beds of unconsolidated quartz sand; thin laminae of siliceous, light-gray clay; and small quantities of pyrite, phosphate, and glauconite. Limonite occurs as staining along fractures in the limestone.

The upper surface of the Ocala Limestone is highly irregular, and beneath most of the study field it ranges from about 35 to 52 ft below land surface.

Residuum. At the study site the Ocala Limestone is overlain by sand and clay derived from weathering of the limestone. The stratigraphic section in Figure 3 shows the lithologic variation of the upper 20 ft of the residuum. The residuum consists of layers and lenses of poorly sorted, very fine to very coarse, angular to subrounded quartz sand and gravel that interfinger with lenses of clayey sand, sandy clay, and clay. The clay content of the residuum ranges from about 10 to 70 percent. The highest percentage of clay occurs in the form of lenses that vary in thickness and lateral extent. Fragments of silicified limestone and chert occur throughout the section and are abundant near the base of the residuum. The average thickness of the residuum at the study site is about 44 ft.

A few thin lenses of tight, gray, plastic clay also occur in the residuum beneath the study field. The clay is dense and contains small fragments of chert and silicified limestone. The maximum observed thickness of the clay lenses is 5 ft in the southeastern corner of the field. Although they are areally restricted, the clay lenses act as semiconfining zones.

Soils

Soils in southeastern Lee County are classified as Ultisols. They are moist soils that have argillic horizons and base saturations of less than 35 percent. Ultisols develop in areas that have long frost-free seasons, abundant rainfall, and adequate ground-water supplies.

The subsurface horizons are commonly red or yellow owing to the accumulation of free oxides of iron, although some weatherable minerals are retained in the soil profile. Ultisols are acidic and have low fertility and low base status. Ultisols are not naturally fertile, but have enormous potential for agricultural productivity when properly managed through applications of lime and fertilizer (20, 21).

Ultisols form on relatively old geologic terranes, where abundant precipitation produces deeply weathered soils. Extensive leaching and warm soil temperatures over prolonged periods result in rapid and nearly complete alteration of weatherable minerals into secondary clays and oxides. Soils in the study site are classified as Paleudults, in which "udult" refers to the suborder of Ultisols and "pale" means "old development". Udult soils have low organic-matter content. They form in humid climates where dry periods are of short duration and the water table remains below the solum throughout most of the year (21).

Two common features of soils in the study site are plinthite and fragipan. Both are cemented materials that tend to restrict water movement and root penetration. Plinthite forms in subsoils that have developed in the oldest areas of the landscape.

Fragipans are common indicators of poor drainage. A fragipan is a loamy, brittle, subsurface horizon that is low in porosity, organic-matter content, and clay content, but high in silt or fine sand. Fragipans can, but do not necessarily, occur in the presence of plinthite. Where they occur, fragipans can create a perched water table (16, 21).

The three major soil series found at the site are the Clarendon, the Ardilla, and the Tifton (Figure 4). A fourth soil series, the Lucy, is present in the field; however, due to its limited areal extent (less than 0.5 acre) and the physical similarity between the Lucy series and the Tifton series, it is not recognized as a separate and distinct soil series in the site.

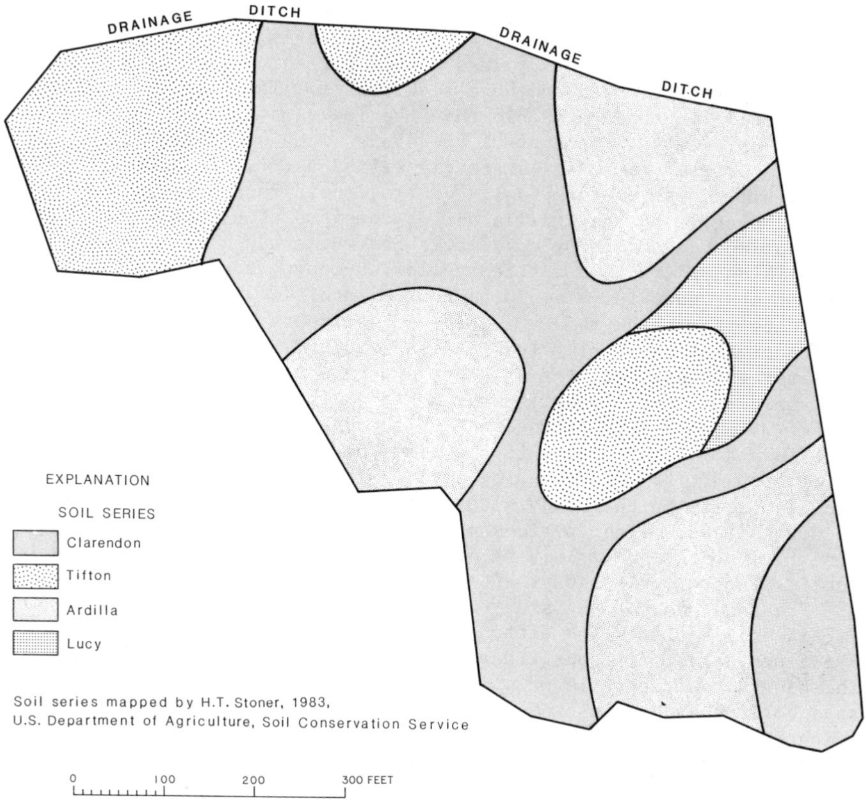

Figure 4. Major soil series in the study site.

The Clarendon series covers slightly more than 45 percent of the area in the field. Clarendon soils are classified as members of the fine-loamy, siliceous, thermic family of Plinthaquic Paleudults. The typical Clarendon profile consists of a dark grayish-brown, sandy horizon underlain by a yellowish-brown, sandy clay loam horizon, which is underlain by a mottled yellowish-brown, red, and gray, sandy clay loam horizon that contains 5 to 30 percent plinthite. Clarendon soils originate from unconsolidated sediments of medium texture. They are typically found on uplands and have a slope gradient of less than 2 percent. The soils are moderately well drained and have slow runoff. The upper part of the soil profile is more permeable than the lower part.

Approximately 29 percent of the study site is covered by the Ardilla series. These soils are classified as members of the fine-loamy, siliceous, thermic family of Fragiaquic Paleudults. The typical profile of the Ardilla series consists of a fine, sandy loam horizon underlain by a yellowish-brown, sandy clay loam horizon which is underlain by a mottled yellowish-brown, red, and gray horizon that contains 5 to 30 percent plinthite. Ardilla soils originate from thick beds of weathered sediments of marine origin, and consist of sandy clay loam and sandy clay. They commonly occur on level to gently sloping uplands, and have a slope ranging from 0 to 5 percent. Ardilla soils are poorly drained and have slow runoff. Permeability is moderate in the upper part of the soil profile and moderately low in the lower part. Ardilla soils receive runoff from surrounding soils. The areal extent of the Ardilla series in the study site is restricted to topographic lows or depressions. After periods of increased rainfall, water tends to pond in low areas underlain by Ardilla soils, and these areas tend to remain flooded several days after the rest of the field has dried.

The remainder of the study site (26 percent) is covered by the Tifton soil series. The Tifton soil is classified as a member of the fine-loamy, siliceous, thermic family of Plinthic Paleudults. The general soil profile consists of a dark grayish-brown, loamy sand horizon underlain by a yellowish-brown, sandy clay loam horizon which is underlain by a coarsely mottled brown, red, gray, and yellow horizon containing 5 to 15 percent plinthite. Tifton soils originate from weathered sediments of marine origin that were predominantly loamy. The soils occur on level to gently undulating uplands and have a slope ranging from 0 to 8 percent. Tifton soils are moderately permeable, well drained, and have medium runoff. The occurrence of Tifton soils in the study site is restricted to topographic highs; these areas drain faster than other areas of the field. Small ironstone nodules are common surface features in areas underlain by Tifton soils.

Hydrology

The primary hydrologic units of interest in this study are, in descending vertical order: (1) the residuum which forms the water-table aquifer and acts as an upper confining unit, (2) the Upper Floridan aquifer (formerly the principal artesian aquifer), which consists of the Ocala Limestone, the Clinchfield Sand, and the upper part of the Lisbon Formation, and (3) the basal part of the Lisbon

Formation which hydraulically separates the Upper Floridan aquifer from underlying sediments and serves as the lower confining unit (Figure 3).

Residuum Water-Table Aquifer. Because of its varied lithology the residuum is a heterogeneous, anisotropic geologic unit. The properties of anisotropy and heterogeneity refer specifically to variations in the permeability or hydraulic conductivity of a geologic unit. An anisotropic, heterogeneous unit is one in which the values of hydraulic conductivity vary directionally at a point and from point to point within the unit. Because lenses of sandy clay and clay interfinger with sand and clayey sand throughout the thickness of the residuum, values of hydraulic conductivity vary directionally and from point to point within the residuum. Consequently, water infiltrating the residuum migrates in varying directions and at differing rates, depending on the hydraulic conductivity of the material through which the water is migrating.

In the area surrounding the study site, Hayes (22) reported a wide range of estimated values for hydraulic conductivity and for transmissivity of the residuum. The hydraulic conductivity was calculated from aquifer tests, borehole geophysical logs, and sieve analyses of drill cuttings. The transmissivity was calculated using an average value of saturated thickness, which was based on measured, seasonal water-level changes at each individual well. The estimates of transmissivity represent only average conditions (22).

Estimated values of transmissivity and average hydraulic conductivity were determined for well TW1 at Piedmont Plant Farm which is 4 mi northwest of the study site, and for well TW1 at Stocks Farm which is 3 mi southwest of the study site (Figure 1).

The estimated average vertical and horizontal hydraulic conductivities at Piedmont Plant Farm were 0.003 and 0.02 ft/d, respectively. The ratio of the average horizontal to vertical hydraulic conductivity was 7.

The residuum is 47 ft thick at Piedmont Plant Farm. The water level was measured from January 1980 to September 1981, and the average residuum water level at Piedmont Plant Farm was 32.9 ft below land surface. Consequently, the average saturated thickness was 14.1 ft. Based on a saturated thickness of 14.1 ft, the estimated transmissivity of the residuum at Piedmont Plant Farm was calculated to be 0.3 ft^2/d.

South of the study site, at Stocks Farm, estimated values of hydraulic conductivity and transmissivity were much greater than at Piedmont Plant Farm. The estimated average vertical and horizontal hydraulic conductivities at Stocks Farm were 9 and 30 ft/d, respectively. The ratio of the average horizontal to vertical hydraulic conductivity was 3.

The residuum is 50 ft thick at Stocks Farm. Residuum water-level measurements made from January 1980 to September 1981 showed that the average residuum water level at Stocks Farm was 13.0 ft below land surface. In contrast to Piedmont Plant Farm, the average saturated thickness at Stocks Farm was 37.0 ft, which resulted in a higher estimated transmissivity. The transmissivity of the residuum at Stocks Farm was estimated to be 1,000 ft^2/d, based on the saturated thickness of the residuum.

During 1983, precipitation that infiltrated through the residuum at the study site either percolated vertically downward to recharge the Upper Floridan aquifer or moved laterally to points of discharge. The drainage ditch bordering the north side of the field and a low, swampy area along the west-central side of the field acted as intermittent discharge points for the water-table aquifer during winter and spring when the water table was high (Figure 2).

The residuum water-table aquifer responded rapidly (within hours) to seasonal variations in precipitation. Figure 5 shows the 1983 hydrograph for well OW3, which fully penetrates the residuum, and a graph showing accumulated rainfall for the same period. Fluctuations of the water table were directly related to variations in the amounts and intensity of rainfall, so rapid fluctuations in the water table were common. The water table generally began rising within 8 hours after the start of heavy rainfall. During winter and early spring 1983, when evapotranspiration was low, increased precipitation caused the water table to rise. Although precipitation was generally heavy from April through September 1983, water lost to evapotranspiration was great and the reduced amount of water available for recharge caused the water level to decline. Because the water table responded rapidly to changes in precipitation and because the water level in the residuum fluctuated seasonally, the thickness of the unsaturated zone also varied seasonally.

Beginning in April 1983, the water level was measured semimonthly in 16 temporary observation wells finished in the residuum (Figure 2). The water table underwent an average decline of 12.1 ft from April to November 1983 (Figure 6). Total rainfall at the study site during 1983 was 50.4 in.

During the two subsequent years following 1983, evaporation data were collected daily from an evaporation pan that was installed as part of a weather station at the study site. From mid-fall to early spring, the average 2-year value for evaporation ranged from about 0.07 to 0.17 in/d, and values of less than 0.05 in/d were common. From late spring to early fall, the average 2-year pan evaporation value ranged from about 0.2 to 0.26 in/d, but values as high as 0.34 in/d were common during both July and August of 1984 and 1985.

During winter and spring 1983, precipitation increased, resulting in a total accumulation of 21.7 in. from February to mid-May 1983. The average 2-year pan evaporation for the same period was about 0.11 in/d or a total of about 13.2 in. The surplus water infiltrated downward causing the water level in the residuum to rise. Vertical leakage through the base of the residuum also increased, which resulted in recharge to the Upper Floridan aquifer.

During summer and fall 1983, total precipitation decreased to about 17.5 in. from mid-May to mid-November 1983. The 2-year average value for evaporation was about 0.23 in/d or a total of about 33.2 in. During this period, evaporation exceeded precipitation, and in response, the water level in the residuum declined. The decreased infiltration of water into the residuum caused a decrease in the vertical leakage through the base of the residuum, and so recharge to the Upper Floridan aquifer decreased. Water levels continued to decline in both the water-table aquifer and the

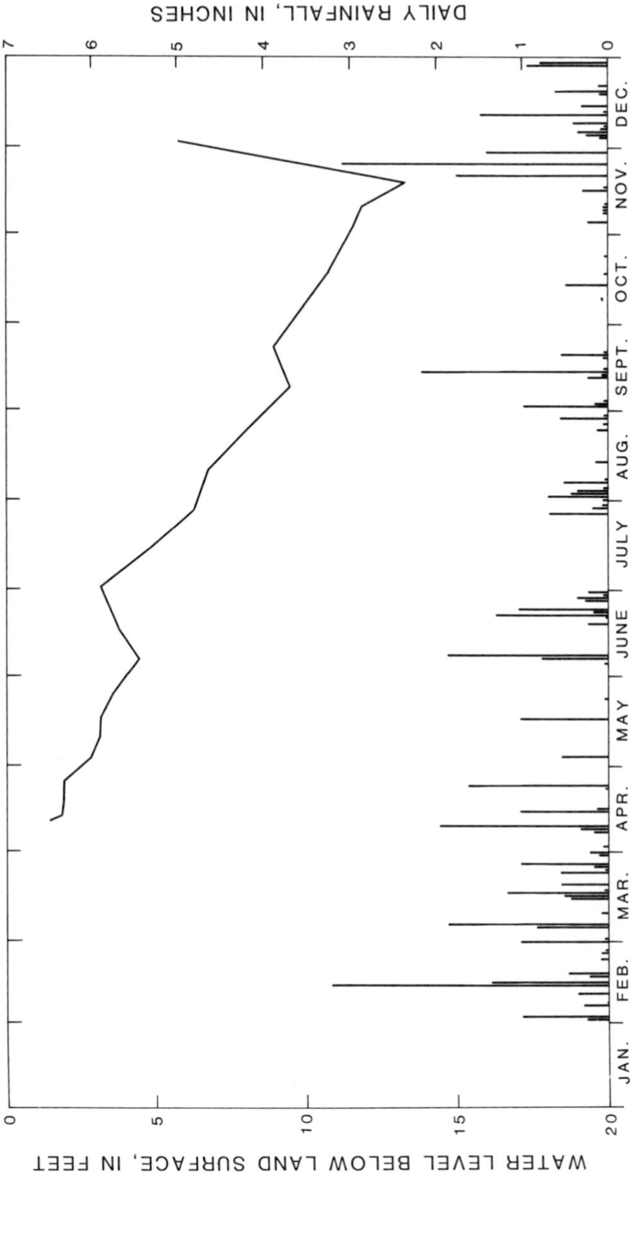

Figure 5. Water-level fluctuations in the residuum and daily rainfall at the study site, 1983.

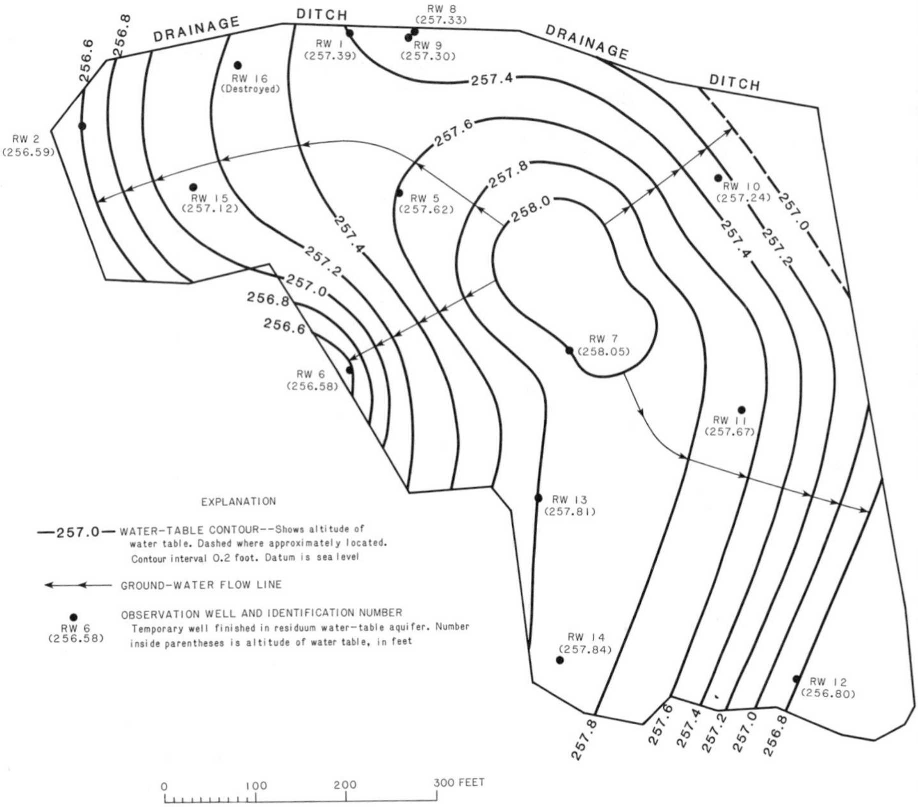

Figure 6. Altitude of water table in the residuum at the study site, January 1984.

Upper Floridan aquifer until significant precipitation occurred in late November 1983 and water lost to the atmosphere through evaporation and transpiration diminished.

The configurations and the gradients of the water table are influenced not only by geohydrologic properties of the residuum, but also by changes in the amount of precipitation and the rate of evapotranspiration. Increased precipitation and decreased evapotranspiration during winter and spring 1983 caused the water table to rise and ground water to temporarily mound in the central part of the field (Figure 6). During January 1984, the direction of ground-water flow was away from the mound. During summer and fall 1983 when precipitation decreased and evapotranspiration increased, the water table declined and the direction of ground-water flow shifted to the east-southeast (Figure 7). The hydraulic gradient was nearly flat in the north-northwest corner of the field, while the gradient steepened in the southeastern part of the field.

Small quantities of water were obtained from the shallow wells finished in the residuum. The well yields ranged from less than 3 gal/min to 5 gal/min.

Upper Floridan Aquifer. The Upper Floridan aquifer is formed by the Ocala Limestone, the Clinchfield Sand, and the upper part of the Lisbon Formation which are, for the most part, highly permeable. The aquifer is confined above by the basal part of the residuum and below by the lower part of the Lisbon Formation (Figure 3).

In the area surrounding the study site and away from streams, Hayes (22) estimated that the transmissivity of the Upper Floridan aquifer ranged from about 4,000 to 6,000 ft^2/d. The effective hydraulic conductivity of the aquifer in this area was estimated to be about 100 ft/d, and the effective porosity was estimated to be about 20 percent (22). Hayes (22) reported that the hydraulic gradient in the northern part of the Dougherty Plain was about 2 ft/mi, and so the average velocity of ground-water flow in the vicinity of the study site is about 0.2 ft/d.

Large quantities of water have been obtained from nearby wells that partially penetrate the Upper Floridan aquifer. Hayes (22) reported that the yield from several wells in Lee County, Ga., ranged from 150 to 225 gal/min. Well yields at the study site exceeded 100 gal/min.

Geologic and hydrologic data from the study site indicate that the Upper Floridan aquifer includes upper and lower permeable zones that are hydraulically separated by a middle, semiconfining unit. Borehole geophysical logs run in test hole DH3, which fully penetrates the Upper Floridan aquifer, show the two permeable zones and the semiconfining unit (Figure 3).

The upper permeable zone consists of 52 ft of coarsely crystalline, fossilferous limestone. Permeability is mainly of secondary origin and includes openings developed by fracturing and dissolution of the limestone. The lower permeable zone is 102 ft thick, and consists of sand that intertongues with sandy limestone. The sand in the lower part of the aquifer is more permeable than the limestone in the upper part and it has a higher water-bearing potential. The high permeability of the sand is mainly from primary

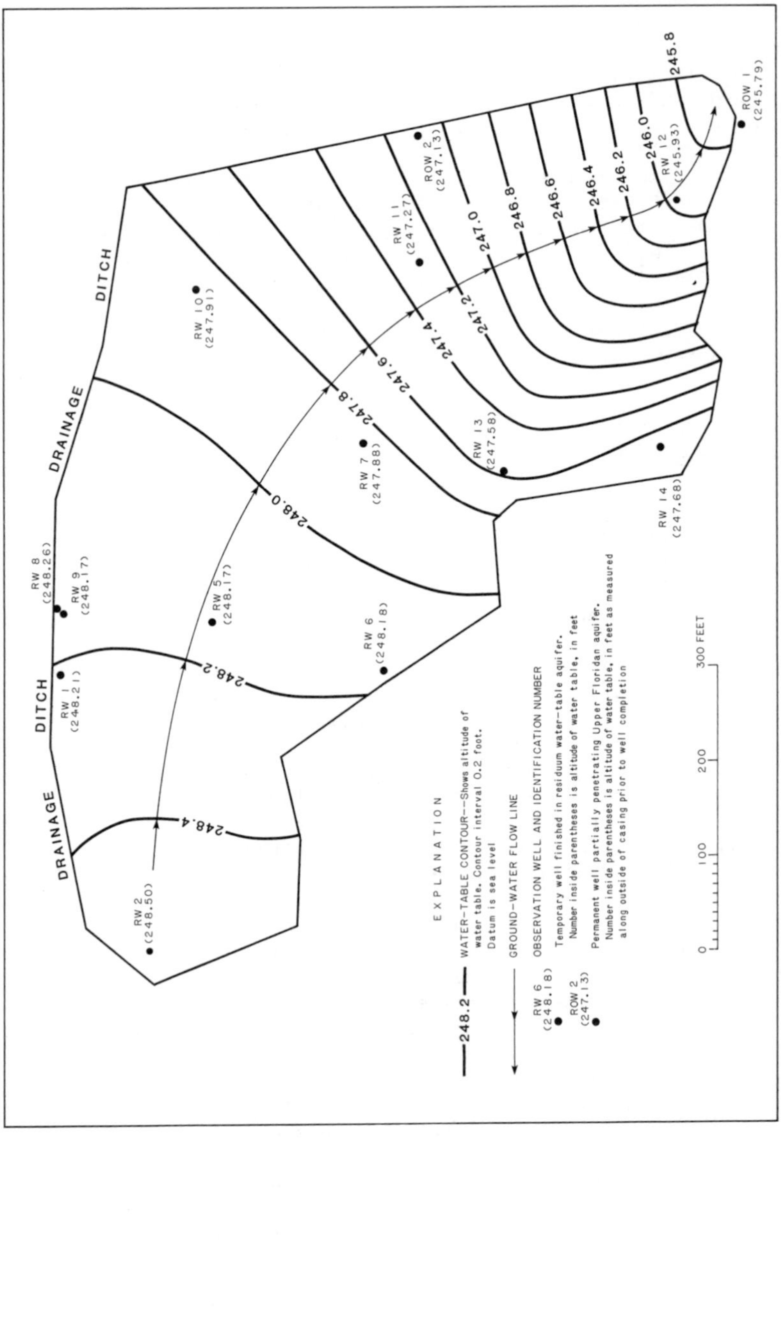

Figure 7. Altitude of water table in the residuum at the study site, November 1983.

porosity. The semiconfining unit near the middle of the aquifer is 47 ft thick and has low permeability owing to high clay content; therefore, it permits only minor vertical leakage between the two permeable zones.

The upper permeable zone is connected hydraulically to the residuum which allows vertical leakage through the base of the residuum, and thus response by the upper permeable zone to changes in precipitation is rapid. Precipitation infiltrating the residuum recharges the upper permeable zone, and the rate of recharge is dependent upon such factors as the head differential between the water table in the residuum and the water level in the upper permeable zone, variations in hydraulic conductivities between the residuum and the upper permeable zone, and the thickness of the confining unit, which is the residuum.

The hydrographs in Figure 8 reveal that a head differential exists between the upper and lower permeable zones, indicating that the two are hydraulically separated. During 1983, the water level in well OW3, which partially penetrates the upper permeable zone, ranged from about 4 to 16 ft below land surface, whereas the water level in well OW1, which taps the lower permeable zone, ranged from about 18 to 34 ft below land surface.

Table 1 shows water-level measurements made at the same time in the water-table aquifer, the upper permeable zone, and the lower permeable zone.

Table 1.--Water levels and head differences between the water table and the two permeable zones in the Upper Floridan aquifer [WT = water table; UPZ = upper permeable zone; LPZ = lower permeable zone]

Date	Water-level measurements (ft)			Head differences (ft)	
	WT	UPZ	LPZ	WT and UPZ	UPZ and LPZ
Apr 1983	3.19	2.34	18.27	0.85	15.93
May 1983	4.24	3.65	20.38	.59	16.73
Nov 1983	12.50	12.43	29.38	.07	16.95

Based on a comparison of the head differences shown in Table 1, the degree of hydraulic separation is much greater between the upper and lower permeable zones in the Upper Floridan aquifer than between the upper permeable zone and the water-table aquifer.

The small head differential, exhibited throughout the year, between the water-table aquifer and the upper permeable zone indicates that although the residuum acts as a confining unit for the Upper Floridan aquifer, vertical leakage does occur through the base of the residuum, and there is some degree of hydraulic connection between the two hydrologic zones.

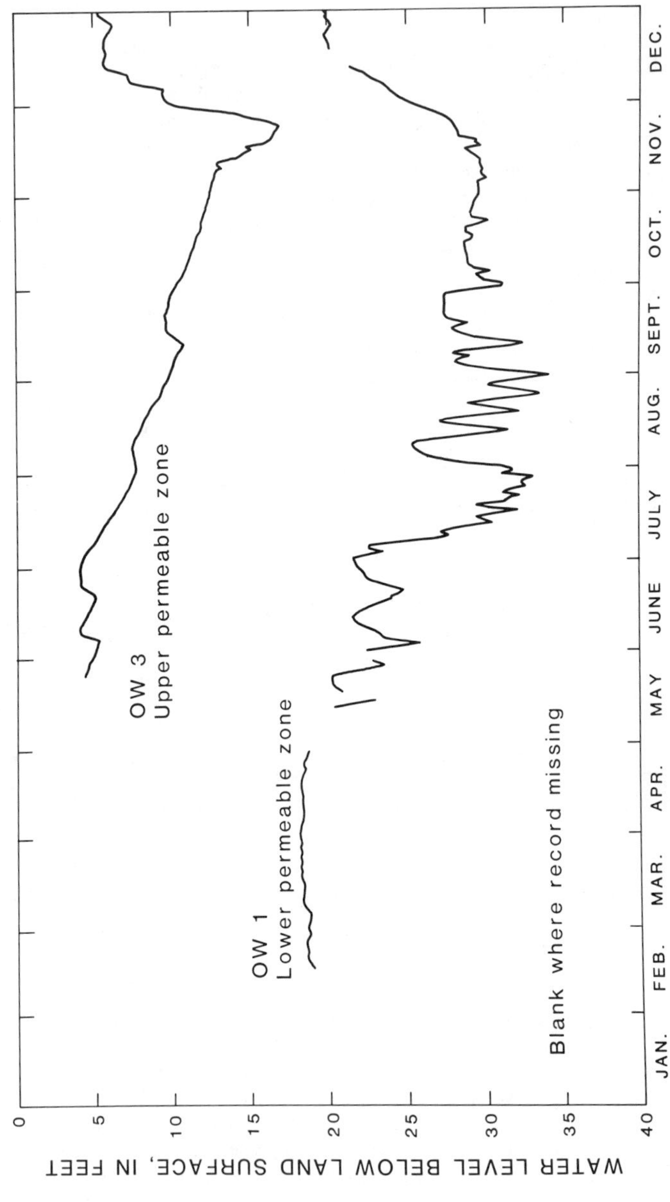

Figure 8. Daily mean water levels in the Upper Floridan aquifer at wells OW1 and OW3, 1983.

A large head differential exists between the upper and lower permeable zones indicating that the 47 ft of semiconfining material (Figure 3) separating the two permeable zones allows only minor vertical leakage downward to recharge the lower permeable zone.

The hydrographs in Figure 8 show that the upper and lower permeable zones respond differently to precipitation and pumping. Climatological variations and water-level changes in the residuum and the upper permeable zone have little effect on the lower permeable zone. Moreover, when nearby irrigation wells that tap only the lower permeable zone are pumped, the water level in the lower permeable zone responds rapidly, but there is almost no change in the water level in the upper permeable zone.

The 1983 hydrograph of well OW1 shows that between mid-May and late October, the water level in the lower permeable zone fluctuated in response to nearby agricultural pumping. By contrast, during the same period, the water level in the upper permeable zone declined steadily and showed minor fluctuations, owing to periods of reduced precipitation and increased evapotranspiration. Well OW3 exhibited little, if any, response to nearby irrigation pumping, thus, once again indicating that leakage through the semiconfining unit was probably minor.

Summary and Conclusion

This study was designed to investigate the migration and degradation of the pesticide aldicarb in a field environment. To evaluate the potential for the pesticide to leach through the soil profile into the ground-water system, detailed information was gathered concerning the physical nature of the unsaturated and saturated zones and the relation between them.

The 10-acre study site is underlain, in descending order, by the undifferentiated residuum, which forms the water-table aquifer, and the Ocala Limestone, the Clinchfield Sand, and the Lisbon Formation, which form the Upper Floridan aquifer. Geohydrologic data indicate that in the area of the study site, the Upper Floridan aquifer includes upper and lower permeable zones. The upper permeable zone is hydraulically connected to the residuum water-table aquifer and is recharged directly by water infiltrating through the residuum. A semiconfining unit separates the lower permeable zone from the upper permeable zone. There is, however, minor vertical leakage through the semiconfining unit into the lower permeable zone.

Because the residuum water-table aquifer and the upper permeable zone are hydraulically connected, they respond rapidly (within hours) to seasonal fluctuations in precipitation. During winter and spring of 1983, the water levels in the residuum and the upper permeable zone rose in response to increased precipitation and decreased evapotranspiration. Water levels in the residuum and the upper permeable zone declined during the summer and fall as precipitation decreased and evapotranspiration increased. During late fall, decreased evapotranspiration and increased precipitation caused the water levels to rise.

Seasonal water-level changes in the residuum and the upper permeable zone have little effect on the lower permeable zone. The water level in the lower permeable zone does, however, respond rapidly (within hours) to the pumping of irrigation wells that tap the lower permeable zone, whereas the water level in the upper permeable zone shows almost no change.

From the initial background field data, the following conclusions were drawn concerning the hydrogeologic environment of the field site:

1. The thickness of the unsaturated zone varies, depending on the water level in the residuum which is directly affected by fluctuations in precipitation and evapotranspiration.
2. The residuum is a thin, surficial layer of variable lithology. Sand, gravel, and clayey sand interfinger with lenses of sandy clay and clay throughout the residuum, but the lenses of clay and sandy clay are restricted in both their lateral and vertical extent. As a result, permeability or hydraulic conductivity in the residuum is variable, which causes water infiltrating through the residuum to migrate in varying directions and at differing rates.
3. The water-table aquifer is hydraulically connected to the upper permeable zone of the Upper Floridan aquifer, and water infiltrates through the residuum to recharge the upper permeable zone.
4. The lower permeable zone is hydraulically separated from the upper permeable zone by a semiconfining unit that allows only minor vertical leakage into the lower permeable zone.

In general, water infiltrating the soil profile migrates vertically downward from the unsaturated zone to the saturated zone and eventually into the ground-water system. Consequently, the potential exists for a pesticide to migrate from the crop-root zone into the water-table aquifer and the Upper Floridan aquifer.

As the study moves into the second phase when field data will be collected for analysis after application of the pesticide aldicarb, background data will continue to be collected to better define and characterize the local ground-water system and its relation to the unsaturated zone.

Literature Cited

1. Supak, J. R. Ph.D. Thesis, Texas A & M University, College Station, Texas, 1972.
2. Martin, H., Ed. "Pesticide Manual" 2nd ed.; British Crop Protection Council, Worchester, England, 1971.
3. "Initial Scientific and Minieconomic Review of Aldicarb," U.S. Environmental Protection Agency, 1975.
4. Zaki, M. H.; Moran, D.; Harris, D. Amer. J. Pub. Health 1982, 72, 1391-1395.
5. Hebb, E. A.; Wheeler, W. B. J. Environ. Quality 1978, 7, 598-601.

6. Peoples, S. A.; Maddy, K. T.; Cusick, W.; Jackson, T.; Cooper, C.; Frederickson, A. S. Bull. Environ. Contam. Toxic. 1980, 24, 611-618.
7. Rothschild, E. R.; Manser, R. J.; Anderson, M. J. Ground Water 1982, 20, 437-445.
8. Spalding, R. F.; Exner, M. E.; Sullivan, J. J.; Sullivan, P. A. J. Environ. Quality 1979, 8, 374-383.
9. Spalding, R. F.; Junk, G. A.; Richard, J. J. Pest. Mon. J. 1980, 14, 70-73.
10. Todd, D. K.; McNulty, D. E. "Polluted Ground Water: A Review of Significant Literature", U.S. Environmental Protection Agency, 1974.
11. Wehtje, G. R.; Spalding, R. F.; Burnside, O. C.; Lowry, S. R.; Leavitt, J. R. C. Weed Science 1983, 31, 610-618.
12. Fenneman, N. M. "Physiography of Eastern United States"; McGraw-Hill: New York, 1938; p. 714.
13. Owen, V. Jr. "Geology and Ground Water Resources of Lee and Sumter Counties, Southwest Georgia," U.S. Geological Survey Water-Supply Paper 1666, U.S. Government Printing Office, Washington, D.C. 1963.
14. Moon, J. W. "Soil Survey of Lee County, Georgia," U.S. Department of Agriculture, Bureau of Chemistry and Soils, Series 1927, No. 4, 1927.
15. LaForge, L.; Cooke, C. W.; Keith, A.; Campbell, M. R. "Physical Geography of Georgia," Georgia Geol. Surv. Bull. 42, 1925.
16. Pilkinton, J. A. "Soil Survey of Lee and Terrell Counties, Georgia," U.S. Department of Agriculture, 1978.
17. Veatch, J. O.; Stephenson, L. W. "Geology of the Coastal Plain of Georgia," Georgia Geol. Surv. Bull. 26, 1911.
18. Herrick, S. M. "Well Logs of the Coastal Plain of Georgia," Georgia Geol. Surv. Bull. 70, 1961.
19. Herrick, S. M. "Age and Correlation of the Clinchfield Sand in Georgia," U.S. Geological Survey Bulletin 1354-E, U.S. Government Printing Office, Washington, D.C., 1972.
20. Brady, N. C. "The Nature and Properties of Soils" 8th ed.; MacMillan: New York, 1974; p. 639.
21. Buol, S. W.; Hole, F. D.; McCracken, R. J. "Soil Genesis and Classification"; The Iowa State University Press: Iowa, 1973; p. 360.
22. Hayes, L. R.; Maslia, M. L.; Meeks, W. C. "Hydrology and Model Evaluation of the Principal Artesian Aquifer, Dougherty Plain, Southwest Georgia," Georgia Geol. Surv. Bull. 97, 1983.

RECEIVED April 7, 1986

6

Spatial Variability of Pesticide Sorption and Degradation Parameters

P. S. C. Rao[1], K. S. V. Edvardsson, L. T. Ou, R. E. Jessup, P. Nkedi-Kizza, and A. G. Hornsby

Soil Science Department, University of Florida, Gainesville, FL 32611

Data were collected at two field sites, one in Florida and the other in Georgia, to evaluate the variability in pesticide concentrations and in pesticide sorption and degradation parameters. The observed variability can be attributed to intrinsic factors leading to inherent variability, and to extrinsic factors resulting in imposed variability. Data are presented to demonstrate the predominant effects of the method of pesticide application and tillage operations in introducing significant extrinsic variability in pesticide concentrations. Variability introduced at the soil surface may persist as the pesticide leaches to deeper depths. At both field sites, a close relationship was found between the spatial variations in soil organic carbon content and pesticide sorption coefficients. At the Georgia field site, small variations (coefficient of variation < 30%) were noted in measured pesticide degradation half-lives in soil samples collected from several locations and four morphologic horizons. Spatial patterns in pesticide sorption and degradation parameters did not correspond to spatial boundaries of the soil series.

Recent reports of increasing incidence of pesticides and toxic organic pollutants in groundwater (1, 2) have prompted a number of laboratory and field investigations of the processes and factors influencing pesticide behavior in soils and groundwater. Several simulation models have been developed for forecasting pesticide

[1]Current address: Agronomy and Soil Science Department, University of Hawaii, Honolulu, HI 96822

0097-6156/86/0315-0100$06.00/0
© 1986 American Chemical Society

fate in soils (3, 4). To validate these models and to use them for either management or regulatory purposes, several site-specific values for soil and pesticide parameters are needed.

A typical field site, varying in area from about 1 to 10 ha, may include several soil series. The model parameter values may be different not only for each of these soil series, but may also vary considerably within a single series. Such variability in a number of soil hydraulic properties (e.g., soil hydraulic conductivity, soil water flux, etc.) has been widely reported in the literature (5 - 7). The model parameter values for a given location in the field may also vary with profile depth depending upon soil horizonation as well as a function of the soil and environmental factors (e.g., soil aeration, temperature, etc.). Since soil and environmental factors undergo dynamic changes with time, model parameters are also expected to exhibit temporal variability. At present, only limited data are available to characterize such spatial and temporal variability in pesticide sorption and degradation parameters required in several simulation models.

In this paper, we will discuss data collected as a part of two recent field studies, one in Georgia and the other in Florida. The objective of both studies was to monitor the environmental dynamics of pesticides in the crop root zone and to use these data to evaluate the predictive capability of several simulation models. Specific attention will be focused on the data characterizing the spatial variability of pesticide sorption and degradation parameters measured at these two field sites.

Description of the Field Sites

The Georgia field study was a cooperative effort between the U.S. Environmental Protection Agency, the U.S. Geological Survey, and the University of Florida. The 4.5-ha field site, planted to peanuts, was located near Albany, GA. At this site data are being collected to characterize pesticide migration within the unsaturated and the saturated zones. The criteria used in selecting this field site and in designing the study are discussed by Carsel et al. (8). Cooper (9) presented a detailed geohydrologic description of this site. Bulk samples of surface soil (0-20 cm) were collected using a bucket auger from 20 locations, designated as the primary sites (8), and an additional 16 random locations at this field site. Each sample was identified by its location with reference to an arbitrary grid (8). Soil organic carbon content (OC) of these samples was determined by dry combustion method using a LECO Carbon Analyzer. Pesticide sorption coefficients (K_d) were determined using the batch equilibrium techniques (10, 11). The pesticides used were: aldicarb [(methylthio) propionaldehyde O-(methylcarbamoyl) oxime]; metolachlor [2-chloro-N- (2-ethyl-6-methylphenyl)-N- (2-methoxy-1-methyl ethyl) acetamide; and diuron [3-(3,4-dichlorophenyl)- 1,1-dimethyl urea]. At the 20 primary locations, soil cores were also collected from each of the four morphologic horizons (0-20, 25-45, 48-63, and 94-107 cm) using aseptic sampling techniques (12). Aldicarb and metolachlor degradation in these soil samples under aerobic conditions was measured using batch incubation techniques (13). These data were used to calculate the half-lives ($t_{1/2}$) for disappearance of the parent compound.

The second field study was conducted on a 0.8-ha citrus grove, located near Davenport, Florida. This study was a cooperative effort between the University of Florida and the Union Carbide Agricultural Products Company, Inc. The soil at this site is classified as Candler sand (hyperthermic, Typic Qurtzipsamments), which is typical of the deep, sandy, and well-drained soils planted to citrus on the central ridge of Florida. Bulk samples of surface soil (0-15 and 15-30 cm) were collected using bucket auger at 29 locations, 12.5 m apart, on a north-south (N-S) transect. On a west-east (W-E) transect, soil samples were collected in a similar manner at another 17 locations. The N-S and W-E transects shared one soil sample at the point of their intersection. Additional soil samples were taken from 26 sites selected randomly within the field. The exact location of each sampling site was noted and was used to identify the samples. OC values for these soils were determined by the dry combustion method. Aldicarb K_d values were measured using the batch equilibrium technique.

Types of Spatial Variability

Rao and Wagenet (14) have proposed that the total variability observed in a given soil property is the sum of intrinsic and extrinsic variability. The former arises from inherent variability in soil properties due to pedogenic processes, while the latter is the result of various soil and crop management practices at a specific field site. They suggested that the spatial variability in soil properties such as soil hydraulic conductivity or pesticide sorption coefficients might be influenced primarily by intrinsic factors, whereas variability in pesticide concentrations and fluxes might exhibit the combined influence of both intrinsic and extrinsic factors. Given data on total variability, it is not always possible to quantify the contributions of each type of variability. In this paper, we will present data representing both types of variability.

Extrinsic Variability

Pesticides and fertilizers are usually applied at the soil surface either in granular or liquid formulations and are subsequently incorporated by some sort of tillage operation (e.g., discing). This operation can induce extrinsic variability in agrochemical concentrations and fluxes in soils due to variability in application itself (random if broadcast; nonrandom if banded) and the tillage operation. Weed scientists have examined such variability, in particular as related to herbicide efficacy (15 - 21). Such extrinsic variability in pesticide concentrations can also pose serious problems in obtaining representative soil samples for pesticide concentration determinations.

At the Florida field site, we evaluated the significance of tillage-induced extrinsic variability in bromide (Br) and aldicarb's total toxic residues (TTR) variability.
Bromide was applied in a 3-m strip centered between the citrus trees that were planted about 8 m apart. A concentrated KBr

solution was sprayed on the soil surface using a tractor-mounted boom spray rig. The spray nozzles were set such that solution spray from adjacent nozzles overlapped and provided essentially a uniform Br application. On the same 3-m strip to which Br was applied, aldicarb (Temik 15-G; granular formulation) was applied in 16 parallel bands that were spaced 20-cm apart. Br and aldicarb were then incorporated into the soil to a depth of about 10 cm using a tractor-mounted multiple disk implement with 24 disks that were set about 12.5 cm apart. Immediately following discing, soil samples were collected to a depth of 30 cm as shown schematically in Figure 1. Note that a total of 16 samples were taken, 8 each on two parallel transects set 0.67 m apart and perpendicular to the application band. Soil samples were collected in this manner in each of the four quadrants of the field.

The variations in Br and aldicarb TTR concentrations across the application band are shown in Figures 2 and 3. Note that both Br and aldicarb TTR concentrations are highly variable within the application band, with several orders of magnitude difference in concentrations even in adjacent soil samples. Note that even though Br was applied essentially uniformly to the soil surface, the variations in Br concentrations are quite similar to that of aldicarb TTR concentrations. This suggests that post-application discing was the primary source of the observed extrinsic variability. These results are consistent with the findings of other workers (8, 18, 21). We have also observed similar variability patterns in aldicarb TTR concentrations in field studies conducted during 1983 at two other locations in Florida (22). Soil samples collected up to 4-m depth during 1983, and up to a depth of 10 m in 1984 showed that such variability will persist throughout the year and at all depths. We are presently analyzing our 1983 and 1984 aldicarb and Br field data in order to calculate a statistically valid "average" concentrations which can be, in turn, compared with the values predicted by simulation models. A discussion of these model validation efforts is beyond the scope of this paper.

Intrinsic Variability in OC and K_d

Variations in OC values in soil samples collected along the N-S and W-E transects at the Florida site are shown in Figure 4. Note that for both the transects, OC in soils from the 0-15 cm depth were more variable than in samples collected from the 15-30 cm depth. Variograms (14) calculated using these data indicated that for the 0-15 cm depth, OC values in soil samples collected within a separation distance (i.e., lag) of 15 m would be spatially correlated. On the other hand, OC values for the 15-30 cm depth are spatially independent. OC data for both depth increments could be fitted to a normal frequency distribution; the normality was confirmed by the Kolmogorov- Smirnov D-statistic (23). The coefficient of variation (CV) in OC data for both depths was less than 20%.

The measured data for OC and aldicarb K_d were used to generate 3-dimensional plots depicting their spatial variations at the Florida field site. These plots, shown in Figure 5, indicate a close, but not exact, correspondence in the spatial patterns of OC and K_d. The OC data and K_d values for three pesticides measured

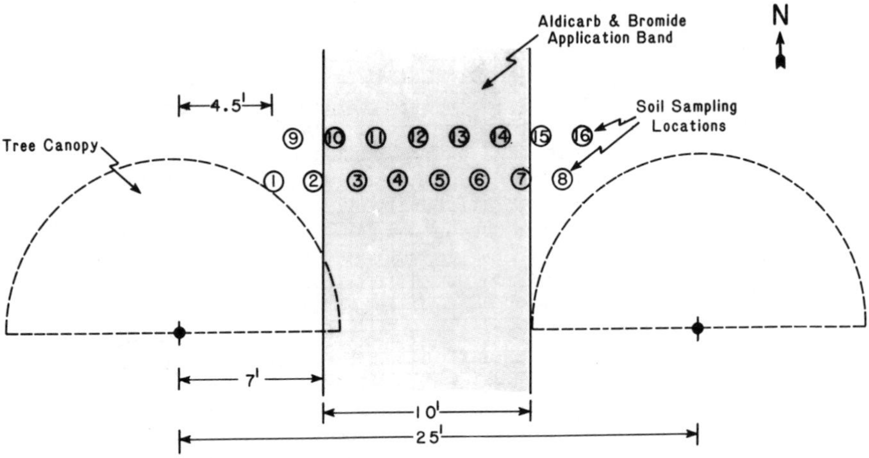

FIGURE 1. Schematic diagram showing the sampling design used to evaluate the extrinsic variability in bromide and aldicarb TTR concentrations at the Florida site.

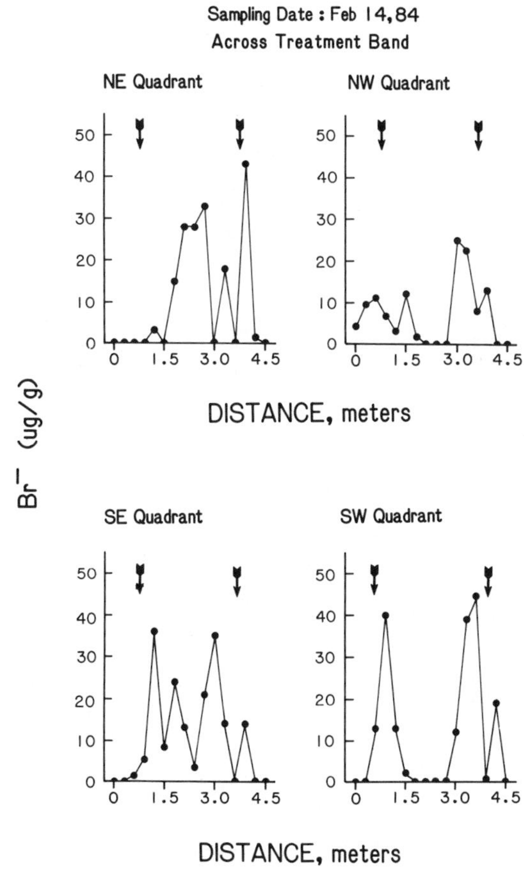

FIGURE 2. Variations in bromide (Br) concentrations in soil samples taken across the application band. Vertical arrows indicate the width of the application band.

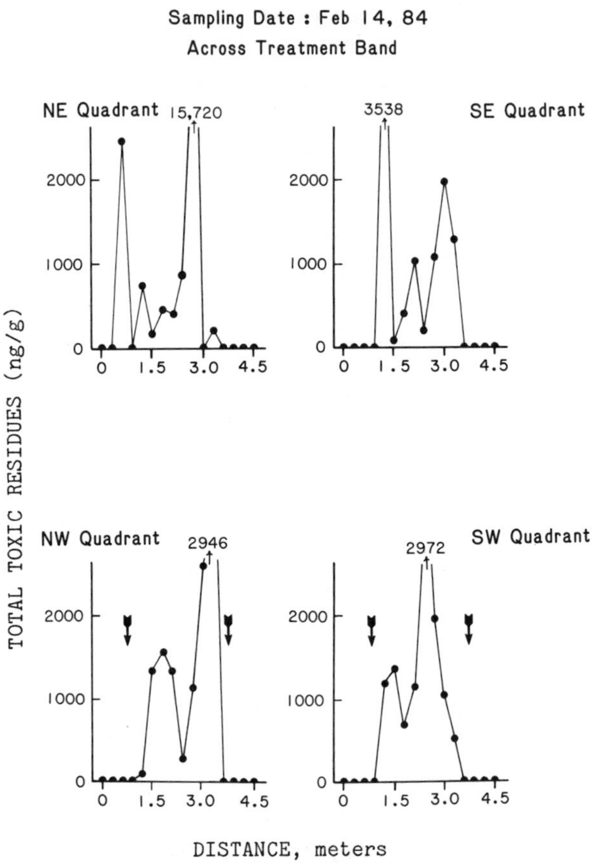

FIGURE 3. Variations in aldicarb TTR concentrations in soil samples taken across the application band. Vertical arrows indicate the width of the application band.

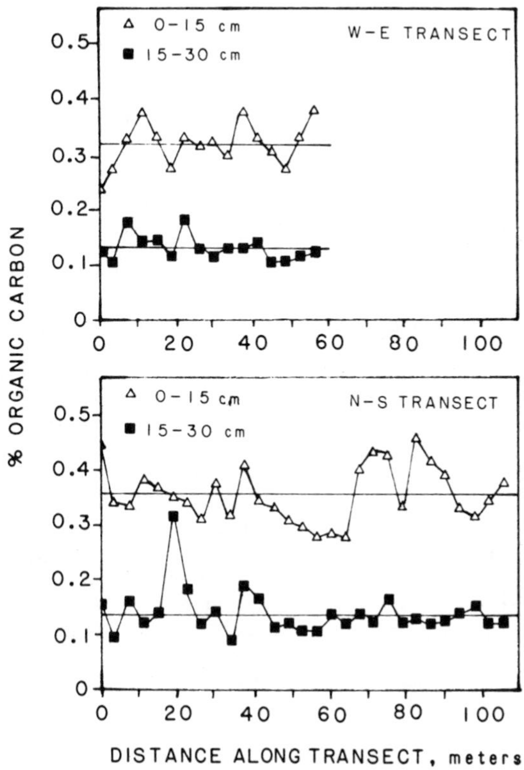

FIGURE 4. Variations in soil organic carbon content (OC) at two depths (0-15 and 15-30 cm) in samples taken along two transects at the Florida site.

using the surface soil (0-20 cm) samples collected at the Georgia site are presented in Figure 6; these plots were generated in manner similar to those shown in Figure 5. As with the data for the Florida site, spatial patterns in pesticide K_d values and OC values are very similar. These results may be anticipated because OC is the single most important determinant of K_d for nonionic pesticides (11, 24 - 26). Thus, it would appear that given a K_d value referenced to OC, denoted as K_{oc}, the spatial variations in K_d for a pesticide at a field site may be estimated by measuring the variations in OC. Green et al. (27) present data for nematicide sorption to support this conclusion.

Intrinsic Variability of Degradation Half-Lives

On the basis of an analysis of published data, Rao and Davidson (11) noted that for several pesticides the variation in pesticide degradation half-lives ($t_{1/2}$) among several soils was surprisingly small (CV < 100%), especially given the range in soil types and the environmental conditions at which degradation was measured. Data collected by Ou et al. (28) for degradation of several pesticides in selected U.S. soils confirmed this observation. More recently, Walker and Brown (29) measured the degradation of two triazine herbicides in soil samples collected from several locations within a 0.64-ha field. They reported a small variation in $t_{1/2}$ values (CV < 25 %) for both herbicides. Data presented by Walker and Zimdhal (30) indicate that the half-lives for pesticide degradation in soils collected from three states in the U.S. did not vary by more than a factor of 2. It should be noted that in all of the above cited studies pesticide degradation was measured in soils that had been air-dried and re-wetted to some desired soil-water content. Inherent differences in the diversity and the size of microbial populations may have been reduced by air-drying the soils, which could explain the absence of significant variability in laboratory-measured pesticide degradation rates.

In contrast to measuring OC or K_d values, the measurement of pesticide degradation half-lives ($t_{1/2}$) in soils is a much more difficult and time-consuming task. Hence, the spatial variations in $t_{1/2}$ could not be assessed in as much detail as we did with OC and K_d. Metolachlor and aldicarb TTR half-lives were measured in a selected number of soil samples collected from the Georgia field site. Soil samples were selected to represent the three major soil series present at the site (Clarendon, Ardilla, and Tifton) and the four major morphologic soil horizons (0-20, 25-46, 48-63, and 94-107 cm) within the crop root zone. Soil samples collected from 4 depths at 10 sites were used to characterize aldicarb TTR degradation rates. Metolachlor half-lives were measured in soils taken from four depths at one site for each soil series and also in samples collected at two depths (0-20 cm and 94-107 cm) at 6 sites.

The data for observed variations in metolachlor $t_{1/2}$ are presented in Table I. Of the three sampling sites within the Clarendon series, the shortest distance between two sites (G-13 to K-15) was about 60 m, while the sites K-15 and H-9 were the farthest, located about 100 m from each other. Of the five sampling locations, the greatest separation distance was about 200 m between J-4 (Tifton) and J-18 (Ardilla). Metolachlor $t_{1/2}$ values measured

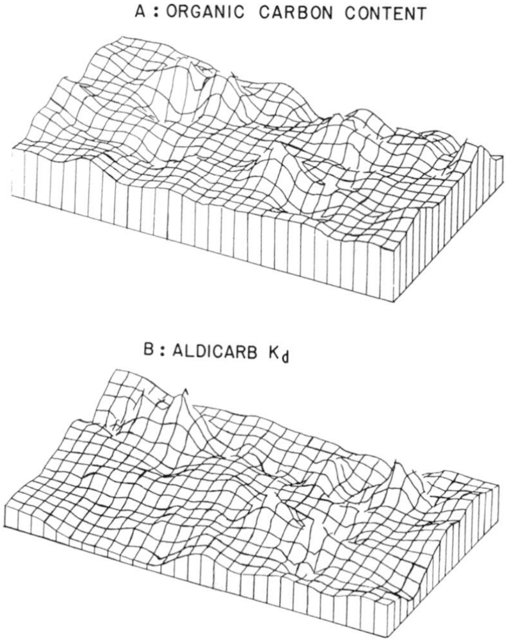

FIGURE 5. Spatial variations in soil organic carbon content (OC) and aldicarb sorption coefficient (K_d) at the Florida site.

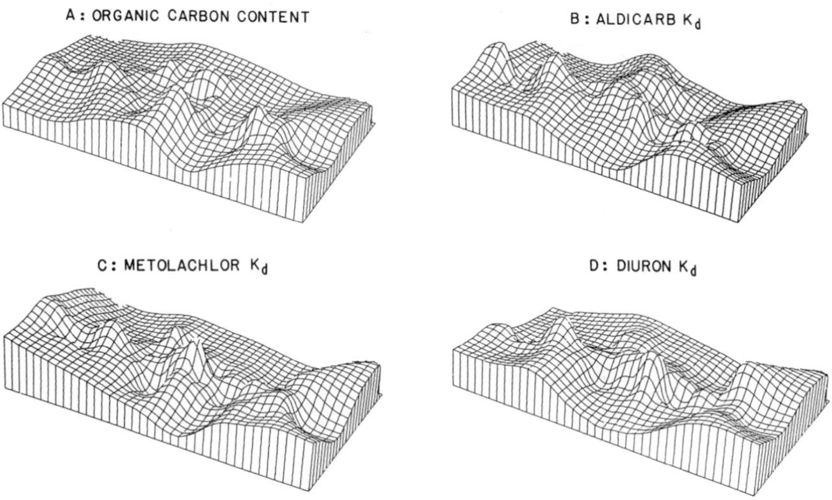

FIGURE 6. Spatial variations in OC and K_d values for three pesticides (aldicarb, metolachlor, and diuron) at the Georgia site.

in all surface soils (0-20 cm) were essentially identical. The $t_{1/2}$ values are larger by about a factor of 2 for the subsoils (25-46, 48-63, 94-107 cm) and tend to be more variable (CV < 38%). The metolachlor half-lives shown in Table I are smaller than those reported by earlier workers (30, 31).

Table I. Variations in Metolachlor Degradation Half-Lives (days) Measured in Soils from the Georgia Site

Soil Series	Soil Depth Increment (cm)			
	0-20	25-46	48-63	94-107
Clarendon (G-13)*	19	17	44	41
Clarendon (H- 9)	19	N.D.**	N.D.	35
Clarendon (K-15)	17	N.D.	N.D.	20
Ardilla (J-18)	18	34	44	53
Ardilla (B-13)	14	N.D.	N.D.	68
Tifton (J-4)	19	31	39	41
Average	18	27	42	43
CV (%)	11	33	7	38

* Code for sampling location on an arbitrary grid
** Not determined

Table II. Variations in Aldicarb TTR Half-Lives (days) Measured in Soils from the Georgia Site

Soil Depth (cm)	Soil Series			Average
	Clarendon	Ardilla	Tifton	
0-20	42 (26)*	45 (24)	37 (11)	42 (21)
25-46	51 (26)	44 (46)	28 (25)	42 (38)
48-63	51 (26)	47 (36)	39 (13)	47 (26)
94-107	67 (21)	54 (30)	63 (5)	62 (21)
Average	53 (28)	48 (30)	42 (34)	48 (31)

* numbers in parenthesis are % CV

The measured variability in aldicarb TTR degradation half-lives are summarized in Table II. Data shown for Clarendon soil were averaged over 4 sites, and the values for Ardilla and Tifton soils are averages for 3 sites. Among the sampling sites, J-4 and J-18 were the farthest apart (about 233 m), whereas the sites I-15 and K-15 were the closest (about 30 m). The greatest distance between two sampling sites within a soil series was: 126 m for Clarendon; 157 m for Ardilla; and 195 m for Tifton. The variations in aldicarb TTR $t_{1/2}$ values with depth and location are somewhat larger (CV < 50%) than those for metolachlor $t_{1/2}$. However, the variation in TTR $t_{1/2}$ values within a soil series are equal to or larger than the variations among the series. An overall average $t_{1/2}$ for aldicarb TTR degradation in these soils (Table II) was 48 days (CV = 31 %).

It should be noted that in our study, soil samples collected from the field were stored at the same water content at which they were sampled. Pesticide degradation was measured in these soils under identical environmental conditions (i.e., substrate concentration, aeration status, temperature, and soil-water content). Thus, the lack of spatial variability in pesticide degradation rates suggests that the inherent capacity of the soil microorganisms to degrade these pesticides might be similar at the Georgia site. This does not necessarily imply that the actual *in situ* degradation rates would not vary spatially or temporally within a field. Local soil environmental factors may be expected to vary as a result of spatial and temporal variations in other soil properties as well as variations in crop/soil management practices (irrigation, rainfall, fertilization, etc.). Thus, a knowledge of the relationship between pesticide degradation rates and soil environmental factors and the variability of these factors in a field is necessary for predicting the spatial variability of pesticide residue concentrations. We are not aware of any published data for assessing *in situ* spatial variations in pesticide degradation rates. Such data have been collected for spatial variations in denitrification rates (32, 33).

Soil-water content (Θ) and temperature (T) are the two major soil environmental factors that control pesticide degradation rates. Walker (34) and Walker and Barnes (35) have proposed a model for pesticide persistence in soils. In their model the Θ-dependence of $t_{1/2}$ is described by a power function [$t_{1/2} = A\,\Theta^{-B}$, where A and B are constants], while the relationship between $t_{1/2}$ and T is described by the Arrhenius equation. Walker and co-workers (36 - 40) determined the values of the necessary model parameters from laboratory measurements of pesticide degradation under controlled conditions. These parameter values and weather data were used as model inputs to predict the persistence of several pesticides in field plots at several locations. Walker et al. (41) summarized the results of an international cooperative study, in which simazine herbicide degradation was measured in soils collected from 21 locations in 11 countries. As in earlier studies (36 - 40), model parameters were estimated on the basis laboratory studies, and were used to predict simazine persistence in field plots at 16 locations. They also simulated simazine degradation in field plots at 5 other locations for which companion laboratory data were not available. In all these studies (36 - 41), the model generally underestimated the amount of pesticide residues remaining in the soil. Walker and co-workers considered the model to be sufficiently accurate for practical applications given the uncertainties in measured pesticide concentrations and the possibility that processes other than microbial degradation (e.g., leaching past the sampling depth; volatilization; and chemical or photolytic degradation) may have been responsible for pesticide dissipation.

Summary

We have attempted to show that both intrinsic and extrinsic factors contribute to the observed spatial variability in pesticide concentrations. The method of pesticide application and subsequent tillage operations can have a major impact on the extrinsic variability in the measured pesticide concentrations and fluxes. Such variability needs to be taken into account in designing soil sampling strategies and in interpreting the field data. Intrinsic variability in pesticide sorption and degradation parameters appears to be small (CV < 30%). Because pesticide sorption coefficients are strongly correlated to soil organic carbon contents, there is a close correspondence between the measured spatial distributions of OC and K_d. Thus, a field-average K_d value may be estimated given the average OC value. Pesticide degradation rates in soil samples collected from different locations and subjected to identical environmental conditions were not variable. However, in situ degradation rates may vary as function of variations in soil environmental conditions.

Acknowledgments

Financial support for this study was provided, in part, by the Cooperative Agreement No. CR-810464 between the U.S. Environmental Protection Agency and the University of Florida; and a grant from the Union Carbide Agricultural Products Co., Inc. Assistance of R.F. Carsel and C.N. Smith, AERL, USEPA, Athens, GA; Ms. S.C. Cooper, USGS, Albany, GA; and J.F. McNabb, RSKERL, USEPA, Ada, OK in collecting the soil samples at the Georgia site is acknowledged. R.L. Jones and R.R. Romine, Union Carbide, assisted in collecting the soil samples at the Florida site and analyzed the samples for bromide and aldicarb TTR. We are grateful to Union Carbide for providing ^{14}C-aldicarb and to Ciba-Geigy for supplying ^{14}C-metolachlor used in the sorption and degradation studies. Finally, we appreciate Ms. Linda Lee's assistance in data analysis and in drafting the figures.

Literature Cited

1. Cohen, S.Z.; Creeger, S.M.; Carsel, R.F.; Enfield, C.G. In "Treatment and Disposal of Pesticide Wastes"; Kreuger, R.F.; Seiber, J.N., Eds.; Symposium Series No. 259; American Chemical Society: Washington, 1984; pp. 297-325.

2. Pye, V.; Patrick, R.; Quarels, J. Groundwater Contamination in the United States; University of Pennsylvania Press: Philadelphia, 1983; 315 p.

3. Carsel, R.F.; Smith, C.N.; Mulkey, L.A.; Dean, J.D.; Jowsie, P. EPA-600/3-84-109; U.S. Environmental Protection Agency: Athens, 1984; 216 p.

4. Wagenet, R.J.; Rao, P.S.C. Weed Sci. 1986, 34, In Press.

5. Nielsen, D.R.; Biggar, J.W.; Erh, K.T. Hilgardia 1973, 42, 215-260.

6. Biggar, J.W.; Nielsen, D.R. Water Resour. Res. 1976, 12, 78-84.

7. Nielsen, D. R.; Bouma, J. (Eds.). Soil Spatial Variability; Proc. Workshop of ISSS and SSSA; Pudoc: Wageningen, 1985; 243 p.

8. Carsel, R.F.; Smith, C.N.; Parrish, R.S.; Mulkey, L.A.; Payne, Jr., W.R. In "Evaluation of Pesticides in Groundwater"; Honeycutt, R.; Ed.; Symposium Series; American Chemical Society: Washington, 1986.

9. Cooper, S.C. In "Evaluation of Pesticides in Groundwater"; Honeycutt, R.; Ed.; Symposium Series; American Chemical Society: Washington, 1986.

10. Green, R.E.; Davidson, J.M.; Biggar, J.W. In "Agrochemicals in Soil"; Banin, A.; Kafkafi, U.; Eds.; Pergamon Press: New York, 1980, pp. 73-80.

11. Rao, P.S.C.; Davidson, J.M. In "Environmental Impact of Nonpoint Source Pollution"; Overcash, M.R.; Davidson, J.M.; Eds.; Ann Arbor Science Publishers: Ann Arbor, 1980; pp. 23-67.

12. Wilson, J.T.; McNabb, J.F.; Balkwill, D.L.; Ghiorse, W.C. Groundwater 1983, 21, 134-142.

13. Ou, L.T.; Edvardsson, K.S.V.; Rao, P.S.C. J. Agric. Food Chem. 1985, 33, 72-78.

14. Rao, P.S.C.; Wagenet, R.J. Weed Sci. 1986, 34, In Press.

15. Clay, D.V.; Scott, K.G.; Weed Res. 1973; 13, 42-50.

16. Horrman, W.D.; Kardhuber, B.; Ramskiner, K.A.; Eberle, D.O. Proc. European Weed Res. Symp. 1973, pp. 129-140.

17. Polzin, W.J.; Brown, Jr., I.F.; Manthey, J.A.; Probst, G.W. Pestic. Monit. J. 1971, 4, 209-215.

18. Robinson, E.L. Weed Sci. 1976, 24, 420-422.

19. Wauchope, R.D.; Chandler, J.M.; Savage, K.E. Weed Sci. 1977, 25, 193-196.

20. Taylor, A.W.; Freeman, H.P.; Edwards, W.M. J. Agric. Food Chem. 1971, 19, 832-836.

21. Thompson, Jr., L.T.; Skroch, W.A.; Beasley, E.O. Pesticide Incorporation: Distribution of Dye by Tillage Implements; North Carolina Agricultural Extension Service: Raleigh, 1981, 32 p.

22. Hornsby, A.G.; Rao, P.S.C.; Nkedi-Kizza, P.; Wheeler, W.B.; Jones, R.L. In "Characterization and Monitoring of the Vadose (Unsaturated) Zone"; Nielsen, D.M.; Curl, M.; Eds.; National Water Well Association: Worthington, 1983; pp. 936-958.

23. Rao, P.V.; Rao, P.S.C.; Davidson, J.M.; Hammond, L.C. Soil Sci. Soc. Amer. J. 1979, 43, 274-278.

24. Karickhoff, S.W. Chemosphere 1981, 10, 833-846.

25. Rao, P.S.C.; Nkedi-Kizza, P.; Davidson, J.M.; Ou, L.T. In "Agricultural Management and Water Quality"; Schaller, F.; Bailey, G.; Eds.; Iowa State Univ. Press: Ames, 1983; pp. 126-140.

26. Kenaga, E.E.;Goring, C.A.I. In "Proc. Third Aquatic Toxicology Symposium"; Eaton, E.G.; Parrish, P.R.; Hendricks, A.G.; Eds.; American Society for Testing and Materials: Philadelphia, 1980; pp. 78-115.

27. Green, R.E.; Cheng-Tsu, M.Y.; Lee, C.C. Agron. Abstr. 1985, p. 25.

28. Ou, L.T.; Rao, P.S.C.; Wheeler, W.B. In "Estimation of Parameters for Modeling the Behavior of Selected Pesticides and Orthophosphate"; Rao, P.S.C.; Berkheiser, V.E.; and Ou, L.T.; Eds., EPA-600/3-84-019; U.S. Environmental Protection Agency: Athens, 1984; pp. 48-88.

29. Walker, A.; Brown, P.A. Crop. Prot. 1983, 2, 17-25.

30. Walker, A.; Zimdhal, R.L. Weed Res. 1981, 21, 255-265.

31. Bouchard, D.C.; Lavy, T.L.; Marx, D.B. Weed Sci. 1982, 30, 629-632.

32. Ryden, J.C.; Lund, L.J. J. Environ. Qual., 1980, 9, 387-393.

33. Ryden, J.C.; Lund, L.J. Soil Sci. Soc. Amer. J., 1980, 44, 505-511.

34. Walker, A. J. Environ. Qual. 1974, 3, 396-401.

35. Walker, A.; Barnes, A. Pest. Sci. 1981, 123-132.

36. Walker, A. Pest. Sci. 1976, 7, 41-49.

37. Walker, A. Pest. Sci. 1976, 7, 50-58.

38. Walker, A. Pest. Sci. 1976, 7, 59-64.

39. Smith, A.E.; Walker, A. Pest. Sci. 1977, 8, 449-456.

40. Walker, A. Weed Res. 1978, 18, 305-313.

41. Walker, A.; Hance, R.J.; Allen, J.G.; Briggs, G.G.; Chen, Y-L.; Gaynor, J.D.; Hogue, E.J.; Malquor, A.; Moody, K.; Moyer, J.R.; Pestemer, W.; Rahman, A.; Smith, A.E.; Streibig, J.C.; Torstensson, N.T.L.; Widyanato, L.S.; Zandvoort, Z. Weed Res. 1983, 23, 373-383.

RECEIVED April 7, 1986

GROUND WATER MONITORING TECHNIQUES

7

Applications of Surface Geophysical Methods to Ground Water Pollution Investigations

Nicholas De Rose

Roy F. Weston, Inc., Raritan Center, Edison, NJ 08837

Several surface geophysical survey methods are presented and the principles of operation for each technique discussed. These methods include seismic refraction, resistivity, electromagnetic conductivity, ground penetrating radar and magnetometry. Applications of these geophysical methods to groundwater contamination investigations include; determining lateral and vertical variations in soil, rock and groundwater characteristics; mapping the extent of groundwater contaminants present within aquifers; and locating buried objects. The suitability and potential application of surface geophysical surveys to monitor the migration of field-applied pesticides, and pesticides from waste disposal sites within the unsaturated and saturated zones is also discussed. An example to illustrate the use of geophysical techniques at a hypothetical pesticide field-application site is outlined. The example details the application of seismic refraction and electromagnetic conductivity techniques to define erratic subsurface conditions at a site where groundwater contamination is suspected. The site is underlain by an unconsolidated deposit of varying thickness and composition which is underlain by cavernous limestone. A groundwater monitoring system is designed based upon interpretation of the geophysical data obtained. By utilizing surface geophysical methods, extensive site coverage can be completed cost effectively and groundwater monitoring systems designed efficiently, thereby increasing the quality and success of site groundwater investigations.

The science of geophysics was originally developed as the study of the physics of the earth's structure and shape. Geophysics includes measuring gravitational, electrical, electromagnetic and magnetic fields, and recording seismic vibrations to identify compositional and structural features of the earth. Developments from the 1920's

0097–6156/86/0315–0118$06.75/0
© 1986 American Chemical Society

through 1950's resulted in the application of geophysical methods to locate potential mineral resource deposits including petroleum-bearing geologic structures and metallic ore bodies. More recent advancements in applied geophysics have produced equipment and technology which are easily accessible to professional consultants, industry and government through several reliable manufacturers and scientific equipment suppliers.

The field portability, cost-effectiveness, and ability to collect data which does not require obtaining actual samples of subsurface material provides several considerations for incorporating surface geophysical surveys into groundwater contaminant investigation programs (1). As part of a preliminary site investigation, surface geophysical surveys may be conducted to identify potential trends or anomalous areas prior to selecting the locations of sampling points for the monitoring system. Utilizing geophysical surveys to screen areas of investigation will increase the effectiveness of the monitoring systems and site evaluation by providing a geophysical-data base for correlation with samples obtained in the field and analytical data from the laboratory (2).

Surface Geophysical Survey Techniques

This paper discusses five of the most widely used techniques for conducting groundwater contaminant investigations. They are characterized by relatively simple operational principles, field procedures, and data analyses and interpretation. The five surface techniques include: seismic refraction, ground penetrating radar, electrical resistivity, electromagnetic conductivity and magnetometry. Seismic refraction and ground penetrating radar both identify subsurface interfaces which are used to determine the thickness and depth of materials and to locate isolated buried objects. Electrical resistivity, electromagnetic conductivity, and magnetometry identify lateral and vertical variations in subsurface geologic formations by locating lateral and vertical variations in bulk electrical, and magnetic properties of the subsurface materials.

Valid interpretations of geophysical data collected in the field require correlation with information obtained from both conventional sampling and analytical programs. The types of geophysical and conventional data required to describe the subsurface environment will vary, depending upon the physical, chemical, and structural complexity of the subsurface, and the intended application of the field study.

Seismic Refraction. Seismic methods are useful tools in determining the thickness and depth of geologic units. In addition, the velocity with which seismic waves are either reflected or refracted is an indication of the physical properties of the subsurface materials. Seismic refraction surveys consist of transmitting a wave into the subsurface by means of an acoustic source. The wave travels through the subsurface in all directions and at differing velocities until contact is made with a geologic interface. The wave is refracted to the surface and received by an array of geophones. The travel time of the refracted wave is recorded on a seismograph. Figure 1 presents a schematic cross section of all the components used to conduct a seismic refraction survey to determine the thickness of layer V_1 which overlies layer V_2.

The seismograph produces a seismic record which graphically depicts all the vibrations received at each geophone during the survey. The seismic record is interpreted to identify the "travel time" of the seismic wave from the acoustic source to each geophone. Travel times of refracted waves are identified on the seismic record as "first arrivals" at each geophone, and they are dependent upon the depth to which the refracted wave traveled, and the seismic velocity of the wave (3).

First arrival times at each geophone are plotted on arithmetic graph paper as time vs distance plots for data interpretation (Figure 2). The travel times and corresponding distances at each geophone will generally plot along one of several linear segments on the graph. The slope of each linear segment corresponds to the seismic velocity of each seismic layer. The velocity and either the intercept time or the critical distance are used to calculate the depth to the interface between the layers (Figure 2).

Seismic velocities are directly related to material density. Bulk densities may vary as a result of composition, water content, age and depth, weathering, fracturing, and degree and type of consolidation. Different soil and rock types are characterized by certain overlapping ranges in seismic velocity. By correlating observed data with information obtained from conventional studies, the types and distribution of geologic units present may be interpreted (2). The most common application of seismic refraction surveys in groundwater contaminant investigations is to define the thickness of the overburden (soil) and to map the stratigraphy.

Limitations to consider when evaluating the suitability of the seismic refraction method for a given site include the following: (1) subsurface layers of limited thickness (generally less than five feet) are not detected; (2) sufficient contrast in seismic velocities (bulk density) between subsurface layers must be present for the accurate location of interfaces; (3) bulk densities of subsurface materials must increase with depth.

Electrical Resistivity. Bulk electrical resistivities of subsurface materials are determined by injecting an electric current into the ground by a pair of electrodes and measuring the resulting drop in voltage through the ground by a second pair of electrodes (Figure 3). The magnitude of the voltage drop depends upon the lithology, moisture content and concentrations of dissolved solids in pore water of the subsurface materials. By varying the spacing and position of the electrode array, lateral and vertical resistivity trends can be determined. Results from a resistivity survey can be used to (1) map lateral and vertical variations in the thicknesses of subsurface materials with similar resistivities; (2) delineate resistivity boundaries associated with variations in groundwater quality; and (3) identify waste burial areas or locate buried tanks associated with local resistivity anomalies (2).

Field surveys are conducted to complete either lateral profiles or vertical electrical soundings (VES). Resistivity profiles are completed by utilizing a fixed electrode spacing and obtaining an electrical resistivity value at each selected station along the

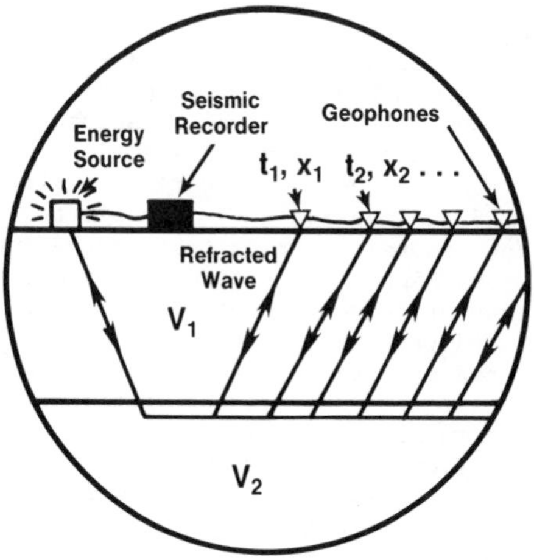

Figure 1. Cross sectional view of seismic refraction survey.

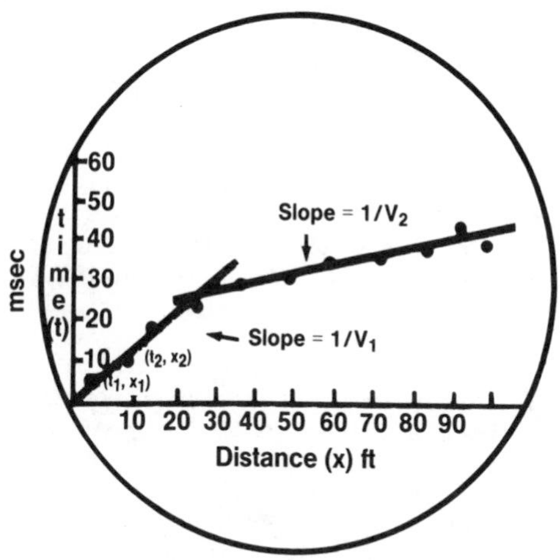

Figure 2. Time vs. distance plot of "first arrivals" for seismic refracted waves at corresponding geophones.

profile traverse (Figure 4). A number of profiles may be completed to construct a resistivity map of an entire site or of selected features at a site. Lateral variations in subsurface electrical resistivity may be used to assess: (1) changes in groundwater quality; (2) groundwater contaminant plumes and saltwater intrusion; and (3) variations in the composition of subsurface materials (2).

Vertical electrical soundings are completed by measuring a series of electrical resistivity values at one location. Each measured value corresponds to one of a series of selected electrode spacings. The resulting set of electrical resistivity values obtained at each VES station are generally plotted against electrode spacing to produce sounding curves that may be empirically evaluated to identify trends (Figure 4). The data may also be solved graphically by using master curves to determine resistivity layer thicknesses, depths, and true resistivities.

Electrical resistivity surveys have been successfully used to locate groundwater aquifers, delineate groundwater contaminant plumes, map salt water intrusion, determine subsurface stratigraphy, map thicknesses of surficial deposits, map fracture zones in bedrock areas, map karst features, locate buried man-made objects and to implement in-place electrical leak detection systems for surface impoundments and underground storage tanks (2,4,5,6,7). In addition, research is on-going to evaluate the ability of resistivity for detecting and mapping organic plumes (8).

The field application of electrical resistivity techniques can be affected by the presence of nearby power lines, fences, railroad tracks, and buried pipes and cables. These cultural features may create electrical interference or alter the subsurface pattern of current flow distribution. In addition, in order to complete electrical resistivity surveys you must be able to "seat" the electrodes in the ground to establish electrical continuity with the subsurface materials to be studied.

Data analysis and interpretation of electrical resistivity data may be limited because: (1) resistivity values may be associated with any one of several geologic units (i.e. a silty sand unit may have similar resistivity values as a sand unit saturated with salt water); (2) thin beds of lower resistivity will be masked when they are sandwiched between two layers of higher resistivity; and (3) the interpreted layer thickness will be greater than the actual thickness due to the anisotropic nature of the individual layers, which are generally characterized as having greater vertical resistivity values than horizontal (9).

Electromagnetic Conductivity. The electromagnetic conductivity survey is similar to the electrical resistivity survey in the sense that both methods quantify electrical characteristics of subsurface materials. However, there are vast differences between the field techniques and the data and analyses which characterize the individual methods. The electromagnetic conductivity survey consists of directing an electromagnetic field into the ground from an above ground source to create a secondary electromagnetic field that is measured by a receiver. As a result, direct measurements of bulk subsurface conductivities are obtained (10) (Figure 5).

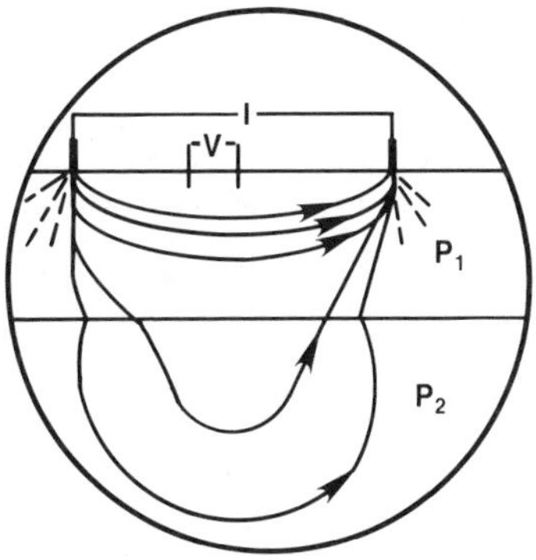

Figure 3. Cross sectional view of electrical resistivity survey. 'I' represents current injection electrodes, 'V' represents electrodes used to measure resulting voltage drop.

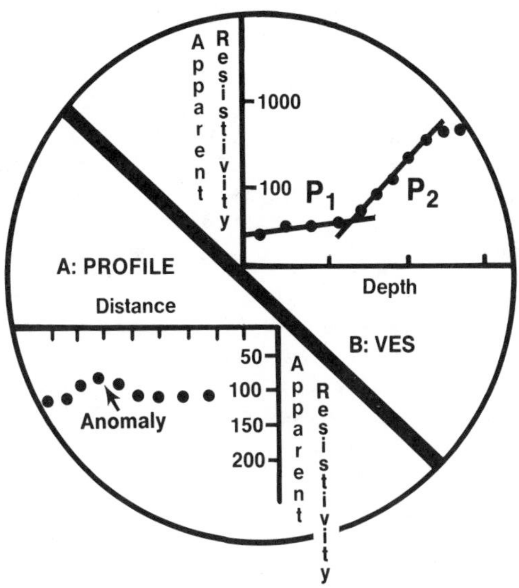

Figure 4. Resistivity plots for (A) lateral profile and for (B) vertical sounding.

Field equipment required for a survey consists of a low frequency transmitter coil and a receiver coil. The receiver coil intercepts a portion of the secondary electromagnetic field and produces an output voltage which is linearly related to subsurface conductivity (11).

Inducing an electromagnetic field eliminates the need for an electrode array and, therefore, the electromagnetic field method is a more rapid tool for surveying. The elimination of electrodes to measure electrical properties of subsurface materials also enables electromagnetic surveys to be conducted in many areas where resistivity surveys cannot be considered (i.e. pavement areas, very dry sandy soils, frozen ground, railroad tracks, etc.). Electromagnetic conductivity surveys may also be used to produce rapid continuous profiles up to depths of 15 meters.

Electromagnetic (EM) surveys are most widely used to create profiles or maps of subsurface conductivity. Lateral variations in conductivity, at a given depth, may be interpreted as contaminant plumes, sand and gravel deposits, clay deposits, karst features, salt water intrusion or buried objects (Figure 6). Because field surveys may be completed rapidly, EM techniques are often implemented to map large sites and to provide detailed maps of variable site subsurface features such as isolated karst features, and buried tanks or 55 gallon drums.

Computer-data processing techniques can be utilized to interpret and present results obtained from site surveys in order to filter out unwanted cultural interferences, and to evaluate plume characteristics. Correlation of EM maps with field-collected geologic information is essential for the final interpretation of the survey data. Two and three layer conductivity models can be constructed using quantitative techniques. The resulting models are generally less detailed in the vertical resolution of the subsurface layers than those produced from vertical electrical soundings.

Electromagnetic surveys may not be conducted in areas characterized by unusually high or low values of subsurface conductivity. In addition, electromagnetic methods are subject to interference from many cultural features and the presence of nearby electrical fields. Finally, the ability of electromagnetic conductivity surveys to identify groundwater contaminant plumes requires that a significant electrical conductivity contrast exists between contaminated and natural groundwater (2).

Ground Penetrating Radar. Ground penetrating radar (GPR) is an impulse radar system that provides a continuous profile of subsurface conditions by radiating electromagnetic pulses into the subsurface and displaying the reflections from surface and subsurface "interfaces" on a strip chart recorder (Figure 7). The term "interface" in this geophysical method refers to any discontinuity in electrical properties such as soil or geologic boundaries or imbedded objects such as drums, or boulders.

The GPR system consists of five major components, these include: 1) power distribution unit; 2) radar control unit; 3) antenna transceiver; 4) graphic recorder; and 5) tape recorder. The power distribution unit provides proper AC/DC voltages to all the GPR equipment. The radar control unit triggers the antennae transceiver to

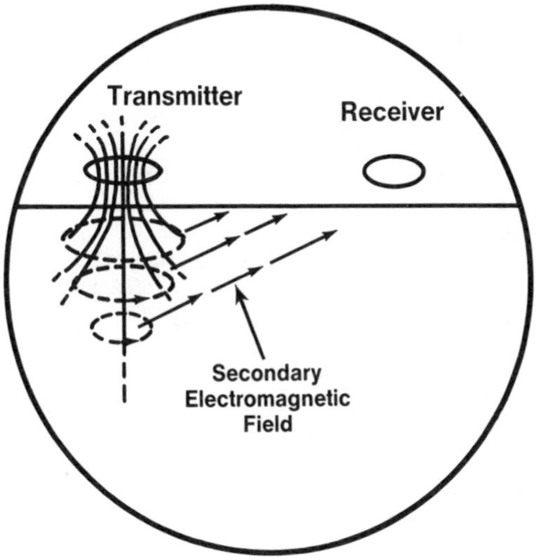

Figure 5. Cross-sectional view of transmitted electromagnetic field and generated secondary electromagnetic field measured by a receiver for completing electromagnetic conductivity surveys.

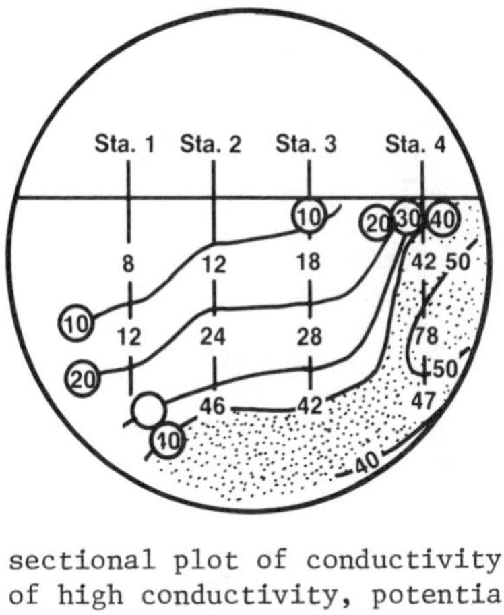

Figure 6. Cross sectional plot of conductivity values. Anomalous values of high conductivity, potentially representing the extent of an ionic groundwater contaminant plume, shaded with stipled pattern.

produce the electromagnetic pulse. The reflected portions of the transmitted pulse are received by the antennae transceiver, converted to the audio frequency range and processed by the radar control unit. Processing is an operator-initiated procedure and allows for the selection and enhancement of that portion of the data which is of greatest importance to the objective of the survey. The processed data is sent to the graphic recorder which produces a permanent chart profile of the subsurface interfaces. Processed data may also be sent to the tape recorder for storage, reprocessing, or printing at a later date. Data may be collected at a rate as much as 16 times faster than the graphic recorder by tape recording the data (12).

Field surveys may be completed on foot or by utilizing vehicles for mounting equipment and towing the antennae transceiver. The resulting GPR profile which is produced by the graphic recorder prints strong signals as black and weak signals as white (Figure 8). The result is a display of dark bands extending across the profile at varying depths. These bands represent the reflection from an interface. The horizontal scale of the profile is dependent upon the travel time of the antennae transceiver. The vertical scale is dependent upon the travel time of the GPR pulse. The travel time of the GPR pulse may be converted to depth if either the dielectric constant of the medium being profiled or the depth to a specific interface is known.

Typical applications of GPR surveys include, mapping depth to bedrock, and mapping interfaces including changes in soil type, geologic formations, and depth to water table. Buried objects and excavations may also be located and in some cases identified, as well as buried cultural features including pipes, cables, and conduits (13).

The effective penetration depths of the GPR system is dependent upon the bulk conductivity of the subsurface materials being profiled. The GPR signal is rapidly attenuated within highly conductive materials (i.e. clays, highly ionic groundwater) which severely limits the penetration depths.

Other limitations of GPR profiles include: 1) masking of a reflection as a result of overlapping interfaces, such as, where the surface of the water table is level with the top of a buried drum; 2) variation in identifying reflectors depending upon the orientation of the traverse to the buried object; and 3) the presence of vegetative cover, surficial debris and irregular surface topography which may limit the penetration depth of the return signal or the ability to traverse the site with the GPR system.

Magnetometry. Magnetometers measure with accuracy and precision, the intensity of the earth's magnetic field. By mapping lateral and/or vertical magnetic gradients, subsurface features such as geologic ore deposits, bedrock features, buried drums and buried pipes may be located (Figure 9). Nonferrous metals such as aluminum, copper, tin and brass cannot be detected by magnetometers (14).

Several types of magnetometers are presently available, however, the two most widely referenced in engineering and groundwater publications are the proton and the fluxgate magnetometers. Proton magnetometers measure the earth's total magnetic field intensity by utilizing the precession of spinning protons in a sample of hydro-

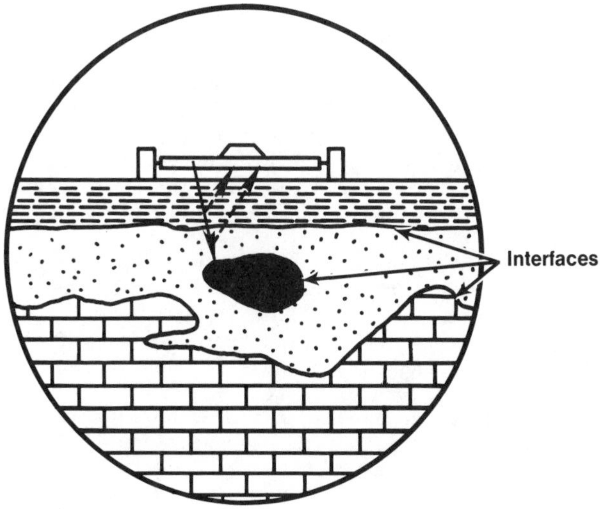

Figure 7. Cross sectional view of transmitted electromagnetic signal generated from ground penetrating radar antenna transceiver.

Figure 8. Ground penetrating radar profile record. Hyperbolic reflectors (dark bands) represent outline of buried drums.

carbon fluid to generate a small signal whose frequency is precisely proportional to the total magnetic field intensity. Fluxgate magnetometers measure variations in the intensity of the total earth's magnetic field by measuring changes in the magnetic saturation level of an iron core sensor. Unlike the proton magnetometer, the signal output of a single fluxgate magnetometer is extremely sensitive to orientation. To overcome this sensitivity, the gradient of the earth's magnetic field may be measured by mounting two fluxgate sensors together to form a gradiometer. The gradiometer is the most common method of obtaining measurements with the fluxgate magnetometer (14).

Proton magnetometers are generally more sensitive than fluxgate magnetometers, however, they are also more sensitive to interference from unwanted magnetic objects (i.e. fences, debris and automobiles). Fluxgate magnetometers also provide the advantage of obtaining continuous measurements and data readout for conducting total site surveys.

The magnitude of an anomaly produced by a buried object is directly related to the magnetic mass and magnetic intensity of the body, and inversely related to the depth of the body. Magnetic anomalies, such as the contact between a non-magnetic sedimentary deposit and a magnetic igneous rock body, that are identified on magnetic profiles or contour maps can be located using qualitative techniques (Figure 10). Computer data processing and analytical techniques can determine the depth of burial, the mass of the buried objects, the geologic structure, and the depth to bedrock.

Magnetometer surveys are limited to identifying or locating subsurface magnetic features or objects. Magnetometer surveys may be hindered by the presence of unwanted local magnetic fields associated with power lines, railroad tracks, etc. Diurnal variations, which are natural changes in the earth's magnetic field over time, must be compensated for during a magnetometer survey. During magnetic storms, which may occur as often as several times a month, significant variations in the earth's magnetic field will result, making completion of a magnetic survey impractical. Finally, identification of local anomalies are affected by the presence of many cultural features including steel fences, vehicles, buildings and iron debris. The search for buried objects may also be limited by the presence of natural iron ore deposits and bog irons, or by other variations in subsurface geology.

Groundwater Applications

Currently, there are several publications which document applying surface geophysical techniques to groundwater contaminant investigations (2,8,15,16,17). Two conclusions drawn in all of the above referenced publications are: (1) geophysical surveys provide several useful nondestructive test methods for evaluating subsurface geohydrologic conditions; and (2) these methods may often improve the quantity and quality of the data base obtained from conventional destructive test methods, while reducing overall investigation and monitoring costs.

Advantages of geophysical surveys include the cost effective benefits provided for by the rapid acquisition of data. Geophysical

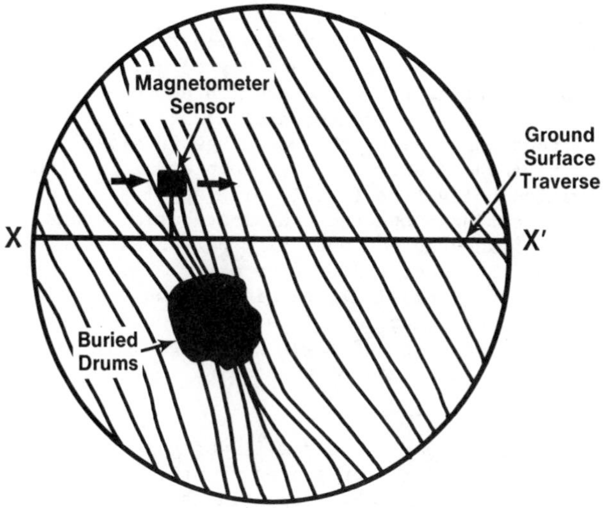

Figure 9. Cross sectional view of ground surface traverse, X-X', and the distortion in the magnetic field intensity created by several buried drums.

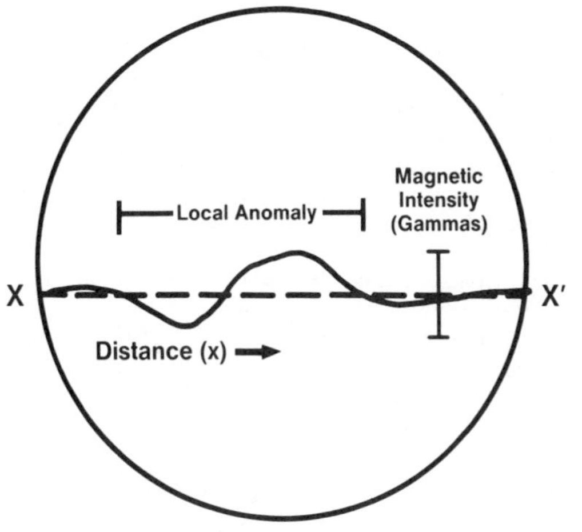

Figure 10. Magnetic profile along ground surface traverse, X-X', and the local anomaly created by buried drums.

survey techniques can be implemented during preliminary project stages in order to establish a site data base, and thereby identify anomalies which require more detailed evaluations. In addition, continuous data collection techniques may be utilized for mapping features to a greater degree of resolution not generally attainable by using conventional sampling techniques.

At hazardous waste sites the use of geophysical survey techniques would be advantageous prior to completing on-site test drilling or excavation activities (1,13). The application of geophysical techniques in locating suspected container burial areas would reduce the likelihood of unexpected contact by personnel or equipment with buried hazardous, toxic or radioactive materials. Site characterization with surface geophysical techniques also helps in identifying the proper procedures and precautions that should be implemented during the conventional phases of an investigation, in order to eliminate the possible spread or release of toxic or radioactive materials.

Selected Illustrative Models. Three illustrative models of geophysical surveys are briefly presented to provide examples of the results that can be obtained by completing surface geophysical surveys. The models include: 1) a vertical electrical resistivity sounding program (VES) that evaluates salt water intrusion; 2) an electromagnetic conductivity survey to identify groundwater contaminant plumes and contaminant sources; and 3) an evaluation of limestone terrain with seismic refraction and electromagnetic conductivity surveys.

Vertical Electrical Resistivity Sounding. The first example illustrates the ability of VES techniques to characterize regional subsurface geology into resistivity layers that have similar resistivity values at similar depths. Sediment lithologies and groundwater quality characteristics were determined for each unit based on observed resistivity values. The survey was conducted in Northampton County, Virginia which is located at the south end of the Del-Mar-Va Peninsula (9) (Figure 11).

Data was collected at six VES sites selected at the study area. The VES data were analyzed using computer processing techniques to create 5 and 10 layer subsurface models (18,19). Figure 12 presents a three-dimensional isometric projection prepared from the computer-generated, 5-layer VES curves.

Three general lithologies describe the subsurface conditions. The upper layer, which produces high resistivity values, consists mostly of unsaturated sand that ranges in depth from about 0 to 150 feet.

The second resistivity layer, which is characterized by moderately low resistivity values, occurs at depths generally ranging between 150 to 160 feet and consists of fresh water-bearing sands. The Manokin and Pocomoke acquifers are present within this layer. In addition, lenses of clay and/or brackish water occur within this layer. Values of resistivity within the second layer decrease eastward, indicating an increase in the thickness and/or frequency of the clay lenses and/or brackish water zones.

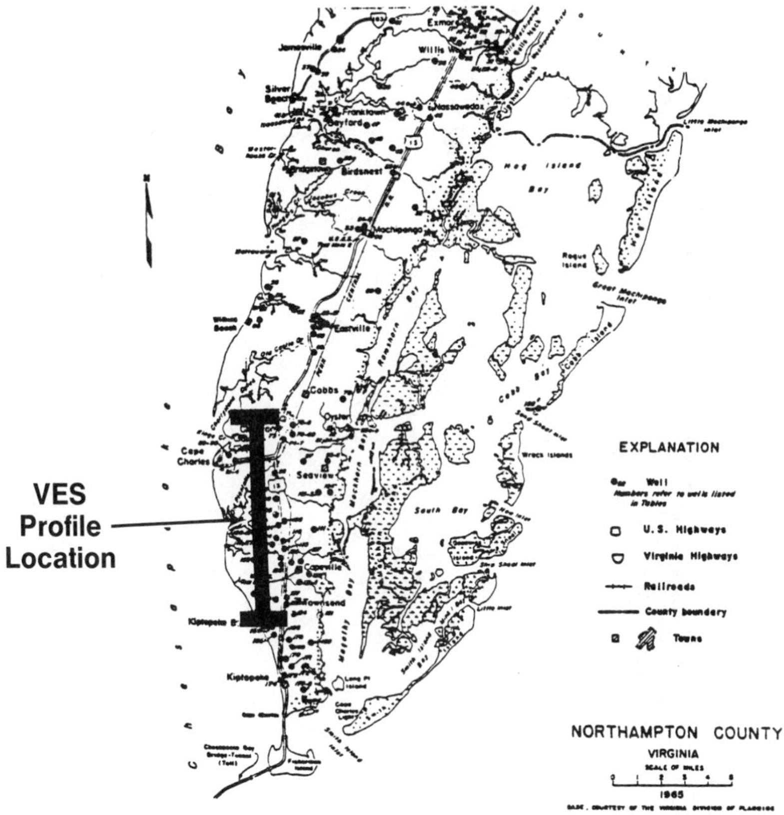

Figure 11. VES site area location in Northampton County, Virginia. Reproduced with permission from Ref. 9.

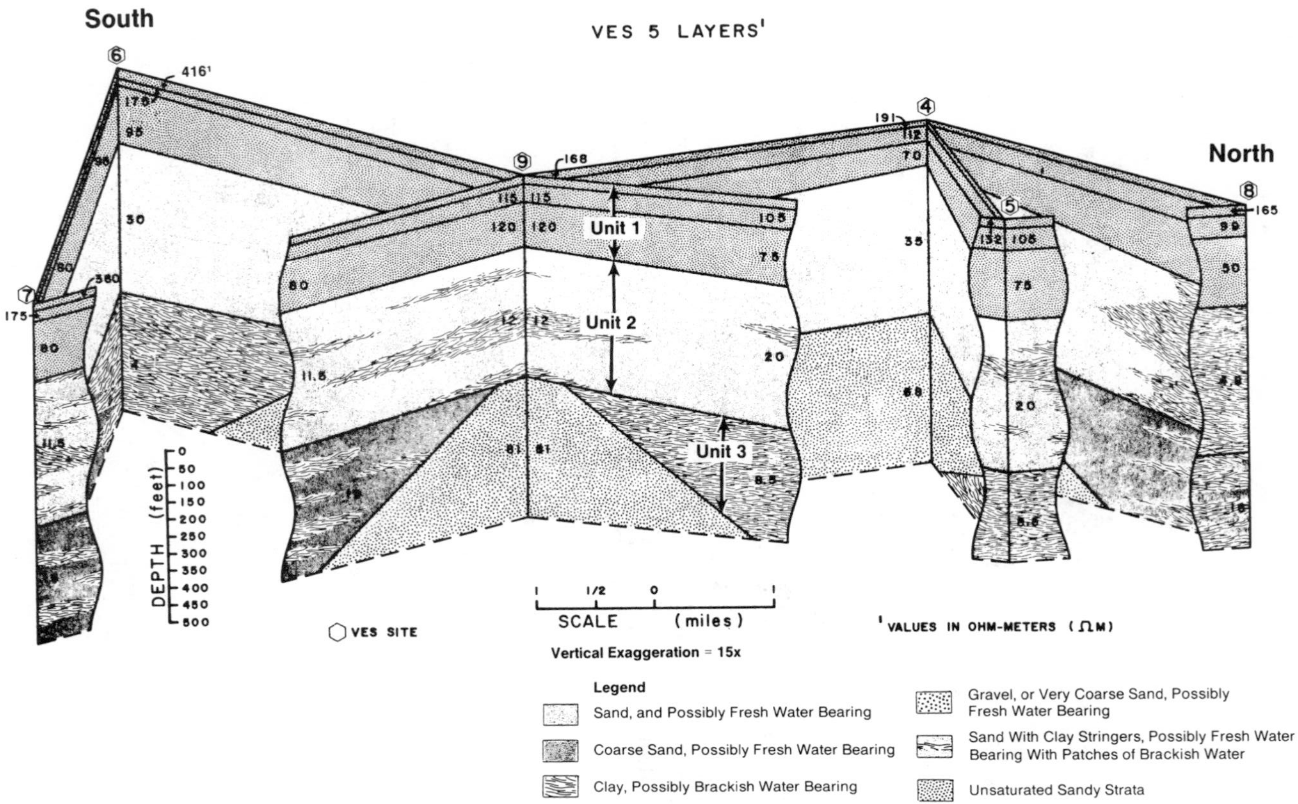

Figure 12. VES five layer, three-dimensional isometric projection. Reproduced with permission from Ref. 9.

Very low resistivity values characterize the deepest layer, indicating the presence of either brackish, or saline water-bearing sediments. The vertical extent of this layer is well defined by sudden large scale decreases in resistivity, however, the lateral extent of the layer is not well defined.

Electromagnetic Conductivity Survey. The results of an electromagnetic conductivity survey conducted at a uranium mill tailings site are shown in Figures 13 and 14. The purpose of this survey was to identify subsurface conductivity anomalies that could be associated with variations in groundwater quality. Once the anomalies were identified, test borings and monitoring wells were completed to determine their significance.

A survey grid was established at the site by measuring 150 and 300 foot intervals across a 6,000 foot by 3,900 foot area. The survey area is predominantly flat, with the major topographic features being tailings piles and levees around the evaporation ponds. The subsurface soils are composed of silty clay and gravel deposits. The depth to groundwater is generally less than 5 feet below land surface.

Electromagnetic conductivity measurements were made at 316 survey stations. Two sets of subsurface conductivity values were obtained at each survey station. These consisted of one set of readings to an effective measurement depth of 25 feet with a large portion of the total reading contributed by near surface materials (Figure 13), and one set of readings to an effective measurement depth of 50 feet with a small portion of the total reading contributed by near surface materials (Figure 14). Conductivity contour maps were computer generated using each data set. The contour maps were then used to identify anomalous areas of subsurface conductivity (Figures 13 and 14).

A comparison of the resulting two conductivity contour maps shows anomalously high conductivity values in an area extending from the site coordinates N8800, E4700 to N6700, E3800. In addition, higher values were consistently recorded at the greater effective measurement depth in comparison to those values obtained at the shallower effective measurement depth. This indicates that subsurface materials within the anomalous area have higher conductivity values at depth rather than near the surface.

Those anomalies that are shown in Figure 13, but are not evident in Figure 14, are related to near surface variations in soil type and/or the accumulation of evaporite deposits. The isolated anomaly highlighted on Figure 14, is probably related to one of several geothermal springs.

Electrical resistivity measurements were also taken at 28 locations selected across the site to verify the data collected by electromagnetic methods. The data obtained was in agreement with the results of the electromagnetic conductivity survey.

The entire geophysical survey of the field site was completed in four days. Later investigations, including the installation and sampling of monitoring wells and analysis of groundwater samples from those wells, confirmed that the anomaly of high conductivity was the result of groundwater contamination occurring downgradient from the tailings piles.

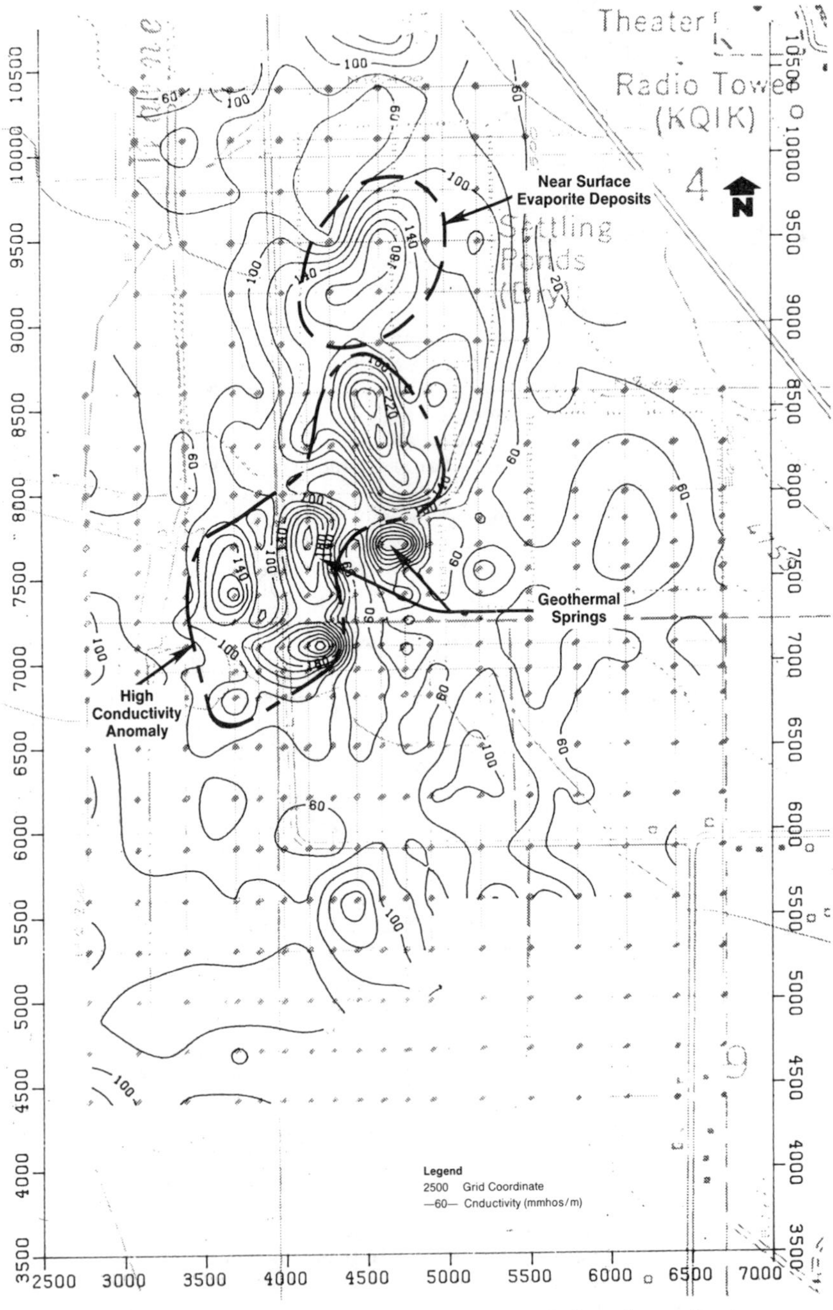

Figure 13. Conductivity contour map resulting from electromagnetic conductivity survey completed at an effective measurement depth of 25 feet.

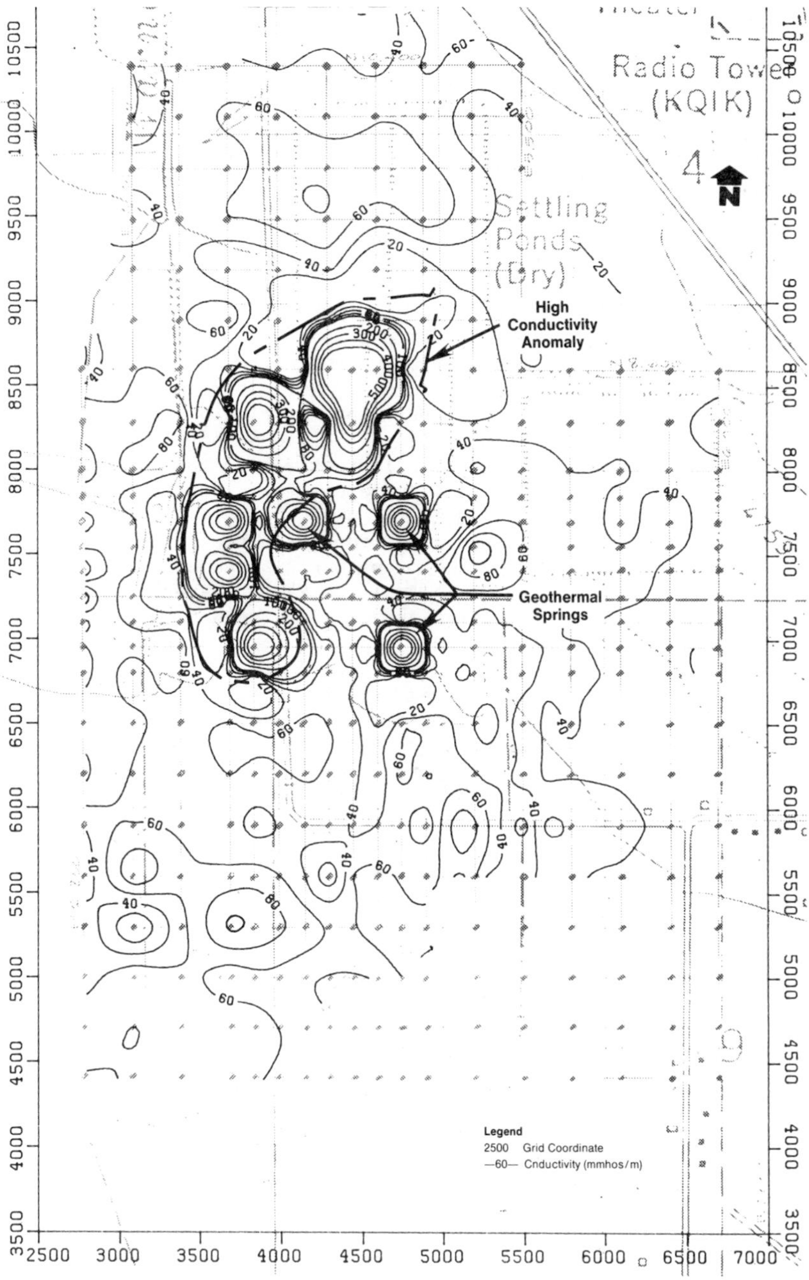

Figure 14. Conductivity contour map resulting from electromagnetic conductivity survey completed at an effective measurement depth of 50 feet.

Geophysical Investigation at a Potential Field-Application Site. Based upon several successful site evaluations that employed geophysical methods combined with conventional exploratory techniques in investigating sites underlain by limestone, the following potential site evaluation was developed. The intent of this hypothetical situation was to illustrate the concept of an integrated site evaluation in which geophysical techniques are utilized to increase the effectiveness of the groundwater monitoring system.

The area of the proposed site for the application of pesticides is about twenty-five acres and is underlain by residual soil of variable lithology which is underlain by the Leithsville Formation and the Triassic Border conglomerate. The Leithsville Formation is a limestone of Cambrian age. The site is also characterized by a relatively shallow water table (Figure 15).

The initial site investigation to define subsurface conditions included seismic refraction profiles and a limited number of test borings. Interpretation and correlation of seismic data with results obtained from the test borings identified areas underlain by limestone at shallow depths (less than 10 feet), by limestone and at moderate depths (15 to 25 feet), and areas where depth to limestone is irregular (Figure 16). Within the area characterized by moderate depth to limestone, based upon comparison with the results of the seismic refraction profiles, four test borings encountered limestone at great depths associated with localized karst features. These data points were therefore identified as anomalous values. In addition, conglomerate was encountered underlying a portion of the site area at depths of 15 to 25 feet.

Based upon test borings, the site was underlain at shallow depths by coarse, light gray limestone and sand, with minor amounts of silt and clay. The site area characterized by moderate depth to limestone was underlain by fine, red dolomitic limestone and silty clay, with minor amounts of sand. The remaining site area was underlain by coarse sandstone and conglomerate and silt and clay with moderate amounts of sand.

An electromagnetic survey was run at the site at stations placed along a 50 foot grid. The results of the survey correlated favorably with the results obtained from the initial investigation. A Phase I Geologic Map was prepared for the site based upon the results of the seismic refraction profiles, test borings and electromagnetic conductivity survey (Figure 17). High areas of conductivity were generally associated with the central part of the site which is underlain by moderately deep limestone and silt and clay. Low conductivity values were generally associated with those areas of the site underlain by limestone at shallow depths. In addition, two linear trending areas of variable conductivity were mapped in the area between the shallow and moderately deep limestone areas.

Follow-up test borings were completed for the final site characterization and implementation of a groundwater monitoring system. Based upon evaluation of the test boring data and the geophysical data, the site was found to be underlain by three different limestone members (20). The two contact zones which occur between the three members were associated with the two linear trending areas of variable conductivity. These zones were characterized by the development of deep karst zone that resulted from the dissolution of the lime-

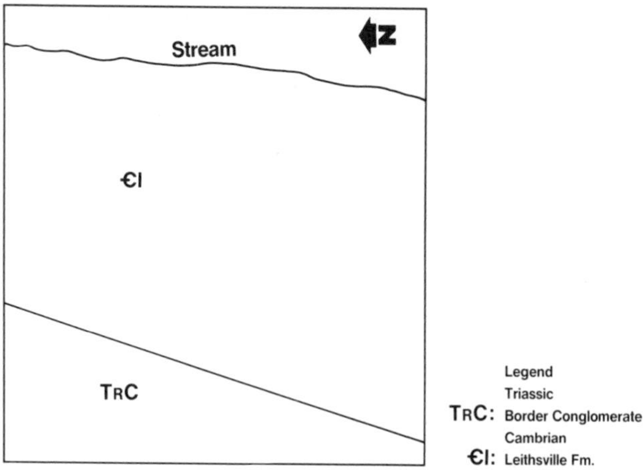

Figure 15. Geologic map of proposed field-application site.

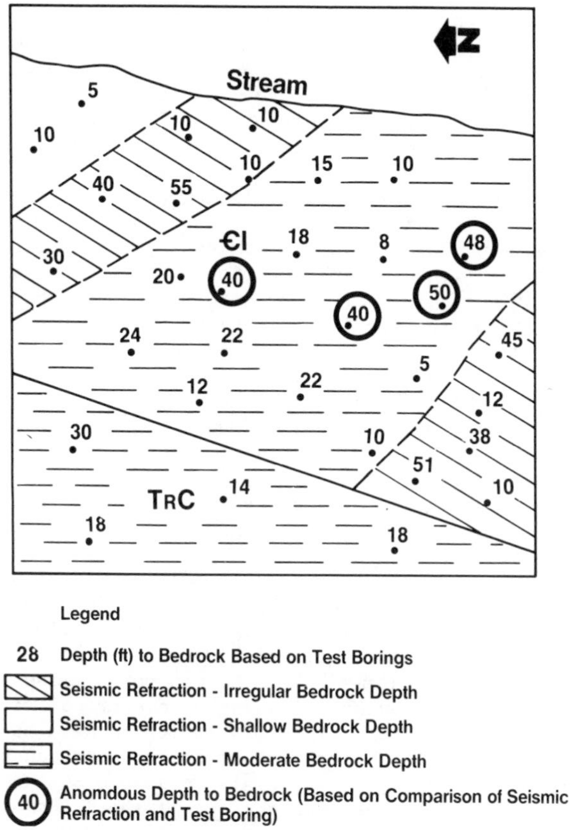

Figure 16. Geologic map of proposed field-application site summarizing bedrock depths determined from seismic refraction surveys and test boring data.

stone by groundwater flowing along the contact zone (Figure 17). Karst features including pinnacles, cavities, and voids were found extending to depths greater than 100 feet along the contact zones. In order to map the extent of the minor solution cavities found in other areas of the site, detailed electromagnetic conductivity surveys were completed on a grid spacing of twenty feet.

Monitoring well locations were chosen based on results from the geophysical surveys and the test drilling which characterized the subsurface conditions underlying the site (Figure 18). Without utilization of geophysical techniques, characterization of the site would have been impractical without incurring excessive drilling costs.

Summary and Potential Applications to the Pesticide Industry

Five cost-effective surface geophysical survey techniques have been described which are widely used in groundwater site investigations. By employing geophysical techniques to complement conventional methods of subsurface exporation, a monitoring network can be designed that has an overall increase in efficiency and is cost-effective (2,6).

The application of these surface geophysical techniques to groundwater contaminant investigations include the following: 1) regional and local evaluations of subsurface geohydrologic conditions; 2) feasibility studies of proposed field application sites and manufacturing, storage and disposal facilities; 3) design of monitoring systems; 4) characterization of existing field application sites and manufacturing, storage and disposal facilities; and 5) design of leak detection systems utilizing electrical resistivity techniques.

The detection of pesticide contaminant plumes, originating from generally non-ionic organic chemicals, presents a difficult situation to effectively monitor by utilizing applicable geophysical techniques such as electrical resistivity and electromagnetic conductivity. However, this reportedly has been successfully accomplished where contaminant concentrations in groundwater are greater than 10 mg/l. At high contaminant concentrations the detection of organic plumes may be possible as a result of the formation of a conductivity low which is associated with the non-ionic contaminants in contrast to the surrounding highly ionic groundwater. Hydraulic processes occurring in the unsaturated zone of fine grained soils may also result in the accumulation of organic compounds above the water table forming a low conductivity anomaly overlying the organic contaminant plume. It is also possible that degradation by-products of pesticides may create a mappable ionic contaminant plume.

Several reports have been published which describe the design and implementation of electrical and electromagnetic early warning and/or leak detection systems (7). The general premise by which the systems operate is that by establishing an electrical resistivity or conductivity map of a site or facility in its natural state the effects of potential site discharges to groundwater can be monitored by continued site surveying. It is also possible that for certain pesticide field application sites the establishment of a baseline conductivity map at selected depths, for example 1.0 meters, 3.0

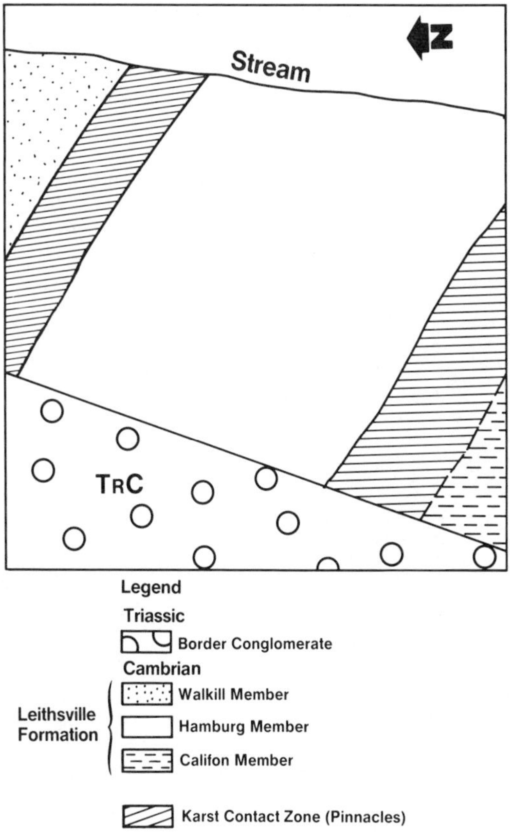

Figure 17. Detailed geologic map of proposed field-application site based upon results of seismic refraction and electromagnetic conductivity surveys and test boring data.

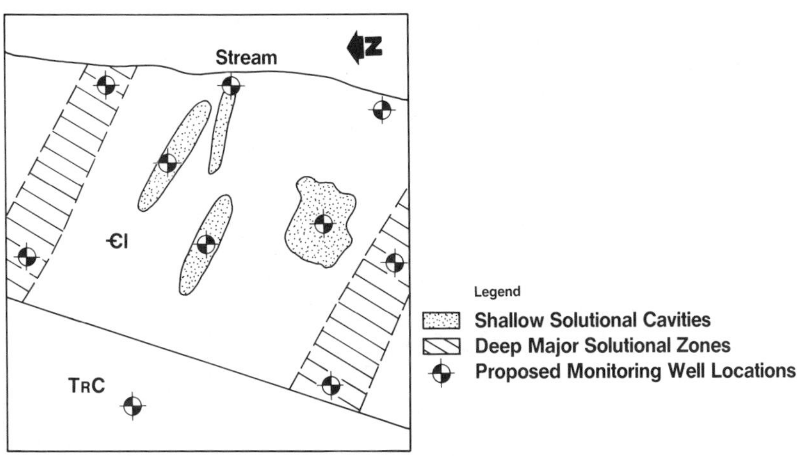

Figure 18. Geologic map of field application site showing karst features and proposed monitoring well locations.

meters, and 7.5 meters, may prove to be useful in determining areas where subsurface migration of pesticides or their degradation products is occurring. Such techniques may be applicable to unsaturated and saturated zone monitoring programs.

Literature Cited

1. Lord, A. E., Jr.; Koerner, R. M.; Freestone, F. J. J. of Hazardous Materials 1982, 5, 221-233.
2. Benson, R. C.; Glaccum, R. A.; Noel, M. R. "Geophysical Techniques for Sensing Buried Wastes and Waste Mibratum"; U. S. Environmental Protection Agency: Las Vegas, Nev., 1984; p. 236.
3. Griffiths, P. H.; King, R. F. "Applied Geophysics for Engineers and Geologists"; Pergammon Press: Elmsford, N. Y., 1975; p. 223.
4. Kelly, W. E.; Frohlich, R. K. Groundwater 1985, 23, 182-189.
5. Alao, D. A. Bull. of the Assoc. of Eng. Geologists 1985, 22, 95-100.
6. Schwartz, F. W.; McClymont, G. L. Groundwater 1977, 15, 197-202.
7. Schultz, D. W.; Duff, B. M.; Peters, W. R. Proc. Intl. Conf. on Geomembranes, 1984, p. 445-449.
8. Evans, R. B.; Schweitzer, G. E. Environ. Sci. Technol. 1984, 18, 330A-339A.
9. Cross, J. W.; Baine, E. H.; Baine, P. M.; Carroll, R. I.; DeRose, N.; Foster, M. A. "Industrialization of Virginia's Eastern Shore: An Environmental Impact Study in Northampton County" Old Dominion Univ.: Norfolk, Va.; 1977, p. 271.
10. McNeill, J. D. Technical Note TN-5 Geonics Ltd.: Ontario, Canada, 1980, p. 22.
11. ______. Technical Note TN-6 Geonics Ltd.: Ontario, Canada, 1980, p. 15.
12. "Ground Penetrating Radar - Summary of Capabilities", Roy F. Weston, Inc., West Chester, PA.; 1984.
13. Koerner, R. M.; Lord, A. E., Jr. Proc. 5th Natl. Conf. on Magmt. of Uncont. Haz. Waste Sites, 1984.
14. Breiner, S. "Applications Manual for Portable Magnetometers"; Geometrics: Sonnyvale, Calif., 1973; p. 58.
15. Sendlein, L. V. A.; Yazicigil, H. Groundwater Monitor Rev. Fall, 1981.
16. Yazicigil, H.; Sendlein, L. V. A. Groundwater Monitor Rev. Winter, 1982.
17. Kaufmann, R. F.; Gleason, A.; Ellwood, R. B.; Lindsey, G. P. Groundwater Monitor Rev. Fall, 1981.
18. Zohdy, A. A. R. Computer Program, Ntl. Tech. Inf. Ser., No. PB-232703, 1973.
19. Zohdy, A. A. R. Computer Program, Ntl. Tech. Inf. Ser., No. PB-232056, 1974.
20. Markewicz, S. J., et al. Pennsylvania Geologists Field Conference Annual Meeting, Guide Book 42, 1977, p. 117.

RECEIVED April 7, 1986

DRASTIC: A System to Evaluate the Pollution Potential of Hydrogeologic Settings by Pesticides

Linda Aller, Truman Bennett, Jay H. Lehr, and Rebecca Petty

National Water Well Association, Worthington, OH 43085

A methodology is described which allows the pollution potential of any area to be systematically evaluated anywhere in the United States. The system, which optimizes the use of existing data, has two major portions: the designation of mappable units, termed hydrogeologic settings, and the superposition of a relative ranking system called DRASTIC. Hydrogeologic settings incorporate the major hydrogeologic factors which are used to infer the potential for pesticides to enter ground water. These factors, which form the acronym DRASTIC, include depth to water, net recharge, aquifer media, soil media, topography, impact of the vadose zone and hydraulic conductivity of the aquifer. The relative ranking scheme uses a combination of weights and ratings to produce a numerical value, called the DRASTIC Index, which helps prioritize areas with respect to pollution potential.

Only in the past few years has the nation become aware of the dangers of ground-water contamination and of the many ways in which ground water can become contaminated. The application of pesticides is one such practice which may result in ground-water contamination. The potential for contamination to occur is affected by the physical characteristics of the area, the chemical nature of the pesticide and the rate, frequency and method of application. Ground-water contamination can be minimized by controlling the pesticide application rate and frequency, by the choice of appropriate pesticides and by evaluating and managing pesticide application in vulnerable areas.

This paper presents a standardized system which can be used to evaluate the ground-water pollution potential of any hydrogeologic setting in the United States. The system has been designed to use information which is available through a variety

0097-6156/86/0315-0141$06.00/0
© 1986 American Chemical Society

of sources. Information on the parameters including the depth to water in an area, net recharge, aquifer media, soil media, general topography or slope, vadose zone media and hydraulic conductivity of the aquifer is necessary to evaluate the ground-water pollution potential of any area using hydrogeologic settings.

This system has been prepared to assist planners, managers and administrators in the task of evaluating the relative vulnerability of areas to ground-water contamination by pesticides. It has been assumed that the reader has only a basic knowledge of hydrogeology and the processes which govern ground-water contamination. This methodology is neither designed nor intended to replace on-site inspections. Rather, it is intended to provide a basis for comparative evaluation of areas with respect to potential for pollution of ground water.

Classification Systems

One of the fundamental needs of any natural science is the development of an effective system to group similar entities into categories. Well-established systems exist in the fields of botany, geology, and many other sciences (1). These systems permit an appropriately trained person to gain certain insight about an entity simply by knowing the appropriate category in which it is grouped.

This systematic and logical way of imposing an artificial system on natural entities has long been used in the field of geology also. For example, rocks have been classified according to origin and minerals grouped according to crystal systems. However, as a science expands and changes, so must the types of systems used to describe those characteristics which need to be studied. The field of hydrogeology is one area of geology which has only been overtly recognized since the term was coined by Lucas in 1879 (2). Since that time hydrogeology has expanded, from a discipline devoted to water occurrence and availability, to include the broad aspect of water quality and solute chemistry. Definition of water quality is fundamental to the protection of the ground-water resource from pollution.

The idea of an organized way to describe ground water systems is not new. Meinzer (3) prepared a small-scale map of the United States showing general ground-water provinces. Thomas (4) and Heath (5) prepared similar but more detailed maps and descriptions which grouped aquifers mainly on their water-bearing characteristics within certain geographic areas. Blank and Schroeder (6) attempted to classify aquifers based on the properties of rocks which affect ground water. Of all these systems, geographic ones have been more widely accepted as ways to describe the quantity of water which is available in various regions.

Some Existing Systems Which Evaluate Ground-Water Pollution Potential

Within the last 20 years the need to expand these systems or to create a new system to address ground-water quality has become evident. Many different systems have been developed to address site selection for waste disposal facilities such as sanitary landfills or liquid waste ponds. Among these, the LeGrand System (7) and the modified version used by the U.S. EPA in the Surface Impoundment Assessment (SIA) are probably the most well known. The LeGrand system uses numerical weighting to evaluate ground-water pollution potential from a given waste disposal site. By evaluating the site through a series of four stages, a description of the hydrogeology of the site, the relative aquifer sensitivity combined with the contaminant severity, the natural pollution potential presented at that site, and the engineering modifications which might change that potential are all evaluated.

The LeGrand system presupposes only a limited technical knowledge but encourages the user to become familiar with the concepts presented in the manual so that skilled judgements can be made in the subjective portion of the system. The similarities between sites are emphasized and the uniqueness of each site is downplayed.

The U.S. EPA methodology (8) uses the basic LeGrand System to define the hydrogeologic framework, but modifies the system to place emphasis on establishing a monitoring priority for the facility. Once the hydrogeologic characteristics have been rated, a table is used to define the monitoring priority. This priority may be adjusted by the rater using prescribed techniques. Once again, only a limited technical knowledge is presupposed.

Other systems have been designed to tailor the results to more specific purposes. Thornthwaite and Mather (9) and Fenn et al. (10) developed water-balance methods to predict the leachate generation at solid waste disposal sites. This approach is based on the premise that by knowing the amount of infiltration into the landfill and the design of the cell, the leachate quantity for the landfill can be determined. The system is intended to be used as a tool by engineers in the early design phase of a facility.

Gibb et al. (11) devised a rating scheme to establish priorities for existing waste disposal sites with respect to their threat to human health by ground water. Via ranking the site through four factors, (1) health risk of the waste and handling mode, (2) population at risk, (3) proximity to wells or aquifers, and (4) susceptibility of aquifers, a number that ranges from 0-100 was used to display the relative risk. The system was used in a specific 2-county assessment by technically qualified individuals. Another rating scheme, developed by the Michigan Department of Natural Resources (12), is designed to rank larger numbers of sites in terms of risk of environmental contamination. By evaluating the five categories: (1) release potential, (2) environmental exposure, (3) targets, (4) chemical hazard, and (5) existing exposure, the user obtains a number ranging from 0 to 2000 points which evaluates the relative hazard of that site with respect to other sites in Michigan.

Seller and Canter (13) evaluted seven empirical methods to determine their usefulness in predicting the ground-water pollution effects of a waste disposal facility at a particular site. The methods they reviewed included rating schemes, a decision tree approach, a matrix and a criteria-listing method. They determined that each method took into account the natural conditions and facility design and construction, but that each method was best applied to the specific situation for which it was designed.

This brief review of selected existing systems reveals that there are a number of methods that can be applied to site-specific situations or to evaluation of the pollution potential of existing sites. However, a planning tool is needed for use before the site-specific methods are employed. The system must (1) function as a management tool, (2) be simple and easy-to-use, (3) utilize available information , and (4) be able to be used by individuals with diverse backgrounds and levels of expertise. This document contains a system which attempts to meet these needs and to provide the planning tool necessary before site-specific evaluations.

The DRASTIC System

The system presented herein is part of a more complete system developed for the United States Envorinmental Protection Agency. A complete description is contained in the draft document EPA # - 570/9-84-002 (14). The methodology has two major portions: the designation of mappable units, termed hydrogeologic settings; and the application of a scheme for relative ranking of hydrogeologic parameters, called DRASTIC, which helps the user evaluate the relative ground-water pollution potential of any hydrogeologic settings. Although the two parts of the system are interrelated, they are discussed separately in a logical progression.

Hydrogeologic Settings

This document has been prepared using the concept of hydrogeologic settings. A hydrogeologic setting is a composite description of all the major geologic and hydrologic factors which affect and control ground-water movement into, through and out of an area. It is defined as a mappable unit with common hydrogeologic characteristics, and as a consequence, common vulnerability to contamination by introduced pollutants. From these factors it is possible to make generalizations about both ground-water availability and ground-water pollution potential.

In order to assist users who may have a limited knowledge of hydrogeology, the entire standardized system for evaluating ground-water pollution potential has been developed within the framework of an existing classification system of ground-water regions of the United States. Heath (5) divided the United States into 15 ground-water regions based on the features in a ground-water system which affect the occurrence and availability of ground water (Figure 1). These regions include:

1. Western Mountain Ranges
2. Alluvial Basins
3. Columbia Lava Plateau
4. Colorado Plateau and Wyoming Basin
5. High Plains
6. Nonglaciated Central Region
7. Glaciated Central Region
8. Piedmont and Blue Ridge
9. Northeast and Superior Uplands
10. Atlantic and Gulf Coastal Plain

11. Southeast Coastal Plain
12. Alluvial Valleys
13. Hawaiian Islands
14. Alaska
15. Puerto Rico and Virgin Islands

For the purposes of the present system, Region 12 (Alluvial Valleys) has been reincorporated into each of the other regions and Region 15 (Puerto Rico and Virgin Islands) has been omitted. Since the factors that influence ground-water occurrence and availability also influence the pollution potential of an area, this regional framework is used to help familiarize the user with the basic hydrogeologic features of the region.

Because pollution potential cannot be determined on a regional scale, smaller "hydrogeologic settings" were developed within each of the regions described by Heath (5). These hydrogeologic settings create units which are mappable and, at the same time, permit further delineation of the factors which affect pollution potential.

Each hydrogeologic setting is described in a written narrative section and illustrated in a block diagram. Figure 2 shows the format which is used. The descriptions are used to help orient the user to typical geologic and hydrologic configurations which are found in each region and to help focus attention on significant parameters which are important in pollution potential assessment. The block diagram enables the user to visualize the described setting by indicating its geology, geomorphology and hydrogeology.

A set of hydrogeologic settings has been developed for each ground-water region. To date, hydrogeologic settings have been identified and described in the United States. Although similar hydrogeologic settings may appear in more than one ground-water region, the document is designed so that once the broad geographic area is located the user does not have to refer to other hydrogeologic settings in other regions. This means that although similar hydrogeologic settings may appear more than once in the document, they have been tailored to reflect the typical hydrogeologic conditions within each individual region.

Factors Affecting Pollution Potential

Inherent in each hydrogeologic setting are the physical characteristics which affect the ground-water pollution potential. Many different biological, physical and chemical mechanisms may actively affect the attenuation of a contaminant and, thus, the pollution potential of that system. Because it is neither practical nor feasible to obtain quantitative evaluations of intrinsic mechanisms from a regional perspective, it is necessary to look at the broader parameters which incorporate the many processes. After a complete evaluation of many characteristics and the mappability of the data, the most important mappable factors that control the ground-water pollution potential were determined to be:

D - Depth to Water
R - (Net) Recharge

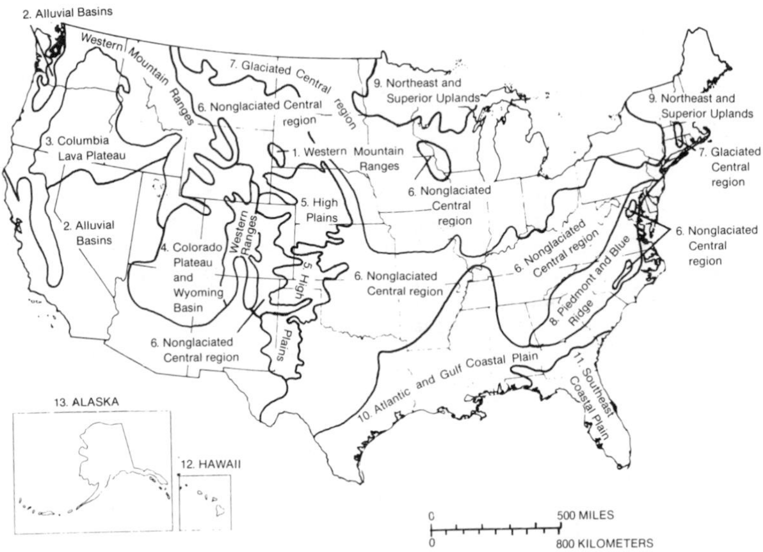

Figure 1. Ground-water regions of the United States: "Reproduced from Ref. 5"

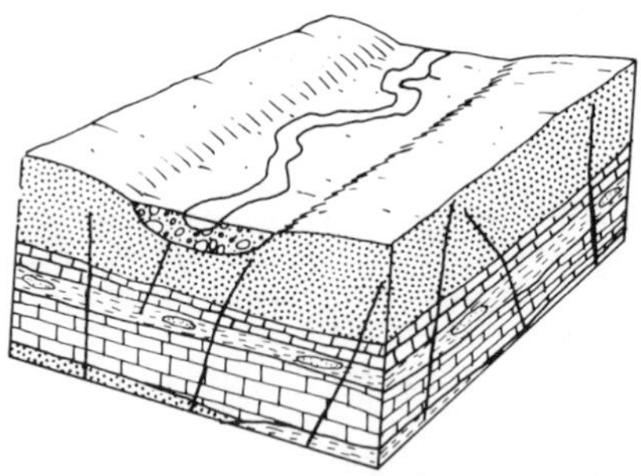

Figure 2. Format of hydrogeologic setting. This setting is identical to 10Ba River Alluvium with Overbank except that no significant fine-grained floodplain deposits occupy the stream valley. This results in significantly higher recharge and sandy soils at the surface. Water levels are typically closer to the surface because banks of fine-grained deposits are not present. Throughout much of this region, there is an abundance of course-grained material, which limits this setting for water supply. These materials however, provide a good source of recharge to the underlying consolidated and semiconsolidated bedrock. Reproduced with permission from reference 14.

A - Aquifer Media
S - Soil Media
T - Topography (Slope)
I - Impact of the Vadose Zone
C - (Hydraulic) Conductivity of the Aquifer

These factors have been arranged to form the acronym, DRASTIC for ease of reference. While this list is not all inclusive, these factors, in combination, were determined to include the basic requirements needed to assess the general pollution potential of each hydrogeologic setting. The DRASTIC factors represent measurable parameters for which data are generally available from a variety of sources without detailed reconnaissance. It is recognized that many of the factors may be considered to be overlapping. However, great care has been taken to try to separate the factors for purposes of the developed system. A complete description of the important mechanisms considered within each factor and a description of the significance of the factors follows.

A Description of the DRASTIC Factors

Depth to Water. The water table is the expression of the surface below the ground level where all the pore spaces are filled with water. Above the water table, the pore spaces are filled with water and air. The water table may be present in any type of media and may be either permanent or seasonal. Depth to water refers to the depth to the water surface in an unconfined aquifer. Depth to water does not include saturated zones which have insufficient permeability to yield significant enough quantities of water to be considered an aquifer.

This document also can be used to evaluate unconfined aquifers. For purposes of this document, depth to water refers to the top of the aquifer where the aquifer is confined. In this case, depth to water may include saturated zones above the top of the aquifer.

The depth to water is important primarily because it determines the depth of material through which a contaminant must travel before reaching the aquifer, and it may help to determine the amount of time during which contact with the surrounding media is maintained. The depth to water is also important because it provides the maximum opportunity for oxidation by atmospheric oxygen. In general, there is a greater chance for attenuation to occur as the depth to water increases because deeper water levels infer longer travel times.

Net Recharge. The primary source of ground water is precipitation which infiltrates through the surface of the ground and percolates to the water table. Net recharge indicates the amount of water per unit area of land which penetrates the ground surface and reaches the water table. Although this amount of water may vary seasonally or annually, a water balance is usually achieved or long-term trends can be established. This recharge water is thus available to transport a contaminant vertically to the water table and horizontally within the aquifer. In addition, the quantity to water available for dispersion and dilution in the vadose zone and in the saturated zone is controlled by this parameter. In areas where the aquifer is unconfined, recharge to the aquifer usually occurs more readily and the pollution potential is generally greater than in areas with confined aquifers. Confined aquifers are partially protected from contaminants introduced at the surface by layers of low permeability media (aquitards) which retard water movement to the confined aquifer. In parts of some confined aquifers, head distribution is such that movement of water is through the confining bed from the confined aquifer into the unconfined aquifer. In this situation, there is little opportunity for local contamination of the confined aquifer. The principal recharge area for the confined aquifer is often many miles away. Many confined aquifers are not truly confined and are partially recharged by migration of water through the confining layers. The more water that leaks through, the greater the potential for recharge to carry pollution into the aquifer. Recharge water, then, is a principal vehicle for leaching and transporting solid or liquid contaminants to the water table. Therefore, the greater the recharge, the greater the potential pollution. This is true until the amount of recharge is great enough to cause dilution of any contaminant which would be introduced into the system. At this point, the pollution potential would cease to increase and might actually decrease. For purposes of this document, this phenomenon has been acknowledged but the system does not reflect the dilution factor.

One additional factor which must be considered is augmentation of natural recharge through artificial recharge or by irrigation. When a range for net recharge is assigned, these additional sources of water must be considered.

Aquifer Media. Aquifer media refers to the consolidated or unconsolidated medium which serves as an aqufier (such as sand and gravel or limestone. An aquifer is defined as a water-bearing rock formation which will yield sufficient quantities of water for use. Water is held by aquifers in the pore spaces of granular and clastic rock and in the fractures and solution openings of non-clastic and non-granular rock. Rocks which yield water from pore spaces have primary porosity; rocks where the water is held in openings such as fractures and solultion openings which were created after the rock was formed have secondary porosity. The aquifer medium exerts the major control over the route and path length which a contaminant must follow. The path length is an important control (along with hydraulic conductivity and gradient) in determining the time available for attenuation processes such as sorption, reactivity and dispersion and also the amount of effective surface area of materials contacted in the aquifer. The route which a contaminant will take can be strongly influenced by fracturing or by any other feature such as interconnected series of solution openings which may provide pathways for easier flow. In general, the larger the grain size and the more fractures or openings within the aquifer, the higher the permeability and the lower the attenuation capacity; consequently the greater the pollution potential.

Soil Media. Soil media refers to that uppermost portion of the vadose zone characterized by significant biological activity. For purposes of this document, soil is commonly considered the upper weathered zone of the earth which averages six feet or less. Soil has a significant impact of the amount of recharge which can infiltrate into the ground and, hence, on the ability of a contaminant to move vertically into the vadose zone. Moreover, the attenuation processes of filtration, biodegradation, sorption and volatilization may be quite significant in this zone. Thus, for certain on-land surface practices such as applications of pesticides, soil can be a primary influence on pollution potential. In general, the pollution potential of a soil is largely affected by the type of clay present, the shrink/swell potential of that clay and the soil texture. The quantity of organic material present in the soil may also be a major factor. In general, the less the clay shrinks and swells and the finer the texture, and the greater the organic content, the less the pollution potential. Soil media are best described by referring to the basic soil types as classified by the Soil Conservation Service.

Topography. As used here, topography refers to the slope and slope variability of the land surface. Basically, topography helps control the likelihood that a pollutant will run off or remain on the surface in one area long enough to infiltrate. Therefore, the greater the chance of infiltration, the higher the pollution potential associated with the slope. Topography influences soil development and therefore has an effect on attenuation. Topography is also significant from the standpoint that the gradient and direction of flow often can be inferred by water table conditions from the general slope of the land. Typically, steeper slopes signify higher ground-water velocity.

Impact of Vadose Zone. The vadose zone is defined as that zone above the water table and below the soil media which is unsaturated. For purposes of this document, this strict definition can be applied to all water table aquifers. However, when evaluating a confined aquifer, the "impact" of the vadose zone is expanded to include both the vadose zone and anysaturated zones which overlie the aquifer. The significantly restrictive zone above the aquifer which forms the confining layer is used as the type of medium which has the most significant impact.

The type of vadose zone media determines the attenuation characteristics of the material below the typical soil horizon and above the water table. Biodegradation, neutralization, mechanical filtration, chemical reaction and dispersion are all processes which may occur within the vadose zone with a general lessening of biodegradation with depth. The media also control the path length and routing, thus affecting time available for attenuation and the quantity of material encountered. The routing is strongly influenced by any fracturing present. The materials at the top of the vadose zone also exert an influence on soil development.

Hydraulic Conductivity of the Aquifer. Hydraulic conductivity refers to tha ability of the aquifer materials to transmit water, which in turn, controls the rate at which ground water will flow under a given hydraulic gradient. The rate at which the ground water flows also controls the rate at which a contaminant will be moved away from the point at which it enters the aquifer. Hydraulic conductivity is controlled by the amount of interconnection of void spaces within the aquifer which may occur as a consequence of intergranular porosity, fracturing, bedding planes, etc. For purposes of this document, high hydraulic conductivities are associated with higher pollution potential. This is because the pollutant has the potential for moving quickly away from the point in the aquifer where it is introduced. Obviously, a wide range of hydraulic conductivities may be present in all areas.

DRASTIC

A numerical ranking system to assess ground-water pollution potential in hydrogeologic settings has been devised using the DRASTIC factors. The system contains three significant parts: weights, ranges and ratings. A description of the technique used for weights and ratings can be found in Dee et al. (15).

Weights. Each DRASTIC factor has been evaluated with respect to the other to determine the relative importance of each factor. Each DRASTIC factor has been assigned a relative weight ranging from 1 to 5 (Table I). The most significant factors have weights of 5; the least significant, a weight of 1. This exercise was accomplished by an advisory committee using a Delphi (consensus) approach. These weights are a constant and may not be changed when using the system.

Table I. Assigned Weights for DRASTIC Features
"Reproduced from Ref. 14"

Feature	Agricultural Weight
Depth to Water Table	5
Net Recharge	4
Aquifer Media	3
Soil Media	5
Topography	3
Impact of the Vadose Zone	4
Hydraulic Conductivity of the Aquifer	2

Source: Reproduced with permission from reference 14.

Ranges. Each DRASTIC factor is divided into either ranges or significant media types which have an impact on pollution potential (Tables II-VIII). The media types have been assigned descriptive names to assist the user.

Table II. Ranges and Ratings for Depth to Water

Depth to Water (feet)	
Range	Rating
0-5	10
5-15	9
15-30	7
30-50	5
50-75	3
75-100	2
100+	1

Source: Reproduced with permission from reference 14.

Table III. Ranges and Ratings for Net Recharge

Net Recharge (inches)	
Range	Rating
0-2	1
2-4	3
4-7	6
7-10	8
10+	9

Source: Reproduced with permission from reference 14.

Table IV. Ranges and Ratings for Aquifer Media

Aquifer Media		
Range	Rating	Typical Rating
Massive Shale	1-3	2
Metamorphic/Igneous	2-5	3
Weathered Metamorphic/Igneous	3-5	4
Thin Bedded Sandstone, Limestone, Shale Sequences	5-9	6
Massive Sandstone	4-9	6
Massive Limestone	4-9	6
Sand and Gravel	4-9	8
Basalt	2-10	9
Karst Limestone	9-10	10

Source: Reproduced with permission from reference 14.

Table V. Ranges and Ratings for Soil Media

Soil Media	
Range	Rating
Thin or Absent	10
Gravel	10
Sand	9
Peat	8
Shrinking and/or Aggregated Clay	7
Sandy Loam	6
Loam	5
Silty Loam	4
Clay Loam	3
Muck	2
Nonshrinking and Nonaggregated Clay	1

Source: Reproduced with permission from reference 14.

Table VI. Ranges and Ratings for Topography

Topography (percent slope)	
Range	Rating
0-2	10
2-6	9
6-12	5
12-18	3
18+	1

Source: Reproduced with permission from reference 14.

Table VII. Ranges and Ratings for Impace of Vadose Zone Media

Impact of Vadose Zone Media		
Range	Rating	Typical Rating
Silt/Clay	1-2	1
Shale	2-5	3
Limestone	2-7	6
Sandstone	4-8	6
Bedded Limestone, Sandstone, Shale	4-8	6
Sand and Gravel with significant Silt and Clay	4-8	6
Metamorphic/Igneous	2-8	4
Sand and Gravel	6-9	8
Basalt	2-10	9
Karst Limestone	8-10	10

Source: Reproduced with permission from reference 14.

Table VIII. Ranges and Ratings for Hydraulic Conductivity

Hydraulic Conductivity (gpd/ft)	
Range	Rating
1-100	1
100-300	2
300-700	4
700-1000	6
1000-2000	8
2000+	10

Source: Reproduced with permission from reference 14.

Ratings. Each range for each DRASTIC factor has been evaluated with respect to the others to determine the relative significance of each range with respect to pollution potential. Based on graphs, the range for each DRASTIC factor has been assigned a rating which varies between 1 and 10 (Tables II-VIII). The factors of D, R, S, T, and C have been assigned one value per range. A and I have been assigned a "typical" rating and a variable rating. The variable rating allows the user to choose either a typical value or to adjust the value based on more specific knowledge.

This system allows the user to determine a numerical value for any hydrogeologic setting by using an additive model. The equation for determining the DRASTIC Index is:

$$D_R D_W + R_R R_W + A_R A_W + S_R S_W + T_R T_W + I_R I_W + C_R C_W = \text{Pollution Potential}$$

Where:

R = rating
W = weight

Once a DRASTIC Index has been computed, it is possible to identify areas which are likely to be susceptible to ground water contamination relative to one another. The higher the DRASTIC Index, the greater the ground-water pollution potential. The DRASTIC Index provides only a relative evaluation tool and is not designed to provide absolute answers.

Table IX shows a typical index computed for the hydrogeologic setting, 10Ba River Alluvium without Overbank, which is described in Figure 2. In contrast, Table X and Figure 3 illustrate a very similar hydrogeologic setting with a pollution potential that is significantly lower. These numbers, although not unique values, can be evaluated with respect to one another by knowing that for all hydrogeologic settings evaluated in the United States, DRASTIC Indices ranged from 53 to 224. This relative comparison helps the user evaluate pollution potential with respect to any other area. In areas of widely variable hydrogeology, the pollution potential

Table IX--Computation of the DRASTIC Index for Setting 10 Bb-River Alluvium Without Overbank Deposit

SETTING 10 Bb River Alluvium Without Overbank Deposit		AGRICULTURAL		
FEATURE	RANGE	WEIGHT	RATING	NUMBER
Depth to water table	5-15	5	9	45
Net Recharge	10+	4	9	36
Aquifer Media	Sand and Gravel	3	8	24
Soil Media	Sand	5	9	45
Topography	0-2%	3	10	30
Impact Vadose zone	S & G w/sig. Silt	4	6	24
Hydraulic Conductivity	1000-2000	2	8	16
			Agricultural Drastic Index	220

Source: Reproduced with permission from reference 14.

Table X--Computation of the DRASTIC Index for Setting 10Ba-River Alluvium with Overbank Deposit

SETTING 10 Ba River Alluvium With Overbank Deposit			AGRICULTURAL	
FEATURE	RANGE	WEIGHT	RATING	NUMBER
Depth to Water Table	15-30	5	7	35
Net Recharge	7-10	4	8	32
Aquifer Media	Sand and Gravel	3	8	24
Soil Media	Silty Loam	5	4	20
Topography	0-2%	3	10	30
Impact Vadose	Silt/Clay	4	1	4
Hydraulic Conductivity	700-1000	2	6	12
			Agricultural Drastic Index	157

SOURCE: Reproduced with permission from reference 14.

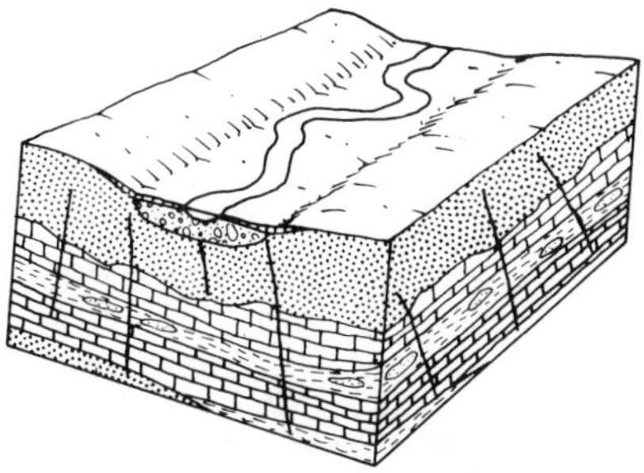

Figure 3. Sample hydrogeologic setting in Atlantic and Gulf Coastal Plain. This hydrogeologic setting is characterized by low topography and thin to moderately thick deposits of alluvium along parts of river valleys. The alluvium is underlain by consolidated and semiconsolidated sedimentary rocks. Water is obtained from sand and gravel layers which are interbedded with finer-grained alluvial deposits. The floodplain is covered by varying thicknesses of fine-grained, silty deposits called overbank deposits. The overbank thickness is usually greater along major streams (as much as 40 feet) and thinner along minor streams. Precipitation in the region is abundant, but recharge is reduced because of the silty soils which typically cover the surface. Water levels are typically moderately shallow. The alluvium may serve as a significant source of water and may be in direct hydraulic connection with the underlying sedimentary rocks. The alluvium may also serve as a source of recharge to the underlying bedrock. Many streams in this setting provide only fine-grained deposits (silts and clays) and as such do not form good aquifers. They still, however, provide a good source of recharge. Reproduced with permission from reference 14.

may also vary widely with an associated spread of DRASTIC Indices. In areas with more subtle changes in hydrogeology, the DRASTIC Indices would reflect more subtle changes. The system does not attempt to define "good" or "bad" areas, but simply offers the user a tool to evaluate the relative pollution potential of whatever areas are desired. The user may wish to then consider additional factors such as importance of the aquifer population or other factors in fully assessing the importance of pollution potential in any area.

Status of the Project

This paper summarizes the efforts to produce a standardized system to evaluate ground water pollution potential. The project was designed to span two years. During the first year, the methodology was developed with the help of an advisory committee. The second phase of the project comprises a year of testing of the system through a series of demonstration mapping projects conducted in ten counties throughout the United States. These counties are being chosen for their wide variability in hydrogeology and their position in different ground-water regions. During this verification stage, the system will be expanded and changed to more fully incorporate the elements which will make the DRASTIC methodology most useful to users.

Conculsions

It is evident that all of the DRASTIC parameters are interacting, dependent variables. Their selection is based on available data quantitatively developed and rigorously applied, but on a subjective understanding of "real world" conditions at a given area. The value of the DRASTIC parameters is in the fact that they are based on information that is readily available for most portions of the United States, and which can be obtained and meaningfully mapped in a minimum of time and at minimum cost. The DRASTIC ranking scheme can then be applied by enlightened laymen for valid comparative evaluations with acceptable results.

Acknowledgments

The work was funded by the United States Environmental Protection Agency through the Robert S. Kerr Research Laboratory, Ada, Oklahoma. The authors hereby extend their thanks to Jack Keeley and Jerry Thornhill of the U.S. EPA for their assistance during the project. Grateful acknowledgment of the contributions of a very able advisory committee is also made:

Michael Apgar, Delaware Department of Natural Resources
Jim Bachmaier, U.S. EPA, Office of Solid Waste
William Back, USGS
Harvey Banks, Consulting Engineer, Inc.
Truman Bennett, Bennett & Williams, Inc.
Robert E. Bergstrom, Illinois State Geological Survey
Stephen Born, University of Wisconsin
Keros Cartwright, Illinois State Geological Survey

Stuart Cohen, U.S. EPA, Office of Pesticide Programs
Steve Cordle, U.S. EPA, Office of Research & Development
George H. Davis, USGS, retired
Stan Davis, University of Arizona
Art Day, U.S. EPA, Land Disposal Branch, Office of Solid Waste
Norbert Dee, U.S. EPA, Office of Ground Water Protection
Donald A. Duncan, South Carolina Department of Health and Environmental Control
Catherine Eiden, U.S. EPA, Office of Pesticide Programs
Grover Emrich, SMC Martin, Inc.
Glen Galen, U.S. EPA, Land Disposal Branch, Office of Solid Waste
Phyllis Garman, Consultant, Tennessee
Jim Gibb, Illinois State Water Survey
Todd Giddings, Todd Giddings & Associates, Inc.
Ralph Heath, USGS, retired
Ron Hoffer, U.S. EPA, Office of Ground Water Protection
George Hughes, Ontario Ministry of the Environment
Jack Keeley, U.S. EPA, Kerr Environmental Research Laboratory
Jerry Kotas, U.S. EPA, Office of Waste Programs Enforcement
Harry LeGrand, Consultant, North Carolina
Fred Lindsey, U.S. EPA, Waste Managment and Economics Division
Martin Mifflin, University of Nevada
Paula Mugnuson, Geraghty & Miller, Inc.
Walter Mulica, IEP, Inc
John Osgood, Pennsylvania Bureau of Water Quality
Wayne Pettyjohn, Oklahoma State University
Paul Roberts, Stanford University
Jack Robertson, Weston Designers & Consultants
Dave Severn, U.S. EPA, Hazard Evaluation Division
Jerry Thornhill, U.S. EPA, Kerr Research Center
Frank Trainer, USGS, retired
Warren Wood, USGS

Literature Cited

1. Joel, A. H. In "Soil Classification"; C. W. Finkl, Jr., Ed.; Hutchinson Ross Publishing Co.: Stroudsburg, Pennsylvania, 1982, pp. 52-59.
2. Davis, S. N.; DeWiest, R. J. "Hydrogeology"; John Wiley & Sons: New York, 1966, pp. 1-463.
3. Meinzer, O. E. "Outline of Ground-Water Hydrology"; USGS WATER SUPPLY PAPER No. 494, United States Geological Survey: Washington, D.C., 1923; pp. 1-71.
4. Thomas, H. E. "Ground Water Regions of the United States - Their Storage Facilities"; Interior and Insular Affairs Committee, U.S. House of Representatives: Washington, D.C., 1952; pp. 1-76.
5. Heath, R. C. "Ground Water Regions of the United States"; USGS WATER SUPPLY PAPER No. 2242, United States Geological Survey: Washington, D.C., 1984; pp. 1-78.
6. Blank, H. R.; Schroeder, M.C. Ground Water 1973, 11, 2, 3-5.

7. LeGrand, H. E., 1983. "A Standardized System for Evaluating Waste-Disposal Sites"; National Water Well Association: Worthington, Ohio, 1983; pp. 1-49.
8. "Surface Impoundment Assessment National Report," United States Environmental Protection Agency, 1983, No. 570/9-84/002.
9. Thornthwaite, S. W.; Mather, J. R. "Instructions and Tables for Computing Potential Evapotranspiration and the Water Balance"; Drexel Institute of Technology: Centerton, New Jersey, 1957; Vol. 10, No. 3, pp. 1-311.
10. Fenn, D. G.; Hanley, K. J.; DeGeare, T. V. "Use of the Water Balance Method for Predicting Leachate Generation from Solid Waste Disposal Sites"; U.S. EPA Report No. 168, United States Protection Agency: Cincinnati, Ohio, 1975, pp. 1-40.
11. Gibb, J. P.; Barcelona, M. J.; Schock, S. C.; Hampton, M. W. "Hazardous Waste in Ogle and Winnebago Counties: Potential Risk Via Ground Water Due to Past and Present Activities"; Illinois Department of Energy and Natural Resources Document No. 83/26, State of Illinois: Champaign, Illinois, 1983, pp. 1-66.
12. "Site Assessment System (SAS) for the Michigan Priority Ranking System Under the Michigan Environmental Response Act," Michigan Department of Natural Resources, 1983.
13. Seller, L. E.; Canter, L. W. "Summary of Selected Ground Water Quality Impact Assessment Methods"; NCGWR Report No. 80-3, National Center for Ground Water Research: Norman, Oklahoma, 1980, pp. 1-142.
14. Aller, L.; Bennett, T.; Lehr, J. H.; Petty, R. J. "DRASTIC: A Standardized System for Evaluating Ground Water Pollution Potential Using Hydrogeologic Settings"; USEPA No. 600/2-85/018, United States Environmental Protection Agency: Ada, Oklahoma, 1985; pp. 1-163.
15. Dee, N.; Baker, J.; Drobny, N.; Duke, K.; Whitman, I.; Fahringer, D. Water Resources Research 1973, 9, 3, 523-535.

RECEIVED April 7, 1986

9

Hydrogeologic Investigations of Pesticide Spills

Ralph E. Moon and Carol D. Henry

Geraghty & Miller, Inc., Ground Water Consultants, 14310 North Dale Malory Highway, Tampa, FL 33618

A major reason for installing monitor wells is to provide early warning of ground-water contamination. If properly designed, monitor-well networks can also aid in determining the effectiveness of ground-water protection measures. The ultimate effectiveness of monitor wells is dependent upon a clear definition of the desired results. For example, a monitoring program designed to supply information for litigation is quite different from one implemented to determine whether local agricultural practices adversely affect ground-water quality. Monitoring based upon a regulatory philosophy of zero discharge to ground water will require a monitoring system substantially different in design from one reflecting a regulatory philosophy of containment within the boundaries of a disposal site. Whatever the reason for installing a ground-water monitoring system, its design should be based on demonstrated hydrogeologic principles and site-specific data.

Contamination of ground water with pesticides can occur in two distinctly different situations, those originating from widespread agricultural use and those attributed to a specific point source loss. Each of these situations requires a different approach to ground-water monitoring.

Cases of widespread pesticide contamination can result from unusually high application rates under shallow water-table conditions, on extremely porous soils, or near gelogic conduits that connect the surface to deeper aquifers. In these cases, sampling of existing public and private water-supply wells generally forms the basis of the ground-water monitoring. In other cases of pesticide contamination on great expanses of land, the ground-water monitoring network has been designed to identify the "clean" potable sources of ground water rather than the extent of the contamination.

The focus of this paper is on spills and point source losses of pesticides, which are generally associated with the manufacture and

0097-6156/86/0315-0159$06.00/0
© 1986 American Chemical Society

disposal of pesticides. Commonly, the ground-water contamination stems from leaky pipes, used drums, lagoons, landfills, and waste streams on plant property. The hydrogeologic investigation of pesticide contamination is essentially no different from the hydrogeologic investigation of most other hazardous chemical constituents. The purpose of the investigation is to identify the vertical and horizontal extent of contamination so that ultimately the contamination can be isolated and the aquifer remediated. The regulatory and hydrologic requirements of the monitoring plan consist of four basic elements: problem identification, health and safety requirements, monitor-well network design, and drilling and well-installation techniques.

Problem Identification

All investigations of ground-water contamination problems involve the same basic considerations, namely the location and nature of the sources of the contamination, the mechanisms for introducing the contaminants into the ground, the paths of flow through the ground-water system, and the locations of wells, springs, and surface-water bodies where the contaminants ultimately leave the ground-water system. A detailed review of the history of the site can prove extremely useful in defining the location and nature of sources of contamination. Installation of monitor wells and water-quality sampling are the methods for defining subsurface transport of contaminants.

Site History. Aerial photographs provide valuable information to help direct the course of a ground-water investigation. It has been said by those in the legal profession that "The palest ink surpasses the value of the finest memory." Access to historical photographs supports this statement, often revealing old activities long since forgotten by the most faithful employees. Personnel interviews can be helpful, but the best way to establish a chronological sequence of disposal and storage practices is through scrutiny of photographs, consulting reports, corporate documents, and governmental records (Federal and state regulatory agencies).

Waste Disposal Practices. Maps showing previous waste disposal sites may help explain the results obtained from the analysis of ground-water samples. Information that locates storage areas and effluent discharge points, characterizes waste streams, and provides chemical analyses of soil, surface water, and ground water can also help to define the waste disposal practices of a facility. In addition, identifying adjacent land owners and land use practices are beneficial when it is necessary to explain the presence of ground-water constituents that could be attributed to upgradient practices.

Geology. Drilling logs of on-site water wells, on-site monitor wells, foundation borings, and water wells owned by adjacent property owners, together with data obtained from the U.S. Geological Survey and state agencies, can be useful in designing a ground-water monitoring plan and/or recovery program. Lithologic descriptions obtained from these sources can define the depth to the water table, characterize the soil types, and establish the degree to which aquifers are isolated from or connected to one another. Geologic information helps to establish the potential risk to underground drinking water supplies and to predict the direction and rate of movement of the pesticide plume.

Field Reconnaissance. Field reconnaissance techniques for evaluating ground-water contamination incorporate several methods other than visual and olfactory detection. Biologic indicators, for example, can demonstrate the presence of pesticides if the plant community is sparse (herbicides) or unusually dense (insecticides). On a visit to the historic location of a South Florida DDT formulation site (15 years old), olfactory detection of pesticide loss was not apparent. Inspection of the plant community, however, showed several large circular patches (20 feet in diameter) of dense, low-lying vegetation at the site. Removal of the plants and soil with a pickax along the periphery of the community resulted in the immediate release of pesticide, which was confirmed with an organic vapor analyzer and by smell.

Hydrogeologic Considerations. The direction and rate of movement of a contaminant plume can change in response to hydrologic conditions. Recharge from precipitation, the molecular density of the pesticide, and the hydraulic gradient (slope of the water table) are predictable conditions that control the horizontal and vertical extent of contamination.

Springs, streams, lakes, and rivers are principal areas where contaminants are discharged from ground-water systems. Stormwater ditches, irrigation wells, community well fields, domestic wastewater sprayfields, infiltration ponds, industrial impoundments, and leaking municipal sewer lines may alter the ground-water flow pattern by creating artificial points of recharge and discharge. These possible hydrologic influences should be taken into account when analyses of ground-water samples show organic compounds that seem unrelated to the location of the inferred sources.

Health and Safety Requirements

Worker health and safety are among the basic considerations in a ground-water investigation of chemicals with harmful properties. However, most instances of pesticide contamination other than losses of free product, involve ground-water and soil concentrations much lower than those posing an immediate occupational health threat.

In instances where free product is lost, vacuum recovery systems and/or absorbents may be used to remove the surface concentrations before the ground-water investigation begins. These actions, normally taken as the first step in the remedial action plan, often render the land surface at the site nearly clean before the ground-water problem is studied.

Protective wear and associated equipment are required to safeguard the worker. In the field where no detectable airborne pesticide emissions are present, but where the potential for skin contact is high, a hard hat with splash protector, safety glasses, chemically resistant outer clothing, gloves and boots with taped wrists and ankles, and inner surgical gloves would provide a conservative dress for field protection. The presence of detectable concentrations of pesticide dust or vapor requires supplemental respiratory protection ranging from a combination particulate-removing filter and organic vapor/pesticide filter to a heavy duty canister similarly equipped. Work conditions that require personal protection above this level may necessitate a reevaluation of the drilling or sampling locations for ground-water monitoring. The specific regulatory requirements of a Health and Safety Plan are described in "Standard Operating Safety Guides." (1)

Monitor-Well Design

The objectives of installing a monitor-well network must be clearly defined before choosing the specific locations, numbers, and dimensions of the wells. The preferred design depends on the proposed uses of the monitor-well network, the dimensions and chemical characteristics of the contaminant plume, and the site hydrogeology.

Location and Number. The location of a monitor well refers to both its areal location in relation to the waste source and its vertical location or depth in the uppermost or other aquifer. The number and locations of the wells depend on several factors, including regulatory constraints and the current level of knowledge of the site hydrogeology and water quality. State and Federal regulations, for instance, generally require that a minimal number of monitor wells be installed at or near the source of ground-water contamination, in both the upgradient and downgradient directions. However, if only scant hydrogeologic data are available, it may not be possible to determine in advance the horizontal and vertical direction of ground-water flow. In these cases, a presumed flow direction, based on the locations of surface water bodies, the topography, and other surface indications, is used to designate the proposed monitor-well locations, both upgradient and downgradient of the waste sources. Additional wells are generally required for characterization of the site hydrogeology.

Wells monitoring different aquifers or different zones within a single aquifer are generally installed for two reasons: (1) to identify hydraulic characteristics of and the relationship between the different zones or aquifers, and (2) to characterize the ground-water quality at different depths. Characterization of site hydrogeology also involves investigating the presence of confining layers. To demonstrate the effectiveness of a confining layer in preventing the downward migration of a plume, for instance, it is often necessary to install wells to a depth beneath the confining layer in order to collect water samples and measure water levels during a pumping test. If the confining layer is relatively impermeable or if the hydraulic gradient is upward, monitoring of deeper aquifers may not be required. Where shallow and deep wells are needed, they commonly are installed at the same location and are referred to as paired wells or cluster wells.

The choice of well locations can be influenced by the types of wastes or contaminants and their propensity to migrate from the waste sources. Heavy organics, for instance, are likely to migrate to deeper aquifers zones, where their movement is not totally dependent on ground-water flow directions but may be partly controlled by the slope of the top of the uppermost confining deposits. Thus, in designing a monitor-well network to detect heavier-than-water solvent plumes, consideration must be given to installing wells at the base of the uppermost aquifer to detect gravity flow of pure product away from the site. Shallow monitor wells also must be installed to detect dissolved components moving in the direction of ground-water flow.

Often, the preliminary characterization wells can be incorporated in the permanent monitor-well network. However, in cases where the initial presumed ground-water flow direction turns out to be incorrect, due to the influences of nearby pumping wells, tidal fluctuations, ground-water mounds, or other hydraulic phenomena, additional regulatory wells may be required.

Dimensions. The dimensions of a well refer to its depth, cased interval, length of the screened or open-hole sampling interval, screen slot size, and borehole, screen, and casing diameters. The length of the sampling interval depends on the purpose of the well, the site hydrogeology, and the ground-water quality. A very preliminary investigation to determine which, if any, contaminants have migrated from a site may involve the installation of only a few shallow wells with a 3 to 10 ft screened interval at the water table. Once contaminants are identified, deeper wells may be installed to determine how those contaminants may have migrated vertically. In determining the length of the sampling interval for the deeper holes, a balance should be sought between sampling a small interval to better define the contaminant plume and sampling a larger interval to avoid missing the plume. If pumping tests are to be performed, wells in the same aquifer should be screened or opened to the same intervals; fully penetrating wells are generally preferred.

Slot sizes of the screens are chosen to maximize well yield so that reliable water samples and water-level drawdown data can be collected during specific-capacity and other field tests. To accomplish this, sieve analyses can be performed on split-spoon samples from the interval to be monitored. A grain-size distribution curve can be field plotted and analyzed to establish the screen slot size and the size of the gravel pack. For wells installed in competent bedrock, it may be wisest not to install a screen but to leave the hole open for maximum yield. In these cases, a surface casing is installed to prevent the collapse of the unconsolidated sediments and possible downward migration of contaminants.

The diameter of a well is dictated by its proposed uses, its depth, and the depth to water. Wells installed for water-level measurements only are commonly only 1-1/4 to 2 inches in diameter. Wells to be sampled can also be this small, but consideration should be given to the depth of the well and the depth to water. Centrifugal pumps and peristaltic pumps can be used for sampling in small-diameter wells, if the depth to water is less than 25 feet (including drawdown caused by the pumping); otherwise, bailing is required. Because bailing is less efficient than pumping, it is often preferable to install a larger-diameter well; a submersible pump then can be used for sampling. If automatic water-level recording instruments are to be installed or if geophysical logging is to be performed on the wells, it is generally advisable to install 4-inch-diameter or larger diameter wells. The diameter of the borehole should be sufficient to install the casing and/or screen and a minimum 3-inch-thick gravel pack. The grout envelope also should be at least 3 inches thick to insure that it completely surrounds the well.

Materials. Polyvinyl chloride (PVC), black carbon steel, and stainless steel are commonly used materials for casings and screens. PVC and black carbon steel are less expensive, but are sometimes not compatible with contaminants found in the ground water. Although generally immune to chemical attack by most naturally occurring compounds, PVC may adsorb or leach organic compounds (2, 3, 4); thus, PVC must be used with care in situations where there is a possibility of organic contamination. Where PVC is deemed appropriate, generally Schedule 40 should be used for wells less than 200 feet deep and Schedule 80 for deeper wells. Cement and other adhesives should be avoided; joints should be threaded and coupled or flush jointed.

Because galvanized steel is not corrosion-resistant, it should not be used in plumes of an acidic nature, or in water with high electrical conductivity due to chlorides or sulfates. Stainless steel is much more resistant to corrosion and can often be used in place of galvanized steel, but is several times more expensive. For this reason, wells can be cased with a cheaper galvanized steel above the water table and with more expensive stainless steel casing or screen below the water table.

The grading of a gravel pack in a well should be based on a sieve analysis of the finest aquifer materials of the screened zone. The screen slot opening should retain 90 percent or more of the gravel pack material. The gravel pack material should be clean, with well-rounded grains that are smooth and uniform. These characteristics increase the permeability and porosity of the pack material.

Drilling and Well Installation Techniques

Several methods are available for the drilling and installation of monitor wells. The advantages and disadvantages relating to ease of construction, character of formations penetrated, well diameter and depth, risks of contamination, and intended use of the well have been reported by several investigators (5, 6, 7). All field personnel should be thoroughly familiar with well specifications, quality assurance/quality control procedures, and safety plans and requirements prior to installation of monitor wells.

Drilling. Common drill rigs available for monitor-well installation include cable tool, air rotary, mud rotary, reverse rotary, and hollow-stem auger. The cable tool rig repeatedly lifts and drops a drill bit, drill stem, drilling jars, and rope socket. The drill bit crushes hard rock or loosens unconsolidated material and mixes the loosened particles with water to form a slurry or sludge. The sludge is removed at intervals by a sand pump or a bailer.

A mud rotary rig cuts a borehole by rotating a bit into the formation materials and removing the cuttings by continuous circulation of a drilling fluid. In a conventional rotary system, mud is pumped down through the drill pipe and out through nozzles in the bit. The mud flows upward in the annular space around the drill pipe to the surface, where it is channeled into a settling pit and a storage basin. Although the wall of the borehole is effectively sealed with the mud, large quantities of fluid can be lost to the formation prior to sealing. In reverse rotary, the drilling fluid is water which flows down the annular space around the drill rods and up inside the drill rods to the surface. To prevent caving, the fluid level must always be kept at ground level.

Hollow-stem auger rigs are commonly used for drilling in unconsolidated material with enough clay so that the borehole will stand without caving. An auger stem is turned and pushed as the auger flights carry material to the surface. Air rotary drill rigs are used for consolidated materials. Air is circulated through the drill pipe, out through ports in the drill bit, and upward in the annular space around the drill pipe.

Table I describes the relative advantages and disadvantages of different drilling rigs. For instance, a hollow-stem auger is very economical for drilling in unconsolidated materials and provides good soil samples, particularly above the water table. Mud rotary is useful for drilling in running sands and silts, is moderately expensive, but provides poor soil samples. Air rotary is expensive and provides poor soil and rock samples, but is very rapid; rock fragments can generally be identified, but other characteristics such as fracturing cannot. Small zones of contamination can be missed. Clearly the type of rig can be chosen only after careful consideration of the formations to be drilled, time and budget constraints, and sampling requirements.

<u>Well Installation and Development.</u> Once boreholes are completed, wells should be installed according to detailed well specifications. The following is a set of good practices applicable to most well installations.

- The gravel pack should be installed by the tremie method to assure proper placement.
- A fine sand pack should be installed above the gravel pack to prevent migration of the overlying cement into the gravel pack.
- If surface casing is installed for a bedrock well, grout should be emplaced by the pressure tremie method to insure a complete seal around the well. Grout should be allowed to set for at least 24 hours.
- A method such as the plumb bob method should be used to determine if the casing and borehole are plumb and true. Centralizers should be attached to wells as necessary.
- All fluid and mud produced by the drilling operations should be retained in portable mud pits for proper disposal.
- All equipment should be cleaned between each well installation to prevent cross-contamination. Specific cleaning procedures should be outlined in the well specifications.
- To the extent possible, drilling of the borehole and installation of the well should be accomplished in a continuous manner. This will minimize the risk of cross-contamination.
- A water-level measurement should be taken upon completion of well installation.
- Wells should be developed until the pumped or bailed water is clear or until successive pH and specific conductance values remain stable.

Table I. Comparisons of Drilling Methods

	Cable Tool	Air Rotary	Mud Rotary (Conventional)	Reverse Rotary	Auger-Hollow-Stem
Cost	Moderate	High	Moderate to High	Moderate	Low
Drillable Formations	Most Materials	Unlimited	Unlimited	Unconsol-idated Formations	Unconsolida-ted Formations
Depth Restrictions	Unlimited	Unlimited	Unlimited	Unlimited	About 100 ft
Hole Size	Unlimited	Unlimited	Unlimited	Unlimited	Up to 12-inch
Drilling Fluids	Water	None or small amounts of water	Mud	Water	None
Installing Grout or Gravel Pack	Easy	Easy	Easy	Sometimes Difficult	Difficult if annular space is limited
Quality of Soil/Rock Samples	Excellent	Poor	Fair to Poor	Good	Good Above Water Table
Drilling Rate	Slow	Rapid	Slow	Moderate	Rapid

- o Elevations of measuring points on wells should be determined by a qualified surveyor to the nearest 0.01 foot.

Recordkeeping. Keeping complete, accurate drilling logs is a critical aspect of well drilling and installation techniques. Table II lists items that may be included in a well log, depending on the type of well and the specifications for sampling. This is a detailed list and in many cases, it may not be necessary to record all the items listed. Generally, as much detail as the drilling technique can provide should be recorded by the field inspector. Some items that may not seem critical at the time of drilling may provide insight to unanswered questions or ambiguities that may arise during analysis of the data. Items such as length of casing and number of casing lengths should be checked by the field inspector. In some cases, samples may be numbered by a random numbering system; these values should be recorded in field notes. A record of field or trip blanks should also be kept. Any checking in the field of water-quality parameters, i.e., pH and conductivity, should be recorded. In general, the field inspector should understand the critical nature of his or her task, and specific guidelines for recording field notes should be established prior to the drilling program.

Table II. Information to Include on Drilling Logs

- o Name of contractor, driller, inspector
- o Location of site and specific borehole location
- o Borehole number
- o Date - start/finish
- o Soil Characteristics
  - depth
  - grain size and texture
  - angularity
  - moisture content
- o Drilling operations
  - rate
  - loss of circulation
  - use of water/mud
  - core diameter
  - color of circulating water
  - core recovery
  - diameter of auger or core
  - type core barrel
  - size of drive hammer and free fall distance to drive hammer
  - blows per foot to drive sample
  - force to push thin wall samples
- o Well materials/characteristics
  - casing - type, diameter, interval
  - screen - type, diameter, interval, slot size
  - grout - type, interval
  - total drilled depth
  - open hole interval
  - sand pack - type, interval
  - protective casing - type, diameter

Continued on next page

Table II--Continued

- o Samples - number, depth, type, size
- o Stabilized water level
- o Rock Characteristics
  - depth unit/formation and member names
  - color
  - hardness
  - fracturing
  - coatings or fillings (or lack of) in joints or seams
  - angle of bedding, schistosity, or other planar features
  - luster
  - thickness of bedding planes
  - degree of weathering
  - decomposition
  - strength
  - weathering
  - rock alteration other than by weathering
  - induration
  - rock quality
  - cavities or voids

Literature Cited

1. "Standard Operating Safety Guides." Office of Emergency and Remedial Response, Hazardous Response Support Division, Environmental Response Team, November, 1984.
2. Miller, G.D. Proc. 2nd Natl. Sym. Aquifer Restoration and Ground-Water Monitoring, 1982, p. 236.
3. Houghton, R.L.; Berger, M.E. Proc. 3rd Natl. Sym. Aquifer Restoration and Ground-Water Monitoring, 1983; p. 203.
4. Barcelona, M.J.; Gibb, J.P.; Millar, R.A.; "A Guide to the Selection of Materials for Monitor Well Construction and Ground Water Sampling;" Illinois Water Survey; Champaign, Illinois; 1983; p. 39.
5. Briggs, G.F.; Fieldler, A.G., Eds.; In "Ground Water and Wells;" Johnson Division, UOP, Inc.; St. Paul, Minnesota, 1980 edition; p. 209.
6. Richter, H.R.; Collentine, M.G; Proc. 3rd Natl. Sym. Aquifer Restoration and Ground-Water Monitoring, 1983, p. 194.
7. Minning, R.; Proc. 3rd Natl. Sym. Aquifer Restoration and Ground-Water Monitoring, 1983, p. 194.

RECEIVED April 7, 1986

FIELD AND LABORATORY GROUND WATER MONITORING

10

Monitoring Ground Water for Pesticides

S. Z. Cohen, C. Eiden, and M. N. Lorber

Office of Pesticide Programs (TS-769C), U.S. Environmental Protection Agency, Washington, DC 20460

At least 17 pesticides have been found in ground water in a total of 23 states as a result of agricultural practice. These results have been obtained through three different types of monitoring studies: (1) large-scale retrospective, (2) small-scale retrospective, and (3) small-scale prospective. The first two types of studies survey areas where the pesticide(s) in question has already been used. The third type of study is an intensive field study where the pesticide is applied and monitoring begins at time zero. Often, soil core data are at least as important as ground-water data. The ability to draw meaningful conclusions from large-scale studies is greatly diminished unless the studies have a statistical, stratified design. The purpose of this paper is threefold: to describe the three study types; suggest guidelines for ground-water sampling, soil sampling and well construction; and update the data summary of pesticides in ground water from agricultural practice.

In 1984, Cohen et al. reviewed leaching and monitoring data on 12 different pesticides found in ground water in a total of 18 different states as a result of agricultural practice (1). They also established criteria for predicting whether certain pesticides could leach to ground water as a result of normal use. In the less than 2 years since that paper was published, the numbers of pesticides found in ground water have increased significantly as have the number of states found to have pesticides in ground water. Seventeen pesticides have now been found in the ground water of 23 states as a result of agricultural practice (Figure 1, Table I). (As described in the "Occurrence" section at the end of this paper, many additional findings can be attributed to poor disposal practices, mixing-loading operations, etc. and are not included in this count.) This significant increase is more likely due to an increase in the quality and quantity of studies rather than an increase in

This chapter not subject to U.S. copyright.
Published 1986, American Chemical Society

Figure 1. Numbers of pesticides found in ground water as a result of agricultural practice.

Table I. Typical Positive Results of Pesticide Ground-Water Monitoring in the U.S.†

Pesticide	Use*	State(s)	Typical Positive, ppb
Alachlor	H	MD, IA, NE, PA	0.1-10
Aldicarb (sulfoxide & sulfone)	I, N	AR, AZ, CA, FL, MA, ME, NC, NJ, NY, OR, RI, TX, VA, WA, WI	1-50
Atrazine	H	PA, IA, NE, WI, MD	0.3-3
Bromacil	H	FL	300
Carbofuran	I, N	NY, WI, MD	1-50
Cyanazine	H	IA, PA	0.1-1.0
DBCP	N	AZ, CA, HI, MD, SC	0.02-20
DCPA (and acid products)	H	NY	50-700
1,2-Dichloro-propane	N	CA, MD, NY, WA	1-50
Dinoseb	H	NY	1-5
Dyfonate	I	IA	0.1
EDB	N	CA, FL, GA, SC, WA, AZ, MA, CT	0.05-20
Metolachlor	H	IA, PA	0.1-0.4
Metribuzin	H	IA	1.0-4.3
Oxamyl	I, N	NY, RI	5-65
Simazine	H	CA, PA, MD	0.2-3.0
1,2,3-Trichlor-opropane	N (impurity)	CA, HI	0.1-5.0

†Total of 17 different pesticides in a total of 23 different states.

*H = herbicide
I = insecticide
N = nematicide

the problem. However, our knowledge of the true extent of pesticide occurrence in ground water has increased only slightly. This is not to say that EPA has been idle in the interim. On the contrary, there has been a tremendous increase in activity in this area. The problem is that the task is enormous. The task is to characterize the extent of contamination or extent of potential contamination, in terms of:

1. pesticides in general and the nation as a whole;
2. specific pesticides and groups of pesticides, e.g., nematicides;
3. specific crop growing areas, e.g., citrus;
4. areas as defined by specific field conditions, e.g., soil permeability and ground-water depth;
5. areas as defined politically, e.g., counties;
6. individual pesticides, the relationship between their properties, field condition, and the mechanisms of leaching.

These characterizations are necessary in order to make better informed regulatory decisions on pesticide regulation and drinking water contaminant regulations. Clearly, no single monitoring or predictive exercise can address all of these issues. What needs to be done, in the case of monitoring studies, is to clearly state and limit the objectives of the studies, and then design these studies so that they contain considerations for hydrogeology, pesticide usage, and statistics. This paper will discuss three study types which, collectively, can be used to characterize the problem in the six different ways described above. Large-scale retrospective probability, small-scale retrospective, and small-scale prospective surveys are described below. The two other purposes of this paper are to offer some guidelines for the conduct of these studies and to update the data summary of pesticides in ground water from agricultural practice. But first it would be helpful to list some of the tools available to design these studies.

It is important to note that potential pesticide leachers can be and have been identified (1, 2). Criteria are available to assess ground-water vulnerability for pesticide application (1, 3, 4). The acronym "DRASTIC" describes a scheme for combining weighted evaluations of seven different hydrogeological parameters into a score, which is an indicator of relative ground-water contamination potential for a county or subcounty region (3). Also, regional pesticide sales data (5) and county level crop data (6) are available to the public, while more specific and confidential sales or use data can be provided to the EPA by the pesticide industry ("registrants"). Finally, statistical survey methods are available (7-11). Thus various tools are available which can be integrated into studies designed to satisfy certain objectives as described above.

Large-Scale Retrospective Studies

The objective of large-scale retrospective ground-water studies is to characterize the extent of occurrence of pesticides in wells over a large area. These statistical surveys cover multicounty or multistate areas and typically involve sampling of more than 100 wells. They are complex and expensive surveys and are usually done for specific pesticides after determination of a human hazard concern and after documented well contamination. This type of survey can also be useful to state or other agencies who would like to make estimates about percentages of wells contaminated by various chemicals using multiresidue methods. A brief discussion of some principles of probability surveys follows. Fuller explanation of the basic methods and theories of survey statistics are given elsewhere (7-11).

Probability surveys are surveys where samples are selected by a specified random process. The population from which the samples will be selected is called the universe or the target population. The data collected from these surveys are used to make inferences about the target population of wells which may be categorized in various ways. Some examples of target populations are wells in:

- all corn-growing counties where pesticide X is applied to >10% of the land area;
- certain ground-water basins in the southeast;
- the United States;
- the State of Florida;
- areas with highly vulnerable hydrogeologies in a certain state, etc.

Once the target population has been defined, two principles should be considered. First, statistical conclusions drawn from the study are not reliably applicable outside the target population. Second, if wells are arbitrarily excluded from sampling at later stages of the survey design or execution, the statistical validity of the study suffers significantly. The converse is true - wells can't be arbitrarily included. This is a corollary of the statement that all sampling units (wells, farm fields, etc.) must have a non-zero probability of selection. Otherwise, water quality results from wells selectively chosen may reflect a particular well-construction or "aquifer-material effect," and will not be representative of all the water in the specific aquifer that is under study (12). Helsel and Ragone make use of an example where wells from two areas are selectively chosen, one set of wells are screened in coarse gravels, the second set being screened in a variety of materials. Residue concentrations may differ between the two sets, reflecting the difference in lithologic materials surrounding the well screens, and not the differences in pesticide use patterns.

After the sample population has been determined and the constraints understood, an overall experimental design can be constructed. The major possibilities are simple random sampling, random sampling within geographical or spatial clusters, and stratified random sampling. (In this paper, "random sampling" is a convenient term which means sampling of equal or proportionate probability and without replacement.) In all cases, a critical step is the determination of the sample size. There are established ways for calculating the required numbers, n (7, 9-11); but basically n is a function of the desired precision of the conclusion (where precision implies a limit on the probability, say .05, that the conclusions differ from the true answer by at most a specified percentage (say 15%) of the true answer, the target population size, and the cost per sample.

Often, probability surveys are designed in stages. This is especially the case when the target population is large and successive stages are necessary to reduce the potential number of wells, farms, etc., to be sampled. Once the target population has been identified, it is carved up into convenient population segments such as counties, enumeration districts, basins, etc. These segments are called primary sampling units (PSU's). The next step in staged design is to randomly select some PSU's for further study. This subpopulation of randomly selected PSU's is referred to as the first stage sample. In a two-stage design, wells are then randomly selected from the sample frame for field sampling, i.e., water analysis.

Before the specific survey types are discussed, some advantages of probability surveys should be mentioned. Such surveys are efficient in that results from a relatively small number of samples can be used to draw reliable conclusions about a large target population. The probability approach also reduces possible selection bias resulting from the 'let's grab a sample here because it looks good' philosophy. Put in different words, "the major obstacle to random selection is convenience - if it is more convenient to select sampling points in a particular fashion, the process is probably not random and should be avoided" (12).

In simple random sampling, samples are randomly selected from the target population, where the target population is usually wells, but may be the ground water under farms, golf courses, etc. This design strategy is generally only recommended in limited circumstances for studies of pesticides in ground water because of expected high variances in the statistical conclusions and the fact that it does not allow one to incorporate knowledge of field conditions into the design. One example where this approach makes sense is when the population to be sampled is very narrowly defined and is known to be relatively homogeneous. For example, a target population could be wells of similar construction in a karst limestone region of a two county area dominated by one soil texture, e.g., a loamy sand. In this case, estimates resulting from simple random sampling would likely have a lower variance compared to other regions containing widely varying types of soil and subsurface hydrogeology. Thus knowledge of ground-water vulnerability was used to limit the target population, but was not a part of the statistical design.

The second type of random sampling, which has some spatial control on the sample distribution, is a special case of stratified random sampling. No examples of this type could be found in the literature pertaining to well sampling design; however, an example of this second type of random sampling pertaining to soil core sampling was found and is discussed in the section on soil sampling under small-scale retrospective studies. As regards well sampling design a possible design strategy is as follows: The wells of the eastern shore of a mid-Atlantic state are identified as the target population. Grids corresponding to 5 mi x 5 mi areas are laid over a map of that region of the state. One or more wells are then randomly sampled within each grid section. This study type may have limited application to areas which contain large hydrogeology and human activity (e.g., disposal sites) data gaps and in the situations when the time required for an appropriate reconnaissance analysis is too long. In such a situation, this may be considered a preliminary study which would lead to the generation of hypotheses and the initiation of additional studies.

The third type of random sampling discussed in this paper is stratified random sampling. The principal weakness of simple random sampling is that it does not use relevant information or judgement that we have about the environment, i.e., about pesticides leaching to ground water. For example, it would incorporate our knowledge that areas with loamy sands overlying solution limestone aquifers are particularly vulnerable to pesticide pollution. This knowledge can be used in stratified random sampling. In this survey approach, the target population is divided into subpopulations or strata that are internally more homogeneous than the target population as a whole (11). Then specified numbers of samples are randomly selected from each stratum. Recommended stratification variables are ground-water vulnerability and pesticide usage. Human or well population density may be used or it may be used as a 'size measure', a weighting factor to increase the probability that certain areas may be selected. The strata should be constructed so that they are mutually exclusive and account totally for the target population.

Following is an example of a stratified random survey. EPA's Office of Drinking Water and Office of Pesticide Programs (OPP) are jointly designing a national stratified random survey with the help of statistical and hydrogeology contractors. The survey is being designed in three stages - select counties, select county segments, select wells. As of this writing, the design of the second and third stages is uncertain. However, in the first stage the stratification variables are pesticide usage (5, 6) and ground-water vulnerability (3). All counties in the U.S. are being categorized according to whether they have high, medium, low or uncommon pesticide usage and high, medium, or low ground-water vulnerability. Thus all counties in the U.S. will be placed into one of the 12 (3 x 4) strata. The number of counties will probably be near 200 and will be determined using the criteria described above (desired precision, etc.)

There are three advantages of stratified random sampling for large target populations, when properly done, relative to simple random sampling. The overall variance for the estimates should be significantly lower. Theoretically, there is an inverse square relationship between the number of strata and the variance, although practically the results are usually less dramatic (7, pp. 133-5).

The second advantage is that estimates can be provided for certain subdivisions of the population. For example, EPA staff desires to draw meaningful conclusions about the high-high stratum in the national survey - the counties with high pesticide use and high ground-water vulnerability. The third advantage is that the possibility that certain segments of the target population may not be sampled at all can be eliminated. For example, if there is a small but significant cotton-growing area in a state, which is of interest, that area can be made into its own stratum, with similar treatments to other crop areas.

This discussion has emphasized the theory and basic principles of probability sampling of wells in a large area. The next two sections address studies of areas of a few acres or less. The guidelines for well sampling in those sections are appropriate for this section as well.

Small-Scale Retrospective Studies

For a small-scale retrospective study, participants enter and study a field in which a pesticide has been used over a period of time. The purposes of small-scale retrospective studies are to determine whether the pesticide(s) in question has leached to ground water in certain fields, and to characterize the leaching pattern in the soil profile at a given point in time. This section contains discussions on site selection, site characterization, the number and location of observation wells, well construction and sampling, and soil sampling.

Site Selection. The following criteria are important in choosing a site (13, 14).

1) There should be documented prior usage of the pesticide in the field site, preferably for several previous years. If possible, the field should be separate hydrogeologically and physically from other fields where the pesticide under study has been previously used. This avoids the possibility of runoff from other fields contaminating the study fields and leaching from other fields contaminating the underlying aquifer.

2) Land owner/farmer cooperation is essential since he/she will be closely involved in the study.

3) It is preferable to install observation wells specifically designed for the purposes of the study. Guidelines for well construction are given in the well construction section. However, existing wells can be used if sufficient information is available to characterize the wells. If existing wells are used, the history of the well-construction will be necessary. The well should be rejected from the study if these data are not complete. Some pertinent questions to ask are: What is the well depth? At what depth is it screened? With what material is the well annulus filled, i.e., gravel, cement? Was the well sealed properly? Does the well have a device to prevent back-siphoning installed in it? Of what material is the well casing constructed? There should be only one short screen or perforation in the casing. Well casings with several screens or perforations at different depths should be rejected from the study. The well casing may be inadequate for monitoring certain organics if it is constructed with any type of plastics or plasticizers known to adsorb organics.

4) Uniform soil characteristics on the field are desirable.

Uniformity in this context refers to slope and texture. Ideally, the field should have a slope of less than 2%, and if possible, there should be only one soil series on the field. If more than one soil series is on the field, all the series should be of the same texture, i.e., all should be sandy loams. This criterion does not preclude the possibility of layered soils (soils with distinct horizons) as long as the entire field is characterized by this layered soil. The more uniform the field, the easier will be the interpretation of results. Soil Conservation Service soil surveys give sufficiently detailed information to characterize a field.

Characterizing a Site. The soil in the field must be characterized. A minimum of the following kinds of information should be recorded:

- permeability of the individual soil layers as a function of depth down to the water table;
- organic matter;
- bulk density;
- particle size distribution of sand, silt, and clay as a function of depth, if distinctly different horizons exist in the soil column down to the water table;
- available water holding capacity, as defined by the difference of field capacity and wilting point, both of which should also be ascertained for the soil;
- SCS soil series classification for the soils of the study site, if information is available.

The hydrogeology of the field must be characterized. In many cases, only semiquantitative analyses are necessary. Some points to consider are:

- types, depths, and extent of layers of reduced permeability such as clay pans and silt-clay lenses;
- water table depth below the surface as a function of time; the water table may fluctuate seasonally with rainfall/runoff events and snowmelt;
- the direction of ground-water flow (construction of a flow-net);
- type of aquifer (confined, unconfined, artesian);
- geologic materials comprising the aquifer (sand and gravel, glacial till, carbonates, etc.);
- hydraulic conductivity(s) (K m^2/day) - whether the local portion of the aquifer is characterized by one K (isotropic) or several K values (anisotropic);
- transmissivity;
- whether the aquifer has recharging or discharging characteristics;
- any man-made activities that may affect the water level of the aquifer such as ground-water pumpage.

Climatological information for the site will be necessary. Precipitation data, air temperature and pan evaporation data, and any use of irrigation water must be carefully recorded. Irrigation and natural recharge is a critical aspect of pesticide leaching.

A water balance detailing the water inputs and outputs from the field can be used to estimate recharge. Inputs to this balance include precipitation and irrigation. The outputs which can be estimated are runoff and evapotranspiration, and the remainder can be assumed to be recharge.

Number and Location of Observation Wells. The precise amount of pesticide which was applied to the soil surface and is available for leaching is known as the source term. For pesticides which are applied directly to the soil, such as granules or liquid formulations sprayed only onto the soil surface, the source term can be assumed to be equal to the total amount of application.

However, if the pesticide is sprayed onto a developing canopy, then some of the pesticide will be intercepted by the plant and will be unavailable for leaching. The source term obviously cannot be estimated accurately without post-application sampling, which is why landowner records are important. Estimation of the source term is critical for the interpretation of all subsequent data collection efforts, as well as for modeling purposes (if that will be part of the use of field data).

Another critical information need for either a retrospective or a prospective study is the direction of ground-water flow. If it can be determined from preexisting hydrogeological information on the study site, then the design and placement of the monitoring wells can be determined. If the direction of ground-water flow cannot be determined from existing data, and previously installed wells are present on the property, they can be used to try to establish the ground-water flow net.

Understanding the direction of ground water flow is necessary to determine the location of monitoring wells. The minimum number of well sites is suggested to be 4 for one study site. (Federal Regulations in 40 CFR Part 256, subpart F of the Resource Conservation and Recovery Act suggest at least four monitoring wells as the minimum number for landfills.) A "well site" can be defined as a cluster of three wells, located very near each other, which penetrate three depths of the aquifer, i.e., each well is screened at a different depth to obtain 3 dimensional sampling of the aquifer at each well site. The first well is placed centrally with respect to the study site up-gradient and outside of the study site boundary; the second well is placed centrally with respect to the study site and down-gradient and outside of the study site boundary; and the third and fourth on either side of the center of the study site within the the study site boundaries.

Once the wells are in place, ground-water samples can be taken for water quality measurements. Assuming the field has had several years of seasonal pesticide use, pesticides may be detected in well samples any time of the year. However, the two optimal times for sampling beneath and just down-gradient of the field are not long following application in late spring and early summer, and during the winter-spring snowmelt period (ca. March). Experience has shown that leaching pesticides typically contaminate very shallow ground water beneath the field with the first major period of recharge following application; i.e., with spring recharge following application in the spring for pesticides in the Northeast, or with summer rain for spring applied pesticides in the Southeast. Because of the temporal changes in ground-water quality (15-17), wells should be sampled beginning at the onset of the sampling program and continuing through the winter and into the next season.

Well Construction. The selection of a drilling method for this phase of work should be based on the ability of a particular method to achieve the objectives of a drilling program, which include the following points from Luhdorff and Scalmanini (18):

1) the ability to penetrate all anticipated formations and materials, to penetrate at a desired rate, and to construct a borehole of desired diameter for the anticipated well, as well as for the placement of a gravel or sand pack and necessary formation sealing material such as bentonite or cement;

2) identification of lithology or development of a geologic log of all formations and materials penetrated, including physical characteristics and visual description of color, texture, etc.;

3) collection of samples of aquifer fluids during the drilling process and prior to well construction, while at the same time minimizing potential for cross-contamination ("cleanliness" is a key issue in drilling - there should be no contamination from surface soils and water, in addition to cross-contamination between layers);

4) collection of "undisturbed" soil samples from the center line or sidewall of the borehole (this objective often requires the drilling to be halted while soil samples are taken from the bottom of the incomplete borehole);

5a) completion of the borehole into a monitoring well during the initial construction process, i.e., constructing a well as the borehole is drilled or constructing a well in the borehole immediately after the drilling tools are removed;

or

5b) completion of a monitoring well in the borehole following a time lapse for interpretation of geologic or geophysical data from the borehole.

Though geophysical logging of the borehole is desirable in many situations, the additional expense and equipment necessary may be prohibitive. For the smaller, more shallow wells used in small-scale ground-water studies this is not necessary.

There are many drilling techniques available for well construction: hollow stem; flight auger; direct circulation rotary drilling; cable tool drilling; reverse circulation rotary drilling; and air rotary drilling. Barcelona, Gibb and Miller (1983) provide a more detailed discussion of each (19).

For the purposes of a small-scale retrospective or prospective study, the hollow stem, continuous-flight auger is recommended. The hollow-stem auger is mobile and inexpensive to operate. It is capable of drilling approximately 150 feet into unconsolidated material. (Practical experience indicates that 100 feet may be successfully drilled.) The rig is equipped with a removable plug that helps in the drilling process, but that is inserted inside the hollow stem of the auger and is, therefore, easily removable. The drilling procedure uses no drilling fluids, thereby minimizing contamination problems between the borehole materials and the drilling process (20). Once the borehole has been drilled to the desired depth, the plug is removed from inside the hollow stem and a small-diameter well casing, 1 1/4"-2", (3-5 cm) can be inserted inside the hollow stem. The hollow-stem auger can then be pulled out of the borehole leaving the well casing in place, which can then be easily grouted in. Soil core samples can also be obtained during the drilling process by inserting a Shelby Tube or a split spoon (split barrel) sampler inside the hollow stem, lowering the assembly to the bottom of the hole and driving the sampling tube into the undisturbed profile (21). These core samples will be used for lithologic identification.

Cross-contamination of drilled materials and soil samples collected for lithologic identification can be minimized during drilling by installing temporary casing as the drilling proceeds, and reversing the drill spin in place. After the first well has been drilled from the soil surface to the desired depth, usually the water table and below, the soil cores from this site may be used for lithologic identification, as mentioned. For any other wells drilled on the same plot, in which a Shelby Tube or split spoon sampler are not used, the first 18" of earth can be removed with a shovel reducing the possibility of soil from these upper zones from contaminating the lower drilling depths. This is desirable for retrospective studies as the first 18" of soil often contains the higher concentrations of pesticides.

In most cases the hollow-stem auger will produce a sufficiently deep borehole for a small-scale ground-water study designed to detect the leaching of pesticides from normal agricultural use. Where deeper wells are needed or where consolidated formations are encountered other techniques may be researched for their applicability.

Once a new well is in place it needs to be developed. Well development refers to the procedure used to clear the well-screen of fine silts and clays produced during drilling. Pumping the well until sediment free-flow is established or by using a surge block to loosen clogged material are recommended (21). For the purposes of small and large scale ground-water studies, all wells should be constructed with a single, short screen at a known depth (12).

Once a screened casing has been lowered into the borehole to the depth of interest, quartz sand, frac sand or pea gravel should be filled in around and a few inches above the screened interval. Expandable grout should be used to fill in the annulus above the screen to the surface around the well. The entire length of the casing from surface to the screened interval should be grouted in to prevent seepage from the surface down along the casing (21). The permeability of grout seals is further discussed by Kurt and Johnson (1982) (22). It is recommended that cement seals be used, and bentonite seals be avoided.

The recommended monitoring well diameter is 2 inches (5 cm) (23). The type of well casing is important, and will be determined by the types of compounds sampled for. The following materials were ranked by the U.S. Geological Survey (USGS) as to their inertness and suitability as casing materials: glass, Teflon, stainless steel, galvanized steel, PVC, black pipe, fiberglass. From the aforementioned USGS memorandum a combination of materials for well casing is recommended, specifically, "a Teflon or stainless steel screen and casing in the water bearing zone and PVC casing for the remainder of the hole." This procedure is recommended for both volatile and non-volatile pesticides of expected low concentration levels in a non-corrosive environment, i.e., pH greater than 5, no iron precipitation, and low concentrations of organic solvents. Whenever PVC is used, no organic-based solvents or sealers should be used because of the possibility of contamination. Joints of casing should be threaded and screwed together, not glued together.

As with well construction and drilling method precautions, all sampling devices should be carefully cleaned prior to use. There are many suggested methods. Current practice at the U.S. Geological Survey recommends a hot detergent scrub with an inorganic detergent, preferably a sulfonated detergent, followed by a distilled water rinse for plastic and Teflon parts and an organic solvent rinse for metal parts (24). All sampling devices should be pre-rinsed with the well water to be sampled.

Well Sampling. Before a well is sampled, it must be purged of its standing water or storage water until the well yields representative aquifer water upon pumping. Storage water is water that "does not come into contact with the flowing ground water" (25, 26). In the past, the most common method used to obtain a representative aquifer sample was to flush the well-bore by pumping a specified number of well-bore volumes of water. This procedure is considered outdated, and is not advised. The following excerpt is a compilation of the information compiled by Gibbs and others outlining recommendations for the collection of ground-water samples (25).

1) A two or three hour pumping test should be conducted on each monitoring well to be sampled. Analysis of the pump test data and other hydrologic information should be used to determine the frequency at which samples will be collected and the rate and period of time each well should be pumped prior to collecting the sample. If pumping tests cannot be conducted, slug tests may be substituted to provide the needed hydrologic information.

2) In general, peristaltic and Teflon pumps are recommended for small diameter, 2" (5 cm) wells. The peristaltic pump is used to help evacuate the casing.

The U.S. Geological Survey requires that the specific conductance and temperature stabilize before taking a water sample that is considered representative of the aquifer; this is the OPP/EPA policy, as well. The pH of the sample should be recorded to $\pm$ 0.1. There are no set number of well casing volumes to be pumped.

These small 2" wells should not be overpumped to the point of dryness. A pumping rate slow enough to allow the well to recharge or recover is recommended. Overpumping can cause excessive silt and clay fines to be drawn from the well (27).

Measurement of chemical parameters is best accomplished with an in-line closed measurement cell (25). When the values of the indicator parameters are observed to vary less than $\pm$ 10% over three consecutive well-bore storage volumes, the well may be presumed to have been adequately flushed for representative sampling. When in-line measurement cells are not practical, standard pH and conductivity meters and thermometers are used. All containers used for measurements must be rinsed 3 times with representative well water.

3) Once the water has been determined to be representative of the aquifer, the peristaltic pump should be removed. A Teflon bladder pump or positive displacement pump should be used to collect the actual water samples. At least 1 liter of representative well water should be pumped through the bladder pump before sampling. The bladder pump (3 feet long) is inserted into the well and fills passively with water. The Teflon bladder inside of the pump inflates with N_2 gas and displaces the water sample up a Teflon tube and into the sample container.

   It is important that there is enough water to completely cover the pump in order to prevent the introduction of air into the water sample. Sampling devices should be selected that minimize the introduction of air and gas bubbles into the sample (28, 29). For wells without enough water to cover the bladder, a Teflon or stainless steel bailer resembling a long, narrow bucket may be used (27). All collection containers should be rinsed 3 times with representative well water prior to collection.

Most pesticides still in use are not volatile organics. However, for volatile organics, the following procedures are recommended (30).

1) A silicon/Teflon septum capped vial is recommended for the collection of volatile organic compounds. This allows the laboratory conducting the chemical analyses to withdraw a sample through the septum top, minimizing the sample's exposure to air.

2) Fill the vial until a positive meniscus forms at the top. Avoid aerating the sample while sealing the container.

3) Look for air bubbles by turning the container upside down. Any bubbles are an indication of trapped air and the sample should be discarded and a new sample collected.

Because pesticides may degrade after collection, the following steps may be considered to minimize that breakdown. 1) Follow collection procedures outlined above. 2) Wrap the glass sample container in aluminum foil or use amber glass bottles or use Teflon sample containers to shield pesticides from the effects of sunlight. 3) Deaerate samples with N_2, He or Ar gas when they arrive in the lab. 4) Store samples at 4°C until analysis, or if samples are collected in Teflon bottles they can be frozen until analysis. 5) Extract the samples as soon as possible, preferably within 2 weeks. DO NOT freeze samples in glass containers. Freezing may crack glass containers. Although Teflon containers can be shipped with frozen samples, freezing and thawing procedures may disturb the chemical equilibria of solutes in solution, and is therefore not recommended (31).

Soil Sampling. Soil sampling provides information on unsaturated zone pesticide migration, and as such, is a critical component of retrospective and prospective studies. For this reason, soil sampling for a retrospective field study should adequately describe pesticide fate horizontally (over the entire field surface) and vertically (through the various soil horizons to the ground water). Unlike water sampling, this can be accomplished with only a single sampling date. This requirement results in a significant savings of resources in contrast to a prospective field study, which requires several soil sampling dates over time. Obviously, the timing of this soil sampling date is critical. It should occur following a time period when the previous application would have spread out in the profile and leached, if that is what it would do. For a spring application, this is usually in the fall, for example.

Few examples of experimental designs for soil sampling were found in the literature. Tourtelot and Miesch (32) discuss the estimation of the total geochemical variation of a shale unit using a design approach called a "hierarchial analysis of variance." This is one approach which could be applied to the variance of pesticide residues in a soil plot study in the following manner. For a plot of variable soil characteristics:

a) determine areas of variable soil types (textures, % organic matter) on the field, and consider these different "areas" level 1 of the hierarchial analysis of variance;

b) determine, by grid placement, sections within each "area" to be sampled, and consider these "sections" level 2 of the design; and

c) determine, randomly, the number of samples per grid section to be taken, and consider this level 3 of the design.

Several procedures may be used as to where to take samples within a grid placement: at intersection points of the grid, in the center of the grid sections, and at opposite corners of the grid sections. The number of samples taken per grid section should be at least two for the purposes of sample comparison within a given grid section. Levels 1, 2, and 3 allow for comparison of the variance in the parameter in question (here, pesticide residue concentrations in soil) as affected by differences between samples from "areas" of different soil characteristics, differences between samples from grid "sections" of the same soil characteristics, and differences between samples from the same grid "section". The percentage of variance contributed by each level of the design for the pesticide residue involved of interest can be determined (32). Krumbein and Slack (33) illustrate details of the calculations involved in computing the variance components. The sample variance contributed by analytical technique, level 4, can be estimated by duplicate analyses of the samples.

There are then 4 levels of variance which are additive and their sum is equal to the total variance of the soil plot as a whole, if sampling and analyses biases are absent throughout the experimental design procedure (32).

For a plot where all soil characteristics are homogenous, level 1, described above, would not be necessary, and there would be 3 levels of variance to compare.

The use of a grid to select sections to be sampled has been referred to as "regularly spaced" sampling, as opposed to random selection procedures, (32). A systematic sampling procedure, such as a grid is recommended, (34, 35).

The preceding discussion is illustrative and is one strategy for soil core sampling. EPA continues to investigate this issue. Basically, the guidance which the authors give to pesticide companies and others is 1) locate soil coring sites in an X pattern or other regularly spaced pattern as described above, and 2) choose at least 5 sites per 1 to 3-acre plot.

Generally, soil probes and hand augers are used to collect soil cores from depths near the surface. Probes are stainless steel tubes, approximately 2.5 cm in diameter sharpened and bevelled on one end, and fitted with a T handle that is used to push the probe into the soil. Augers are used where probes will not penetrate the soil. Augers are twisted or screwed into the ground. They are 3-20 cm in diameter. Augers collect a disturbed sample by virtue of the twisting motion used to insert the tube into the soil. Probes and hand augers are designed to sample the upper 2 meters of soil (36).

Shelby tubes and split barrel samplers are used for deeper soil core sampling as mentioned earlier under well construction. Cross contamination of soil layers during the coring process should be avoided. This can be accomplished either by installing temporary casing as drilling proceeds, or by drilling to a point, reversing the spin of the auger flight to carry loose soil out of the hole, and ensuring that the soil at the surface is stamped down.

Soil samples should be taken to the depth of pesticide penetration at each sampling event, and then a little further to ensure that all the pesticide is accounted for. This depth can be estimated with a computer simulation model, by judgement and experience, or the sampling can be to a certain depth, three feet for example. When the pesticide nears that level, the sampling depth can be increased by increments such as a foot at a time. The location of the ground water should be considered when determining the total depth of soil sampling, with deeper sampling for a deeper groundwater level. Sampling should be done in 6-inch increments.

Each individual sample should originate from a continuous core which has been taken to the predetermined total depth of sampling. Resources permitting, there should also be at least one set of deep cores to characterize the profile to the ground water, if the soil profile has not already been established from past or current well drilling procedures.

Small Scale Prospective Studies

These studies are done by pesticide registrants and other scientific researchers, and are usually associated with the pesticide-use registration process. The primary objective of this study type is to characterize the subsurface fate of a particular pesticide, i.e., establish the leaching potential in a controlled study. A prospective study design attempts to follow the pesticide from the time it is applied to the surface and until it has degraded, reached the saturated zone, or reached a depth significantly greater than the root zone. As such, particular emphasis is placed on soil coring.

Since many of the details of retrospective and prospective studies are similar, such as constructing wells and extracting samples, only the differences between the two will be covered in the following sections.

Site Selection and Characterization. The most important criteria for site selection is that there should be no prior use of the pesticide in question on the field site. This demands a landowner/farmer who keeps careful records of his field and method of application for the pesticide of interest. All other criteria, including hydrogeologic considerations, soil homogeneity, existing well information, and new well construction are similar in the prospective and retrospective study. Characterization of a prospective field site is similar to retrospective field site characterization.

Collection of Soil and Water Samples. Soil sampling provides information on unsaturated zone pesticide migration, and as such, is the most critical data that will come out of a prospective field study. For this reason, soil sampling programs should adequately describe pesticide fate horizontally, vertically, and temporally (over time).

One set of soil samples should be taken prior to pesticide application, to ensure that the field is free of residues. Another set of samples should be taken on the same day as pesticide application or on the day after application. Three to four more sets of samples should be taken over the course of the season, particularly after rainfall events to demonstrate the effects of water infiltration on pesticide leaching. One set should be taken after the winter fallow period at springtime during snowmelt. Soil cores will need to be timed with irrigation events in fields using irrigation practices.

One set of well samples should be taken prior to pesticide application to ensure that the well water is free of residue. Water sampling should then occur following major recharge events, i.e., major storms producing recharge. Placement and number of well sites is similar to retrospective field sites.

One important difference between a prospective and retrospective study is the application and analysis of tracers. Davis (1980) discussed types of tracers that are available (37). The addition of tracers along with the pesticide is valuable for the following reasons:

1) tracers allow the researcher to track the "water front" as it moves down the soil profile.

2) tracers allow for the determination of the "retardation factor", R, of the pesticide as it traverses down the unsaturated zone. R is defined as a ratio of the distance of travel of the water front to the distance of travel of the pesticide front.

3) tracers allow for the determination of a phenomena known as "macropore flow", which is the rapid flow of a small portion of water through macropores, e.g. worm holes, rapidly through the soil profile at the onset of a storm, directly to the aquifer. Tracers appearing at the well following a storm can allow for macropore flow analysis.

4) tracers allow for the determination of the hydraulic conductivity (K) in the saturated zone.

Chloride and bromide are the tracers of choice. Tracers must be applied at the time of pesticide application, so that the movement of the tracer can be directly correlated to the movement of pesticide. However, in a more detailed study, Tennyson and Settergren (38) found that bromide moved more rapidly in soil-water than would be expected from the laboratory determined soil hydraulic conductivity (K). Radioactive materials, organic dyes, gases, and fluorocarbons are not recommended for use as tracers. The fluorocarbons have been used successfully, though they have not been used extensively to date (39). After the application of pesticide and tracer, the procedures previously outlined should be followed for monitoring soil and ground water.

Occurrence of Pesticides in Ground Water

At least 17 pesticides have been found in ground water in a total of 23 states as a result of agricultural practice. These data are summarized in Table I and Figure 1 as well as in the discussion below. Two of these pesticides, 1,2-D and 1,2,3-TCP, are not usually considered to be part of the active ingredients but are synthetic by products of the active ingredients. Unlike the 1984 summary (1), the term "at least" has to be used. This is because of the significant increase in the number of data generators as well as in the data generated. It is difficult to keep track of so much knowledge. The next update of this statistic will require either additional coauthors or a qualifier such as "...many more than X pesticides...". Hopefully, the next update will be more conclusive as a result of more systematic and rigorously designed surveys.

At present, we cannot yet determine the extent of the problem. The EPA is designing a national statistical survey which will attempt to answer this question (described below). In addition, EPA's Office of Pesticide Programs (OPP) continues to involve itself in various field projects around the U.S. in collaboration with state, county, and USGS staff. These studies and others which are being done or have been done since 1984 are very interesting and are worth high-lighting below. The authors regret if key, current studies are not mentioned in the synopsis below. The intent was to list only the most recent and exciting studies of which the authors have personal knowledge. They are listed in no particular order by chemical, state and agency. Following this summary of recent project highlights is a more complete summary of monitoring results organized by chemical.

Aldicarb. Union Carbide Corporation continues to monitor for aldicarb around the U.S. with the assistance of local government officials. As well, more mechanistic studies continue on its subsurface persistence and mobility in Long Island and Florida (40-43). OPP is in the midst of a regulatory decision on this chemical, a public use-by-use risk-benefit analysis.

Regional Assessments. OPP is collaborating with the Water Resources Division of USGS and local universities on a multi-year study of certain areas of the High Plains or Ogalalla Aquifer as well as certain areas in the San Joaquin Valley. The objective is to study the relationship between agricultural land use on ground-water quality in these areas.

National Survey. EPA's Office of Drinking Water and OPP are designing a national statistical survey of pesticides in existing drinking water wells. The goals are to make statistical estimates about nationwide occurrence, to relate the occurrence to field conditions (pesticide usage and hydrogeology), and to make broad estimates of the population exposed. Technical details were described above under "Large-Scale Retrospective Studies." A final report is expected in 1989.

Dibromochloropropane (DBCP) in California and Hawaii. DBCP has now been found in 2500 private domestic, public, and irrigation wells in California, and was recently found to have leached 400 feet downward through the unsaturated or vadose zone (44). California State Assembly Bill 1803, passed in 1984, requires the state health department to monitor 40 priority pesticides and other organics. The remaining use of DBCP on pineapples in Hawaii was cancelled (45). In a collaborative effort between EPA and USGS, a field by field ground-water vulnerability reconnaissance analysis was performed on Maui (46).

Iowa. The State of Iowa Geological Survey and the Department of Water, Air, and Waste Management continue to aggressively monitor for pesticides in ground water in many areas of the state (47-50). EPA/OPP is collaborating with the University of Iowa and the IGS in a two-year stratified random survey of wells and farm fields in four types of hydrogeologic environments.

Ethylene Dibromide (EDB). Results of a collaborative study of EDB between USGS and EPA/OPP in southwest Georgia were recently published (51). A follow up project is continuing. The State of Florida has been successfully struggling with detections of EDB in approximately 1000 private and public wells (11% positives), resulting in exposure to greater than 50,000 people (52). Laboratory research on the subsurface fate of EDB continues (53, 54).

Cape Cod. EPA/OPP is collaborating with the Cape Cod Planning and Economic Development Commission and the USGS to design and conduct a small-scale retrospective study of ground-water quality under golf courses treated with pesticides. The USGS district office is also lending some technical support to this effort.

Two other recent California reports are worth highlighting. The California Department of Food and Agriculture recently completed and reported on a useful computerized well inventory data base for agricultural pesticides in ground water from nonpoint sources (55). A 1983 report by Ramlit Assoc. showed that the number of pesticides found in ground water as a result of point source pollution greatly outnumbered those from nonpoint sources (56). This latter report underscores the need to determine contamination sources so that proper remedial and regulatory actions are taken.

Following is a chemical-specific summary of ground-water monitoring results. This is an update of the 1984 tabulation, which also contained environmental chemistry data (1). In order for a pesticide to be included on this list, there must be analytical confirmation, there must be lab and/or controlled field data which demonstrate some leaching potential in certain environments, and the findings must be tied to agricultural practice. The same criteria applied to the 1984 tabulation. The chemicals are listed in alphabetical order. The data are summarized by chemical and by state in Table I and Figure 1, respectively. There are additional reports of pesticides in ground water, but the reports only satisfy two of the three criteria for inclusion on the list. The three pesticides in question are arsenic in Texas and EDB and atrazine in Hawaii.

These reports all focus on nonpoint sources of pesticides leaching to ground water. However, the extent of occurrence of pesticides in ground water from agricultural and industrial point sources is also not known. A separate investigation into this topic would be warranted.

Alachlor. Alachlor has been found recently in Iowa (47-50), Pennsylvania (57), and Maryland (58). The studies would all be classified as large-scale retrospective. All study designs had strong hydrogeological components, but had little, if any, probabilistic components. In the more localized areas the latter components would not be necessary. The Iowa findings have been in private wells, public wells, and observation wells in various areas around the state and hydrogeologic environments ranging from alluvial to karst limestone. In Pennsylavania, the USGS reported six positives in a network of approximately 82 observation and residential wells in central Pennsylvania sampled for pesticides. Likewise, it was found in four public and private wells out of 30 wells sampled in various locations on Maryland's Eastern Shore. Most positives fall in the 0.1-10 ppb range. Alachlor has now been found in four states' ground water.

Aldicarb. As noted previously (1), the aldicarb species usually found are the biooxidized metabolites, aldicarb sulfoxide and aldicarb sulfone. The only new state with a report of aldicarb in ground water is Rhode Island. The finding resulted from a water quality investigation by the USGS in the vicinity of potato fields (59). The well screens generally ranged between 15 and 50 feet (4.6 - 15.2 m) deep. As Table I indicates, aldicarb has now been found in the ground water of 15 states. The findings have been in observation, irrigation, and private domestic wells but apparently not in public wells. Based on work cited previously (1) and other more recent work (42, 60), roughly 2000 wells nationwide have been found to contain aldicarb residues, at levels typically ranging from 1 to 50 ppb total aldicarb residues. These study designs would be classified under all three study types. The site selection process has been purposive in nature with a heavy emphasis on soil type, well depth, and pesticide application timing with respect to recharge events and crop growth.

Atrazine. Atrazine was found in the same Pennsylvania survey mentioned in the alachlor discussion above. Atrazine was reported in 21 of those wells. It was also reported in three public wells in the Maryland survey mentioned in the alachlor discussion. It has now been reported in five states' ground water at levels typically ranging between 0.3 and 3.0 ppb.

Bromacil. There are no new positives of bromacil to report, although it is seldom analyzed for in well surveys.

Carbofuran. The only new state with a report of carbofuran in ground water is Maryland. The finding was beneath test plots on Maryland's Eastern Shore in a small-scale prospective study, and positives typically exceeded 5 ppb. A tracer was applied. More details cannot be given due to the fact that the information is considered by the registrant, FMC, to be confidential business information (60). Carbofuran has now been reported in three states' ground water, at levels typically ranging between 1 and 50 ppb. It should be noted that carbofuran ground-water monitoring has rarely attempted to detect the toxic metabolites, 3-OH and 3-keto carbofuran. This is a deficiency which should be addressed in future monitoring.

Cyanazine. This chemical had not been found in ground water as of the time the 1984 list (1) was written. The new findings are in northeastern Iowa (47-49) and central Pennsylvania (57). In the Iowa studies, cyanazine was found at low levels in three wells and a spring in incipient karst and karst areas. Cyanazine was reported at 1.1 ppb in one of the 82 wells in the central Pennsylvania study where the water level in the well was approximately 75 feet below land surface.

DBCP. No additional states have reported DBCP findings. In 1984, DBCP was reported in wells of five states (1).

DCPA (and acid products). No additional states have reported DCPA findings. It was reported in New York ground water (1). These chemicals are not usually included in ground-water monitoring programs.

1,2-Dichloropropane. 1,2-D was recently reported in at least seven shallow wells in western Washington in association with soil injection in strawberry fields (61). The purposive well selections in this retrospective study were done with strong bases in hydrogeology and pesticide usage information. 1,2-D has now been reported in four states' ground water at levels generally ranging between 1 and 50 ppb.

Dinoseb. No additional states have reported dinoseb findings. It had been found in New York ground water (1). This chemical is not usually included in ground-water monitoring programs.

Dyfonate. This chemical was not on the 1984 list. It was found at 0.11 ppb in a spring draining a solution limestone aquifer in northeast Iowa (49). It is a marginal leacher; i.e., its persistence and mobility are less than most of the other ground-water contaminants (2).

EDB. Four additional states have reported EDB in ground water. EDB was found in 14 public, private, and observation wells, out of 95 wells sampled, tapping shallow unconfined and deeper confined aquifers in western Washington (62, 63). EDB has been found in several wells in southwest Arizona at sub-ppb levels, including community wells in Phoenix (64). Contamination of over 220 public and private wells was reported in western Massachusetts and western Connecticut where it had been used in the Connecticut River Valley (65, 66). In many cases these large-scale retrospective studies have involved saturation sampling. That is, every possible threatened well was sampled. In such situations, a statistical/probabilistic design is meaningless and not required. EDB has been found in the ground water of eight states at widely varying levels which usually range between 0.05 and 20 ppb.

Metolachlor. Both state reports of metolachlor are new. It was found in 4 out of 82 wells sampled in the USGS central Pennsylvania study (57). Metolachlor was also found in a well in an incipient karst area and two springs draining a solution limestone aquifer in northern Iowa (48, 49). Concentrations typically range between 0.1 and 0.5 ppb.

Metribuzin. This pesticide is a recent addition to the list. It was found in three observation wells in northern Iowa at levels ranging from 0.09 to 4.35 ppb (48).

Oxamyl. The only new state reporting oxamyl in ground water is Rhode Island (59). It was found in at least four observation wells out of 11 sampled. Typical levels of oxamyl in ground water range from 1 to 60 ppb, with the Rhode Island results closer to 1 ppb. Oxamyl has now been found in the ground water of two states.

Simazine. Simazine was recently found in eight out of 82 wells sampled in central Pennsylvania (57). It was also found in one well at the detection limit in the previously mentioned Maryland study (58). Simazine has been found in the ground water of three states at levels typically ranging between 0.2 and 3.0 ppb.

Trichloropropane. 1,2,3-Trichloropropane (TCP) had been a by-product of the manufacture of dichloropropene/dichloropropane nematicides. It is uncertain whether this persistent and mobile impurity is still present in pesticide formulations. TCP has been found in small-scale and large-scale retrospective studies in California and Hawaii soil and ground water (67, 68). It was found at least 10 feet down in the soil profiles in Hawaii, and in wells in Oahu and the Central Valley of California. Positives typically range between 0.2 and 2 ppb in the soil and in ground water.

Concluding Remarks

The extent of the problem of pesticides in ground water is not known. What we have is a collection of monitoring studies, conceived with different objectives and using different design strategies, although most are various scales of retrospective studies. (Most of these have either been cited in the present work or previously (1)). Therefore the reader is cautioned about making generalizations about the results so far. For example, it would be easy to look at Figure 1 and conclude that Iowa has a worse ground-water problem than, say, Missouri. However, Figure 1 may only reflect the fact that the Iowa State government has a very active monitoring program, especially when one considers that Missouri has a potential high use of nematicides in permeable soils over solution limestone aquifers. Perhaps the next update of this general assessment will be more conclusive and will cite additional good monitoring studies which contain several of the key elements described earlier in this paper. Likewise, improvements in our predictive capabilities should enable us to be more conclusive in our assessments.

Acknowledgments

The authors gratefully acknowledge the assistance of the Statistics Team of OPP's Toxicology Branch for several helpful suggestions. Likewise, George DeBuchananne of DeBuchananne & Associates and Les McMillion of EPA-Las Vegas were invaluable sources of information on well construction and hydrogeology.

Literature Cited

1. Cohen, S.Z.; Creeger, S.M.; Carsel, R.F.; Enfield, C.G. In "Treatment and Disposal of Pesticide Wastes"; Krueger, R.F.; Seiber, J.N., Eds.; ACS SYMPOSIUM SERIES No. 259, American Chemical Society: Washington, DC, 1984; pp. 297-325.
2. Cohen, S.Z. "List of Potential Ground-Water Contaminants," Office of Pesticide Programs (TS-769C), Environmental Protection Agency, Washington, DC 20460, August 28, 1984.
3. Aller, L.; Bennett, T.; Lehr, J.H.; Petty, R.J. "DRASTIC: A Standardized System for Evaluating Ground Water Pollution Potential Using Hydrogeologic Settings," EPA/600/2-85/018, Robert S. Kerr Environmental Research Laboratory, Environmental Protection Agency, Ada, Oklahoma 74820.
4. Helling, C.S.; Gish, T.J. "Soil Characteristics Affecting Pesticide Movement Into Ground Water," 189th ACS National Meeting, PEST 2, Florida, 1985.
5. "1984 Herbicide Market Study," Doane Marketing Research, Inc., Princeton, New Jersey 08540; 1984 (also available are reports on insecticides and specialty crops).
6. "1982 Census of Agriculture, Volume 1, Geographic Area Studies," Bureau of the Census, Department of Commerce, 1984.
7. Cochran, W.G. "Sampling Techniques Second Edition"; John Wiley & Sons, Inc.: New York, 1967; Chaps. 2, 5.
8. Nelson, J.D.; Ward, R.C. Ground Water 1981, 19, 617-625.
9. Mason, R.E.; McFadden, D.D.; Iannachione, V.G.; McGrath, D.S. "Survey of DBCP Distribution in Ground Water Supplies and Surface Water Ponds," EPA 68-01-5848, Research Triangle Institute, Research Triangle Park, North Carolina 27709.
10. Williams, B. "A Sampler on Sampling"; John Wiley & Sons, Inc.: New York, 1978.
11. Snedecor, G.W.; Cochran, W.G. "Statistical Methods Seventh Edition"; The Iowa State University Press: Ames, Iowa, 1980; Chap. 21.
12. Helsel, D.R.; Ragone, S.E. "Evaluation of Regional Ground-Water Quality in Relation to Land Use: U.S. Geological Survey Toxic Waste-Ground-Water Contamination Program", 1984, Water Resources Investigations Report 84-4217, 33 pages.
13. "RCRA Permit Writer's Manual Ground-Water Protection 40 CFR Part 264, Subpart F," U.S. EPA, 1983.
14. "Field Agricultural Runoff Monitoring (FARM) Manual," Smith, C.N.; Brown, D.S.; Dean, J.D.; Parrish, R.S.; Carsel, R.F., U.S. EPA, EPA-600/3-85/043, 1985, 300 pages.
15. Porter, K. S.; Trautmann, N.M. "Seasonality in Ground Water Quality," EPA Office of Research and Development, Las Vegas, NV; in press.

16. Keith, S.J.; Wilson, L.G.; Fitch, H.R.; Esposito, D.M. Ground Water Monitoring Review 1983, 2, 21-32.
17. Hallberg, G.R.; Libra, R.D.; Bettis, A.E.; Hoyer, B.E. "Hydrologic and Water Quality Investigations in the Big Spring Basin," 1984, Iowa Geological Survey, Open-File Report, 231 pages.
18. Luhdorff, G.; Scalmanini, J. "Ground Water and the Unsaturated Zone Monitoring and Sampling," National Water Well Assoc.; unpublished training manual, 359-365.
19. Barcelona, M.J.; Gibb, J.P.; Miller, R.A. "A Guide to the Selection of Materials for Monitoring Well Construction and Ground-Water Sampling," 1983, Ill. State Water Survey, Contract Report 327.
20. Scalf, M.R.; McNabb, J.F.; Dunlap, W.J.; Cosby, R.L.; Fryberger, J. "Manual of Ground Water Sampling Procedures", National Water Well Association/Environmental Protection Agency Series.
21. "Quality of Water Branch Technical Memorandum No. 85.09," U.S. Department of the Interior Geological Survey, April 22, 1985.
22. Kurt, C.E.; Johnson, R.C. Ground Water. 1982, 20(4), 415-419.
23. Nacht, S.J. Ground Water Monitoring Review. 1983, 3, 23-8.
24. Feltz, H. U.S. Geological Survey, Reston, VA., July 1, 1985, personal communication.
25. Wilson, L.C.; Rouse, J.V. Ground Water Monitoring Review. 1983, 1, 103-108.
26. "Ground Water and the Unsaturated Zone Monitoring and Sampling", National Water Well Assoc.; unpublished training manual, pp. 257-262.
27. ibid. pp. 143-148.
28. Schuller, R.M.; Gibb, J.P.; Griffin, R.A. Ground Water Monitoring Review. 1981, Spring, 42-46.
29. Gibb, J.P.; Barcelona, M.J. J. Am. Water Works Assoc. 1984, May, 48-51.
30. Holden, P.W. "Primer on Well Water Sampling for Volatile Organic Compounds," Water Resources Research Center, University of Arizona, U.S. Department of the Interior, 1978.
31. Steinheimer, T. U.S. Geological Survey, Central Analytical Lab, Aravada, CO, July 9, 1985, personal communication.
32. Tourtelot, H.A.; Miesch, A.T. The Geological Society of America-Special Papers. 1975, 155, 107-119.
33. Krumbein, W.C.; Slack, H.A. Geol. Soc. Am. Bull. 1956, of 67(6), 739-762.
34. Hormann, W.D.; Karlhuber, B.; Ramsteiner, K.A. Proc. Eur. Weed Res. Coun. Symp. Herbicides-Soil. 12-3, 129-140.
35. Reed, J.F.; Rigney, J.A. J. Am. Soc. Agron. 1947, 39, 26-40.
36. Mason, B.J. "Preparation of Soil Sampling Protocol: Techniques and Strategies," U.S. Environmental Protection Agency, 1982.
37. Davis, S.N.; Thompson, G.M.; Bentley, H.W.; Stiles, G. Ground Water. 1980, 18, 14-23.
38. Tennyson, L.C.; Settergren, C.D. Water Resources Bulletin. 1980. 16, 433-437.
39. Thompson, G.M.; Hayes, J.M. Water Resources Research. 1979, 15, 546-554.

40. Jones, R.L.; Back, R.C. Environ. Toxicol. Chem. 1984, 3, 9-20.
41. Miles, C.J.; Delfino, J.J. J. Agric. Food Chem. 1985, 33, 455-460.
42. Porter, K.S.; Lemley, A.T.; Hughes, H.B.; Jones, R.L. In "Second International Conference on Ground-Water Quality Research Proceedings", National Center for Ground Water Research, Oklahoma State University, Stillwater, OK, in press.
43. Lemley, A.T.; Zhong, W. J. Agric. Food Chem. 1984, 32, 714-719.
44. Cohen, D.B. "Ground Water Contamination By Toxic Substances: A California Assessment," 189th ACS National Meeting, PEST 91, Florida, 1985.
45. "Dibromochoropropane; Intent to cancel Registrations of Pesticide Products Containing Dibromochloropropane (DBCP)", EPA Fed. Reg. 50, 1122-1130, 1985.
46. Cohen, S.Z. "DBCP Use on Certain Pineapple Fields on Maui - Implications for Potential Drinking Water Contamination," TS-769C, U.S. EPA, Washington, DC 20460, 1985.
47. "Hydrogeology, Water Quality, and Land Management in the Big Spring Basin, Clayton County, Iowa," Iowa Geological Survey Open File Report 83-3, Iowa City, IA, 1983.
48. "Ground Water Quality and Hydrogeology of Devonian-Carbonate Aquifers in Floyd and Mitchell Counties, Iowa," Iowa Geological Survey Open File Report 84-2, Iowa City, IA, 1984.
49. "Hydrogeologic and Water Quality Investigations in the Big Spring Basin, Clayton County, Iowa 1983 Water-Year," Iowa Geological Survey Open File Report 84-4, Iowa City, IA, 1984.
50. Kelly, R., Iowa Department of Water and Waste Management, Des Moines, IA, 1984 and 1985, personal communications.
51. McConnell, J.B.; Hicks, D.N.; Lowe, L.E.; Cohen, S.Z.; Jovanovich, A.P. "Investigation of Ethylene Dibromide (EDB) in Ground Water in Seminole County, Georgia," USGS Circular 933; Reston, VA, 1984.
52. Reich, A. Florida Dept. Health & Rehab. Serv. - Env'l Epi. Sect.; personal communications, 1985.
53. Jungclaus, G.; Cohen, S.Z. "Hydrolysis of Ethylene Dibromide," Environ. Sci. Technol., in review.
54. Weintraub, R.A.; Jex, G.W.; Moye, H.A. "Degradation of 1,2-Dibromoethane in Florida Ground Water and Soil," American Chemical Society 189th National Meeting, PEST 110, Florida, 1985.
55. "Agricultural Pesticide Residues in California Well Water: Development and Summary of a Well Inventory Data Base for Non-Point Sources", Environmental Hazards Assessment Program, Cal. Dept. Fd. Ag., Sacramento, CA, 1985.
56. Litwin, Y. "Groundwater Contamination by Pesticides: A California Assessment", Ramlit Assoc., Berkeley, CA, 1983.
57. Buchanan, J.W.; Loper, W.C.; Schaffstall, W.P.; Hainly, R.A. "Water Resources Data Pennsylvania Water Year 1983 Volume 2. Susquehanna and Potomac River Basins," USGS/WRD/HD-84/060, U.S. Geological Survey; Harrisburg, PA, 1984.

58. "Results of a Maryland Groundwater Herbicide Survey Fall 1983," Office of Environmental Programs, Dept. Health and mental Hygiene; Annapolis, MD, 1984.
59. Johnston, H.E., United States Geological Survey, Providence, RI, July 24, 1985, personal communication, and subsequent contacts with that office.
60. EPA/Office of Pesticide Programs confidential business information files.
61. "EDB Contamination Survey, Skagit County, WA," EPA - Drinking Water Branch, Seattle, WA; 1984.
62. "Results and Implications of the Investigation of Ethylene Dibromide in Ground Water in Western Washington," Washington Department of Social Services and Health, Water Supply & Waste Section, Olympia, WA, 1985.
63. Plews, G.; Baum, L., ibid., personal communication.
64. "Results of Ethylene Dibromide (EDB) Sampling in Arizona," Ambient Water Quality Surveys and Data Management, Division of Environmental Health Services, Phoenix, AZ, in press.
65. Wing, S.; Robinson, B.R., Connecticut Dept. Env. Protection; personal communications, 1984.
66. Higgins, J., Massachusetts Dept. Env. Qual. Engg. - Western Region, Springfield, MA; personal communication, 1984.
67. Wong, W., Arizumi, T., Hawaii Department of Health, personal communication, 1983.
68. Cohen, D., California State Water Resources Control Board, personal communication, 1985.

RECEIVED April 1, 1986

Field, Laboratory, and Modeling Studies on the Degradation and Transport of Aldicarb Residues in Soil and Ground Water

Russell L. Jones

Union Carbide Agricultural Products Company, Inc., Research Triangle Park, NC 27709

Research conducted on the movement and degradation of aldicarb residues in the unsaturated and saturated zones has shown that it is a complex process affected by soil and hydrogeological properties, climatic conditions, and agricultural practices. This paper presents the results of unsaturated and saturated zone field studies conducted in 16 states over a period of six years in which approximately 20,000 soil and water samples have been collected. Results from laboratory degradation studies are also included. Computer modeling has been used to illustrate the effects of variables such as soil field capacity, soil organic matter, pesticide application timing, and climatic conditions on the potential for aldicarb residues to reach groundwater. These experimental and modeling studies show that in most areas aldicarb residues degrade in the upper portion of the unsaturated zone. In the relatively few areas where aldicarb residues reach groundwater, the primarily lateral movement of groundwater and the continuing degradation usually limit the presence of aldicarb residues to shallow groundwater near treated fields.

The development of modern analytical methodology has made possible the detection and quantification of extremely low concentrations of chemicals. The application of this technology to groundwater analyses made possible the detection of trace levels of aldicarb residues in Long Island groundwater in 1979 (1). This finding prompted research on the environmental fate of water-soluble pesticides by scientists in universities, regulatory agencies, and agricultural chemical manufacturers. Much of this research focused on aldicarb and two of its metabolities (aldicarb sulfoxide and aldicarb sulfone). Although much independent research, especially laboratory studies, has been performed (samples of such work are included in references 2-7), this paper will focus on laboratory and field research work in which Union Carbide scientists have participated often in cooperation with scientists from universities or regulatory agencies.

0097-6156/86/0315-0197$06.50/0
© 1986 American Chemical Society

Summary of Experimental Methodology

Unsatured Zone Studies. These experiments began in 1980 at special interest areas in the United States. In each area, two to five locations, which were grower-treated fields, were sampled once or twice after application. Soil cores were taken by hand auger at three sites per location with sampling of strata from 0-0.3, 0.3-0.6, 0.6-1.2, 1.2-1.8, and 1.8-2.4 m. In 1981, the number of sampling intervals at each location was increased to provide a better estimate of degradation and movement. To reduce the effects of the inherent variability associated with field studies, several sampling changes were made beginning in 1982. Test fields were carefully selected in high use or potential use areas. Application was supervised by university or company personnel. Each treatment was replicated in four subplots and four cores were taken in each subplot by a special bucket auger technique designed to reduce contamination between soil strata. In a typical study, about 100 soil samples would be taken at a given sampling interval in each plot representing a single treatment. Field studies included from three to seven sampling intervals after application. On the average, one area would include two treatments and yield 800 soil samples per year to provide sufficient information for estimation of degradation and movement parameters. However, in 1984 a study was conducted to assess the effect of spatial variability on measurement of degradation and movement. This study involved 64 cores per interval and the collection and analysis of over 3,100 soil samples.

Saturated Zone Studies. Beginning in 1983, saturated zone studies were conducted using several clusters of test wells screened at various depths and located upgradient, within, and downgradient of the treated area. A minimum of five clusters of three wells each is needed at any one study area to effectively assess movement and degradation of residues in the groundwater. At one test site in Florida, approximately 160 test wells were installed and have been sampled on a monthly basis. Two limited saturated zone studies have also been performed by analyzing water taken from the outlet of tiles draining treated fields.

In addition to saturated zone research studies, about 30,000 potable well water samples from 32 states have been collected. Although not specifically covered by this paper, insights gained from these analyses are included in the discussion sections.

Laboratory Studies. A variety of laboratory studies have been performed including degradation rate studies with actual samples from the saturated and unsaturated zones, oxidation mechanism studies, saturated zone degradation mechanism studies, potential for sulfoxide or sulfone reduction studies, and distilled water hydrolysis studies. The experimental methodology in these experiments varied according to the study objectives.

Degradation

Degradation Pathway. The degradation pathway of aldicarb (Figure 1) is the same in plants, animals, and soil. First, aldicarb is oxidized to aldicarb sulfoxide. A portion of the aldicarb sulfoxide

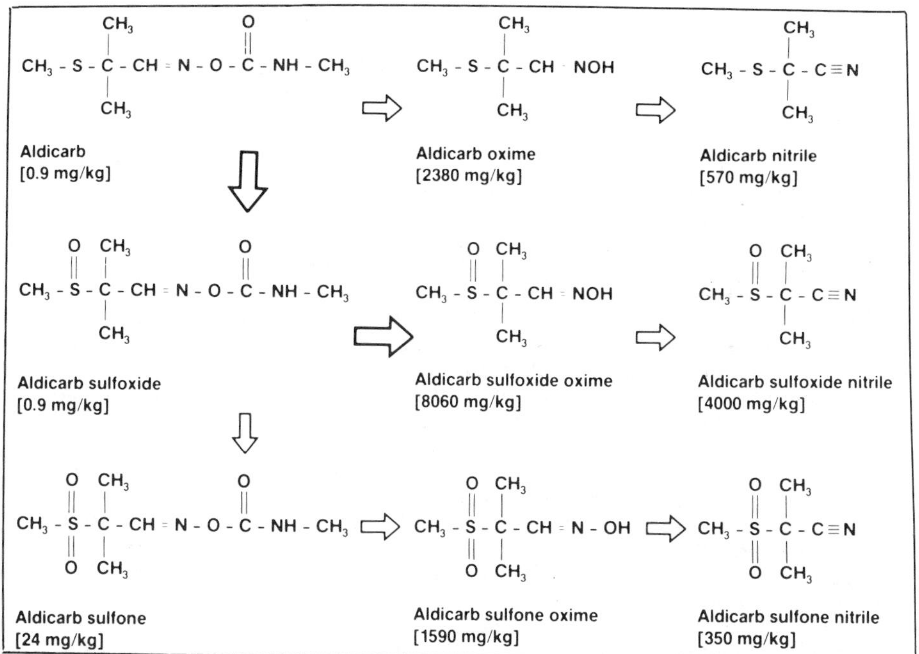

Figure 1. Degradation pathway for aldicarb (values in parentheses are acute oral LD_{50}'s for rats).

is further oxidized to aldicarb sulfone. Both aldicarb sulfoxide and aldicarb sulfone, biologically active carbamates, are concurrently degraded to low toxicity (non-carbamate) compounds. Although reduction of aldicarb sulfoxide has been demonstrated under laboratory conditions (8), field data demonstrate that neither the sulfoxide nor the sulfone are reduced back to parent aldicarb or to sulfoxide in the saturated or unsaturated zones. In this paper, the term aldicarb residues refers to the sum of aldicarb, aldicarb sulfoxide, and aldicarb sulfone.

Many factors influence the degradation of aldicarb residues to biologically inactive compounds. Some of the more important include temperature and pH, and the presence of moisture and microbial populations. The following paragraphs discuss these in more detail for degradation occurring in both the unsaturated and saturated zones.

Unsaturated Zone Degradation. After application, soil moisture rapidly dissolves aldicarb in the granule. Once in solution, the degradation processes begin immediately. The oxidation process is relatively rapid and little or no parent aldicarb exists a month after application. The degradation of aldicarb sulfoxide and aldicarb sulfone is the result of both microbial and chemical action. Especially in acidic soils, microbial degradation is believed to be the most important mechanism in the root zone, or the upper strata of the unsaturated zone. Factors which tend to increase the degradation rate of aldicarb residues include microbial populations, high temperatures, high soil pH and high soil moisture content.

The degradation rate of aldicarb residues in the unsaturated zone has been determined from field studies conducted since 1982 in ten different states (9-15) (Table I) where half-life ranged between two weeks and three months. Data from these sites indicate that soil temperature appears to be the most dominant variable affecting half-life of aldicarb residues under normal agricultural conditions in the unsaturated zone, with increasing temperatures resulting in faster degradation. Once below the root zone, no decrease in the degradation rate with depth was observed except in acidic, sand subsoils. Because of the decrease in microbial population with depth, this relatively constant degradation rate as a function of depth indicates that soil catalyzed chemical hydrolysis may be an important degradation mechanism even in the root zone of many soils.

Saturated Zone Degradation. Because of the rapid oxidation process in the root zone, parent aldicarb is rarely detected in the saturated zone. In rare instances where transport from the soil surface is rapid, aldicarb may be present at less than five percent of the total residues found. In the saturated zone, residues are usually a mixture of aldicarb sulfoxide and aldicarb sulfone in an average ratio of 3:2.

If aldicarb residues leach into the saturated zone, or groundwater, degradation of the residues continues, mainly by chemical hydrolysis in cold areas and by both chemical hydrolysis and microbial degradation in warm areas. Factors which tend to increase the degradation rate are high temperature and high pH. Microbial populations may provide a significant contribution to saturated zone degradation, especially in shallow groundwater in warm areas. Half-life degradation rates in the saturated zone, as measured in the laboratory (Table II), range from about three days for east coastal Florida to several years for eastern Long Island, New York. In these experiments, the measured degradation rates are 3 to 40 times faster than would be predicted on the basis of distilled water hydrolysis.(16) This indicates that even in deeper groundwater some catalytic factors (microbes, metals, surface sorption effects) are present. Preliminary data from an ongoing mechanism study indicate that the rate increase is mostly due to the presence of soil rather than microbes. Work by the University of Wisconsin has indicated a half-life of six months to one year in Wisconsin groundwater (5-10°C) which agrees with temperature-extrapolated laboratory data using actual field samples.

Movement

The driving force for movement of aldicarb residues is the movement of water. Therefore, the movement of water must be defined before the movement of aldicarb residues can be quantified. Factors influencing the movement of water include rainfall, irrigation, evapotranspiration and soil field capacity in the unsaturated zone. The amount of organic matter determines the rate of aldicarb movement relative to water movement in both the unsaturated and saturated zones.

Movement of Aldicarb from the Granule. Aldicarb is released from the applied granules upon contact with soil moisture. If the ground is extremely dry, no aldicarb will be released until the ground is moistened by rainfall or irrigation. In arid climates, irrigation after aldicarb application is recommended. Under most agricultural conditions, aldicarb is released from the granule within hours by normal soil moisture.

Unsaturated Zone Movement. Because aldicarb, aldicarb sulfoxide, and aldicarb sulfone do not significantly bind to inorganic soil (10,17,18) and are soluble in water, aldicarb residues move with soil water in both the unsaturated and saturated zones. Surface soil, the upper layer of the unsaturated zone, 0.3 to 1.5 m deep depending on the crop, is often termed the root zone. In the root zone, aldicarb residues move downward with water from rain or irrigation. Conversely, plant transpiration of water will tend to retain residues in the root zone. Evaporation of soil moisture from the land surface will draw soil moisture and aldicarb residues from the root zone toward the surface. Movement of aldicarb residues in the lower strata of the unsaturated zone is generally in a vertical direction. However, soil structures such as clay lenses or hardpans may result in some horizontal movement. Aldicarb residues do sorb to organic matter and in some high organic matter soils the rate of residue movement may be up to a factor of ten slower than the rate of water movement. Factors which increase downward movement are a high recharge rate (rainfall plus irrigation minus evapotranspiration), low soil field capacity, and low soil organic matter. In many cases, movement in the unsaturated zone is the result of a few relatively heavy rains.

Table I. Aldicarb and Aldicarb Sulfone

	Crop
Arizona	
1982 Maricopa	Cotton
Aldicarb at Emergence	
Aldicarb at Planting and Emergence	
Aldicarb Sulfone at Planting	
Aldicarb Sulfone at Emergence	
California	
1980 Exploratory Studies at Two Sites (Aldicarb)	Cotton
1984 Manteca (Aldicarb)	Tomatoes
Livingston (Aldicarb)	Grapes
Fresno (Aldicarb)	Grapes
Colorado	
1984 Exploratory Studies at Three Sites (Aldicarb)	Potatoes
Florida	
1980 Exploratory Studies at Five Sites (Aldicarb)	Citrus
1981 Exploratory Studies at Five Sites (Aldicarb)	
1982 Indiantown (Aldicarb)	
1983 Lake Hamilton (Aldicarb)	
Oviedo (Aldicarb)	
Alcoma (Aldicarb)	
Indiantown (Aldicarb)	
Lutz (Aldicarb)	
Lake Buena Vista (Aldicarb)	
DeLeon Springs (Aldicarb)	
Fort Pierce (Aldicarb)	Tomatoes
1984 Davenport (Aldicarb)	Citrus
Lake Hamilton (Aldicarb	
Alcoma (Aldicarb)	
Lutz (Aldicarb)	
DeLeon Springs (Aldicarb)	
Monitoring Wells (Aldicarb)	
1985 Lake Hamilton (Aldicarb)	
Lutz (Aldicarb)	
DeLeon Springs (Aldicarb)	
Monitoring Wells (Aldicarb)	

*Average concentration in soil cores or water samples

Field Research Programs - 1980-July, 1985.

No. of Samples Soil	Water	Application Rate kg/ha	Unsaturated Zone Half-Life (months)	Maximum Leaching Depth (m)*
1176				
		2.24	0.5	1.8
		1.12,2.24	0.3, 0.8	1.8
		3.36	0.5	1.8
		2.24	0.3	1.8
28	4	0.61-2.5		1.2
584		3.36	1.5	1.8
351	247	4.48	1.5	1.2
759		4.48	1.5	3.0
40	3	3.36	--	1.2
141	23	5.6-11.2	--	2.4
445	88	5.6-11.2	--	2.4
	31	7.56		
575	298	11.2	0.6	3.0
354	171	11.2	0.6	1.2
	146	--		
	91	7.56		
	55	--		
	18	--		
	50	--		
224	20	3.36	--	0.6
3127		5.6	--	6.0
	1187	--		
	212	--		
	49	--		
	29	--		
	44	5.6		
	820	--		
	12	--		
	13	--		
	73	5.6		

is less than 5 ppb below this depth.

Continued on next page

Table I. Aldicarb and Aldicarb Sulfone

	Crop
<u>Indiana</u>	
1983 Bluecast (Aldicarb)	Corn
<u>Maine</u>	
1983 Presque Isle (Aldicarb)	Potatoes
Planting Application	
Emergence Application	
<u>Michigan</u>	
1983 Blissfield (Aldicarb)	Corn
<u>Nebraska</u>	
1985 Bartlett (Aldicarb)	Corn
<u>New York</u>	
1983 Phelps (Aldicarb)	Potatoes
Planting Application	
Emergence Application	
<u>North Carolina</u>	
1983 Harrellsville	Tobacco
Aldicarb at Transplanting	
Aldicarb Sulfone at Transplanting	
<u>Oregon</u>	
1980 Exploratory Studies at One Site (Aldicarb)	Potatoes
<u>South Carolina</u>	
1985 Edisto (Aldicarb)	Soybeans Bare Plot
<u>Texas</u>	
1980 Exploratory Studies at Three Sites (Aldicarb)	Citrus

*Average concentration in soil cores or water

Field Research Programs - 1980-July, 1985 (continued).

No. of Samples Soil	Water	Application Rate kg/ha	Unsaturated Zone Half-Life (months)	Maximum Leaching Depth (m)*
208	20	1.68	1.1	0.6
1440				
		3.36	3.3	1.5
		2.24	2.8	1.2
192	16	1.68	0.7	0.6
111	51	1.7	--	--
456				
		3.36	1.0	0.3
		2.24	0.9	0.3
656				
		3.36	1.3	0.6
		3.36	0.9	0.6
15	1	3.36	--	1.2
135		3.4		--
136	143	3.4	--	--
41	7	5.6	--	1.8

samples is less than 5 ppb below this depth.

Continued on next page

Table I. Aldicarb and Aldicarb Sulfone Field

	Crop
Virginia	
1980 Exploratory Studies at Two Sites (Aldicarb)	Potatoes
1981 Exploratory Studies at Three Sites (Aldicarb)	Potatoes
1983 Blackstone	Tobacco
Aldicarb at Transplanting	
Aldicarb Sulfone at Transplanting	
Washington	
1980 Exploratory Studies at Two Sites (Aldicarb)	Potatoes
1983 Pasco (Aldicarb)	
Wisconsin	
1980 Exploratory Studies at Five Sites (Aldicarb)	Potatoes
1981 Exploratory Studies at Six Sites (Aldicarb)	
1982 Hancock (Aldicarb)	
Emergence Application with Moderate Irrigation	
Planting Application with Moderate Irrigation	
Emergence Application with Heavy Irrigation	
Planting Application with Heavy Irrigation	
Cameron (Aldicarb)	
Emergence Application	
Planting Application	
1983 Hancock	
Aldicarb at Emergence	
Aldicarb at Planting	
Aldicarb Sulfone at Planting	

*Average concentration in soil cores or water samples

Research Programs 1980 - July, 1985

No. of Samples Soil	No. of Samples Water	Application Rate kg/ha	Unsaturated Zone Half-Life (months)	Maximum Leaching Depth (m)*
39	3	3.36	--	1.2
118	27	3.36	--	
664				
		3.36	1.1	0.6
		3.36	1.3	0.6
24	5	3.36	--	0.6
448		6.72	1.7	1.8
127	31	3.36	--	2.4
505	65	3.36	--	2.4
1230	10			
		2.24	1.2	1.8
		3.36	1.7	2.4
		2.24	--	1.8
		3.36	--	3.0
338				
		2.24	1.5	1.2
		3.36	2.0	1.2
1094	21			
		2.24	1.3	1.8
		3.36	0.9	0.6
		3.36	1.1	2.4

is less than 5 ppb below this depth.

TABLE II. Results of Laboratory Studies Measuring Degradation Rates of Aldicarb Residues in Saturated Zone Samples

Sample Location	Compounds Studied	Approximate pH of Sample	Study Temp. °C	Calculated Half-Life in Days [1]
New York				
Long Island	Sulfoxide & Sulfone	6	13	800[2]
			25	360(243-704)[2]
Wisconsin Central Sands	Sulfoxide & Sulfone	7	25	14 (12,16)
	Sulfone only	7	25	13 (10,16)
North Carolina Coastal Plain	Sulfoxide & Sulfone	5	25	137(117,165)[2]
	Sulfone only	5	25	108(63,368)[2]
Florida				
Lutz	Sulfoxide & Sulfone	6	25	49 (41,61)
Lake Hamilton	Sulfoxide & Sulfone	6	25	47 (40,55)
Alcoma	Sulfoxide & Sulfone	6	25	49 (38,63)
Oveido	Sulfoxide & Sulfone	7	25	3 (1,5)
Fort Pierce	Sulfone only	7	25	52 (40,74)
California Livingston	Sulfoxide & Sulfone	7	25	8 (7-10)
Colorado San Luis Valley	Sulfoxide & Sulfone	8	25	23 (17,35)

(1) Numbers in parenthesis are the 95% confidence limits.

(2) Variation in later sampling intervals makes estimation of these values difficult.

The amount, if any, of aldicarb residues leaching to a specified depth depends both on the rate of degradation and movement. For example, the same amount of leaching could occur in a situation where the degradation rate and the rate of movement are relatively slow, as in a situation where both rates are relatively rapid. As discussed earlier, the rate of degradation of aldicarb residues in the unsaturated zone is relatively fast (half-life of 0.5 to 3 months). Therefore, leaching of residues through the unsaturated zone to groundwater occurs only occasionally and then only in areas with relatively shallow water tables (where transport to the water table may be accomplished quickly) or in areas with acidic sand subsoils where degradation rates are slower (such as portions of Long Island or central Florida with relatively deep water tables).

The depth to which detectable leaching of aldicarb residues occurred in field studies is listed in Table I. Leaching beyond three meters occurred in the ridge area of Florida (specifically Lake Hamilton, Alcoma, Davenport) and with planting applications in the Central Sands of Wisconsin (Hancock). Other work has shown leaching below three meters on Long Island. (16,19) The depth to which aldicarb residues may leach in a given site may vary from year to year depending on the amount and distribution of rainfall and irrigation. The timing of an application relative to weather (rainfall and soil temperature) may affect both the rate of movement and the degradation rate, as demonstrated in Wisconsin (9), Florida (20), and the northeastern United States (11,12).

Saturated Zone Movement. The movement of aldicarb residues in the saturated zone is in the same direction as groundwater movement. In most areas, this means that if aldicarb residues reach the saturated zone, they will move primarily in a horizontal direction. Since there is usually little organic matter in the saturated zone, aldicarb residues travel at the same speed as the groundwater, generally 0.03 to 0.5 m per day. The residues tend to remain near the top of the saturated zone and vertical movement is slow.

In only a few areas where aldicarb is used have residues actually been found to move through the unsaturated zone before being completely degraded. In these areas, the potential for presence of residues in drinking water wells depends on the amount of residues entering groundwater, the degradation rate in groundwater, the rate of groundwater movement and the location, casing depth and integrity of drinking water wells near treated fields. In most areas, continued degradation in the saturated zone and the slow horizontal movement accompanied by dispersion tend to result in aldicarb residues being constrained to shallow groundwater near treated areas.

Model Simulation

Unsaturated Zone Models. Various models exist for estimating the movement and degradation of pesticide residues in the unsaturated zone (20). Perhaps the best model available is PRZM developed by the U.S. EPA (21). The applicability of PRZM to the modeling of aldicarb residues has been demonstrated using data from Long Island (22), Florida (20), Wisconsin (23,23,24), and North Carolina (24).

One of the most important uses of an unsaturated zone model is to illustrate the effects of variables. In Figure 2, PRZM has been used to illustrate the effect of organic matter, soil hydraulic properties, and degradation rate on the amount of leaching. The input parameters listed in Table III have been used as a basis for these simulations. The environmental conditions of this example are similar to those of the ridge citrus growing area in central Florida. The base case illustration chosen was one where potential for movement to the water table is high. Downward movement is rapid due to high rainfall, extremely low field capacity for the soil, and low sorption to soil. As Figure 2 shows, even small changes in percent organic matter or field capacity will significantly affect the distance residues move before being decomposed. Small changes in degradation rate have a significant impact as well.

These types of simulations help explain the differences seen in Florida field studies in the magnitude of residues found in shallow groundwater under treated citrus groves in coastal and ridge areas. The higher soil field capacity and higher organic matter content typical of eastern coastal areas results in much lower residues entering shallow groundwater compared to the ridge area. Simulations using the actual parameters for coastal citrus (and other southeastern locations) coupled with experimental data confirm that the central Florida ridge area is more sensitive to leaching of aldicarb residues to groundwater compared to other southeastern agricultural areas.

Because both the amount and distribution of rainfall and evapotranspiration change from year to year at a specific site, modeling can help generalize experimental results. One method for illustrating the effect of year-to-year weather variations is to perform a simulation with several consecutive years of rainfall. The results of this simulation can then be summarized in a cumulative probability distribution curve. The results of a 17-year simulation (1965-1982 with 1978 excluded because of missing weather data) for the Lake Hamilton, Florida, location using the input parameters in Table III are shown in Figure 3. This figure shows that 50 percent of the time the amount of aldicarb residues leaching below 450 cm is less than 7 percent of that applied when the application date was February 15. When the aplication data was June 15, 30 percent of the applied residues leached below 450 cm. These differences are a result of the uneven distribution of rainfall during the year. Therefore, an application on February 15 (during the less rainy portion of the year) will, on the average, result in less residue leaching below a specified depth, than a June 15 application (during the rainy portion of the year).

Saturated Zone Models. Results from unsaturated zone simulations can be used as inputs to saturated zone models to predict concentrations of aldicarb residues in groundwater. The saturated zone model used by the author takes the pesticide inputs into groundwater, as predicted by PRZM, and calculates the concentration and movement of aldicarb residues in the upper portion of the saturated zone. The core of the saturated zone model is a finite element solute transport calculation procedure developed at the University of Wisconsin (25). The accuracy of this model in estimating pesticide movement in groundwater is (as with other

A

B

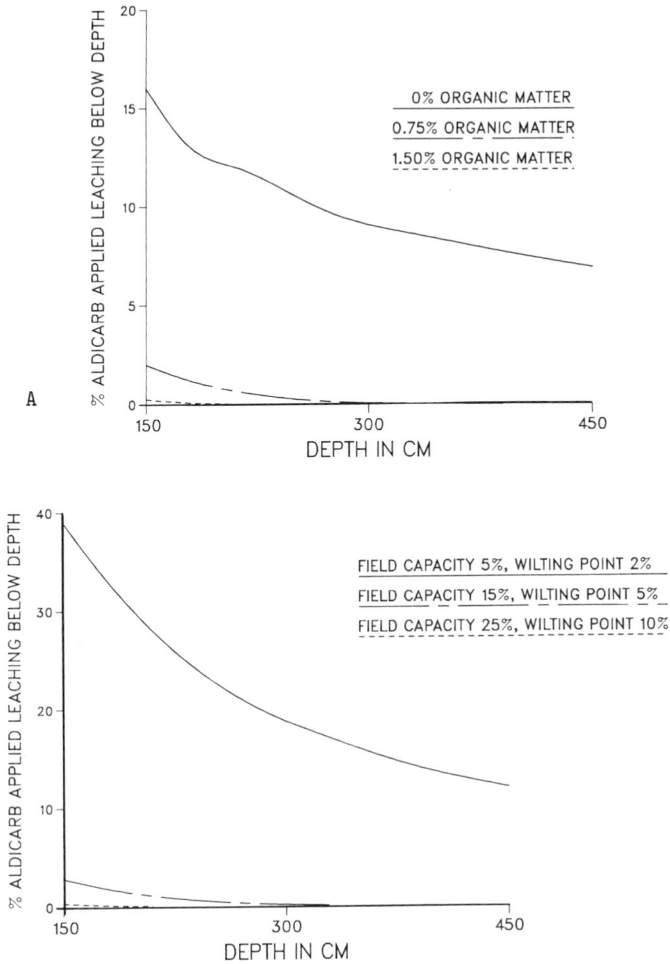

Figure 2(A,B). Simulated effect of organic matter (A) and soil hydraulic properties (B) on the leaching of aldicarb residues under Florida Ridge conditions. Continued on next page.

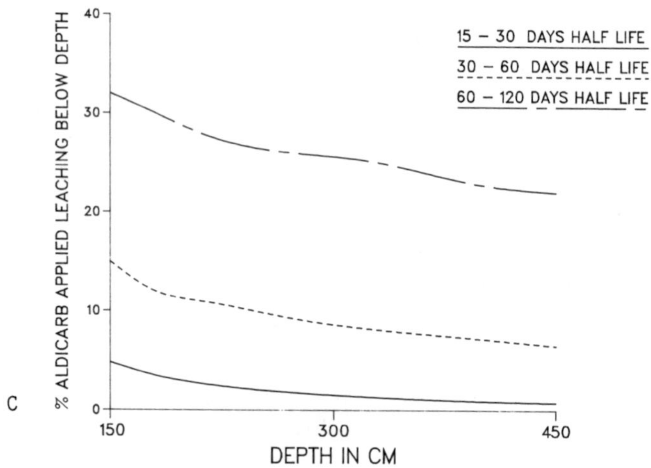

Figure 2(C). Simulated effect of dehydration rate on the leaching of aldicarb residues under Florida Ridge conditions.

TABLE III. PRZM Base Case Using Florida Ridge Conditions

Daily Rainfall Data:		1983 Lake Alfred Station
Daily Evaporation Data:		1983 Lake Alfred Station (Pan Factor of 1.0)
Soil Hydraulic Properties:		
Soil Strata (cm)	Field Capacity (vol. percent)	Wilting Point (vol. percent)
0-15	6.7	1.8
15-30	5.3	1.3
30-45	6.6	1.2
45-60	4.3	0.9
60-75	4.9	0.9
75-90	4.1	0.6
90-450	3.8	0.8
SCS Curve Numbers:	47, 67, 83, 59, 77, 89, 40, 60,78	
Root Zone Depth (cm):	150	
Pesticide Degradation Rate (half life in days):		
0-150 cm	30	
150-300 cm	30-60 (interpolate with depth)	
Below 300 cm	60	
Pesticide Soil Sorption Coefficient:		
0-15 cm	0.08	
15-30 cm	0.08	
30-60 cm	0.03	
Below 60 cm	0.00	
Plant Uptake of Pesticide:	Not considered	
Pesticide Application:	11.21 kg/ha on February 16, 1983	

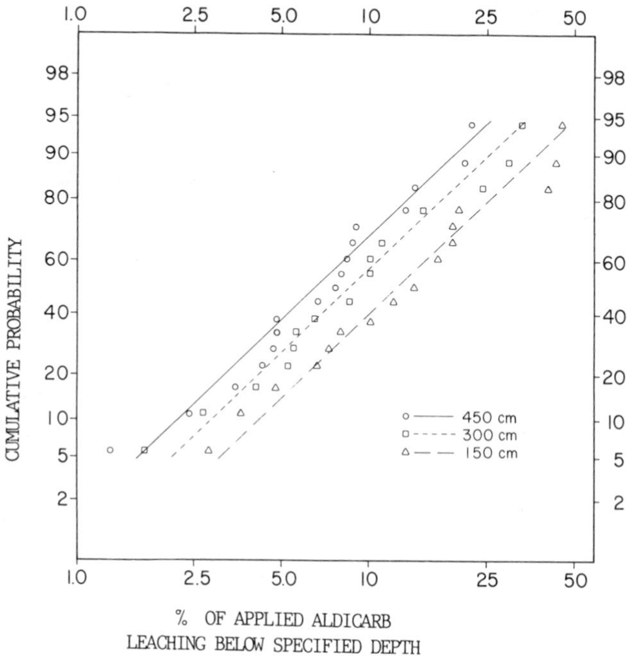

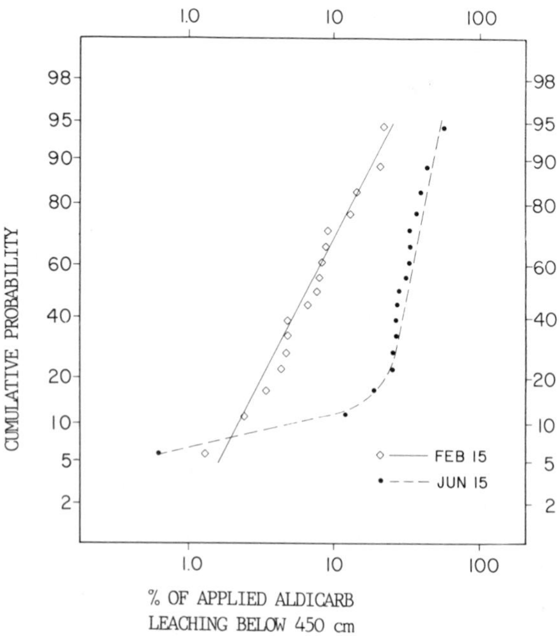

Figure 3. Simulated cumulative probability distributions showing the effect of weather (as a function on depth) and application timing on the leaching of aldicarb residues.

saturated zone models) largely dependent on the accuracy of the input parameters.

A simplified example is presented in Figure 4 which helps illustrate how a saturated zone model can be used in simulating aldicarb residue concentrations. This case assumes a starting concentration of 100 ppb in groundwater under a treated area, a groundwater flow rate of 0.15 m/day, and a degradation rate corresponding to a half-life of six months. These conditions are similar to those encountered in much of the ridge area of Florida. Under these conditions, 18 months later, aldicarb residues will be present up to 82 m downstream of the treated area at a concentration of about 12 ppb.

The results of these model simulation projections are consistent with results obtained from field experimental studies not only in Florida, but at other locations. For example, in the Florida ridge citrus region, aldicarb residues have not been detected in shallow groundwater at distances of more than 200-300 m downgradient of treated groves. Experimental results also demonstrate that under these conditions, the aldicarb residues are constrained to the upper 3-6 m of groundwater due to degradation and lateral groundwater flow. In the coastal regions of Florida, aldicarb residues are constrained to the shallow groundwater within the perimeter of the treated fields or groves due to the more rapid rate of degradation.(10,26) The results of these experimental and model simulation studies helped provide a basis for state regulations on the use of aldicarb in Florida.

Conclusions

The environmental data base for aldicarb is quite extensive, consisting of about 20,000 soil and water samples from 38 research studies and 30,000 potable well analyses. These data indicate that aldicarb residues degrade in both the unsaturated and saturated zones with the rate dependent on a variety of environmental parameters. In the unsaturated zone, the degradation rate corresponds to a half-life of about two weeks to three months with the slower degradation rates occurring in colder areas. The degradation rate in the saturated zone corresponds to half-lives ranging from a few days to a few years. Degradation rates are most rapid in warm, alkaline groundwater.

The movement of aldicarb residues is quite complex, depending on a number of interacting factors. In most aldicarb use areas, residues degrade completely before moving through the unsaturated zone and into the saturated zone. In the few areas where aldicarb residues have entered the saturated zone, residues are usually located in shallow groundwater near treated fields.

When required input parameters are available, unsaturated and saturated zone models are good tools for illustrating the effects of various combinations of variables, applying experimental data to field situations, selecting potential worst case situations for further assessment, and determining the effect of management practices (such as changes in application timing) on residue movement. The applicability of the unsaturated zone model to situations involving aldicarb has been demonstrated over a wide

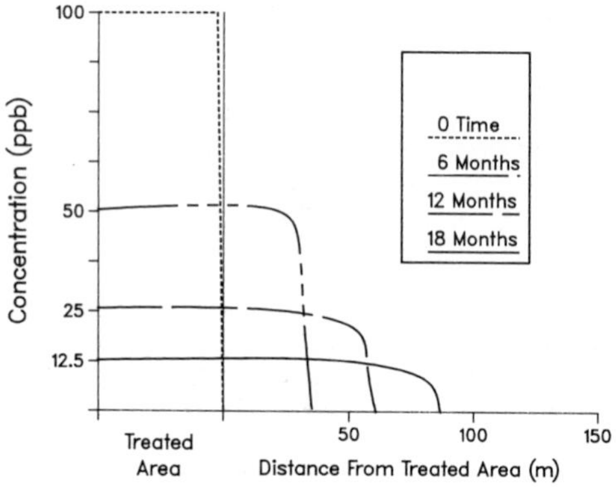

Figure 4. Illustration of saturated zone modeling under Florida ridge conditions.

range of conditions and the accuracy of both unsaturated and saturated zone models predictions is enhanced by the availability of extensive data on sorption and degradation rates.

Acknowledgments

Many scientists from universities and regulatory agencies have contributed to research programs that have been described in this paper. I would like to specifically acknowledge the contributions of: M. P. Anderson, R. F. Carsel, K. S. Porter, P. S. C. Rao, J. A. Wyman, D. Curwin, J. M. Harkin, G. Chesters, A. G. Hornsby, W. B. Wheeler, P. Nkedi-Kizza, J. McNeal, R. J. Wagenet, H. B. Hughes, R. V. Rourke, E. N. Lightfoot, P. S. Thorne, A. P. Lemley, D. A. Komm, D. Powell, R. O. Hegg, and P. Stoffela. I would also like to acknowledge the work of the many people at Union Carbide who participated in these studies.

References

1. Zaki, M. H., D. Moran, D. Harris, Am. J. Public Health 1982, 72, 1391-1395.
2. Smelt, J. H., M. Leistra, W. H. Houx, A. Dekker, Pestic. Sci. 1982, 13, 475-483.
3. Nicholls, P. H., R. H. Bromilow, T. M. Addiscott, Pestic. Sci. 1982, 13, 475-483.
4. Pacenka, S., K. S. Porter, "Preliminary Assessment of the Environmental Fate of Potato Pesticide, Aldicarb, in Soil and Ground Water" Eastern Long Island, New York, U.S. EPA Contract 8006876-02.
5. Lemley, A. J., W. Z. Zhong, J. Environ. Sci. Health 1983, B 18 (2), 189-206.
6. Rothschild, E. R., R. J. Manser, M. P. Anderson, Ground Water 1982, 20, 437-445.
7. Dierberg, F. E., "Aldicarb Studies in Groundwaters from Citrus Groves in Indian River County, Florida," Florida Water Resources Research Center Publication No. 76, 1984.
8. Miles, C. J., J. J. Delfino, J. Agric. Food Chem., 1985, 33, 455-460.
9. Wyman, J. A., J. O. Jensen, D. Curwen, R. L. Jones, T. E. Marquardt, "Effects of Application Procedures and Irrigation on Degradation and Movement of Aldicarb Residues in Soil." Accepted for publication, 1985.
10. Hornsby, A. G., P.S. C. Rao, W., B. Wheeler, P. Nkedi-Kizza, R. L. Jones, "Fate of Aldicarb in Florida Citrus Soils: 1. Field and Laboratory Studies," presented at Characterization and Monitoring of the Vadose (Unsaturated) Zone, Las Vegas, December 8-10, 1983.
11. Jones, R. L., R. V. Rourke, J. L. Hansen, "Effect of Application Methods on Movement and Degradation of Aldicarb Residues in Maine Potato Fields." Accepted for publication, 1985.
12. Wagenet, R. J., K. S. Porter, R. L. Jones, T. E. Marquardt, "Aldicarb Movement in the Field Soil of Upstate New York." Submitted for publication, 1985.

13. Wyman, J. A., J. Medina, D. Curwen, J. L. Hansen, R. L. Jones, "Movement of Aldicarb and Aldoxycarb Residues in Soil." Submitted for publication, 1985.
14. Jones, R. L., J. L. Hansen, R. R. Romine, T. E. Marquardt, "Unsaturated Zone Studies on the Degradation and Movement of Aldicarb and Aldoxycarb Residues." Submitted for publication, 1985.
15. Jones, R. L., "Central California Studies on the Degradation and Movement of Aldicarb Residues," July, 1985 (to be submitted for publication).
16. Porter, K. S., A. T. Lemley, H. B. Hughes, R. L. Jones, "Developing Information on Aldicarb Levels in Long Island Groundwater," paper presented at the Second International Conference on Groundwater Quality Research, Tulsa, March 26-29, 1984.
17. Supak, J. R., A. R. Swoboda, J. B. Dixon, Soil Sci. Soc. Am. J., 1978 42, 244-247.
18. Bromilow, R. H., Ann. Appl. Biol. 1973, 75, 473-479.
19. Martin, J. M., "Long Island Deep Soil Boring for Aldicarb Residues," Suffolk County Department of Health, 1981.
20. Jones, R. L., P.S.C. Rao, A. G. Hornsby, "Fate of Aldicarb in Florida Citrus Soil: 2. Model Evaluation," paper presented at Characterization and Monitoring of the Vadose (Unsaturated) Zone, Las Vegas, December 8-10, 1983.
21. Carsel, R. F., C. N. Smith, L. A. Mulkey, J. D. Dean, P. Jowise, "Users Manual for the Pesticide Root Zone Model (PRZM) Release 1", EPA-60013-84-109, December 1984.
22. Carsel, R. F., L. A. Mulkey, M. N. Lorber, L. B. Baskin, "The Pesticide Root Zone Model (PRZM): A Procedure for Evaluating Pesticide Leaching Threats to Groundwater," accepted for publication, Ecological Modeling 1985.
23. Jones, R. L., "Movement and Degradation of Aldicarb Residues in Soil and Groundwater," presented at the Environmental Toxicology and Chemistry Conference on Multidisciplinary Aproaches to Environmental Problems, Crystal City, VA. 1983.
24. Lorber, M. N., "A Method for the Assessment of Groundwater Contamination Potential Utilizing PRZM - A Pesticide Root Zone Model for the Unsaturated Zone, presented at the ACS Symposium on Evaluation of Pesticides in Groundwater, Miami, April, 1985.
25. Wang, H. F., M. P. Anderson, "Introduction to Groundwater Modeling," W. H. Freeman, San Francisco, 1982.
26. Union Carbide, unpublished Florida unsaturated and saturated zone monitoring data, 1983 - 1985.

RECEIVED April 1, 1986

12

Fate of Aldicarb in Wisconsin Ground Water

John M. Harkin, Frank A. Jones, Riyadh N. Fathulla, E. Kudjo Dzantor, and David G. Kroll

Department of Soil Science, University of Wisconsin–Madison, Madison, WI 53706

Small amounts of aldicarb sulfoxide and sulfone leach through sandy soils to shallow groundwater in Central Wisconsin fields where aldicarb is applied to irrigated potatoes. Leachate residue concentrations decrease with application rates and later application time. The pattern of leaching through soil and the levels detectable in groundwater are invariably highly erratic. Leaching occurs mainly during the late summer and fall of years of application; highest residue concentrations in groundwater occur later in a zone near the water table. These levels dissipate as this zone is overlain with uncontaminated leachate, but residues disappear at rates faster than can be ascribed to dilution and dispersion due to natural groundwater movement. The latter is complicated by water table fluctuations and flow perturbation caused by water extraction for irrigation. Residue attenuation is apparently accelerated by hydrolysis in the alkaline groundwater and degradation by a variety of groundwater-inhabiting bacteria. The groundwater alkalinity increases with depth and fluctuates seasonally, rising in summer with CO_2 in the water, which dissolves dolomitic limestone in the aquifer sediments. Rates of breakdown presumably vary with the groundwater temperature, which ranges from 3-4°C in February-March to 20-21°C in August near the water table but remains at 12-14°C 10 m deep in the aquifer. This complex interplay of factors prevents accurate prediction of the fate of aldicarb residues in groundwater by mathematical modeling on the basis of available data. Sensitivity analysis indicates that degradation rates are the crucial determinant of occurrence and persistence in groundwater.

The introduction in the early 1970's of the soil-incorporated, water-soluble systemic insecticide/nematicide aldicarb was regarded as a major advance in the technology of pest control. It eliminated the once prevalent hazard of inadvertent exposure of applicators or inhabitants of nearby dwellings, wildlife, and useful insects to toxic chemicals during spraying or aerial application of pesticides.

0097-6156/86/0315-0219$10.25/0
© 1986 American Chemical Society

It was 1979 before intrusion of aldicarb residues into a domestic well located close to irrigated potato fields in Suffolk County on Long Island, New York was reported in August, 1979 (1). This discovery precipitated an extensive survey of other wells in the vicinity (1,2) and regulatory actions which eliminated the use of the insecticide on Long Island. Aldicarb was being used on a variety of crops in several states, and concern arose whether residues of the pesticide might also be encountered in other areas. On the basis of information then available, this was deemed unlikely, because of the special circumstances of aldicarb use on Long Island. First, application rates were high: 5-7 pounds of active ingredient per acre (lb a.i./A = 5.6-7.8 kg/ha) to ensure adequate control of two major potato pests, the Colorado potato beetle (*Leptinotarsa decemlineata*) and the golden nematode (*Heterodera rostochiensis*); in other areas such as Wisconsin application rates of only 2-3 lb a.i./A (2.2-3.4 kg/ha) were being used on potatoes. Secondly, on Long Island, potatoes are grown on irrigated sandy soils over high water tables. While similar conditions exist in other areas, such as the Central Sands region of Wisconsin, the average annual precipitation rate is much higher for Long Island (50 versus 30 inches = 127 v. 76 cm per year). Third, the contaminated Long Island aquifer is largely a shallow confined aquifer, and the pH and alkalinity of the water are low. In Wisconsin, in contrast, much of the uppermost aquifer in the potato-growing area is close to discharge areas, either drainage ditches or the Wisconsin River and its tributaries, and the pH and alkalinity of the water are high (3,4,5). It was known that aldicarb residues have only brief persistence in surface water, and are more stable under acid than alkaline conditions. Indeed, on the basis of a geographic scale model (cf. 6) projections were made using a numerical rating system which at the time ranked the Long Island aquifer first in sensitivity toward aldicarb residue contamination, Wisconsin's Central Sand Plain a distant second, and all other areas of aldicarb use--especially those in warmer southern states--as relatively safe, regardless of application rates.

However, following press reports claiming that evidence for the occurrence of several pesticides, including aldicarb, had been found by gas chromatographic analysis of six Wisconsin groundwater samples, the manufacturer and the State of Wisconsin examined fresh samples from the same wells. The State tests indicated no contamination by any of 70 pesticides for which analyses were conducted. However at the time, no laboratory in Wisconsin had facilities for conducting aldicarb residue analyses in the parts-per-billion range. Union Carbide's analyses indicated the presence of aldicarb residues in one sample from a monitoring well at a potato processing plant wastewater disposal area. This provoked the company to examine samples from several doemstic wells, and to report in January 1981 that aldicarb residues had been found in 12 samples. As more wellwater samples were analyzed, further contaminated sites were detected. Obviously, the geographic scale model was inadequate to predict pollution incidents for Wisconsin.

As a result of the initial press reports, research was begun by the University of Wisconsin in the Central Sands area of Wisconsin to examine groundwater for the presence of pesticide residues. Because of restricted funding, efforts were concentrated exclusively on aldicarb, as representing a worst-case situation, because this was the only material for which the presence of residues at levels of concern had been confirmed by reliable analyses.

The major goal of the research was to determine which factors affected the amount of intrusion of pesticide residues into the groundwater and their persistence and distribution in the groundwater. Such information could enable the manufacturer and state and federal regulatory agencies to decide whether the material could be used safely, and if so, under what conditions in Wisconsin. A further goal was to predict how long the residue concentrations already in the water would remain at levels of concern and whether their movement to domestic wells could be forecast.

The first step toward attaining these goals was to collect aldicarb residue concentrations and other water quality parameters for water samples withdrawn from various depths in the aquifer in and around several fields in which irrigated potatoes had been treated at differing times with aldicarb. The next step was to describe the mechanisms of aldicarb disposition following application and to calibrate and test a model using data for aldicarb and other chemical species accumulated from the literature and from recent field analyses. It was hoped this approach would permit satisfactory prediction of the distribution and fate of aldicarb applied at a field-size scale, and the potential for entering domestic wells within 0.5-5 miles (0.8-8.0 km) at levels above the recommended safety level of 10 μg/L. Typical 160-acre (65-ha) irrigated potato fields in Wisconsin are square and have sides 0.5 mile (0.8 km) long, and private wells serving farmhouses or rural homes are often located in the corners of or immediately adjacent to such fields or among a series of such fields.

Procedures

Site Selection. Nine 160-acre (65-ha) irrigated fields in two of the main potato-growing counties of Wisconsin (Portage and Waushara, 6.4 and 2.2 million cwt. or 3.2 and 1.1 million quintals in 1981) were selected for study. Soils in each field were well drained sands or loamy sands, mainly of the Plainfield series (Typic Udipsamments), containing 1-2% organic matter in the A horizon. Seven fields had previously been examined (5,7).

The eighth field was located immediately southwest of one field already being monitored in Waushara County, the ninth field was directly northwest of the eighth field. All fields had known histories of aldicarb treatments (Table I). The direction of undisturbed groundwater flow in the area is approximately west-northwest and the rate is approximately 1 foot (30 cm) per day (8,9). When central-pivot irrigators are in use, groundwater flow is perturbed by the cone of depression created by water withdrawal through high-capacity irrigation pumps (5,8,9). The cone of depression does not extend to the edge of the field and recovery is rapid once pumping ceases.

Table I. Aldicarb Application Rates and Frequencies in the 9 Fields under Study

Field	Application rate in pounds of active ingredient per acre (kg/ha)	Years treated
1	3.0 (3.4)	1977, 1979, 1981
2	3.0 (3.4)	1976, 1978, 1980
3	2.4 (2.7)	1976, 1977
	2.5 (2.8)	1978
	2.7 (3.0)	1979
	3.0 (3.4)	1980
4	3.0 (3.4)	30 rows west of center pivot, 1982 (5 acres)
5	3.0 (3.4)	1979
	3.0 (3.4)	East half, 1980
6	2.25 (2.5)	1981*
7	3.0 (3.4)	1980
	1.95 (2.1)	1982*
	1.8 (2.0)	1984*
8	2.7 (3.0)	1978, 1980, 1981
9	3.0 (3.4)	1979, 1980

*Treated at plant emergence; all other treatments were at planting.

The fields were originally selected to allow comparison of aldicarb residue movement and persistence under the conditions pertinent to its prescribed (label) use and pre- and post-1982 maximum application rates [3 versus 2 lb a.i./A (3.4 v. 2.2 kg/ha)]. At the start of the study:

1. Fields 3, 8 and 9 had received aldicarb applications at planting several years in sequence;
2. Fields 1 and 2 had received aldicarb applications at planting in alternate years;
3. Field 5 had received aldicarb at planting only once;
4. Field 4 had never been treated with aldicarb and was downgradient in the water table from a treated field;
5. Field 6 had been treated only once at plant emergence 1 year before monitoring;
6. Field 7 had been treated once at planting, then 2 years later at plant emergence during the first year of monitoring.

Well Installation. In seven fields, monitoring wells were installed at eight or more positions around the perimeter of the field and one or more locations close to the center pivot. Only two perimeter well sites were chosen in the other two fields (to comply with the owner's wishes). The wells were placed as close as possible to the treated areas. The well casings were Schedule 40 polyvinyl chloride (PVC) pipe; the well points were made of PVC with 1 or 3 feet (30 or 91 cm) of 0.006 inch (0.15 mm) slits (Timco Mfg., Sauk City, WI 53578). Shallow wells were installed by hand-augering through the soil to or slightly below the water table using a 4-inch (102 mm) diameter Soil Conservation Service stainless steel bucket auger (Art's Machine Shop, American Falls, OR 83211). The drillings were stored in sequence on a sheet of plastic alongside the hole. A driving cap was placed in the end of the pipe and the well driven to the desired depth using a post driver or sledgehammer. The emergent pipe was cut off at a suitable elevation above ground and a rubber vacuum pump hose inserted into the well. Water pumped from the well using a battery-driven peristaltic pump (ISCO Model 1680 Wastewater Sampler Superspeed Pump, ISCO, Lincoln, NB 68528) was squirted down the outside of the casing as the drillings from the hole were returned in the reverse of the sequence in which they had been removed. This reconstitutes the original soil profile without channeling or arching. The top 1 m of the drilled hole was sealed with dry, pulverized bentonite clay (Quick-Jell, N.L. Baroid Industries, P.O. Box 4350, Houston, TX 77210). After installation, the well was developed by pumping until the withdrawn water contained no visible sediments. Each well was capped with a screw-on top secured with a long-shank bicycle padlock inserted through holes drilled diagonally through the cap and pipe fitting.

A commercial truck-mounted well drilling rig was used to install deeper wells. Drilling was performed using hollow-stem continuous flight augers, which were left in place while a preassembled PVC well was driven by hand to embed it in the aquifer sediments at the preselected depth. The hollow-stem auger was then removed, the drillings washed in as before and the well sealed with bentonite at the top and developed by pumping.

In each field, a rain-gauge was attached to one monitoring well to collect rainfall and irrigation water. The location of the wells and samplers in Fields 1-7 are shown in Figures 1 and 2.

Monitoring wells were also placed in four test plots sown to potatoes at the University of Wisconsin-Madison Experimental Farm, near Hancock, Wisconsin. These plots were treated with aldicarb at a rate of 2 lb. a.i./A (2.24 kg/ha) at planting and at plant emergence. One set of plots was irrigated at a rate corresponding to the measured evapotranspiration rate for the region, the second set at 1.6 times the evapotranspiration rate (21).

Multilevel Samplers. Multilevel samplers (10) assembled in the laboratory were installed at selected positions in five fields. A bundle of 0.5-inch (12.7 mm) o.d. high-density linear polyethylene tubes (Central Plastics Distributors, Madison, WI 53703) were clustered around a central 0.75-inch (19.1 mm) diameter PVC pipe at distances 18 inches (45.7 cm) apart. The ends of each sampler were surrounded by a screen of Style 3401 Typar spunbonded polypropylene (DuPont de Nemours Chemicals Corp., Wilmington, DE 19898) taped to the tubing. The bundle of tubes was secured around the central PVC pipe using duct tape and installed through a hole drilled to the desired depth with the drill rig. The bottom end of the hollow-stem auger was closed with a large rubber stopper to prevent sand from filling the center of the auger during drilling. With the auger in place at a depth such that the uppermost sampling port would be at or a little below the water table, the inside of the auger stem was filled with water to a level above the water table. The multilevel-sampler bundle was inserted through the auger's hollow stem and allowed to sit until the tubes had filled with water. The rubber stopper was then dislodged from the end of the auger by tapping sharply on the end of the central PVC pipe. Unless the auger stem is filled with water before the rubber stopper is displaced, sand is swept up into the hollow stem when the stopper is dislodged. This jams the multilevel sampler in the hollow-stem auger when it is removed from the hole. By filling the hollow stem with water the auger can be withdrawn leaving the multilevel sampler in place at the desired depth. After positioning the samplers, the drill hole around the bundle was refilled by washing in removed material and sealing the surface with bentonite. Each sampler was developed using the peristaltic pump until the water delivered contained no sediment. The ends of the tubes and central pipe emerging from the ground were covered with a capped PVC pipe which was locked to the central PVC pipe with a bicycle padlock.

Field Lysimeters. In addition, potatoes were planted in each of two field lysimeters (11,12) and treated with aldicarb at plant emergence at a rate of 3 (1982) and 2 (1983-5) lb a.i./A (3.36 and 2.24 kg/ha). Each lysimeter was spray-irrigated with 0.75 inch (19.3 mm) of water twice a week throughout the growing season. The soil in each lysimeter was maintained under a constant suction using an electric pump (Gast, Benton Harbor, MI 49022) with a bleed. Drainage from each lysimeter was passed through a separate water meter fitted with an automatic counter and printer to measure incremental and cumulative flow. Portions of drainage were diverted periodically from each lysimeter to a plexiglass sampler in parallel to the drainage system and bottled and frozen for future analysis.

Sampling. Water samples were collected periodically from each monitoring well, each sampling tube on the multilevel samplers, and from the irrigation wells in the fields under study. Before sampling, water was pumped from each monitoring well or sampling tube until a clear sample could be obtained or until 3 to 5 times the volume of the well had been removed, to assure that the sample came from the groundwater around the well and not from stagnant water in the well. Duplicate or triplicate samples were collected in new 500-ml Nalgene bottles or 5-liter brown glass bottles with Teflon-lined caps, each previously washed with acetone, distilled water, and rinsed three times with water from the well or sampler being sampled. The temperature of each sample was measured to the nearest 0.1°C and the pH measured with a Digi-Sense digital pH meter (Cole-Palmer Instrument Co., Chicago, IL 60648), previously calibrated in the laboratory against a Corning Model 130 pH meter (Corning Medical, Medfield, MA 02052). Each sample was chilled in ice water in an insulated picnic cooler and later frozen until analysis.

Before each monitoring well was sampled, the elevation of the groundwater table was measured using a "popper" (a plastic cylinder with a concave base which emits a popping sound when it impinges on a water surface) suspended at the end of a steel measuring tape. This device gave accurate measurements of water table depths without contaminating the well water.

In April 1982, before crops were planted, the drill rig was used to collect water samples from spots directly beneath the growing area of Field 5. A 30-cm metal well point with stainless-steel screen welded to the end of a section of hollow-stem auger was drilled to various depths below treated areas. Water samples were withdrawn through Teflon tubing inserted down the hollow stem into the point using the peristaltic pump. Sediment-free samples were bottled and stored for analysis as previously described.

Periodically, batches of samples were removed from the freezer and thawed for analysis. Parameters examined included laboratory pH, conductivity, total alkalinity, calcium, magnesium, sodium, potassium, nitrate, chloride, sulfate, and aldicarb residues.

The pH was determined using a Corning Model 130 digital pH meter; carbonate and bicarbonate alkalinities were determined titrimetrically, according to Method 403; nitrate-nitrite by the automated cadmium reduction method (Method 418 F) and sulfate by the turbidimetric method (Method 426c) of Standard Methods for the Examination of Water and Wastewater (13); and chloride using a Buchler-Cotlove Model #4-2000 Chloridometer titrator (14). Na and K were determined by flame photometry on a Coleman Model 21 instrument according to Methods 322B and 325B (13); Mg and Ca were determined by atomic absorption on a Perkin-Elmer Model 306 instrument according to Method 303A (13).

Aldicarb Residue Analysis. For analysis of aldicarb residues, a 100-ml aliquot of each thawed sample was oxidized with 1 ml of 40% peroxyacetic acid, the excess acid neutralized with 25 ml of 10% sodium bicarbonate, and the aldicarb sulfone produced by the oxidation extracted with two 100-ml portions of methylene chloride. The extract was dried over 100 g of sodium sulfate, evaporated to dryness, redissolved in 1 ml of acetone:ether and chromatographed on a column 12.5 mm diameter x 10 cm of florosil (Florodin Co., Berkely Springs, WV 25411) using 1:1 acetone:ether as eluent. The purified sulfone fraction was again evaporated to dryness, dissolved in 1 ml of acetone and subjected to gas chromatography in an H-P 5880 instrument (Hewlett-Packard, Palo Alto, CA 94304) on a 2 meter 2 mm i.d. column packed with 10% SP-1000 on 80-100 mesh Supelcoport (Scientific Products, McGraw Park, IL 60085) using a Tracor flame photometric detector (Tracor Instruments, Austin, TX 78721) or an H-P nitrogen/phosphorus specific detector.

The injector temperature was 260°C, the detector temperature 300°C. Analyses were run isothermally at 180 or 200°C using helium as carrier gas at a flow rate of 25 ml/min. The hydrogen flow rate for the detector was 3 ml/min and the air flow rate 60 ml/min using an attenuator setting of 2^4. The limit of detection by this method--as statistically defined by the American Chemical Society's Committee on Environmental Improvement (14)--was below 1 μg/liter or ppb but values below 1 ppb printed out by the instrument processor were reported as non-detects (ND). For quality control, standards were analyzed following every tenth sample. Samples of pure aldicarb, aldicarb sulfoxide and aldicarb sulfone were donated by Union-Carbide Agricultural Products Company, Raleigh, NC. Recoveries of over 90% were consistently obtained with 1, 10, and 100 ppb levels of the individual compounds or mixtures of the pure standard solutions spiked into distilled or Madison well water or field samples and subjected to the extraction, cleanup and analysis. Unless recommended criteria for recoveries from fortified samples (16) and blanks were met, data were discarded and all reagents, solvents and procedures were reviewed. As a further quality control measure, some duplicate samples were sent frozen in insulated containers to Union Carbide for analysis for interlaboratory comparison.

Water samples from rural residences close to the experimental sites were also analyzed. Three shallow domestic wells contaminated by aldicarb residues above the suggested no adverse response level of 10 ppb were deepened to examine whether the water quality would be improved.

Bacterial Analysis. Groundwater and aquifer sediments samples were collected aseptically for bacterial analysis; the sediments were obtained using hollow-stem augers (17). With the augers in position in the soil or groundwater at a preselected depth, samples were collected inside a rigid, clear plastic tube (Acker Drill Co., Scranton, PE) previously sterilized using ultraviolet light passed through the lumen. This tube was inserted inside a split-spoon sampler sterilized by dousing with ethanol and flaming. The sampler was fed through the hollow stem with an extender rod and forced into the undisturbed soil/groundwater below the bore hole. After removal from the split-spoon sampler, the ends of the plastic cylinder were cut off with a sterilized knife and capped with sterilized plastic caps. Subsamples removed under sterile laboratory conditions from the centers of the soil cores were examined for bacteria.

Samples of [S-$^{14}CH_3$] aldicarb, aldicarb sulfoxide and aldicarb sulfone were incubated with samples of groundwater and aquifer sediments to examine the potential for microbial degradation *in situ* in groundwater. The [S-$^{14}CH_3$] aldicarb was provided by Union Carbide Agricultural Products Company; the labeled sulfoxide was prepared from this by oxidation with hydrogen peroxide, the sulfone from the aldicarb by oxidation with peroxyacetic acid. These oxidants afforded better yields of cleaner product than the *m*-chloroperoxybenzoic acid previously used (18).

Results

The results of the monitoring clearly indicate that regardless of application rate, timing and frequency, residues of aldicarb are leached beyond the rooting zone of potatoes grown in irrigated sandy soils into underlying groundwater. The water table depth is not the critical significant factor in restraining aldicarb leaching: residues were found where the water table was only 3.9 feet (1.2 m) or as deep as 17.3 feet (5.3 m) below the soil surface. This is not surprising, since neither aldicarb nor its sulfoxide or sulfone is strongly absorbed by soils, especially coarse-textured soils (19,20). The groundwater table in the area is subject to seasonal and long-term fluctuations. The water tables in the fields studied dropped by 2-4 feet (61-122 cm) during the growing season due to water removal for irrigation but recovered to more or less the original level by the start of the next season. The water table at the U.W.-Madison Experimental Farm near Hancock, WI fluctuated by ± 5 feet (152 cm) around a mean value during the years 1951-1984.

Erratic Residue Distribution. However, the pattern of aldicarb residue concentrations encountered were astoundingly variable given a uniform application rate across the 127-acre (51-ha) treated central portion of each 160-acre (65-ha) field. The water loadings from irrigation plus precipitation may be assumed to be uniform over this relatively small area, and the soils are relatively uniform. Consequently, aldicarb residue penetration to the groundwater should be uniform. The sand-and-gravel aquifer is relatively homogeneous and largely stratified (8,9), so that mainly lateral movement of residues with little vertical mixing might be expected. It was therefore surprising that in each field no aldicarb residues were found in some wells downgradient from treated areas, low levels in others, and high levels in others, all at approximately the same depth (Table II). Occurrences and concentrations were also erratic in the multilevel samplers (Table III). Concentrations of aldicarb residues in the soil solution, i.e. values based on the water content of soil samples taken from the vadose zone, were also erratic (Table IV) and concentrations in leachate from the lysimeters also varied widely over large ranges (Table V). Aldicarb residues apparently do not penetrate to groundwater by bleeding through the vadose zone at a constant rate, but rather in pulses of varying concentration. Peak concentrations appear to be associated with minor wetting fronts passing through the soil. So far we have been unable to determine clearly whether highest loadings (volume x concentration) penetrate with sharp or diffuse wetting fronts under conditions of unsaturated flow through the vadose zone, i.e. whether the loading is proportional to the flow.

Table II records the concentrations of aldicarb residues found in water samples from monitoring wells which had detectable residues present at some time during the sampling period. No significant residues were detected at any time in any other wells (12 more wells in Fields 1 and 2, 13 in Field 3, 16 in Fields 4 and 5, 10 in Field 6, 11 in Field 7, and 1 in each of Fields 8 and 9). Well depths in the groundwater increase in the sequence indicated by the designation A, B or C associated with the location number for any field. The wells which initially contained contaminated or uncontaminated water in any field were at similar depths in the water table; thus, the pattern of residue occurrences cannot be ascribed to inappropriate well placement or depth.

These data give some idea of the erratic areal distribution of aldicarb residue contamination even within a single field. For example, on Field 5, treated in its entirety in 1979 and in the eastern half in 1980, significant concentrations of residues were initially encountered only at two spots--at and directly north of the center pivot (sites 7 and 2, Fig. 1). Surprisingly, no residues were initially encountered in water from wells at positions 3, 4 or 5, or at any time in samples from the neighboring Field 4, all directly downgradient in the aquifer from the treated area. Residues (12 ppb) were found in June 1981 in water from the irrigation well which was cased to a depth of 36 feet (11 m), or 28 feet

Table II. Aldicarb Residue Concentrations (ppb) in Water Samples from Monitoring Wells (in which Residues were Statistically Quantifiable (15).

Well No.	Well depth (cm)	Mean water table (cm)	Date (year, month) 1980	1981						1982						1983								1984						1985
			12	2	3	4	5	8	11	3	6	8	9	10	12	2	3	5	6	7	8	9	10	1	2	3	5	6	7	9
Fields 1,2																														
1B	754	478		ND		1	2	2	1	6	7	7	6			4	9	12	11	8	17	24	8	21	19	11	5	19	16	16
2B	549	480		49		70	10		4	2	30	115	6	121			118	63	58	36	11	13	9	2	ND	1	ND	11	6	ND
2C	1072	476		ND		1	2	5	ND	ND	ND	1	8		1	6	10	ND	4	ND	1	4	4	13	ND	19	6	17	28	ND
3B	572	424		6		11			48	47	11	23	7			10	8	4	3	ND	ND	ND	1	1	ND	ND	ND	ND	1	ND
4A	552	333		4		26			42	65	8	49	22			81	29	90	36	2	ND	88	100	64	51	25	5	ND	7	-
7C	1052	459		ND		1	ND	ND	ND	ND	ND	3	ND		3	ND	2	9		1	6	7	2	3	3	3	ND	8	6	2
8A	531	420		15		30	9	10	2	18	9	3	7	7		11	24	21	6	11	20	14	5	13	10	18	ND	10	15	-
9C	1069	344		22		17	3	10	12	8	6	2	11	7	7	9	21	18	8	15	32	14	23	14	17	19	8	19	29	15
10B	526	342		47		30			68	40	55	30	35		40	39	36	24	24	16	18	15	7	6	7	7	4	4	8	-
13A	541	338		69		103			69	53	37	14	24	6	6	116	116	91	70	48	48	67	85	72	55	40	33	17	10	ND
Field 3																														
3A	365	225			ND		ND	ND	ND	6	6	13	3	3	4	4	ND	ND	2	13	16	13	18	9	9	14	9	11	13	-
4A	399	221			30		72	67	88	13		ND	2		3	ND	ND	13	3	ND	ND	ND	ND	2	5	5	ND	2	2	-
4B	562	226			18		12		3	1		21	15		7	21	20	33	8	18	ND	ND	ND	6	ND	2	ND	1	2	-
5A	373	212			73		10	21	23	ND		ND	ND	ND	ND															
5B	508	215			ND		ND	ND	18	Z	ND	3	1	ND	3		4	2	5	1	ND	ND	ND		ND	ND	-	ND	ND	-
6A	348	180			116		157	153	100	82	3	3	13		17	48	44	38	5	8	ND	ND	ND	2	ND	ND	2	1	1	-
6B	549	181			2		4		2	1		24	11	19	6	ND	3	ND	ND	ND	7	15	8	2	ND	1	ND	1	ND	-
7A	338	174			35		69	86	180	39	30	ND	4	8	10	ND	ND	ND	ND	ND	ND	ND	ND	ND	ND	ND	ND	ND	ND	ND
7B	531	171			22		16		28	39	37	88	80		39	22	24	7	7	7	7	11	24	6	ND	4	3	4	3	-
9A	553	189			191		7	150	123	83	ND	ND	4	1	3	5	6	5	2	ND	ND	ND	ND	ND	ND	ND	1	ND	ND	1
Fields 4, 5																														
2B	292	128	195	150	171	142	140	66	64	1		13	9		6	ND	ND	ND	ND	ND	ND	ND	ND	ND	ND	ND	ND	ND	ND	ND
2C	427	131	10	12				21			1	ND	ND		ND	3	11	2	ND	ND	ND	ND	ND	ND	ND	ND	ND	ND	ND	ND
4C	338	206	5	ND					27	8	23	13	28		11	ND	ND	ND	ND	ND	ND	ND	ND	ND	ND	ND	ND	ND	ND	ND
7B	323	295	110	79	77	83	86	26	44	78		8	ND		ND	ND	ND	ND	ND	ND	ND	ND	ND	ND	ND	ND	ND	ND	ND	ND
7C	427	270	3	2				18	3	1		3	1		1	2	2	3	ND	ND	ND	ND	ND	ND	ND	ND	ND	ND	ND	ND

7B'	389	249						38	28		ND	ND		ND	ND	ND	ND	ND	ND	ND	ND	ND	ND	ND	ND	ND	ND	ND
9C	422	156	ND					5						24	41	76	64	52	43	48	31	11	1	ND	ND	ND	ND	ND
Field 6																												
1A	221	152								69	71	30	24	16	38	67	7	1	ND	ND	ND	ND						
1B	292	155								122	72	116	58	25	18	36	12	6	2	2	ND	ND						
5B	373	144								122	37	29	93	113	64	57	36	18	2	1	ND	1						
Field 7																												
1A	262	199										ND	15	83	49	3	17		Wells							ND		ND
1B	353	208										1	17	62	49	12	2		Destroyed							3		4
2A	277	192								38		10	57	44		49	16									ND		ND
2B	302	192								ND	1	14					2									ND		ND
3A	176	118								46	ND	15	1	3	3	4	9	ND	ND	ND	ND	ND			ND	ND	ND	ND
8A	249	179									ND	ND		100	76	11	12	2	5	ND	ND	ND			ND	ND	ND	ND
8B	338	179									ND	ND		14	157	20	24	6	2	1	ND	ND	ND	ND	ND	ND	ND	ND
9A	183	153									ND	ND		24	12	7	33	5	2	ND	ND	ND	ND	ND	ND	ND	ND	ND
9B	346	154									25	19	10	6	11	9	28	ND	34	30	34	17	19	2	3	2	2	ND
10B	361	145								ND	ND	2	128	24	28	17	33	19	41	42	27	26	15	ND	ND	ND	ND	ND
11B	354	127								ND	ND	6	8	25	32	22	45	33	48	19	29	22	3	ND	ND	ND	ND	ND
12A	341	250								ND	8	7	11	15	54	61	174	112	198	95	119	42	ND	3	-	3	ND	ND
12B	359	247								ND	4	1	24	7	23	116	23	24	24	65	9	3	1	ND	ND	ND	ND	ND
Field 8																												
1	404	237		2	ND	ND		16	4		ND	5		7	8	3	6	ND	ND	ND	ND	ND			ND			
2	359	262		105	57	62	28	22	6		ND	ND		ND	ND	ND	ND	ND	ND	ND	ND	ND			31			
Field 9																												
2	472	315		210		154	30	83	66	ND	16	7		12	26	ND	ND	4	14	14	3	ND		ND	3			

Table III. Aldicarb Residue Concentrations (ppb) in Groundwater Samples Withdrawn through Multilevel Samplers

Site	Depth below water table (cm)	Date (year, month) 1981	1982						1983								1984					
		11	3	6	8	9	10	12	2	3	5	6	7	8	9	10	1	2	3	5	6	7
Field 3	15	173	138	55	24	17		8	27	36	8	3	ND	ND	ND	ND	2	ND	1	ND	ND	ND
	61	193	130	45		36		21	49	46	28	3	4	ND	ND	1	1	ND	2	ND	ND	ND
	107	158	181	160		81		31	17	18	12	3	16	10	5	5	ND	ND	1	ND	ND	1
11	152	77	25	24		47		9	17	25	16	13	12	9	ND	ND	ND	ND	ND	ND	ND	1
	198	38	19	17		24		16	10	24	20	5	2	ND	4	ND	ND	ND	3	ND	ND	2
	244	14	14			37		7	12	18	18	4	10	ND	14	ND	3	ND	2	ND	1	1
	290	12	13	ND		23		10	12	9	19	12	7	5	7	5	ND	ND	ND	ND	ND	ND
	15	133	170	54		11		1	ND	6	ND	ND	3	1	ND	ND	1	ND	1	ND	ND	ND
	61	78	221	49	17	54		13	8	8	4	3	4	ND	ND	ND	1	ND	1	ND	ND	ND
	107	45	45	19	54	46		64	35	30	10	5	3	ND	ND	ND	ND	ND	1	ND	ND	1
12	152	26	68	22	49	34		34		69	29	9	13	3	2	6	3	ND	3	2	ND	2
	198	16	18	29	19	30		12	31	30	49	17	14	4	5	ND	ND	ND	2	ND	ND	3
	244	12	15	17	22	30		19	13	11	17	14	5	4	ND	2	ND	ND	ND	ND	ND	2
	290		10		29	20		23	19	13	14	5	6	11	5	1	ND	ND	ND	ND	ND	1
	15	36	118	162	82	36		11	ND	ND	ND	ND	4	ND	ND	ND	2	ND	1	ND	ND	1
	61		100	128	24	52		46	30	7	ND	1	9	ND	ND	ND	ND	ND	1	ND	ND	ND
	107		94	16	22	50		23	78	62	38	18	8	ND	5	4	2	ND	1	ND	ND	ND
13	152		29	16	21	40		16	24	34	33	14	8	3	2	ND	2	ND	ND	2	ND	2
	198		10	10	17	35		7	14	15	5	17	5	7	3	1	5	ND	ND	3	ND	2
	244		8		14			7	15	17	9	3	4	3	5	1	ND	ND	4	ND	ND	1
	290		5		10	4		1	15	21	10	12	7	15	5	5	ND	ND	ND	ND	ND	2

Fields 4,5	46, 91, 135 183, 226, 274								(all ND's)													
3	318	ND	ND	ND	ND	2		5	ND	ND	2	2	ND	ND	ND	1	ND	ND	ND	ND	ND	ND
	366	ND	ND	ND		2		3	6	5	7	2	4	5	4	2	ND	ND	ND	ND	ND	ND
	46	2		ND	ND	ND	ND	ND	ND	ND	ND	ND	ND	ND	ND	ND	ND	ND	ND	ND	ND	ND
	91	ND		ND	1	ND	ND	ND	ND	ND	ND	ND	ND	ND	ND	ND	ND	ND	ND	ND	ND	ND
	135	ND	ND	3.8	8	ND	ND	ND	ND	ND	ND	ND	ND	ND	ND	ND	ND	ND	ND	ND	ND	ND
4	183	ND	7	2		33	10	5	1	2	ND	ND	ND	ND	ND	ND	ND	ND	ND	ND	ND	ND
	226	30		20	54	59	55	6	1	4	6	ND	ND	ND	ND	ND	ND	ND	ND	ND	ND	ND
	274	12	69	45	20	34	29	20	5	17	11	15	7	12	ND	ND	ND	ND	ND	ND	ND	ND
	318	ND	14	12	22	29	25	16	34	27	22	22	20	24	8	ND	1	ND	ND	ND	ND	ND
	366	ND	17	11	4	18	34	18	22	19	12	2	10	8	7	13	3	ND	ND	ND	ND	ND
	53	122	ND	ND		ND		ND	2	ND	ND	ND	ND	ND	ND	ND	ND	ND	ND	ND	ND	ND
	91	1	46	36		1		ND	ND	ND	ND	ND	ND	ND	ND	ND	ND	ND	ND	ND	ND	ND
7	244	ND	ND	ND	ND	ND		ND	ND	ND	ND	ND	ND	ND	ND	ND	ND	ND	ND	ND	ND	ND
	488	ND	ND	ND	ND	ND		ND	ND	ND	ND	ND	ND	ND	ND	ND	ND	ND	ND	ND	ND	ND
	732	ND	ND	ND	ND	ND		ND	ND	ND	ND	ND	ND	ND	ND	ND	ND	ND	ND	ND	ND	ND
	46	ND	ND	ND	ND	ND		ND	ND	ND	ND	ND	ND	ND	ND	ND	ND	ND	ND	ND	ND	ND
	91	ND	1	ND	2	1		ND	ND	ND	ND	ND	ND	ND	ND	ND	ND	ND	ND	ND	ND	ND
	135	44		2		11		ND	ND	ND	ND	ND	ND	4	ND	ND	ND	ND	ND	ND	ND	ND
	183	193	45	41	53	12		ND	ND	ND	ND	ND	ND	ND	ND	1	ND	ND	ND	ND	ND	ND
13	226	136	13	10	36	57	11	5	2	ND	ND	ND	ND	ND	ND	ND	ND	ND	ND	ND	ND	ND
	274	97		ND	9	140	80	ND	8	16	3	ND	ND	8	3	ND	ND	ND	ND	ND	ND	ND
	318	67		ND	5	10	-	7	8	3	5	4	6	5	5	1	ND	ND	ND	ND	ND	ND
	366	ND		ND	ND	ND	ND	2	6	7	4	1	3	2	1	ND	ND	ND	ND	ND	ND	ND
	411	ND	ND	ND		3	ND	ND	ND	3	3	3	2	2	2	ND	ND	ND	ND	ND	ND	ND

Continued on next page

Table III--Continued

Field 6																			
2		None detected at any depth (high SOM)																	
	15		ND		ND	ND	ND	ND	ND	ND	ND	ND	ND	ND	ND	ND	ND	ND	ND
	61	ND	ND		1	ND	ND	ND	ND	ND	ND	ND	ND	ND	ND	ND	ND	ND	ND
4	107	1	20	18	16	ND	ND	ND	ND	ND	1	ND	1	ND	ND	ND	ND	ND	ND
	152	3	104	64	55	9	3	5	ND	ND	7	ND	ND	ND	ND	ND	ND	ND	ND
	198	14	20	2	7	28	49	37	ND	2	ND	ND	ND	ND	ND	ND	ND	ND	ND
	244		2	1	1	13	3	34	15	3	ND	1	ND	ND	ND	ND	ND	ND	ND
	290	ND	ND	ND	ND	ND	ND	25	23	5	ND	ND	ND	3	ND	ND	ND	ND	ND
	335	ND	ND	ND	ND	ND	ND	11	14	7	ND	ND	2	5	ND	2	ND	ND	ND
	378	ND	ND	ND	ND	ND	ND	ND	12	4	ND	ND	1	4	ND	2	ND	ND	ND
Field 7																			
	46		2	150	59	81	64	23	29	45	7	ND	ND	ND	ND	ND	ND	ND	ND
	91	ND	ND	57	59	48	77	55	27	23	6	13	7	ND	ND	ND	ND	ND	ND
	135	2	ND	47	8	21	22	39	22	31	27	28	20	2	ND	ND	ND	ND	ND
	183	10	6	-	7	13	42	38	32	38	18	44	8	14	3	9	4	6	ND
9	226	16	50	26	24	26	26	5	1	ND	1	ND	1	16	46	39	37	15	9
	274	50	118	33		23	26	16	ND	ND	ND	ND	2	21	31	27	27	10	8
	318	85	85	3	14	30	40	14	16	5	ND	ND	8	4	2	5	6	6	9
	366	6	7	6	6	14	36	6	18	6	11	13	7	14	ND	9	5	5	2

	15	--Dry--		2	14	ND	27	24	16	ND	3	ND	ND	ND	ND	ND	ND	18	
	61		ND	41	47	91	145	76	55	35	22	12	6	1	ND	ND	ND	1	9
	107	ND	ND	37	46	78	98	102	54	103	20	4	9	3	1	2	1	1	2
	152	ND	1	2	12	19	13	28	52	76	75	57	4	36	29	36	18	60	ND
12	198	1	1	2	ND	1	3	17	13	10	49	47	32	45	53	25	10	75	ND
	244	6	5	ND	ND	ND	ND	ND	ND	6	24	7	9	13	8	26	22	10	1
	290	9	7	2	ND	2	ND	ND	ND	8	5	9	8	5	1	1	3	15	1
	335	5	5	2	1	2	2	ND	ND	1	3	2	9	1	ND	ND	1	12	2
	378	5	14	1	ND	3	3	ND	ND	2	2	11	6	ND	ND	ND	ND	4	3

Table IV. Texture, Alkalinity and Moisture Contents of Soil Samples (Oct. 27-28, 1981) from the Vadose Zone in Potato Fields and Aldicarb Residue Concentrations (ppb) in the Soil and Soil Solution ("H_2O" in Last Column).

Field	Sampling depth (cm)	pH	$CaCO_3$ equiv. (%)	Soil separates, % Sand	Silt	Clay	% H_2O	Aldicarb in Soil	H_2O
1	0-30	6.1	0.46	86.4	10.4	5.0	6.8	5	70
	30-60	6.0	0.38	95.3	2.3	2.4	4.0	ND	0
	60-120	5.4	0.41	98.3	0.6	0.1	1.8	ND	0
	120-180	5.3	0.65	99.9	0.1	0	3.5	ND	0
	180-240	5.2	0.43	100	0	0	7.0	ND	0
1	0-30	6.0	0.42	88.5	7.5	4.0	7.8	11	141
	30-60	5.4	0.37	95.3	3.4	1.3	3.5	ND	0
	60-120	5.2	0.40	98.1	1.5	0.4	4.0	ND	0
	120-180	5.2	0.39	99.3	0.5	0.2	4.0	7	175
1	0-30	6.2	0.48	88.5	7.5	3.0	7.1	11	155
	30-60	5.8	0.30	97.2	2.1	0.7	4.3	ND	0
	60-120	5.3	0.36	98.7	0.9	0.4	3.0	ND	0
	120-180	5.0	0.33	98.9	1.1	0	3.9	ND	0
3	0-30	6.9	0.61	93.8	3.5	2.7	4.3	7	163
	30-60	6.3	0.83	91.8	6.2	2.0	4.5	ND	0
	60-120	5.1	1.02	94.2	4.1	1.7	4.0	ND	0
	120-180	5.0	0.64	95.0	3.0	2.0	6.8	ND	0
3	0-30	6.4	0.59	90.7	4.7	4.6	6.7	7	104
	30-60	6.0	0.45	93.3	5.3	1.4	3.9	ND	0
	60-120	5.1	0.40	92.9	5.3	1.8	3.5	15	429
	120-180	5.0	0.42	97.2	0.9	1.9	6.4	ND	0
3	0-30	6.4	0.66	87.5	9.4	3.1	7.1	11	155
	30-60	5.3	1.47	94.7	3.3	2.0	4.3	ND	0
	60-120	5.0	0.38	98.0	1.3	0.7	3.0	ND	0
	120-180	5.1	0.40	95.1	3.4	1.5	3.9	ND	0

Table IV. (Continued)

Field	Sampling depth (cm)	pH	$CaCO_3$ equiv. (%)	Soil separates, % Sand	Silt	Clay	% H_2O	Aldicarb in Soil	H_2O
10*	0-30	6.3	0.74	88.9	7.4	3.7	8.9	5	56
	30-60	5.3	0.40	90.1	6.7	3.2	6.3	ND	0
	60-120	5.5	0.46	95.2	2.5	2.3	4.8	ND	0
	120-180	5.5	0.46	96.2	2.5	1.3	4.3	ND	0
	180-240	5.5	0.39	98.1	1.0	0.9	5.4	ND	0
10	0-30	6.8	0.67	87.3	9.1	3.6	8.7	5	57
	30-60	5.8	0.47	92.1	5.4	2.5	5.9	ND	0
	60-120	5.5	0.86	97.2	1.9	0.9	4.7	ND	0
	120-180	5.0	0.64	93.6	3.6	2.8	6.0	10	166
	180-240	5.1	0.49	96.8	1.7	1.5	5.1	ND	0
10	0-30	6.0	0.55	85.8	10.0	4.2	8.1	ND	0
	30-60	5.0	0.47	89.6	7.4	3.4	5.6	ND	0
	60-120	5.4	0.54	97.6	1.8	0.6	3.7	ND	0
	120-180	5.2	0.44	97.4	1.5	1.1	4.4	ND	0
	180-240	5.2	0.46	99.6	0.4	0	5.3	11	207
10	0-30	5.8	0.60	84.6	10.6	4.8	8.9	25	281
	30-60	5.6	0.44	88.3	6.6	5.1	6.5	21	323
	60-120	5.2	0.51	88.1	8.5	3.4	6.2	50	806
	120-180	5.2	0.49	85.6	14.4	0.4	7.9	79	1000
	180-240	5.4	0.44	99.2	0.6	0.2	7.6	18	237
10	0-30	6.1	0.90	83.8	10.9	5.3	10.6	17	160
	30-60	5.2	0.48	85.4	6.9	7.7	8.4	ND	0
	60-120	4.7	0.47	88.8	5.7	5.5	7.2	9	125
	120-180	5.0	0.41	96.7	1.2	1.1	4.2	16	381
	180-240	5.3	0.46	88.6	0.5	0.9	11.7	15	128

*Field 10 in Portage County was treated with 3 lb a.i./A (3.36 kg/ha) at planting in 1979 and 1981.

Table V. Concentrations [ppb, Mean/Range (No. of Samples)] of Aldicarb Residues in Lysimeter Leachate; Application Rates: 3 lb a.i./A = 3.36 kg/ha in 1982, 2 lb a.i./A = 2.24 kg/ha in 1983, 1984, at Planting

Month	1982		1983		1984	
	Rectangular	Round	Rectangular	Round	Rectangular	Round
June	ND/- (1)	3/0-7 (3)	ND/- (4)	ND/-4	5/ND-6 (7)	ND/ND-2 (7)
July	434/149-693 (6)	414/148-670 (11)	140/ND-162 (9)	100/2-250 (9)	64/5-148 (14)	19/3-71 (18)
Aug.	146/40-400 (40)	430/80-751 (40)	311/178-434 (11)	425/290-546 (11)	145/142-148 (2)	101/64-162 (6)
Sept.			87/64-143 (6)	210/115-316 (6)	11/- (1)	3/- (1)
Oct.			31/ND-127 (5)	21/ND-123 (4)		

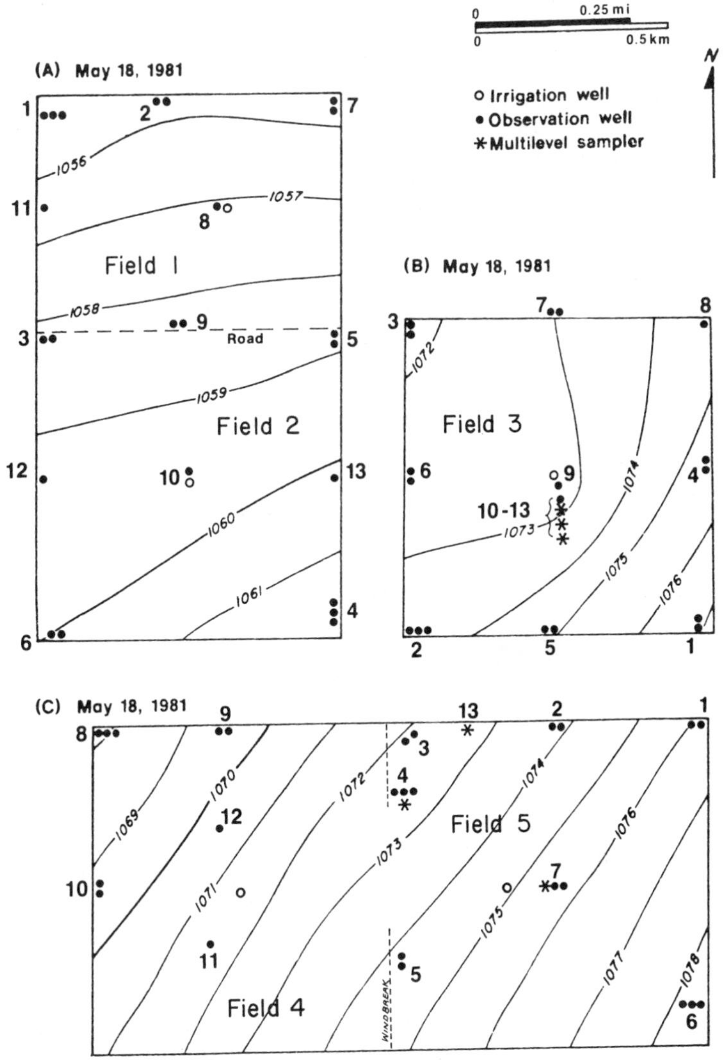

Figure 1. Maps showing water-table contours for Fields 1-5 and locations of monitoring wells, irrigation wells, and multilevel samplers.

(8.5 m) below the water table (5). Similarly, no contamination was initially observed at sites 1, 11 and 12 in Fields 1 and 2, which had each been treated three times, although the monitoring wells were in the path of groundwater flow. Likewise, no residues were detected in one well at position 3 in Field 3 or wells at positions 2 and 4 in Field 6 (Fig. 2).

No aldicarb residues were ever detected in water taken from wells installed in test plots at the U.W.-Madison Experimental Farm near Hancock, WI, even though one set of plots was irrigated at 1.6 times the measured soil moisture evapotranspiration rate (cf. 21).

A reason that no aldicarb residues were detected at many positions could be that the well depths selected were above or below a narrow plume of aldicarb contamination. However, the results observed with the multilevel samplers, with the 45 cm depth intervals between sampling ports (Table III), indicate that this is not likely: where contamination occurs, the plume is usually spread out over a depth of several sampling levels. Also, no significant aldicarb residues were encountered at all in some multilevel samplers (e.g. at locations 3 and 7 in Field 5; location 2 in Field 6).

One explanation for the paradoxical residue distribution apparently lies in the uneven penetration of aldicarb residues to the groundwater and their distribution with depth in the aquifer. Water samples withdrawn from positions directly below the treated area in Field 5 showed a variety of concentration profiles with depth (Figure 3), with maximum concentrations ranging from 18 ppb at position "a" to 108 ppb at position "e". At position "b", residues were found only at a depth of 3 feet (91.5 cm) below the water table (19 ppb).

Soil samples collected from the vadose zone also show a variety of aldicarb residue concentration profiles (Table IV). In view of the low Kd value for aldicarb in sandy soils (22) it can be assumed that any aldicarb residues detected in soil samples are entirely dissolved in the soil moisture. If soil residue values are recalculated on a soil-moisture basis, the resulting peak concentrations are closer to those observed in the groundwater or leachate from the lysimeters. However, the pattern of residue concentrations in the vadose zone is also highly erratic and bears no apparent relation to soil texture, moisture content or alkalinity (Table IV). Equally erratic concentration distributions were observed for samples collected on other dates in the same year. Similarly erratic results were obtained for different soil profiles from Long Island potato fields (22).

Residues in Lysimeter Leachate. The 1982-84 lysimeter studies indicated that breakthrough of aldicarb residues began around mid-July, or 30-40 days after treatment, peaked in mid-August and declined during September through mid-October to <10 ppb or nondetectable levels. No residues were detected in samples taken at any other time of the year. The growing season is the period of the year when water application rates are high by combination of precipitation and irrigation. The mean monthly precipitation for the years 1976 through 1984 is shown in Figure 4. The precipitation of 55.5, 61.1 and 60.2 cm for the months April through September was supplemented by 19 irrigation applications totalling

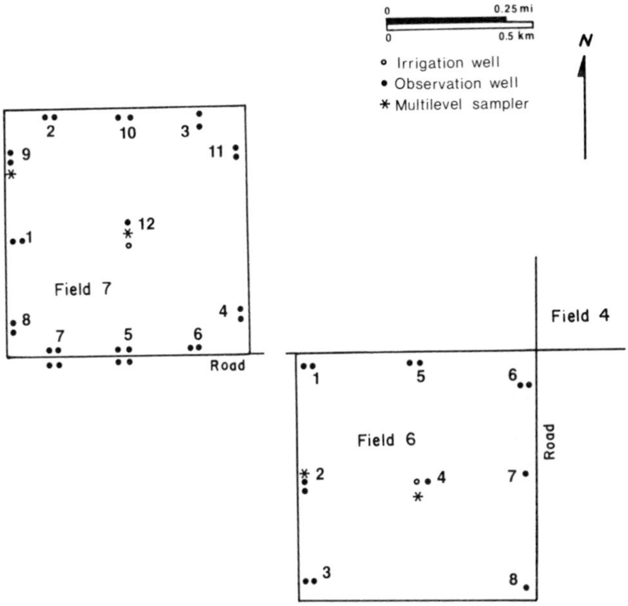

Figure 2. Map showing location of Fields 6 and 7 relative to Field 4, water-table contours and positions of monitoring wells, irrigation wells and multilevel samplers.

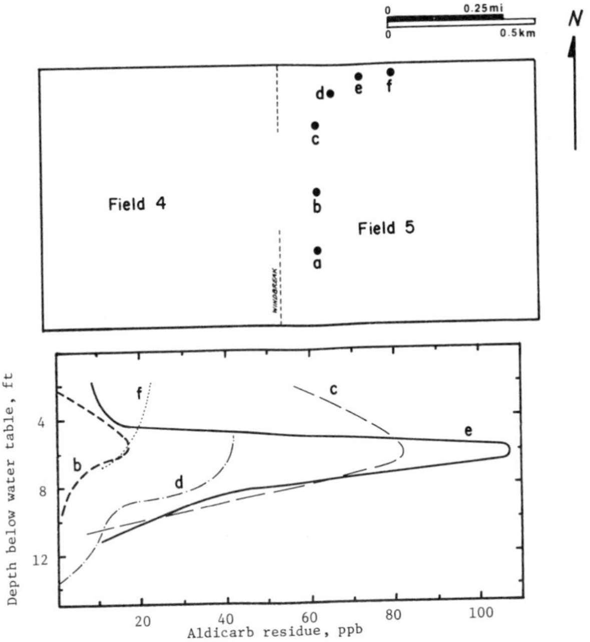

Figure 3. Locations in Field 5 sampled through the hollow-stem auger (April 19-21, 1982) and distribution of aldicarb residues versus depth below the water table.

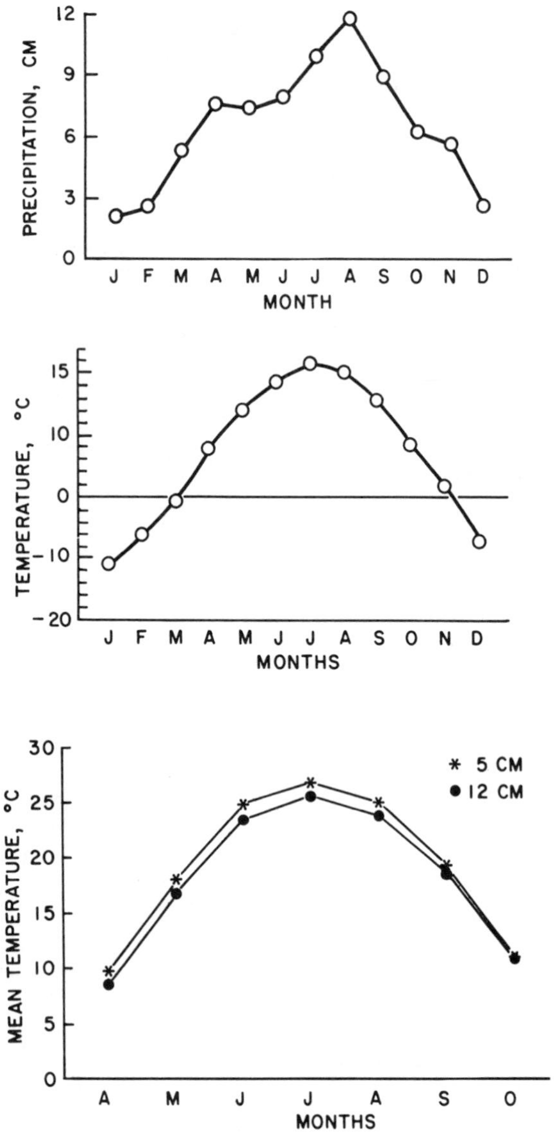

Figure 4. Mean air and soil temperatures and precipitation in study area, 1976-84.

36.7, 32.8, and 22.5 cm for the years 1982, 1983 and 1984. Irrigation water was applied between the beginning of June and mid-September. The average and range of aldicarb residue concentrations for August, September and October in 1982 when the aldicarb application rate was 3 lb/A (3.36 kg/ha) were substantially higher than those for 1983 and 1984, when the rate was 2 lb/A (2.24 kg/ha). Treatment took place in all cases at plant emergence, six weeks after planting. However, in 1982, just after the lysimeters were irrigated immediately following aldicarb application, they were drenched with a heavy (1.9 cm) rainfall.

Because of equipment failures, it was not possible to derive an accurate mass balance for the amount of aldicarb leached. However, by multiplying the average of the concentrations of each consecutive pair of samples by the intervening volume of drainage for 1983, the year with the best data set, a rough estimate of the fraction of applied aldicarb leached was obtained: 3.6% for the rectangular, 13.5% for the round lysimeter. Concentrations were higher for 1982, probably because of the higher application rate, but the fraction leached for the period measured was almost the same as for 1982: 5.4 for the rectangular versus 16.4 for the round lysimeter. For 1983, when the amount of irrigation water applied was lower, the percentages leached and the concentrations were suprisingly low: 3.2% for the rectangular versus 1.6% for the round lysimeter. The volumes of leachate collected in each case were similar to those obtained during previous studies of nitrate leaching with the lysimeters when "improved" (reduced) irrigation rates were used (23). Twenty-six samples of lysimeter leachate analyzed by Union Carbide using an HPLC method which determined aldicarb and its oxides separately established that no "parent" aldicarb leached through the soil; the residues were exclusively sulfoxide and sulfone in a ratio of 2.2 (range 1 to 4). Residues in samples from monitoring and household wells analyzed in the same way by Union Carbide showed that these too were exclusively sulfoxide and sulfone in a ratio of 1.3 (24). Although the lysimeters lie only 30 m apart, and both were treated identically each year, the difference each year in the concentrations and estimated fractions of the residues leached is remarkable; this may reflect non-uniform infiltration observed under potato canopies (25). Non-uniform leaching is also indicated by the variability in residue concentrations in the vadose zone (Table IV); cf. (22).

Residue Distribution with Depth. Wherever persistent aldicarb residues occur in wells or multilevel samplers the plume of contamination is observed at progressively deeper levels with time (Tables II, III). Highest levels are observed in a layer near the water table in years following aldicarb use; peak concentrations represent approximately averages of lysimeter leachate. If aldicarb is not used in the following year, recharge of residue-free water from precipitation and irrigation is gradually superimposed upon this contaminated layer, making it appear to sink. As the plume "sinks," the concentrations decrease dramatically. At most locations, abrupt changes and fluctuations within a few months indicate that mechanisms other than physical dilution and dispersion must be responsible for the concentration changes and disappearance of detectable residues. This suggests that degradation of residues to non-toxic fragments, presumably by chemical or microbial hydrolysis, is occurring in the groundwater. However, in some instances, e.g. sites 3 and 4 in Field 5, a zone of contamination remains and apparently passes below the lowest level in the multilevel samplers (Table III). However, no residues were detected in a monitoring well 6.6 m below the water table at site 4 and no persistent plume was observed at site 7 in the same field.

The same trends were observed with the 3 domestic wells which were deepened. The first well, approximately 1 km west along the road between Fields 1 and 2, was a sand point whose screen began at a depth of only 82 cm below the water table and yielded water with aldicarb residue contents of 90-111 ppb in 1981 and 1982. After deepening to 235 cm, residue levels fell to 26 ppb (12-82), but gradually rose again in 1983 [45 (6-83), 63 (9-83), 72 (12-83] and again declined with fluctuations during 1984-85 [53 (2-84), 63 (3-84), 38 (4-84), 46 (5-84, 51 (6-84), 74 (7-84), 59 (8-84), 42 (9-84), 45 (12-84), 45 (3-85), 39 ppb (6-85)]. The second well, a sand point installed in the northwest corner of Field 1, had initially screen starting 245 cm below the water table and yielded water containing 12-16 ppb of aldicarb residues in 1981 and 1982. After deepening by 305 cm, the water was free from aldicarb residues in 1983 and showed only traces in 1984-85 [1 (1-84), 2 (2-84), 3 (3-84), ND (5-84), 4 (6-84), 8 (7-84), 9 (8-84), ND (9-84), 1 (12-84), 5 (3-85), 6 (6-85)]. The third well, a sand point with screen starting 250 cm below the water table, was located close to Field 4 and had residue leels of 6-11 ppb in 1981 and 1982. Following deepening, no further residues were detected.

Clearly, the persistence and resultant distribution of aldicarb residues in groundwater are far from uniform, even under uniformly treated, uniformly irrigated fields. Residues seem to dissipate to levels of <10 µg/L or disappeared entirely by the time they reach depths of 3-4 m below the water table, but were more persistent in some areas than others.

Relation to Alkalinity. Some explanations for this behavior can be offered. Persistent levels appear to be associated with low pH (4.5-6.5) and alkalinity (<10 mg/L) in the groundwater. Residues disappeared faster from Fields 4 and 5, where pH and alkalinities were high, especially at deeper levels (Figure 5) than in Fields 1 and 2, where pH and alkalinity were low (Figure 6). Rates of chemical hydrolysis of aldicarb and its oxides have been extensively studied (26-29). Alkalinity and pH tend to increase with depth in the aquifer under all fields. However, prediction of rates of chemical hydrolysis based on known rate constants are complicated by two phenomena: 1. fluctuations in groundwater temperature; 2. higher *in situ* pH and alkalinity than those measured in the laboratory.

Shallow groundwater is colder in February and warmer in August than deep groundwater (cf. Figure 7), affecting hydrolysis rates. This reflects the higher air and soil temperatures and precipitation during the summer months (Figure 4). The seasonal differences in temperature with depth are more pronounced at the edges of fields than in the center (Figure 7). Presumably the cone of depression caused by irrigator pumps during the summer causes turbulences which vertically mix the water more at the middle of

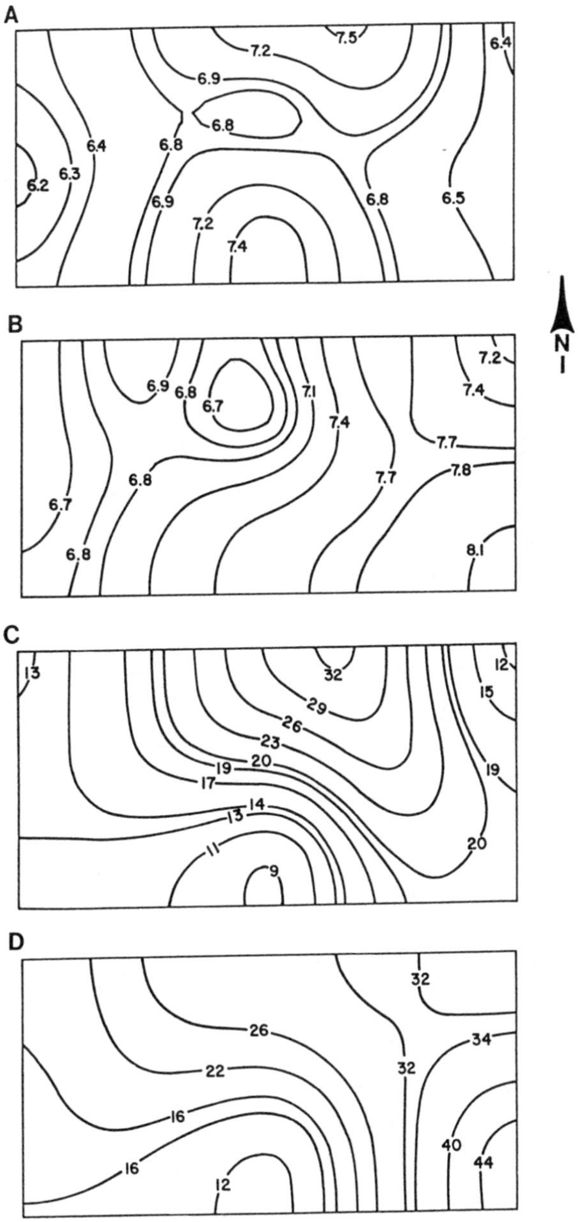

Figure 5. pH/alkalinities in shallow (A/C) and deep (C/D) groundwater under Fields 4 and 5.

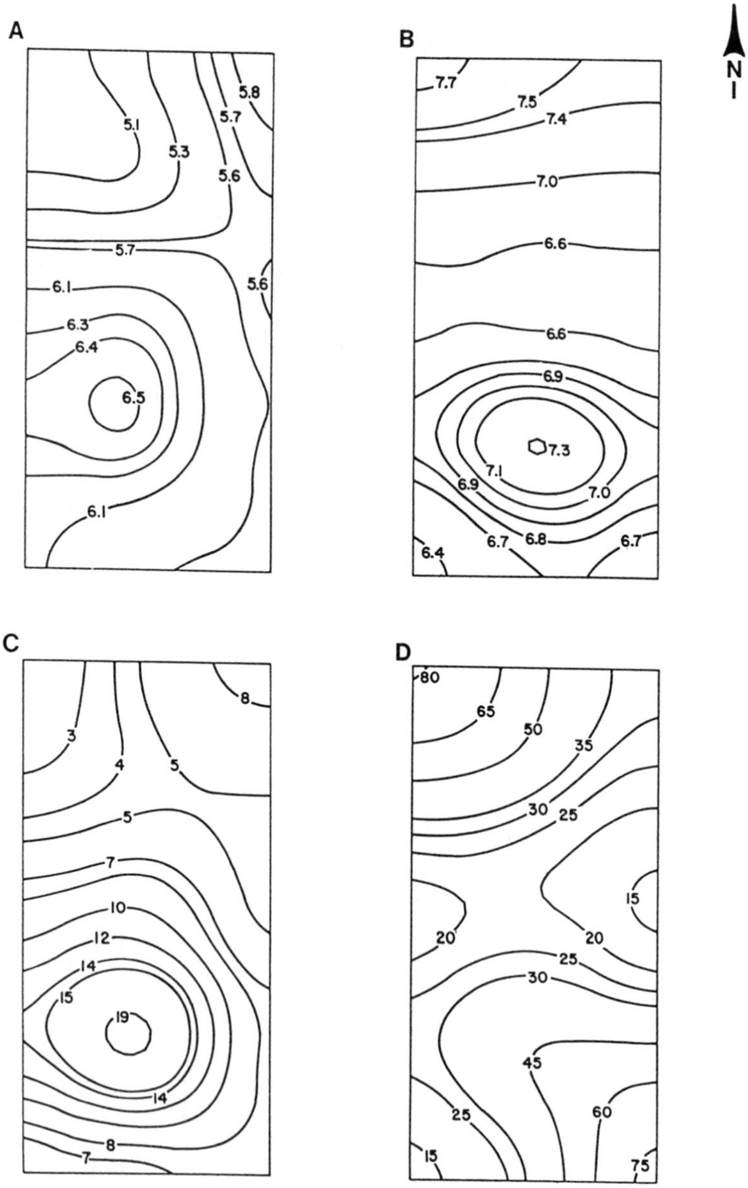

Figure 6. pH/alkalinities in shallow (A/C) and deep (B/D) groundwater under Fields 1 and 2.

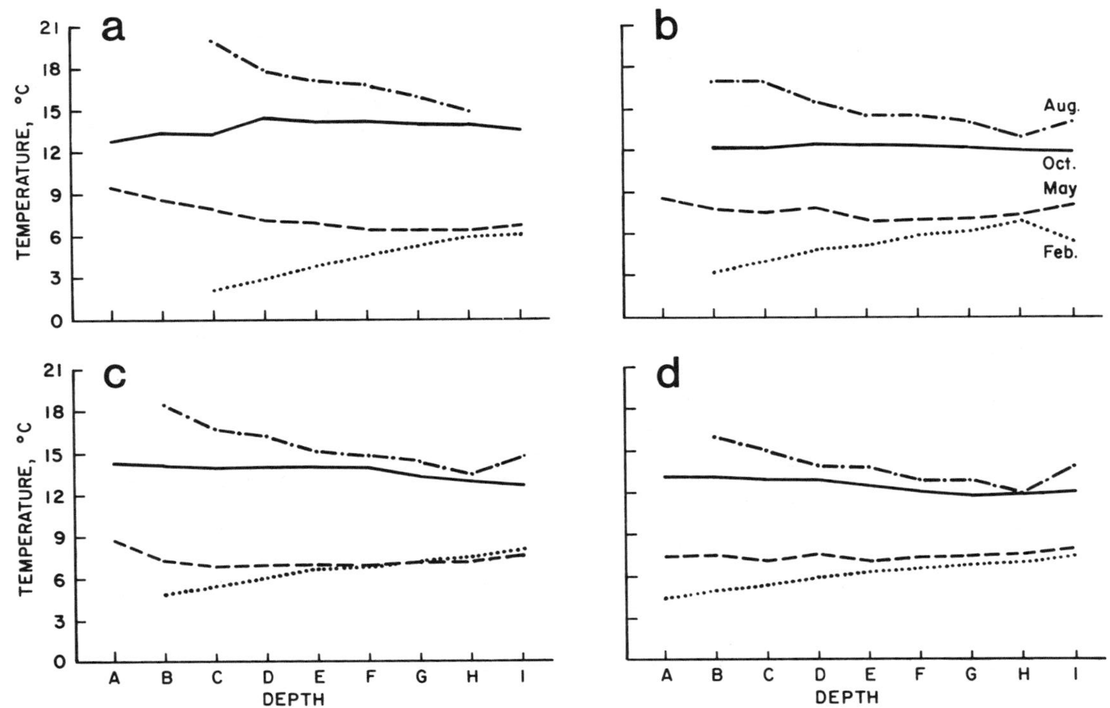

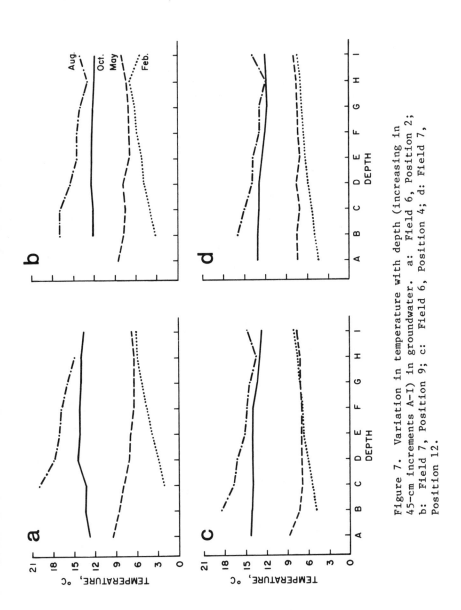

Figure 7. Variation in temperature with depth (increasing in 45-cm increments A-I) in groundwater. a: Field 6, Position 2; b: Field 7, Position 9; c: Field 6, Position 4; d: Field 7, Position 12.

fields. It is not known whether density differences associated with temperature differences promote vertical mixing and affect residue concentrations at aquifer locations beyond the influence of pumping drawdown.

The partial pressure of CO_2 (pCO_2) in groundwater *in situ* may be many times higher than that of extracted samples; this is the case in Central Wisconsin. The excess CO_2 is "lost to the atmosphere during titration" to determine laboratory alkalinity (13, p. 263). The quantity lost depends on the amount of agitation during titration and the degree of supersaturation attained at low pH levels (30). If alkalinity were solely determined by pCO_2 according to the carbonate equilibrium, i.e.

$$[Alk] = [HCO_3^-] + 2\ [CO_3^{2-}] + [OH^-] - [H^+]$$

a higher pCO_2 would tend to decrease alkalinity and pH. However, in the presence of insoluble dolomitic limestone $(Ca,Mg)CO_3$, a higher pCO_2 tends to dissolve more Ca and Mg as moderately soluble bicarbonates, actually increasing pH and alkalinity. If alkalinity is defined as the excess of cations of strong bases over anions of strong acids, a corrected *in situ* alkalinity can be calculated by a balance of all the highly dissociated ion species present in the groundwater (30), i.e.

$$[Alk] = [Na^+] + [K^+] + 2\ [Ca^{++}] + 2\ [Mg^{++}] + \ldots$$

$$-[Cl^-] - 2[SO_4^=] - [NO_3^-]\ \ldots$$

Table VI shows examples of laboratory-measured and the substantially higher calculated *in situ* alkalinities for 10 samples of groundwater from Field 6. The values shown in Figs. 5 and 6 are laboratory, not *in situ* pH and alkalinities. Higher *in situ* pH and alkalinity would lead to faster rates of aldicarb residue breakdown. Projections of aldicarb residue degradation by alkaline hydrolysis should be based on *in situ* pH, alkalinity and temperature. These, like the residue concentrations, vary widely, spatially and temporally.

Table VI. Measured and Calculated Alkalinities for 10 Groundwater Samples from Field 6

Date	Depth in groundwater (cm)	Aldicarb residues (μg/L)	pH/Alkalinity (mg/L $CaCO_3$ equivalents)	
			Measured	Calculated
8-82	107	ND	8.9/64	-/161
8-82	107	1	8.5/36	-/267
10-82	107	18	7.7/33	-/58
10-82	152	64	8.1/50	-/149
7-82	174	ND	6.7/35	-/54
7-82	229	37	5.2/2	-/99
10-82	229	29	5.9/4	-/18
8-82	290	ND	8.8/26	-/257
10-82	290	ND	8.6/45	-/51
10-82	378	ND	8.0/23	-/204

Relation to Microbial Activity. Breakdown of aldicarb residues by biodegradation is presumably superimposed upon chemical hydrolysis. Results of examination of soil and water samples removed aseptically from depths of 20, 75 and 230 cm below the water table in Fields 5, 6 and 7 for microbial content are shown in Tables VII and VIII. Groundwater samples collected in sterilized containers through sterilized equipment were supplemented with the sterilized yeast extract normally used for isolating soil bacteria and colonies enumerated by the pour-plate method. The same method was used for aquifer sediment abstracted aseptically in sterilized plastic cylinders using a sterilized split-spoon sampler (17).

The bacteria are associated primarily with sediment particles. The counts are similar to those obtained in groundwater and saturated sediments by others (31-33). About 36 separate species of bacteria have been isolated; nine of 18 studied in detail were tentatively classified (Figure 8). All are bacteria; no fungi were encountered. All are facultative anaerobes, but their growth into the medium rather than on the suface of semisolid agar deeps suggests their preference for microaerophilic conditions, such as those that occur in Wisconsin groundwater (5 mg/L dissolved oxygen). The well-drained sandy soils clearly permit diffusion of air to shallow groundwater. Several isolates prefer much lower concentrations of substrates in growth media than normal surface soil or wastewater bacteria. Some isolates form pinpoint colonies on aldicarb or its oxides as sole carbon source. Ability of a bacterium and five soil fungi to utilize aldicarb as a sole carbon source or degrade it as a cometabolite has been reported previously (34).

Table IX shows the microbial breakdown with time of [*S*-$^{14}CH_3$]-aldicarb, aldicarb sulfoxide and aldicarb sulfone to $^{14}CO_2$ in groundwater samples. Incubation temperatures of 10°C and 20°C were used to approximate the mean and highest annual groundwater temperatures in Central Wisconsin.

The slow rate of aldicarb breakdown is unimportant, since only the oxides occur in Wisconsin groundwater.

Evolution of $^{14}CO_2$ from aldicarb sulfoxide showed a typical degradation pattern: a lag (acclimation) phase of 2 to 4 weeks followed by rapid mineralization; aldicarb sulfone showed no lag phase. With the sulfone at 20°C, some radioactivity was incorporated into bacterial cells trapped on membrane filters.

In soils, aldicarb and its oxides labeled at the *N*-methyl group were converted more rapidly to $^{14}CO_2$ than [*S*-$^{14}CH_3$] derivatives (35). Total mineralization of aldicarb residues is not needed to detoxify them; hydrolysis of the carbamate ester is enough. Rates of detoxification of aldicarb oxides by microbial hydrolysis in groundwater may be faster than the rates of mineralization to CO_2 observed in the laboratory.

The large variability in temperature, pH, alkalinity, nitrate content, etc. with depth in groundwater we have observed may

Table VII. Enumeration of Groundwater Bacteria by Pour Plate Counts on Groundwater-Yeast Extract Agar

Sampling date	Site/Field	Depth below groundwater feet/cm	Bacterial counts Colony forming units/ml mean (sd) x 10^{-3}
11/82	4D/5	20.5/625	2.7 (0.3)
1/83	1A/6	1.4/43	7.4 (0.2)
	1B/6	3.9/119	6.0 (0.7)
	5A/6	1.8/55	8.4 (0.2)
	5B/6	6.7/205	8.0 (0.3)
2/83	1A/6	1.4/45	4.6 (0.2)
	1B/6	3.9/120	5.3 (0.1)
	5A/6	1.8/55	3.0 (0.2)
	5B/6	6.7/205	4.9 (0.1)
3/83	1A/6	3.4/104	4.8 (0.3)
	1B/6	5.7/174	3.5 (0.2)
	5A/6	3.6/110	5.1 (0.2)
	5B/6	8.5/260	4.8 (0.2)
8/83	5A/7	1.9/58	3.7 (0.1)
9/83	5B/7	5.6/171	0.6 (0.3)
12/83	5A/7	2.4/73	0.9 (0.4)
	5B/7	4.4/134	0.6 (0.4)

Table VIII. Enumeration of Bacteria in Aquifer Sediments by Pour Plate Counts on Groundwater-Yeast Extract Agar

Sampling depth m (ft)	Colony forming units/g of sediment (x 10^{-4})
2.2 (7.5)	9.0
3.1 (10.0)	8.5
4.6 (15.0)	4.7

Table IX. Mineralization of [*S*-$^{14}CH_3$] Aldicarb and its Oxides in Groundwater Samples

Compound	Duration of incubation, days	% of added radioactivity evolved as $^{14}CO_2$ at 10°C	20°C
Aldicarb	70	0.9	0.3
Aldicarb sulfoxide	63	3.6	5.9
Aldicarb sulfone	63	1.3	12.6

affect both bacterial numbers and rates of aldicarb oxides degradation, accounting in part for the erratic pattern of residues concentrations encountered. Similarly, the decreases in residue concentrations in lysimeter leachate from 1982-1984 (Table V) may reflect adaptation of vadose zone bacteria to aldicarb oxide residues.

Variability of Chemical Parameters. Patterns of concentrations of inorganic species derived from fertilizer leachate, e.g. nitrate, chloride and potassium, in monitoring wells and multilevel samplers were also erratic, even though ammonium nitrate and potassium chloride are applied every year. High levels of NO_3, Cl, K or any other parameter measured did not correlate with high levels of aldicarb residues. For instance, K concentrations were often much higher (10-30 mg/L) than Na concentrations (1-10 mg/L), especially in samples taken from Field 3, regardless of whether the samples contained aldicarb residues or not. Concentrations and distribution with depth of nitrate were equally erratic. In extreme cases (Well 4, Field 6) the average NO_3 content was 77 mg/L, the range of concentrations 4-360 mg/L (n = 19). For wells 5A and 5B (Field 6), the average and ranges were 52 (28-96) and 49 (10-134) mg/L (n = 19). Similarly for wells 9C, Field 1 and 7A and 7B, Field 5, the averages and ranges were 21 (6-36, n = 22), 15 (4-44, n = 18) and 35 (7-70, n = 20). In no instance did peaks in NO_3 levels correspond to peaks in aldicarb residues. However, in no case were aldicarb residues ever found in samples that contained no or only little NO_3.

Modelling Efforts. Two different approaches are being used to evaluate the field data: contouring plots with real data and predictive modeling in efforts to match the field data.

Contour plots could be prepared successfully for groundwater elevations and groundwater chemistry data (e.g. pH, alkalinity, NO_3 Cl, K) because real values were available for every sample analyzed. The density of sampling locations was insufficient to contour the aldicarb residue plume because many of the samples did not contain detectable levels.

The contour plots were constructed using three Madison Area Computing Center (MACC) subroutines. The first subroutine runs an interpolation of the data provided to create a uniform grid network. Refining and smoothing subroutines are then used to reduce large numerical variations. The smoothed numerical matrix is subjected to the contouring subroutine which plots isopleths from the numerical data. The data entered into the subroutines include two numbers to give a location in the grid network and a third value representing the parameter being plotted. Variations of depth in the groundwater are plotted on different maps, so that each contour represents a specified depth in the aquifer. The plots of pH and alkalinity for shallow and deep wells (Figs. 6 and 7) were obtained in this way. Similar diagrams for NO_3, Cl, Na and

FIRST STAGE TESTS IN IDENTIFYING BACTERIAL ISOLATES

Gram	+	+	−	−	−	+
Shape	R[a]	R	R	R	R	R
Size						
Width, μm	1.0-1.5	1.0	0.8	0.5-1.0	v[b]	v
Length, μm	2.0-4.0	2.0-3.0	1-5	1.5-2.0	v	v
Spores	+	−	−	−	−	−
Motility	±	−	±	−	±	±
Catalase	+	+	+	+	+	+
Oxidase	−	−	+	±	±	±
Growth in air	+	+	+	+	+	+
Growth anaerobically	+	+	+	+	+ (?)	+ (?)
Glucose O/F	−	−	−	±	v	v
Bacillus	3					
Corynebacterium		2				
Cytophaga			1			
Flavobacterium				3		
Unidentified					4	
Unidentified						5

[a] R = Rods
[b] v = varies

Figure 8. Characteristics of some faculative anaerobic bacteria isolated from Wisconsin groundwater. The boxed numbers indicate the number of species tentatively identified.

K concentrations were prepared but bore no obvious relation to observed aldicarb residue levels.

To try to predict aldicarb movement to and in the groundwater, efforts are being made to combine two models, the U.S. Environmental Protection Agency's Pesticide Root Zone Model (36) and the Random Walk Model (37). The goal is to match the output of the combination of these models with the observed aldicarb residue concentrations listed in Tables II and III.

The Pesticide Root Zone Model (PRZM) is being utilized to calculate a daily concentration of aldicarb leaching into the groundwater. Daily meteorological records and irrigation schedules are input into the model. Other important input parameters include

Table X. Effects of Varying Soil Organic Matter (SOM) and the Half-Life of Aldicarb on the Output of the PRZM Model, within Ranges in the Fields Studied

Unsaturated soil column length (cm/ft)	Percent SOM in rooting/vadose zones	Percent change[a]	Half-life (days) in rooting/vadose zones	Percent change[b]
122/4	2/0.1	6	14/180	11
	3/0.1	11	14/365	21
	5/0.1	19	30/365	44
			60/365	61
305/10	2/0.1	7	14/180	39
	3/0.1	12	14/365	71
	5/0.1	21	30/365	106
			60/365	127
488/10	2/0.1	8	14/180	148
	3/0.1	14	14/365	327
	5/0.1	24	30/365	436
			60/365	496

[a]Compared to "standard" conditions (1% SOM in rooting zone, 0.1% SOM in vadose zone), aldicarb concentrations in groundwater would be decreased by this percentage.

[b]Compared to "standard" conditions (half-life = 14 days in rooting zone, 90 days in vadose zone), aldicarb concentrations in groundwater would be increased by this percentage.

several soil and pesticide properties. PRZM is a compartmentalized mass-balance type model and downward movement of water and solutes in the soil column is assumed to occur whenever the soil moisture content exceeds the field capacity. The field capacity and wilting point in our simulations were 11.2% and 4.8% (21). The soil organic matter content was estimated from county summaries (38) and core samples (39). The soil organic matter content (OM) is used to calculate the soil bulk density (BD) by the following equation (40):

$$BD = \frac{100.0}{\frac{\%\ OM}{OMBD} + \frac{100.0 - \%\ OM}{MBD}}$$

where OMBD and MBD are the bulk densities of the organic matter and mineral fractions of the soil. The soil organic matter content is also used to calculate the soil distribution coefficient (K_d) from the relationship between the partition coefficients between water and octanol (K_{ow}), and water and soil organic carbon (K_{oc}):

$$\log K_{oc} = 1.00 \log K_{ow} - 0.21$$

using the equation (41):

$$K_d = K_{oc} \frac{(\%\ OC)}{100}$$

An octanol/water partition coefficient, log K_{ow}, of 0.69 for aldicarb (42) gives a K_{oc} of 3.1, so that the calculated soil distribution coefficient K_d of 0.0155 indicates very little adsorption of aldicarb. Values for K_d for the aldicarb oxides would be presumably much lower, i.e. these compounds would be sorbed even less by soil.

The final significant paramater is decay rate. A sensitivity analysis of half-lives ($T_{1/2}$) ranging from 14 to 365 days was conducted (Table X). A very significant (>400%) change was seen in the residue leaching as aldicarb below 488 cm (16 feet). The high variability of decay rates in the unsaturated zone (43-6) makes the choice of a single value or single range of decay rates of questionable value. A number of simulations utilizing best and worst case decay rate ranges is being used to determine the concentration of aldicarb that could leach into the groundwater. The output of these simulations will be fed into the Random-Walk Model; movement of aldicarb residues through the saturated zone is being simulated utilizing this model (37). Residue concentrations are input into the model daily and are assumed to act as a conservative solute showing very little retardation. The movement of aldicarb residues is assumed to be governed by the direction and velocity of the groundwater flow. Simulation of hydrogeological parameters such as hydraulic conductivity (K), storage coefficients, dispersion coefficients are similar to those used in an earlier modeling study (9). The thickness of the aquifer and the decay rate are initially being set at 6.1 m and 900 days. These two parameters are quite significant and no single value may perhaps adequately simulate even a field-size model. Consequently, ranges of values are being utilized in a sensitivity analysis.

Our efforts to calibrate the models with the field data have been unsuccessful. Even using the K_d value for aldicarb--which is less water-soluble and more strongly sorbed than the sulfoxide and sulfone, the forms which actually leach--the PRZM model overestimates the amount of residue leaching, compared with measured values (cf. 39); the inputs into the Random Walk Model are therefore too high and the outputs do not correspond to observed data. This suggests that the degradation of residues by alkaline hydrolysis and microbial metabolism in both the vadose zone and groundwater is substantially higher than has been assumed on the basis of rate constants derived from laboratory studies (26-9, 43-6).

Conclusions

Regardless of application rates and dates, some residues of aldicarb, apparently exclusively the sulfoxide and sulfone, leach through sands/sandy loams in irrigated potato fields in Central Wisconsin and into shallow groundwater. However, the penetration patterns are highly erratic in time and space, judging from levels measured in soil samples from the vadose zone, lysimeter leachate and shallow monitoring wells. Other contaminants being leached from agricultural chemicals applied to surface soils, e.g. nitrate, chloride and potassium, exhibit equally erratic concentration distributions. Aldicarb residues encountered in the groundwater tend to be initially more concentrated in a zone near the water table. As this zone is overlain by further drainage, the contaminated zone "sinks" to lower depths in the aquifer, and residue concentrations decrease. The reduction in concentration is greater than can be ascribed to straightforward dilution and dispersion processes. Superimposed upon the normal concentration reductions due to classical dilution/dispersion mechanisms associated with groundwater flow is the seasonal fluctuation in water table and the local fluctuations and perturbations in flow patterns caused by water abstraction by high capacity irrigation wells. Groundwater temperatures increase with depth in winter and decrease in summer; corresponding density differences may promote vertical mixing and accelerate residue dispersion by convection.

Residue persistence in groundwater appears to be associated with low pH and alkalinity. Groundwater pH and alkalinity increase with depth in the aquifer. High CO_2 contents in the groundwater, especially in summer, tend to increase water contents of Ca and Mg by dissolution of dolomitic limestone residues from the aquifer sediments, increasing pH and alkalinity above laboratory-determined values. Such conditions apparently promote chemical hydrolysis of aldicarb residues, reducing their persistence in deeper groundwater.

The ability of bacteria found in the groundwater and aquifer sediments to metabolize aldicarb oxides suggests that microbial degradation is responsible in part for disappearance of residues from groundwater.

The complex interplay of this multiplicity of physical, chemical and biological factors which affect aldicarb residue concentrations in the groundwater at any place and time makes efforts to describe the fate of aldicarb in Wisconsin potato fields by mathematical models an extremely arduous, perhaps futile, task.

Literature Cited

1. Zaki, M. H.; Moran, D.; Harris, D. Am. J. Public Health 1982, 72, 1391-5.
2. Guerrera, A. A. J. Am. Water Works Assoc. 1981, 73, 190-9.
3. Kammerer, P. A. "Ground-water Quality Atlas of Wisconsin." Info. Circ. 39. U.S. Geol. Survey/U.W.-Extension Geol. Nat. Hist. Survey, Madison, WI. 1981, 39 pp.
4. National Uranium Resource Evaluation Program. "Hydrogeochemical and Stream Sediment Reconnaissance Basic Data for Eau Claire NMTS Quadrangle, Wisconsin, Minnesota," Oak Ridge Gaseous Diffusion Plant, Oak Ridge, TN. 1978. (Also corresponding document for Green Bay Quadrangle, WI.
5. Chesters, G.; Anderson, M. P.; Shaw, B.; Harkin, J. M.; Meyer, M.; Rothschild, E.; Manser, R. "Aldicarb in Groundwater" Water Resources Center, University of Wisconsin, Madison, WI, 1982; 38 pp.
6. Back, R. C.; Romine, R. R.; Hansen, J. L. 1984. Envir. Toxicol. Chem. 3, 589-97.
7. Rothschild, E. R.; Manser, R. J.; Anderson, M. P. Ground Water 1982, 20, 437-445.
8. Rochschild, E. R. M.S. Thesis, University of Wisconsin, Madison, WI, 1982.
9. Manser, R. J. M.S. Thesis, University of Wisconsin, Madison, WI, 1983.
10. Pickens, J. F.; Cherry, J. A.; Grisak, G. E.; Merritt, W. F.; Risto, B. A. Ground Water 1978, 16, 322-7.
11. Black, T. A.; Thurtell, G. W.; Tanner, C. B. Soil Sci. Soc. Amer. Proc. 1968, 32, 623-9.
12. Black, T. A.; Gardner, W. R.; Thurtell, G. W. Soil Sci. Soc. Amer. Proc. 1969, 33, 655-61.
13. "Standard Methods for the Examination of Water and Wastewater," American Public Health Association, American Water Works Association, Water Pollution Control Federation, 15th Edit. 1981.
14. Cotlove, E. V.; Trauntham, V.; Bowman, R. L. J. Lab. Clin. Med. 1958, 50, 358-71.
15. American Chemical Society, Committee on Environmental Improvement, "Principles of Environmental Analysis," Anal. Chem. 1983, 55, 2210-8.
16. Horwitz, W.; Kamps, R. L.; Boyer, K. W. J. Assoc. Off. Anal. Chem. 1980, 63, 1344-54.
17. Dzantor, E. K.; Harkin, J. M. Submitted to J. Envir. Qual. 1985.
18. Coppedge, J. R.; Lindquist, D. A.; Bull, D. L.; Dorough, H. W. J. Agr. Food Chem. 1967, 15, 902-10.
19. Smelt, J. H.; Schut, C. J.; Dekker, A.; Leistra, M. Neth. J. Plant Path. 1981, 87, 177-91.
20. Smelt, J. H.; Schut, C. J.; Leistra, M. J. Environ. Sci. Health Bd. 1983, 18, 643-65.
21. Wyman, J. A.; Jensen, J. O.; Curwen, D.; Jones, R. L.; Marquardt. Environ. Toxicol. Chem. 1985, 5 pp.

22. Enfield, C. G.; Carsel, R. F.; Cohen, S. Z.; Phan, T.; Walters, D. M. Ground Water 1982, 20, 711-22.
23. Saffigna, P. G.; Keeney, D. R.; Tanner, C. B. Agron. J. 1977, 69, 251-7
24. Romine, R. R. Union Carbide Agricultural Products Co., Research Triangle, NC. 1985. Personal comunications.
25. Saffigna, P. G.; Tanner, C. B.; Keeney, D. R. Agron. J. 1976, 337-42.
26. Chapman, R. A.; Cole, C. M. J. Environ. Sci. Health 1982, B17, 487-504.
27. Lemley, A. T.; Zhong, W. Z. J. Environ. Sci. Health 1983, B18, 189-206.
28. Hansen, J. L.; Spiegel, M. L. Environ. Toxicol Chem. 1983, 2, 147-53.
29. Miles, C. J.; Delfino, J. J. J. Agric. Food Chem. 1985, 33, 455-60.
30. Stumm, W.; Morgan J. "Aquatic Chemistry;" John Wiley and Sons, New York, 1981.
31. Ghiorse, W. C.; Balkwill, D. L. Dev. Industr.Microbiol. 1983, 24, 213-24.
32. Hirsch, P.; Rades-Rohkohl, E. Dev. Industr. Microbiol. 1983, 24, 183-200.
33. Wilson, J. T.; McNabb, J. F.; Wilson, B. H.; Noonan, M. J. Dev. Industr. Microbiol. 1983, 24, 225-33.
34. Spurr, H. W.; Sousa, A. A. J. Environ. Qual. 1965, 3, 130-3.
35. Heywood, D. L. Environ. Qual. Saf. 1975, 4, 128-33.
36. Carsel, R. F.; Smith, C. N.; Mulkey, L. A.; Dean, J. D.; Jowise, P. "Users Manual for the Pesticide Root Zone Model (PRZM). 1984, EPA-600/3-84-109. U.S. Environmental Protection Agency, Washington, D.C.
37. Prickett, T. A.; Maymik, T. G.; Lonnquist, C. G. "A Random-Walk Solute Transport Model for Selected Groundwater Quality Evaluation. 1981, Illinois State Water Survey Bull. 65, 103 pp.
38. Wisconsin State Soil Testing Laboratory. Summary Reports for Portage and Washara Counties 1976-1980 and 1981-1984. Dept. of Soil Science, University of Wisconsin-Madison, 1981, 1985.
39. Wyman, J. A.; Medina, J.; Curwen, D.; Hansen, J. L.; Jones, R. L. 1985, in the press.
40. Rawls, W. O. Soil Sci. 1983, 135, 123-5.
41. Karickhoff, S. W.; Brown, D. S.; Scott, T. A. Water Res. 1979, 13, 241-48.
42. Rao, P. S. C.; Davidson, J. M. In "Environmental Impact of Nonpoint Source Pollution;" Overcash, M.R.; Davison, J.M. Eds., Ann Arbor Science, 1980. p. 23-36.
43. Smelt, J.H.; Leistra, M.; Houx, N.H.W.; Dekker, A. Pestic. Sci. 1978, 9, p. 279-85.
44. Smelt, J.H.; Leistra, M.; Houx, N.H.W.; Dekker, A. Pestic. Sci. 1978, 9, p. 286-92.
45. Leistra, M; Smelt, J.H.; Lexmond, T.M. Pestic. Sci. 1976, 7, p. 271-82.
46. Kuseke, D.W.; Funke, B.R.; Schultz, J.T. Plant Soil 1974, 41, p. 255-69.

RECEIVED April 17, 1986

13

Complexity of Contaminant Dispersal in a Karst Geological System

David A. Kurtz[1] and Richard R. Parizek[2]

[1]Department of Entomology, The Pennsylvania State University, University Park, PA 16802
[2]Department of Geosciences, The Pennsylvania State University, University Park, PA 16802

The various means of dispersal of environmental contaminants within folded and faulted carbonate (karst) rocks that define a complex karst aquifer system of the Central Appalachian type include surface and subsurface pathways. Chemicals may be exchanged among surface water, soilwater, and groundwater flow systems. Concentrations of a marker compound found in spring waters, Spring Creek surface waters, and groundwater pumped from wells located adjacent to Spring Creek illustrate these exchange processes. Water from residential wells grouped near each other and located downstream from a waste-disposal site contained the marker compound at different levels ranging from 5 to 22 ppt in 1984. Concentrations found in 1977/78 were similar to those found in 1984 in some houses but were different in others. Evidence is presented that show barriers to subsurface water flow exist in the form of geological structures, of subsurface water table divides and gradient directions, and of water flows within groundwater basins of various sizes, all of which channel the potential flow of contaminants and limit dispersion possibilities. The contaminant may have entered the more regional carbonate aquifers by natural streambed leakage, by infiltration induced by groundwater pumpage, or by nearby sewage sludge disposal.

Chemical contamination of some groundwater and surface-water supplies may have occurred as a result of point-source, waste disposal practices located near State College, PA and the Pennsylvania State University. The compound, unidentified, but having a consistent retention time for elution in the analytical method used for tracing and quantification, will be further described as a marker compound in order to properly focus on the dispersal aspects of this problem. It

0097-6156/86/0315-0256$07.50/0
© 1986 American Chemical Society

was found in water areas and may have resulted from production from any of various compounds in a local manufacturing plant. Disposal of waste materials resulting from the manufacture of various chemicals 20-30 years ago was in accordance with accepted procedures of those times in on-site disposal areas. These areas are located on geologically transported and residual soils up to 60 feet thick. They overlie folded and faulted limestone and interspersed dolomitic bedrock that contains secondary solution openings. Unfortunately, the plant was sited on a sequence of tilted layers of limestone whose solution cavities extended deeply from the surface. The marker compound was found first in surface and groundwater some years ago and continues to the present day despite more recent attempts to decontaminate this disposal site. Contaminants are being released slowly from unsaturated soil and bedrock immediately below and adjacent to the plant and are likely to linger for some years to come.

Lined and unlined lagoons were constructed on-site to store liquid wastes. These leaked at various times in the past thus contributing to groundwater pollution. A land treatment operation where liquid wastes were disposed of by flood irrigation methods also added to the pollution. These on-site waste disposal practices have contributed to soilwater and groundwater contamination and are marked "A" in Figure 1. This is the potential initial disposal point for the marker compound and other chemical substances released with plant effluents and sludges.

Surface and subsurface drainage below the plant flows into Spring Creek, a stream known for its fine trout fishing, scenic beauty, and the location of Benner Springs Fish Research Station, operated by the Commonwealth of Pennsylvania (Figure 1). Water samples in this stream near the source of contamination have been found to contain 20 to 90 ng/kg (ppt) of the marker compound during the past 10 years (1).

This paper will discuss the various means of dispersal of compounds disposed in a waste area located in a region of diverse hydrogeological setting.

Chemical Analysis Methods

Sampling. Water samples were collected in gallon glass jugs with teflon-lined stoppers. The jugs had been previously cleaned and silanized with dimethyldichlorosilane and were extracted to check for residues or artifacts of the marker compound prior to use.

Well water samples were collected from residential taps prior to any in-place water softening units. Water samples taken during a spring flood water period were collected from inside taps after a full open flush of 30 minutes. This assured that the sample collected was groundwater withdrawn from the carbonate aquifer some distance from the well. Stream samples were taken at the center of the channel near the stream surface. The cross sectional stream samples were taken equidistant apart at 1/3 the water depth at each lateral position. Water samples were stored less than a week at 5°C.

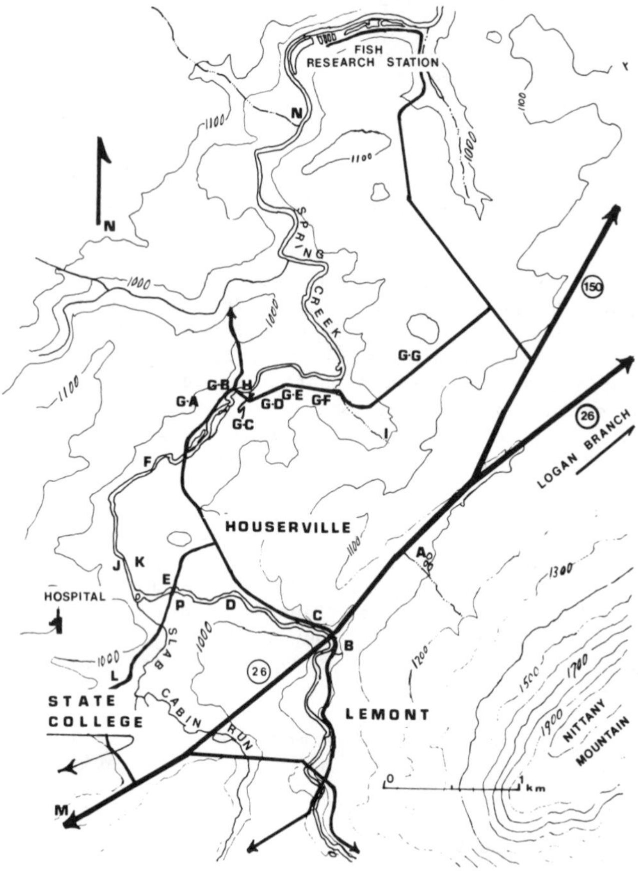

Figure 1. Topographic map showing the area of contamination of the marker compound in groundwater and surface water. Area covers 5 x 6 km. Designated points are described in the text.

Extraction. Accurately measured samples, approximately 3 L in size, were extracted in their original containers four times each with 175 mL of toluene (Baker, Resi-Analyzed). Each layer of toluene was siphoned off the water with an aspirator-driven teflon tube apparatus. The combined solution extract was rotoevaporated to dryness in a silanized flask and was taken up with toluene for chromatographic analysis.

Chromatography. Gas chromatographic analysis (Tracor, Model 220) was performed with a 1.8 m 4 mm i.d. "U" column packed with 1.5% SP-2250/1.95% SP-2401 coated on Supelcoport 100/120 support. The oven temperature was 220°C, and the inlet and detector temperatures were 250 and 350°C, respectively. The gas flow was 60 mL/min with 95:5 argon:methane. Detection was accomplished with a Ni^{63} electron capture detector controlled by a linearized electrometer (Tracor). Output signals were processed electronically (Varian CDS-401).

Samples were also analyzed with a 25 m fused silica capillary column, direct injection, coated with crosslinked 5% phenylmethyl silicone (Hewlett Packard, ultra-performance). The nitrogen carrier gas was 44 cm/sec. Following injection into a 200°C zone, the column was held at 50°C for 6 minutes and then programmed at 3°C/min to 250°C. Peaks were identified relative to a dieldrin internal standard.

Quantification. Residue concentrations were calculated by use of regression calibration graphs on logarithmically transformed data (2). Lack of fit protocol was observed for the calibration graph.

Laboratory recoveries of the marker compound were run at all times when unknown samples were analyzed (Table I). These were always accompanied by a blank recovery of the water used as a spiking medium.

Hydrogeological Setting

The study area is located in Nittany Valley in the central part of Pennsylvania, an area about 100 km long and 10 km wide bounded by Nittany and Tussey Mountains. In this locale the Allegheny Mountains stretch southwest to northeast and are nestled against the Allegheny Front. Bald Eagle Creek flows to the northeast in Bald Eagle Valley against the front and is separated from Nittany Valley by Bald Eagle Mountain. From a position to the south of Nittany Mountain and extending 25 km northward, Spring Creek winds its way across shallow limestone and dolomite aquifers before flowing into Bald Eagle Creek at Milesburg. Figure 1 shows a generalized topographical map of the immediate area of study, and Figure 2 a regional geological cross section of Nittany Valley.

Nittany Mountain is a synclinal mountain. It is cored with resistant sandstones, less resistant shales and siltstones. Tilted shale beds make up the lower, more gentle slopes of Nittany Mountain which stands more than 900 feet above the general upland surface of the valley. This upland is underlain by nearly 8000 feet of folded and faulted carbonate rocks which have been arched upward to form a

Table I. Recovery of the Marker Compound from Spiked Tap Water

Date	Samples Accompanying the Recovery Tests	Spike Charge	Marker Recovery	
		ng	%	std. dev.
Aug. 1981	Single well	6	90	
		23	99	12 (n=4)
Jun. 1983	Spring Creek	150	107	
	Thornton Spring		109	
Jun. 1984	Stream Cross Section	240	92	7 (n=10)
Jan. 1985	House Water Taps	2	99	
			104	
		15	92	
			100	
Feb. 1985	Spring Creek Flood	2.5	127*	
	House Water Taps		127*	
		5	121*	
			141*	

* Peak heights of 1.3 and 2.3 cm, resp. contained unresolvable shoulder peaks.

broad anticline with minor synclinal folds. The region has been weathered to produce distinctive karst terrain. The reference in this discussion to karst refers always to weathered carbonate rocks. It has been cut by at least two zones of thrust faults, one that surfaces near the plant and one that is exposed to the west. Tear faults are recognized that cross cut and displace the upturned karstified rocks and also increase their permeability down valley where marker compound has been detected in groundwater.

Rocks ranging in age from Late Cambrian through Lower Devonian are transected by Spring Creek between the manufacturing plant site and its confluence with Bald Eagle Creek. Limestone units of the Middle Ordovician Series underlie the plant and extend upslope to the base of Nittany Mountain. Together, these limestone units are nearly 900 feet thick and include the more soluble karstified rocks that localize all of the known caves in central Pennsylvania (3). These strata contain abundant solution cavities both above and below the water table near the plant and are responsible for conduit systems that drain groundwater parallel to the surface trend of individual rock layers to springs located on Spring Creek to the southwest of the plant and to the head waters of Logan Branch located northeast of

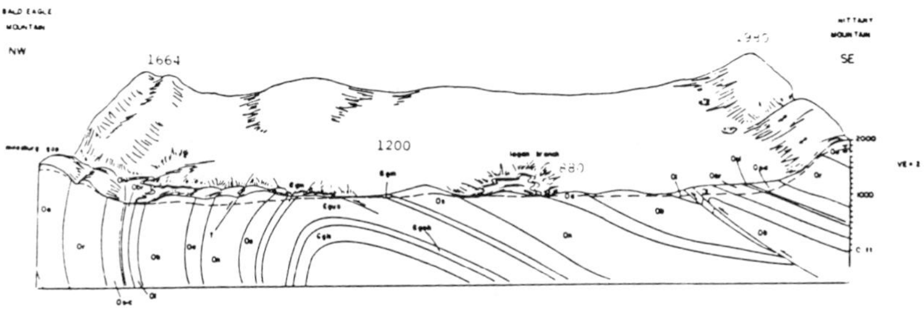

Explanation

Lower Ordovician	Ob Oa On Os	Bellefonte Dolomite Axemann Limestone Nittany Dolomite Stonehenge Limestone	Silurian	St	Tuscorora Formation
			Upper Ordovician	Oj Oo Or	Juniata Formation Oswego Sandstone Reedsville Shale
Upper Cambrian	Cgm Cgus Cgon Cgls	Gatesburg Formation	Middle Ordovician	Os-c Ol-t	Salona-Coburn Limestones Milroy-Nealmont Limestones

Thrust Fault — ∇ — — Water Table

Figure 2. Regional geological cross section of Nittany Valley (4).

the plant (Figures 3 and 4). These conduits are recharged by diffuse infiltration of surface water that enters residual and transported soils, by surface water that enters swallow holes near the base of Nittany Mountain located to the east of the plant, and by groundwater that drains into these conduits from adjacent, less soluble beds of limestone and dolomite located both east and west of the plant.

The Middle Ordovician limestone units are, in turn, thrust above a very fine-grained, relatively insoluble dolomite. This thrust fault causes cavernous limestone units to be exposed along a second narrow belt that is parallel to the northeast-southwest trend of Nittany Mountain and separated from the main belt of Middle Ordovician Series limestones by a thin, repeated section of dolomite. Thornton Springs emerges from the uppermost section of this same dolomite.

The dolomite is more resistant than the overlying Middle Ordovician Series limestone units and an underlying limestone sequence that is exposed further to the northwest of the plant. The dolomite forms a ridge of low relief that parallels Route 26 (Figure 3) and significantly influences local and more regional groundwater flow. A low relief groundwater mound underlies this ridge. Thin shale partings, the rather impure nature of the dolomite, and its inclination toward Nittany Mountain restrict groundwater flow across the direction of bedding. These features also help to support the groundwater mound which serves as a hydraulic barrier to groundwater flow. Water movement along the southeast flank of this groundwater mound is toward the plant and conduit drain system. Along the mountain's northwestern slope, flow is northwestward toward Spring Creek. Besides the hydraulic barrier offered by this groundwater divide, a deep, confined flow system also is present within a thin sandstone bed that is found near the top of the dolomite and crops out along the ridge (Figure 3) (4). Groundwater that circulates through the core of Nittany Mountain appears to be moving updip through this porous sandstone and is discharged at its outcrop along the ridge crest. This deep groundwater flow system also restricts contaminant movement from the plant site through the ridge toward Spring Creek to the northwest. Contaminants first must enter Spring Creek near Thornton Springs to the southwest (point B, Figure 1) before being dispersed within surface and groundwater downvalley. A pumping cone of depression has depressed water levels at one point in this mound (Figure 4) but the marker compound has not been detected in any of the wells responsible for this depression (1).

A transient, groundwater divide separates groundwater flow along the conduit systems developed in the Middle Ordovician Series limestones (point A, Figure 4). Two groundwater subbasins are recognized: one that drains to springs along Logan Branch and that is unaffected by plant wastes, and a second basin that drains to Thornton Spring (Figure 4). This divide is not fixed in its location by rock structure but rather can migrate northeastward or southwestward in response to changes in recharge, discharge, and groundwater pumpage.

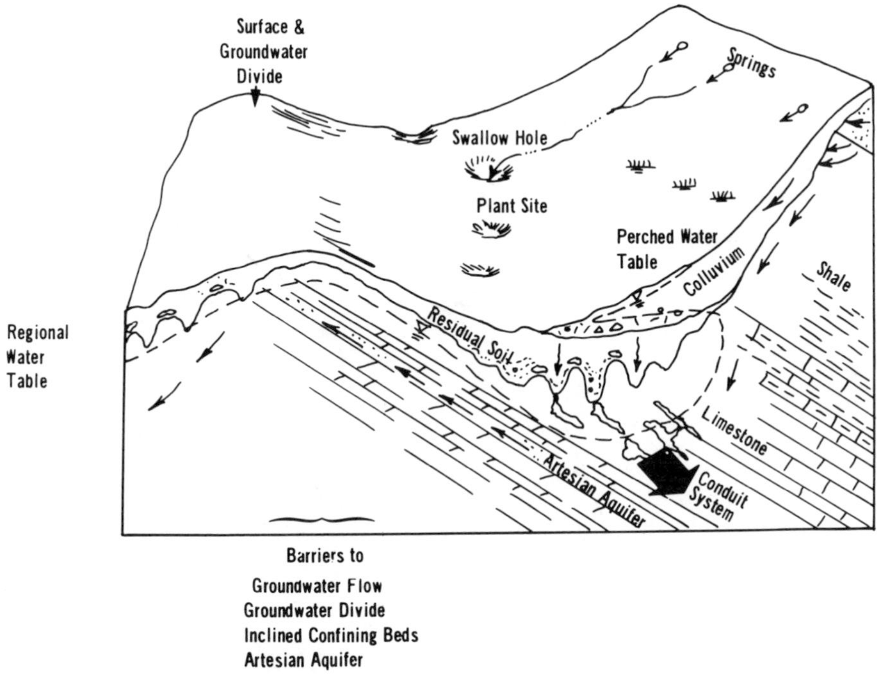

Figure 3. Geological cutaway showing possible entrance, dispersion pathways, and barriers to groundwater flow near the plant site.

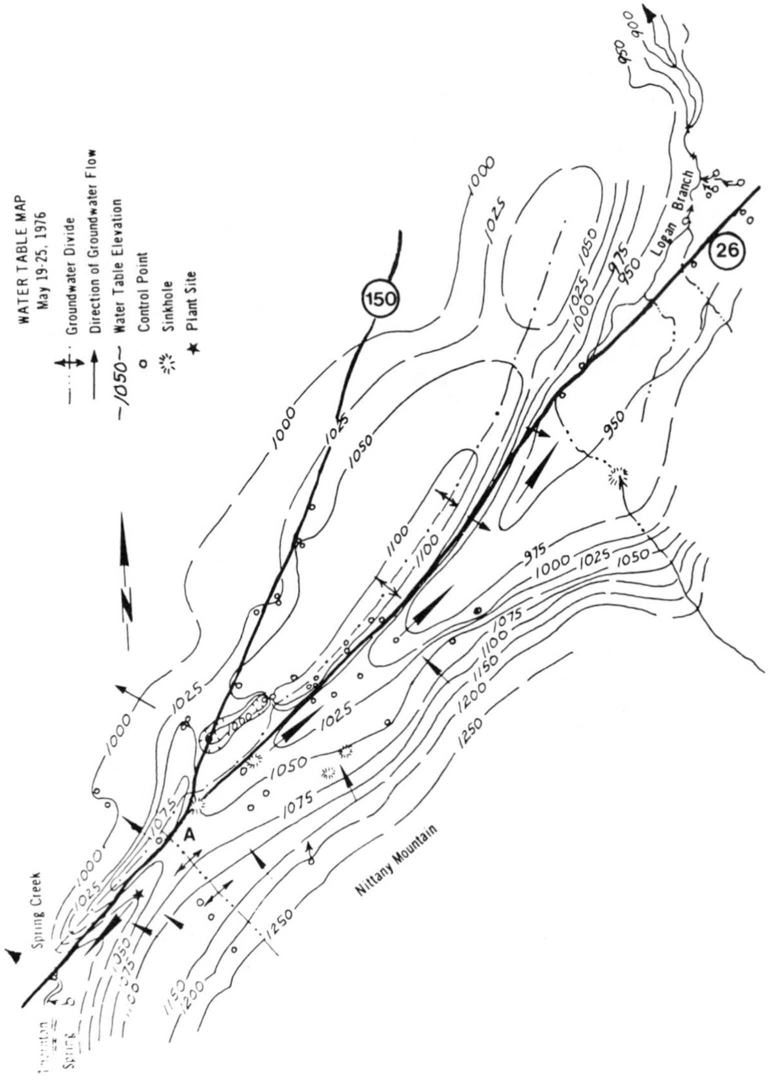

Figure 4. Water table contour map near the plant site.

Karst aquifers downstream from Thornton Spring and located near Houserville can be highly productive to water wells and are relied upon for domestic and farm water supplies. Bathgate (L) and Thompson Springs (M) (Figure 1) along Slab Cabin Run and above the confluence of Spring Creek are used for public water supplies and Benner Spring (N) along Spring Creek, for a state fish hatchery. Three test wells drilled on the flood plain and valley slope of Spring Creek near Houserville penetrated cavernous, carbonate rock that yielded more than a 100 gpm in one well and in excess of 2500 gpm in the two other wells. These wells were intentionally located on fracture-trace intersections as revealed by analysis of aerial photography. Other wells are less productive but, nevertheless, an important aquifer system has been shown to exist along Spring Creek below Thornton Spring.

The upland surface developed on karstified rocks contains a blanket of residual soil derived from the prolonged chemical weathering of soluble bedrock. It varies in thickness from less than 1 to more than 100 feet near the plant. It is highly variable having formed from the differential solution attack on karst rocks. Up to 60 feet of soil underlies the plant site locally, but bedrock outcrops are common along Route 26 just beyond the plant.

Colluvium also makes up some of the unconsolidated overburden deposits that overlie bedrock. These unsorted deposits of sand, silt, clay, cobbles, and boulders were transported down the slopes of Nittany Mountain under a freeze-thaw climate. They vary in thickness from less than 1 to more than 60 feet where they were deposited within pre-existing sinkholes near the base of Nittany Mountain.

Perched groundwater develops in the shallow soil profile of these sloping colluvial deposits which diverts surface water and shallow groundwater down the lower slopes of Nittany Mountain out over the regional water table (Figure 3). Water may perch above a dense shallow soil horizon and near the colluvium-residual soil contact where both soil units are well developed.

Spring Creek and its tributary, Slab Cabin Creek, are perched above the regional water table from Route 26 downstream towards Houserville. Water levels in well J of Figure 1 drilled within less than ten feet of Spring Creek commonly stand from 11 to 13 feet below the level of the stream surface as do water levels in test well K located along its valley slope or wall (Figures 1 and 5). Both wells were cased into bedrock and obtain water from solution openings.

Test holes drilled for the State College bypass across the flood plain of Slab Cabin Creek just upstream from its confluence with Spring Creek show soil thicknesses ranging 5.5 to 30 feet. These rather poorly permeable deposits are composed of silt, sand, clay, and gravel. The water table ranged in depth from 3 to more than 10 feet below land surface in test borings located in the flood-plain of Slab Cabin Creek indicating that water flow is downward through channel and flood-plain sediments into the underlying carbonate aquifer (Figure 5).

An eleven foot deep test pit in channel sand and gravel deposits within 15 feet of the bank of Spring Creek and nine feet below the elevation of the stream surface remained dry until it was backfilled indicating that Spring Creek is perched at this location (point P, Figure 1). In the area between the confluence of Thornton Spring and Houserville, water levels measured in piezometers placed at one foot depths below Spring Creek channel also showed that surface water flow is downward through flood plain deposits.

A downward component of flow is present during periods of high seasonal water table and when Spring Creek and Slab Cabin Run are in flood. During periods of flood, surface water comes into contact with isolated carbonate bedrock outcrops located along the valley slopes of these creeks and provide local pathways for contaminated surface waters to enter the karst aquifer. Elsewhere, flood waters come in contact with residual and transported soil that blankets the lower portion of valley slopes, hence, preventing the direct entry of surface water into the local aquifer.

This perched water system provides a potential natural pathway for the migration of contaminants into the underlying karst aquifer, but these contaminants first must travel through a variable thickness of flood-plain sediments that are rather poorly permeable. No swallow (sink) holes have been recognized along the channel of Spring Creek between where Thornton Spring joins Spring Creek and the vicinity of the downvalley water wells that contain the marker compound. Such openings, if present, would provide a direct connection between the surface-water and groundwater systems.

Man-made connections between surface and groundwater systems may exist along the flood plain of Spring Creek where water wells penetrate flood plain sediments. The annular space between drill holes and well casings are not grouted into bedrock at every home site along the valley and some surface water may gain access to bedrock along these routes. Sewer mains cross the flood-plain and channel of Spring Creek at several locations below Thornton Spring. The mains and their laterals are buried from 5 to more than 12 feet below land surface. These were placed into trenches that were drilled and blasted before being excavated wherever bedrock was penetrated and backfilled with permeable limestone aggregate. Groundwater is free to migrate along or through these gravel envelopes that surround the sewer mains permitting movement into the underlying bedrock wherever the regional water table is below surface water levels in Spring Creek (Figure 6). They may also have allowed contaminated surface water to enter the underlying karst aquifer.

Possible Dispersion Pathways of Land Disposed Wastes

Table II provides a list of the possible pathways in which a land disposed waste can be dispersed and which are believed to be available at the plant site.

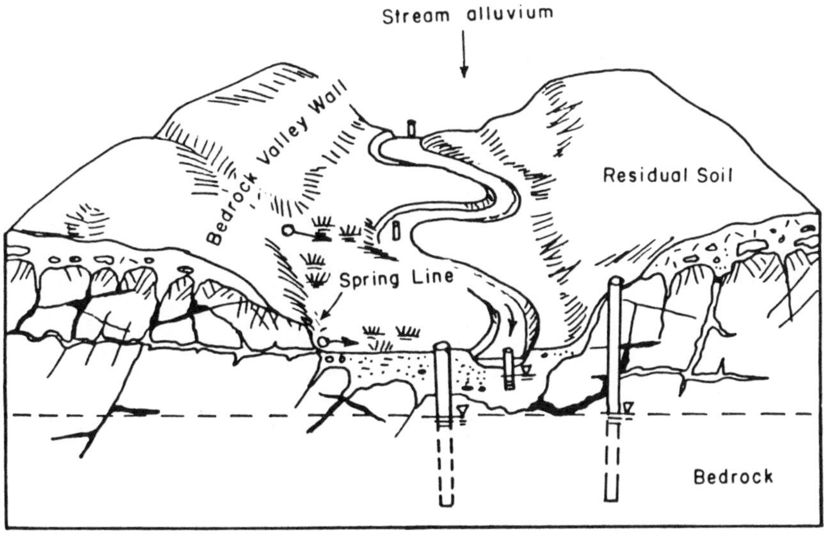

Figure 5. Spring Creek perched above the regional water table near Houserville as revealed by water levels measured in wells and piezometers placed in channel bottom sediments.

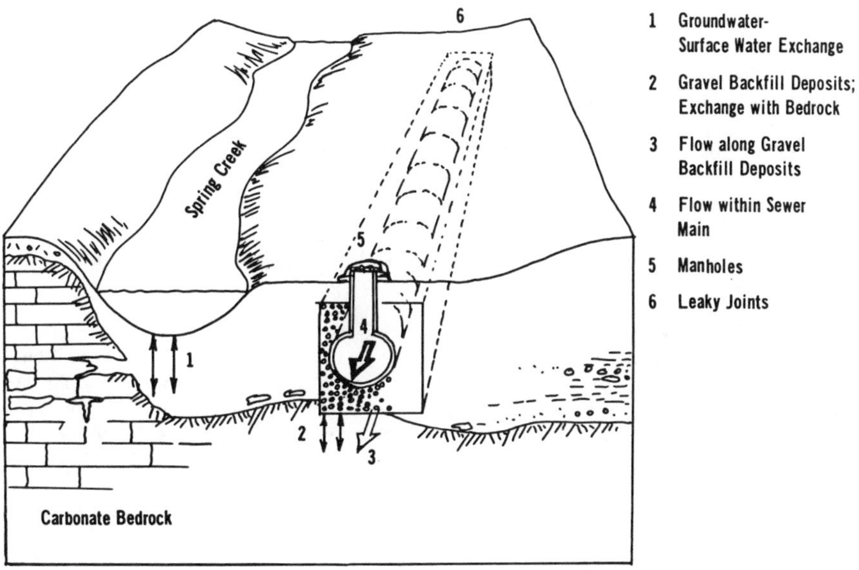

Figure 6. Geological cutaway showing the utility line as a conduit of groundwater and possible included contamination media.

Table II. Possible Pathways for Contaminant Dispersal

Pathway	Marker Compound
Surface Water-Groundwater Flow (downward)	known
Groundwater	known
Groundwater-Surface Water Flow (upward)	known
Surface Water	known
Surface Runoff and Soil Erosion	known (1)
Buried Utility Lines	suspected
Induced Groundwater Flow	suspected
Sewage Sludge Disposal	known (1)
Biota	known (5)
Atmosphere	uncertain

Surface Water to Groundwater Flow. There are four locations in the study area where a surface water to groundwater flow of contaminants could take place: (1) at the burial site where the wastes were disposed of within soil, (2) at various places under the channel and floodplain of Spring Creek where it is perched above the regional water table, (3) where storm runoff occurred at the plant site and entered a drainage ditch along Route 26, and (4) where sewage sludge from the University Area Joint Authority was disposed of on agricultural lands near Houserville.

The surface to groundwater pathway is probably the most important way in which contaminants were initially dispersed into the local environment. Groundwater was contaminated at the plant site by infiltrating surface water and plant effluents, infiltration through soil deposits and unsaturated bedrock, and recharge to the water table.

Thornton Spring, which is downgradient from the plant, is contaminated as are waters obtained in interceptor wells drilled on site as part of an ongoing cleanup program. The waters pumped from the interceptor wells are treated as part of cleanup operations (6).

No testing was done as part of this investigation of interceptor and monitoring wells drilled on-site but a number of tests were made on Thornton Spring water samples 0.9 km below the plant.

Downvalley from Thornton Spring, measurable concentrations of the marker compound were found in various water-supply wells located immediately adjacent to Spring Creek. It is concluded that surface water was the major source and pathway of these organic substances before reentering groundwater downvalley because of the chemical data collected, the perched nature of Spring Creek, and the presence of contaminated fish and plants in the Creek. It is not known, however, if these contaminants entered the carbonate aquifer near Houserville through streambed and floodplain infiltration, leakage along well casings, leakage through gravel backfill deposits placed along sewer mains, infiltration into outcrops along the valley slope of Spring Creek, leaching of sewage sludge disposed of on adjacent uplands, and/or within groundwater that may have migrated beneath Spring Creek valley from the vicinity of Thornton Spring. A combination of these pathways most likely was involved.

Although it has been shown that surface water flows downward through the flood plain of Spring Creek, finding a chemical example of this interaction in a stream bed for the marker compound requires extremely careful work which has not yet been completed.

Groundwater Dispersal. Once a contaminant has found its way into groundwater in a karst aquifer system, it is difficult to predict its exact movement. It depends upon the interconnection of openings afforded by joints, fractures, bedding-plane partings, fault planes, and rock layers with intergranular, primary and secondary openings. Any or all of these planar and intergranular openings may be enlarged to provide an interconnected network of solution voids and conduits. These may be developed as a maze of small openings or be concentrated as a few distinct large openings.

Contaminants may migrate within saturated residual or transported soils, within soil macropores, or within small interconnected pores where they find their way into groundwater. In either case contaminant dispersal will be limited to distinct groundwater subbasins defined by fixed or transient groundwater divides, by the hydraulic-head distribution within the flow system, and by geological barriers which restrict flow such as the presence of poorly permeable soil or bedrock units.

The regional water table shown in Figure 4 provides some insight into the pathways available for subsurface flow. One groundwater divide exists near the top of Nittany Mountain and another along the groundwater ridge that runs parallel to and north of Route 26. These divides preclude shallow groundwater flow from the plant site to any direction except downward, and to the northeast toward springs at the head of Logan Branch or southwest toward Thornton Spring (point B, Figure 1). The short transient divide separating these two groundwater subbasins (Thornton Spring and Logan Branch systems) is located near the Route 26-150 intersection (point A, Figure 4). To date, no marker compound is known to have been recovered from wells northeast of this divide, only in the Thornton Spring subbasin and wells located down valley near Houserville.

The marker compound was detected in an upland well located north of the plant and across the western groundwater divide of this sub-basin. The static water level in this well stands above the elevation of Spring Creek and water levels near the plant. It is unlikely that contaminated groundwater crossed beneath this groundwater divide through upturned, somewhat poorly permeable beds of dolomite and the thin, artesian sandstone aquifer.

The largest amount of initial dispersal had to have been from the plant site down groundwater gradient to Thornton Spring (point B, Figure 1). This spring, located 0.90 km down gradient from the burial site, is the most prominent location of residues of the marker compound in the area. In 1977 concentrations of it ranged from 2000-4000 ng/L (ppt) (7 samples) and in 1978 they were 3000 ppt (2 samples) (3). Measurements were made at this site in June of 1983 as part of the present investigation, fully 20 years after waste burial, and the water was found in a single measurement to contain 6400 ppt at that time.

One well near Houserville drilled and cased to a depth of 200 feet in August of 1981 contained 22 ppt of marker compound (analytically confirmed). The continued presence of the compound in this well was not confirmed with additional sampling.

Other downstream areas, however, have also shown dispersal of the marker compound within the karst aquifer. Concentrations have been found in numerous domestic well-water supplies in testing programs of 1977/78 (1) and in January of 1985 in the region of lower topographic and water table elevation located about 4.0 to 4.9 km downstream of the disposal site (as measured along Spring Creek drainage). Figure 7 shows the concentrations of the marker compound found in these home supplies. All homes are located adjacent to the stream except Number G-G which is located 0.5 km upgradient from the stream.

The chemical data show, first, that the levels found in 1984 generally were comparable to those found in 1977-78 by the Pennsylvania Department of Environmental Resources (1). The 1985 sampling found more for Houses G-B and G-C but less for Houses G-D and G-G. This indicates the long term persistence of the marker compound. The differences indicate possible changing flow patterns in the groundwater system. The high value for House G-C may indicate changes in the hydraulic head of the local groundwater supplies of the area. Heavy pumping of groundwater supplies for the University community, and lesser amounts for farms, and homes nearby may have caused these changing patterns in groundwater movement. This, in turn, may have also diverted away all of the contaminated water as shown for House G-G. Alternately, it may have been the result of long term natural flow of groundwater which has helped to locally flush the aquifer of the marker compound.

The complexity of the groundwater flow system is also seen by comparing the private well analyses with the private well profiles of Figure 8. This figure shows the best estimate of the depth of each well with respect to land-surface elevation. The marker compound

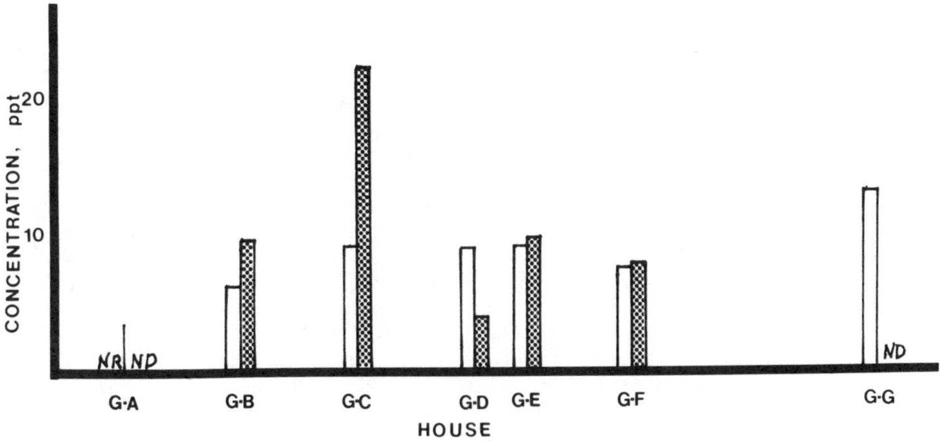

Figure 7. Concentrations of the marker compound found in home water supplies located along Spring Creek for two different sampling times, 1977/78 in open graph and 1984 in darkened graph. NR is not reported; ND is not detected.

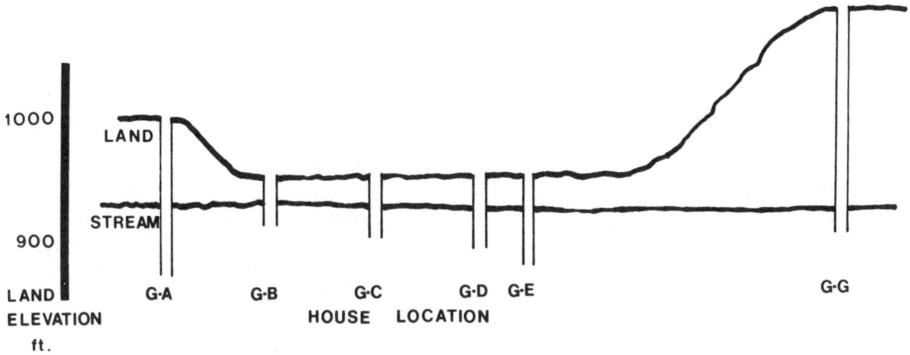

Figure 8. Height-of-land profile of domestic water supplies containing the marker compound showing relative depths of wells below the stream bed of Spring Creek. The house location axis is to scale.

values of 1984 showed large differences in concentration for house wells G-B and G-C even though they are in close proximity and drilled to a similar depth. In contrast, house wells G-B, G-E, and G-F contained similar concentrations of the marker compound but were drilled to different depths. These conclusions are only tentative since the well depths are estimated and the casing depths are not known.

The time frame for comparing movement of surface water and groundwater can be illustrated in the following example. During a heavy rainfall in February of 1985, samples were made of surface water concentrations at point H and groundwater pumped from the wells in the two homes located nearest to that point, G-B and G-C (Fig. 1). The total distance between these homes was only 0.3 km along the stream. Figure 9 shows vials of extracted water samples taken at these three locations and placed in graph form with respect to time. The sample of the surface stream extraction having the darkest color, probably a result of naturally occurring components, occurred at zero hours. The darkest color of the house G-B groundwater flow system samples was found at a delay of 24 hours from the darkest color of the surface samples. It was also noticed that the well water of House G-C had no color to any of the vials despite its close lateral and vertical proximity.

Concentrations of the marker compound (Figure 10) found during the high water table conditions for these three sites discussed above did not appear to correlate with extract color. The marker compound analytical data, however, contain uncertainty due to apparent chromatographic artifacts. The presence of colored compounds in the extractant, however, does provide an independent assessment of the possible interconnection between surface water and groundwater in this locale. The fact that the color patterns of the two houses compared were different also demonstrates potentially different pathways of flow that are reaching these two houses despite the close proximity of their wells.

Groundwater to Surface Water Flow. The possibility of bringing xenobiotic compounds to the surface from contaminated groundwater sources is obvious through the mode of a spring, such as Thornton Spring. It is not so obvious that this mode happens within Spring Creek immediately below Thornton Spring. Direct exchange upward of groundwater to the Spring Creek channel is precluded wherever and whenever the regional water table stands below stream level. Since this hydrological condition has persisted in the flood plain region near Houserville at least since the mid-1960's when the first test well was drilled there, no groundwater upward flow is suspected in this region.

Other springs in the area, aside from Thornton Spring, could bring groundwater to the surface. A spring at location I in Figure 1 was tested in April of 1985 for the marker compound but none was detected at the 1 ppt level.

Surface Water. The spreading of environmental contaminants via surface water is probably the most obvious dispersion pathway. Conclusions based on stream water sampling must be carefully done

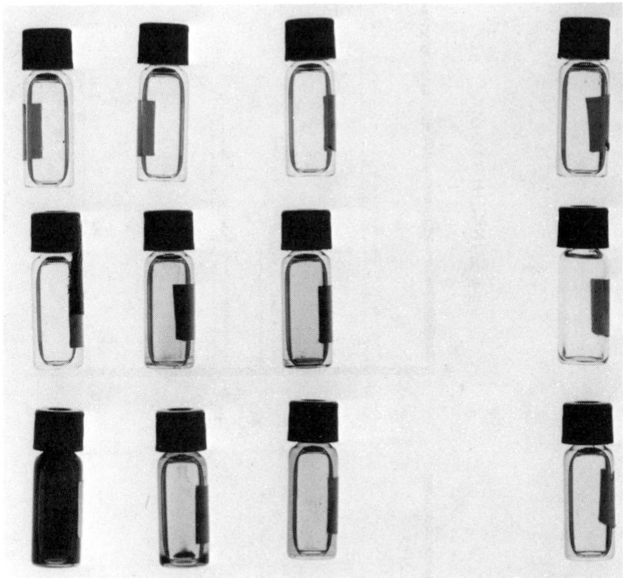

Figure 9. Photograph of vials containing water extraction products from stream and home supplies as a function of time during flood water stream conditions.

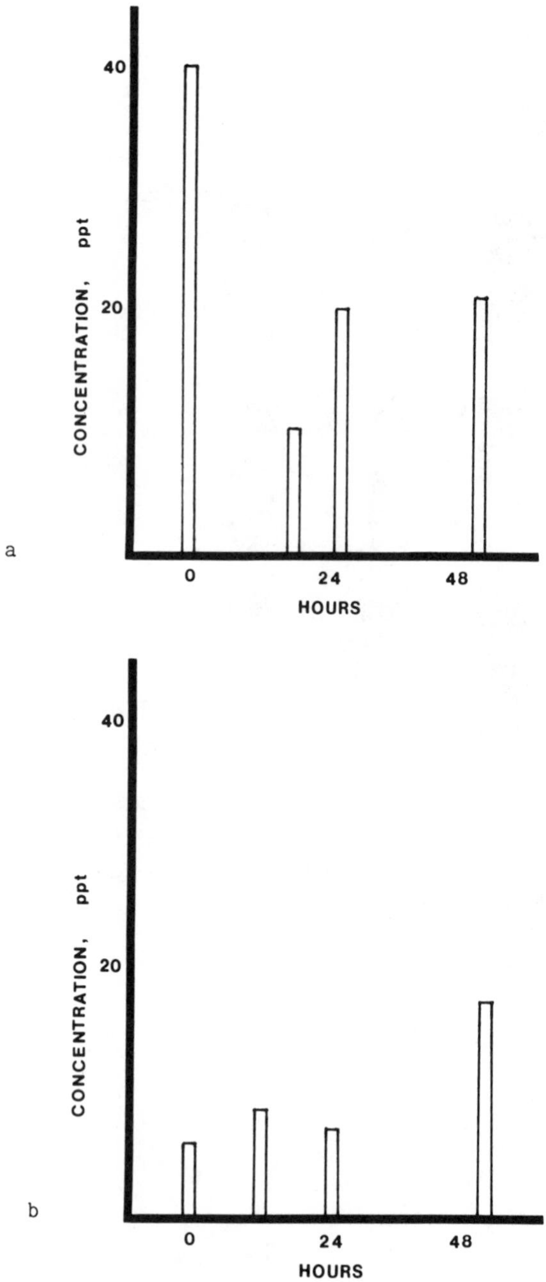

Figure 10(a,b). The marker compound found in water from flooded Spring Creek and two selected homes nearby: stream analysis (a) at point H in Figure 1, and home water analysis (b) in House G-B.

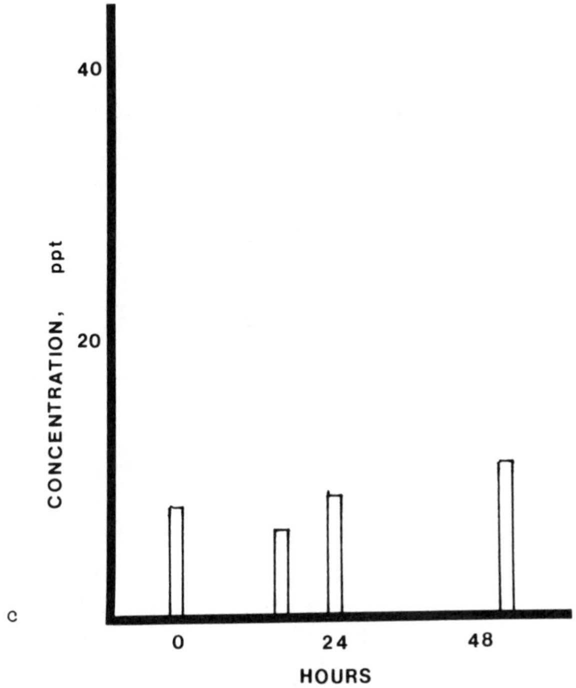

Figure 10(c). The marker compound found in water from flooded Spring Creek at House G-C.

since at any point there may be additions to or subtractions from the water flow. These exchanges can be from groundwater or surface water supplies for water uses to meet industrial or domestic needs.

The marker compound was found in surface waters from the confluence of Thornton Spring to a point some 30 km downstream in June of 1983 (Figure 11). Stream volumes were measured for this entire length and the data correlated fairly well where with larger downstream water volumes at lower stream locations there was obtained the expected lower concentrations.

To determine the sampling accuracy in a stream containing small contaminant influx from one bank, stream cross-sectional sampling was done in June of 1984 at various intervals below the confluence. The data presented in Table III show that at 60 meters downstream the range of the concentrations of the marker compound was wide and the variance of this group of points was large. At both 200 and 1000 m the data are similar with a small variance indicating complete mixing at 200 meters. Hence, for the total watershed study the stream analyses are representative of the whole stream body.

Table III. Cross Section Stream Analysis for the Marker Compound, Spring Creek, 1984

Distance Below Confluence m	Marker Compound Cross Section Analyses[a] ppt					Variance
60	145	40	11	4	0.1	3700
200	45	44	45	44	42	2.0
1000	loss	34	43	36	39	15
	46	43	42	43	43	2.4

[a] Stream width divided into 5 equal segments.

Surface Runoff and Soil Erosion. Surface runoff is overland flow immediately following periods of snow melt and rainfall. It is the main process by which contaminated sediment is eroded and transported from upland settings. Liquids, applied to the land or released to perched groundwater that is discharged back to the surface, and contaminated soils may be transported into surface streams in this manner.

Shortly after the plant landfill wastes were found to be contaminating other areas, mirex, one of the products manufactured at this site, was found in a drainage ditch along Route 26 just 0.18 km

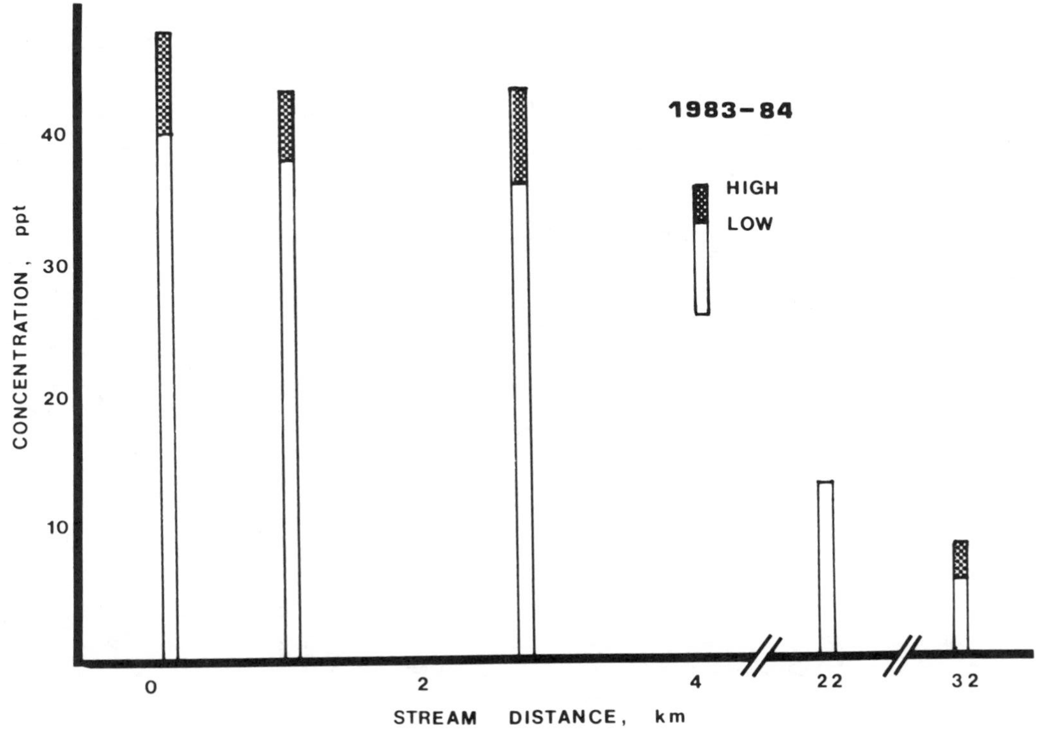

Figure 11. Surface-water measurements of marker compound concentrations at various distances along Spring Creek showing the range of concentrations in shaded tone.

away to the northwest. The material most likely was dispersed there through the overland route. The drainage ditch rarely has flowing water in it except during periods of rainfall or snowmelt. In later tests it was found that concentrations of mirex in fish in Spring Creek, into which the drainage ditch discharges just below Thornton Spring, were less after the surface soil was dug out and removed (1). It is uncertain if there was a causal relationship in this case, however.

Plant effluents applied to surface farm land late in the mid or late 1960's did run off from the spray field. Terraces were later built to contain this runoff, but some sediment and effluents inevitably escaped from the site. It is not known if this surface runoff contains the marker compound.

Utility Lines. Sewer lines are present along Route 26 near the plant and, as well, along and in Spring Creek below Thornton Spring. One sewage line follows the stream for 6.0 km before it heads to a sewage treatment plant. Because of the depth of construction of the mains, this pathway could very well be the main source of flow into the private home well supplies along the stream. With manholes every 100 m there are numerous potential surface entry points of contact. If this dispersal method is proven to be a viable one, then there is a good probability that contaminants would be transported through the sewer mains to the treatment plant and become incorporated in the sewage sludge. The Pennsylvania Department of Environmental Resources has reported (1) that the sludge derived from the University Area Joint Authority Sewage Treatment Plant located below Houserville contained the marker compound. The manner in which this sludge was contaminated has not been further established.

Induced Groundwater Flow. Groundwater can be induced to flow from a nearby stream wherever water levels in the well are drawn down below stream level and the groundwater gradient slopes from the stream to the well. Drawdown is illustrated by the fact that in the early 1960's heavy pumping from wells completed by Penn State University moved a given water table elevation contour outward from the well field almost 2 km over a two-year span. This has resulted in a pumping cone of depression in the Big Hollow area. However, this wellfield is isolated from the Houserville area by a groundwater divide.

The actual waste burial site for these marker compound residues was located only 0.35 km away from another transient water table divide (point A, Figure 4). Heavy pumping from an industrial source located along Route 26 to the northeast and on the other side of the divide could have easily shifted groundwater flow from southwest to northeast along Route 26 (Figure 4). In that case both ends of this groundwater flow system could have been affected by the marker compound, i.e., Thornton Spring side and Logan Branch side.

Domestic wells along Spring Creek also can induce streambed infiltration for the same reason. Induced flow may be a significant pathway for further groundwater dispersal in the future as more groundwater is pumped from existing well fields, or new well fields are completed in the region.

Sewage Sludge Disposal. If a chemically persistent contaminant works its way into a sewage line and ends up in the treatment plant, an additional mode of dispersion will become available: sludge disposal. In many areas the sludge is of such nutrient value that it is spread on farm fields for fertilizer. Although there is probably less use of sludge as fertilizer due to the presence of harmful levels of toxic metals, such as cadmium (7), and organics, such as PCB's (7), pressure has been exerted by both the sewer authority and individual farmers to recycle this nutritive waste. Nonetheless, the marker compound has been found in sludge (1). Once on the fields the materials in the sludge can be dispersed into plants grown there, it can percolate back into the groundwater, or it can be transported as overland flow during periods of storm flow, this time, perhaps, at some distance away from the original sources. This water may enter streams or sinkholes during periods of excessive runoff. The leachate from this sludge could reach home well supplies upgradient from the chemical or sewage treatment plant, or regions with known surface or groundwater contamination. This mechanism may account for the marker compound found in well G-G.

Biota. Biota are another means of dispersal of contaminants in the environment. When the material is biomagnified, such as the case with mirex, DDT, and PCB's and to a lesser extent with more polar compounds, botanical and zoological life can be affected and, as well, serve as a mode of dispersal. Chemical residues from the production plant have been found in fish in Spring Creek resulting in a warning by PA DER to avoid eating the fish.

Atmospheric Route. To complete the dispersion system one cannot leave out the mode of spreading via the atmosphere. Volatile materials can evaporate from land and water sources into the atmosphere. The same is true about compounds that are adsorbed or absorbed into dust that is carried into the atmosphere by the wind or released from stacks. No data exists for this potential pathway at the plant.

Summary and Conclusions

From the range of this study we conclude that the dispersal of waste chemical contaminants, throughout a portion of the karst or carbonate rock aquifer system in Nittany Valley, has proceeded in the following manner: 1. At the disposal site conduit openings in the limestone permitted major entrance to the groundwater system. 2. The dense colluvium and residual soil perched the water system in the disposal site area and contributed to overland flow in times of high water. This resulted in small amounts of contaminated sediment being deposited in ditch areas down gradient from the disposal site and, possibly, transport of contaminants to the receiving stream, Spring Creek. 3. The groundwater mound adjacent to the shallow valley of the plant just to the northeast provided a directional barrier to flow in that direction. 4. The less soluble dolomite ridge with accompanying groundwater mound pressured by a deep, confined aquifer serves as a directional barrier to the northwest. 5. Chemical residues entering the groundwater regions under the plant reentered the surface water at Thornton Spring just downhill from the plant and have persisted to the present. High concentrations of the marker compound, up to 6400 pt, were found in this spring. The contaminant entered Spring Creek at this point. 6. The perched flow system of Spring Creek in the area below the confluence with Thornton Spring provides a natural trough for water to remain on the surface. Only small amounts of water, containing the marker compound could percolate to the groundwater through this limited barrier. Concentrations of the marker compound were found all along this stream in decreasing concentrations from 145 to 5 ppt for a distance of 30 km. The occasional presence of rock outcropping on the margin of the flood plain in segments of Spring Creek in this area does allow some downward flow of surface water and contaminants. 7. Man-made connections between the stream bed and groundwater areas below are allowed via utility sewer line construction along the bed and flood plain of Spring Creek. Minor dispersion can occur within the sewer line through leaks in the joints and manhole covers. 8. There are no known swallow holes in the Spring Creek bed between Thornton Spring and wells containing the marker compound to allow major downflow by this means.

Concentrations of the marker compound were found in shallow domestic water supplies up to 22 ppt in the region of 4-4.8 km below the Thornton Spring confluence. Major contributions to these concentrations could have occurred through the utility line construction or extensive groundwater flow emanating from under the Thornton Spring area and flowing down valley below the flood plain of Spring Creek.

In the subsurface system, however, there is concern about concentrations of contaminant compounds in drinking water. Both shallow and deep wells have been found to contain the marker compound in this system. More information is needed to fully assess the various areas of the aquifer impacted to date and rates of transfer between them especially because changes can occur in aquifer systems when groundwater pumpage increases. The transport of residuals of these substances in flood plain and channel bottom sediments must be better understood to assess the nature of the release into ground and surface water in the future.

Acknowledgments

The authors wish to extend their thanks for the careful technical work done by two students, Christopher Zoky and Richard Hoff, who did all of the analytical work cited in this paper. This paper is published as Journal Series paper No. of the Pennsylvania Agricultural Experiment Station.

References

1. Alters, D. personal communication, unpublished internal reports, Department of Environmental Resources, Bureau of Water Quality Management, Williamsport Regional Office, PA.

2. Kurtz, D. A. Anal. Chim. Acta 1983, 150, 105-114.

3. Parizek, R. R.; White, W. B. "Application of Quaternary and Tertiary Geological Factors to Environmental Problems in Central Pennsylvania. Guidebook of the 50th Annual Field Conference of Pennsylvania Geologists: Central Pennsylvania Revisited"; Department of Environmental Resources, Bureau of Topographic and Geological Survey, Harrisburg, PA, 1985.

Rauch, R. R.; White, W. B. Water Resources Res. J. 1970, 6, 1175-1192.

4. Parizek, R. R. In "Hydrogeology and Geochemistry of Folded and Faulted Carbonate Rocks of the Central Appalachian Type and Related Land Use Problems"; Parizek, R. R., White, W. B., and Langmuir, D., Eds.; Mineral Conservation Series Circular 82, Earth and Mineral Science Experiment Station, The Pennsylvania State University, University Park, PA, 1971; pp. 9-65.

5. Hesser, R. "Contamination of Spring Creek and Foster Joseph Sayers Lake," Environmental Contamination: Fish and Wildlife Concensus; Meeting of PA Chapt. Wildlife Society and Central PA Chapt. Am. Fisheries Society at The Penna. State Univ., Univ. Park, PA, April 19, 1985.

6. Emrich, G. H., S. M. Martin, Inc., Valley Forge, PA, 1985, personal communication.

7. Baker, D. E.; et al. "Criteria and Recommendations for Land Application of Sludges in Northeastern Pennsylvania"; Penna. Agric. Experiment Station, University Park, PA, 1985.

RECEIVED April 17, 1986

14

1,2-Dibromoethane (EDB) in Two Soil Profiles

D. W. Duncan and R. J. Oshima

California Department of Food and Agriculture, Environmental Hazards Assessment Program, Sacramento, CA 95814

Soil properties and Ethylene Dibromide (EDB) concentrations were measured at two locations with histories of EDB applications. The objective was to explain the presence of EDB residues in a well near one location and the lack of residues in a well near the other location. The soil profile was sampled at each location from the surface to groundwater, and groundwater samples were collected at the profile base. A combination of statistical analyses (stepwise linear regression and discriminant analyses) was applied to soil data to interpret differences within each location. Maximum EDB concentrations were found at location 1 between 0 and 0.98 meters and ranged from 0.3 to 12.5 ppb, correlating with organic carbon. Location 1, composed of a silty clay, contained a deeper band of EDB lying between 2.35 and 2.98 meters below the soil surface. Within the band, EDB concentrations ranged from 0.2 to 0.6 ppb. The presence of this deeper band was not correlated with any measured variables, including organic carbon, and may represent migrating EDB. The fumigant was found between 0 and 0.46 meters at location 2 and was correlated with organic carbon. EDB was not found in deeper soil layers of location 2, which was characterized as a more coarsely textured soil than location 1. Depth to groundwater was 5.2 and 4.0 meters at locations 1 and 2 respectively, and EDB was not detected in water sampled at the profile bases.

In 1954, Ethylene Dibromide (EDB) was introduced as a product for the preplant treatment of agricultural fields to control nematodes and it is still used worldwide. EDB is a volatile, halogenated hydrocarbon that is usually marketed as a liquid. The liquid is injected 15 to 30 centimeters beneath the soil surface with a tractor driven chisel tool where the vapors permeate soil air spaces and kill the

0097-6156/86/0315-0282$06.00/0
© 1986 American Chemical Society

parasites. Amounts of applied active ingredient range from 9 to 73 kilograms/hectare depending upon soil characteristics and severity of the nematode problem. Because of high product cost, fumigation is carried out on soil used to grow crops with relatively high monetary yields such as tobacco, vegetable, vineyard and orchard crops (1).

During 1983, EDB residues were identified in well water samples collected in Florida, Georgia, California and South Carolina (2). In addition, a soil coring study in California revealed EDB residues from the soil surface to a depth of 12 meters (3). As a result of water monitoring and soil coring studies, the Environmental Protection Agency (EPA) issued a suspension order in October, 1983 to discontinue the use of EDB as a pesticide (2).

As a regulatory agency, the California Department of Food and Agriculture (CDFA) is concerned with creating a strategy to selectively control the application of pesticides used on soil to reduce the potential for groundwater contamination. Pesticides are currently regulated in California at the county government level with a reporting system based on township, range and section coordinates (1 section = 1 square mile or 2.59 square kilometers). A new regulatory design should incorporate results of laboratory, well sampling, soil coring and computer modeling studies to help estimate the potential for a pesticide to reach groundwater within a section. A CDFA study in progress (4) is statistically comparing well sample data of dibromochloropropane (DBCP) residues with soil types. Preliminary results indicate a very high correlation between wells containing DBCP levels and highly permeable soils. Although seeming to point out an obvious correlation, the data allow statistically based predictions of well contamination as a result of agricultural pesticide use in areas as small as 1 section. This use of well sampling data, in addition to soil coring and variables measured in the laboratory such as sorption, solubility in water and volatility can support local pesticide use decisions perhaps to a resolution of 1 section.

Soil core data can provide real evidence of pesticides leaching to groundwater. This paper describes a soil coring study designed to examine relationships between soil properties of two agricultural locations and the presence or absence of EDB. Sampling locations were nearly a mile apart with similar soils, EDB application histories, and agricultural practices. One location was near a well where EDB residues had been found, and the other location was near a well where EDB residues were not found. Soil cores were taken at both locations from the soil surface to the point at which groundwater was first reached. Soil properties, such as texture, moisture, organic carbon content, pH and electrical conductivity were measured. In addition, clay types and relative amounts of clay types were measured. The study objective was to compare measured variables at both sites and suggest an explanation for the presence of EDB in well water at one location and absence of the chemical in well water at the other location.

Materials and Methods

Sample Site Description. Two soil coring sites were selected in the western portion of Stanislaus County on the west side of the San

Joaquin Valley, California. Selection was based on well sampling survey results conducted during the summer of 1983 (5). Both soil coring locations consisted of agricultural fields as close as possible to one of the previously sampled wells. Location 1 was 62.0 meters from a domestic well 42.7 meters deep where EDB residues were detected. The well had a sanitary seal 6.0 meters deep and was perforated between 31.4 and 37.5 meters. Location 2 was 111 meters from a domestic well 36.6 meters deep where EDB residues were not detected. This well also had a sanitary seal 6.0 meters deep and was perforated from 31.0 to 34.0 meters.

For more than 10 years, EDB was applied at location 1 in 3 out of every 4 years at a rate of 2 to 3 liters/hectare. EDB at location 2, about 1 mile from location 1, was also applied at a rate of 2 to 3 liters/hectare for over 10 years in 3 out of every 4 years. A formulation of 86% active ingredient was used most recently at both locations and applied in the spring of 1983 in preparation for lima bean crops. The fields were historically used for vegetable crop production with water supplied through furrow irrigation. About 50 hectare-centimeters were used in 1983 to irrigate lima bean crops at location 1 and 41 hectare-centimeters were used to irrigate the same crop at location 2. The average precipitation total in the area was 28.73, 36.40 and 52.10 centimeters for 1980, 1981, and 1982, respectively. A total of 51.00 centimeters were recorded in the 9 months prior to soil coring (6).

Soil type at location 1 was Meyer's clay, described as a very deep, well-drained clay soil that cracks when dry to a depth of about 90 centimeters producing an angular, blocky structure (7). Location 2 was characterized in the El Solyo Series which consists of moderately well-drained, fine textured silty clay loams. Subsurface soils are described as compact and consisting of calcareous heavy silty clay loams (7).

Soil Coring. Soil coring occurred in October, 1983, and was accomplished with a truck-mounted Mobile Drill, model B-53 drilling rig and 20.3 centimeter (8 inch) hollow-stem augers. Segments of the soil profile were collected with a split barrel sampler, 50.8 centimeters long and containing three stainless steel cylinders stacked end to end. Each 15.2x6.4 centimeter (6x2.5 inch) cylinder was numbered and weighed before being placed into the split barrel. The interiors of the cylinders were rinsed with ethyl acetate, the solvent used for EDB extraction in the laboratory, and the split barrel was attached to a Moss Wireline sampler and lowered by a steel cable into the hollow auger. Drilling and soil sampling occurred simultaneously but the split barrel did not rotate during the sampling process and was pressed into the soil ahead of the auger. After drilling 50.8 centimeters of soil, the sampler was withdrawn. The cutting tip of the split barrel held a 5 centimeter segment of sampled soil which was discarded. Thus, 45.7 centimeters of soil was sampled for every 50.8 centimeters drilled. Soil samples remained encased in the cylinders when removed from the split barrel. After removal, the cylinder ends were sealed with aluminum foil and tightly fitting plastic caps. The cylinders were weighed again to obtain soil bulk weight and then frozen on dry ice (-70° C).

When water-saturated soil was reached during the drilling process, soil sampling stopped. A 2 mil thick Teflon sheet, about 10x20 centimeters, was attached inside the split barrel with a steel basket insert. The sampler was lowered through the auger and allowed to sink into the saturated layer before being raised to the surface. The Teflon sheet allowed water to move into the sampler, but when raised, weight from the column of water in the barrel caused the Teflon to collapse and sealed the opening. The split barrel was removed and the water poured into 1 liter amber bottles. Two bottles were filled to capacity and sealed with foil-lined caps.

During soil coring, water samples were collected from the well associated with the soil core locations. Fresh recharge water was sampled after operating the well pump long enough to completely purge and replace water in the casing, usually about 30 minutes. All water samples were collected while the pumps were operating. A Teflon tube was attached to a Schrader aeration valve, located between the pump and storage tank. With the core of this valve removed, water from the well was directed into 1 liter amber glass bottles that were filled to capacity and sealed with aluminum foil-lined caps. If the well was not equipped with a Schrader valve, a faucet between the pump and storage tank was used as a sampling port. A Teflon tube was inserted into the faucet opening and pushed into the pipe to reduce aeration before the sample bottles were filled. One well required water to be sampled from a faucet after it had passed through a storage tank. The tank was emptied and filled three times before the sample was collected. Bottles containing water samples were stored on wet ice immediately after filling.

Soil Core Splitting and Analysis. A mechanical sample splitter was constructed to remove frozen soil from the steel cylinders and divide it longitudinally into three portions. This device was an electronically controlled hydraulic pump that pushed the frozen soil past two steel blades and into three shoots which channeled the samples into collection containers: one for moisture determination, one for quantitative chemical analysis and one for physical analyses. Soil samples were kept frozen for about 2 weeks before being split and two weeks more were required to split the samples.

The sample portion used for moisture determination was split into a half-pint jar. A pint jar was used to contain the portion of soil reserved for chemical analysis and was immediately sealed. The pint jars were stored on dry ice and shipped within 3 days to the CDFA Chemistry Laboratory in Sacramento, California for analysis. The third portion of the sample was used to determine texture, organic matter content, pH, electrical conductivity and moisture. This portion was collected in a plastic bag and stored in a refrigerated chamber at 3° C.

Textural analysis was conducted using temperature-controlled water baths and the Bouyoucos hydrometer method (8) with 2 modifications. First, soil samples were agitated for 20 seconds with plastic plungers instead of mixing the soil by inverting the container, and second, the clay suspension was not washed and sieved to determine fractional sizes. Organic matter percentages were determined with a dichromate reduction method (9). Electrical conductivity (ec) and pH were measured from the saturation extract of a 100 gm. sample. A Beckman Solubridge model SD26 was used for the ec measurements and pH was determined with a Corning model 125 and separate pH and reference electrodes. In addition, the weight of the saturated soil paste was used to estimate water-holding capacity for each sample. X-ray diffraction analysis was performed on the clay fraction, obtained from textural analysis, to determine clay types in each sample. A quantitative analysis was performed on each X-rayed sample to determine proportions of clay types. Both analyses were performed by the Department of Geology and Physical Science at the California State University in Chico (10). EDB in both soil and water samples was extracted by refluxing and co-distillation with ethyl acetate followed by drying with sodium sulfate. An aliquot of the extract was then injected into a Perkin-Elmer gas chromatograph equipped with a Ni^{63} electron capture detector (11).

Results

Chemical analysis of the soil profile segments at location 1 revealed EDB residues between 0 and 0.98 meters and between 2.35 and 2.98 meters below the surface (Table I).

Table I. EDB Concentrations (ppb dry wt.) in Soil Segments

Location	Concentration*	Depth (centimeters)	Meters
1	4.3	1.0 - 15.2	
	6.1	15.2 - 30.4	
	12.5	30.4 - 45.7	0 - 0.97
	0.8	66.0 - 81.3	
	0.3	81.3 - 97.0	
	0.6	280.4 - 298.7	
	0.3	304.8 - 323.1	2.35 - 2.98
	0.3	323.1 - 341.4	
	0.2	341.4 - 359.7	
2	5.4	1.0 - 15.2	
	1.3	15.2 - 30.4	0 - 0.46
	0.3	30.4 - 45.7	

* Minimum detectable level = 0.1ppb

The chemical was not detected in the profile basal water or in concurrent samples collected from the domestic well 62 meters away. Chemical analysis of the soil profile at location 2 (Table I) detected EDB between 0 and 0.46 meters below the soil surface while water sampled at the profile base contained no residues. However, EDB was detected in water sampled on the day of coring from the domestic well 111 meters away. Table II summarizes water sample data.

The results of measured soil property variables for both core locations appear in Figures 1, 2 and 3. Portions of the profiles were not sampled and appear as breaks in the figure columns.

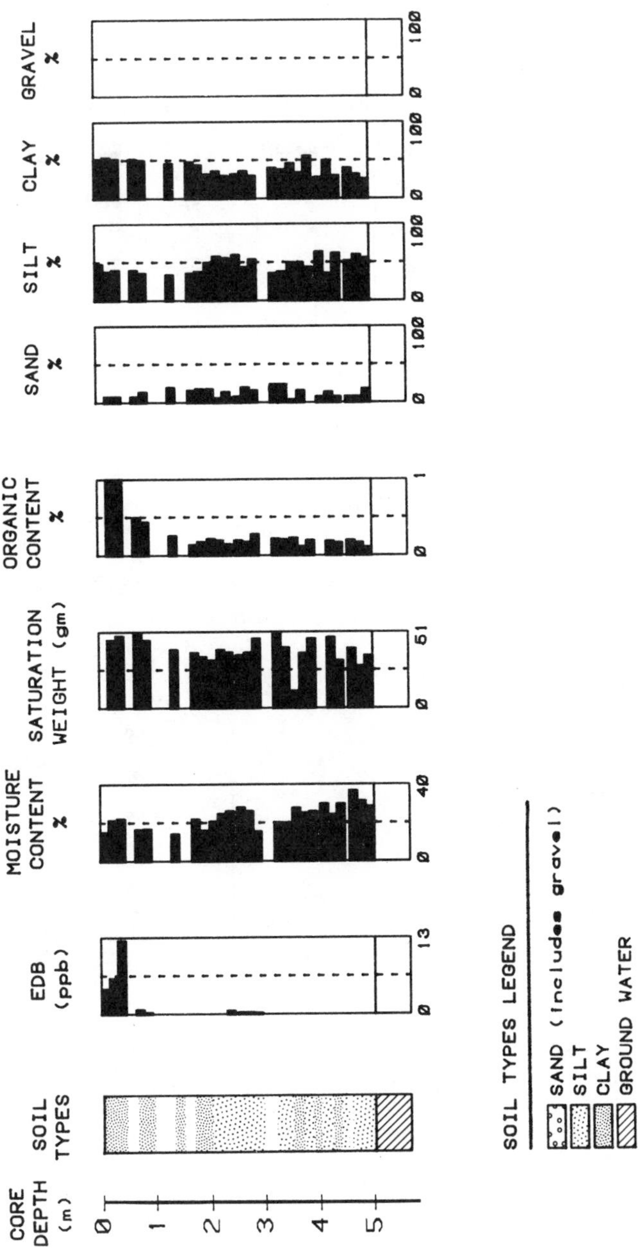

Figure 1. Histograms of measured variables at location 1.

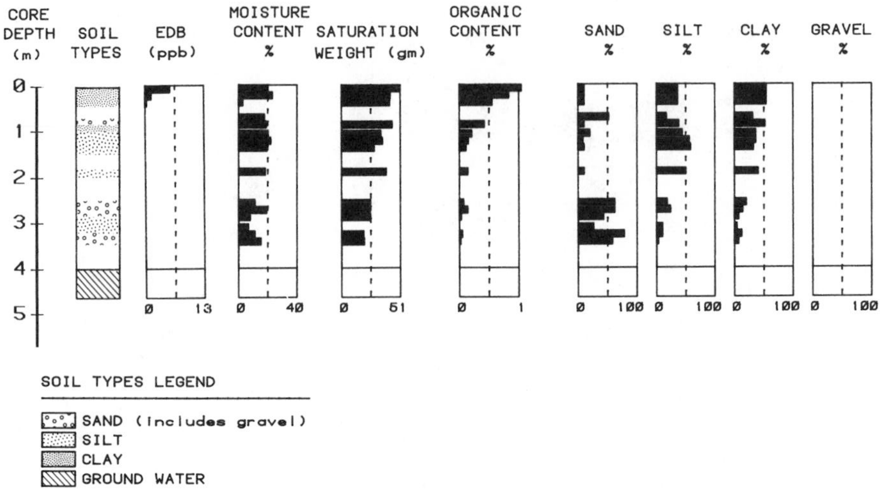

Figure 2. Histograms of measured variables at location 2.

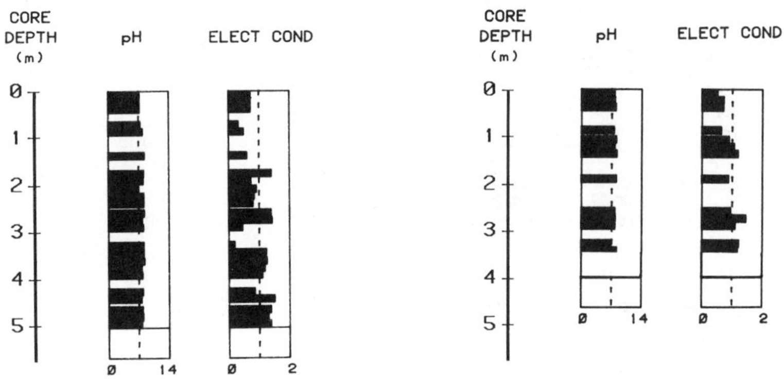

Figure 3. Histograms of electrical conductivity and pH measured at locations 1 and 2.

Occasionally the split barrel sampler would not contain a full allotment of soil although the auger drilled a complete 51 centimeters. Most often the soil was probably compressed, but a portion of the sandy and saturated sample collected near the profile base of location 2 dropped out as the sampler was raised. Another profile section at location 2, between 2 and 2.5 meters was drilled out and not sampled because the split barrel could not penetrate.

X-ray analysis revealed montmorillonite, illite and kaolinite in samples from both locations. Relative proportions are presented in Table III.

Table II. EDB Concentrations (ppb) in Water Samples

	Background Wells	Core Water	Wells
Month	Jun, Jul, Aug	Oct	Oct
Location			
1	N.D.[a]	N.D.	N.D.
2	0.2[b]	N.D.	0.2[c]

a. None detected
b. Mean of 3 replicated samples
c. Mean of 2 replicated samples

Table III. Clay Types and Relative Proportions

Location	Montmorillonite	Kaolinite	Illite
1	59.4	26.6	14.0
2	57.0	27.0	16.0

Statistical Analysis. Stepwise multiple linear regression analysis was used to determine the measured soil properties most correlated with EDB residues and to formulate a predictive equation to estimate chemical concentration. The regression analysis was performed on results from location 1 (Table IV) only because the number of segments containing residues at location 2 were not sufficient to produce an analysis.

Table IV. Regression Analysis for Location 1

F Ratio	Variables	R^2
76.23	Organic Matter	0.7840
	$y = -1.883 + 9.597$ (organic matter)	

The regression of location 1 revealed organic matter ($p>.001$) as the single and only important variable in predicting EDB residues and explained 78% of the variability in the data. The standard error of

1.372 produced 95% confidence intervals of ±2.74 ppb EDB. An additional regression analysis was performed on data from location 1 (Table V) to determine if residues found in the upper portion of the profile were correlated with the same variables as residues in the lower profile portion. The division was made just above the four lower segments containing residues. The analysis excluded any samples that did not have a measurement for each variable. Thus, the low F ratio for the upper portion was probably due to the small sample size, but organic matter still accounted for a significant ($p > .05$) portion of the variability. The lower portion, however, produced no significant F ratio, indicating the presence of EDB residues was apparently unassociated with organic matter and all other measured variables.

Table V. Regression Analysis for Location 1, Divided

	F Ratio	Variables	R^2
Upper Core	14.5	Organic Matter	0.8286
Lower Core	Nonsignificant		

Discriminant analysis created a model to predict the presence or absence of residues in soil segments of location 1 and was used to verify the outcome of the divided regression analysis. The analysis used a stepwise linear regression to determine which variables were important in predicting residues. As expected, organic matter was produced as the predictive or discriminating variable. The analysis assigned probability scores to each sample based on the relationship between the discriminating variable and the presence or absence of EDB in each sample. The probabilities were then used to classify each sample as theoretically containing or not containing residues. The results of this hypothetical classification were compared with experimental results to determine the accuracy of the model (Table VI). EDB in the four profile segments near the soil surface at location 1 was strongly predicted by the model. However, the four profile segments between 2.35 and 2.98 meters that contained EDB were strongly predicted to contain no residues. EDB residues in the lower profile segments were not associated with any of the measured variables.

Table VI. Discriminant Analysis of Location 1

Classification Matrix

	Correct		No. of Cases Classified: No Residues	No. of Cases Classified: Residues
	100.0 %	No Residues	15	0
	50.0 %	Residues	4	4
Total	82.6 %		19	4

Discussion

The correlation between EDB residues and soil organic matter as a result of sorption is a well documented association (12,13). This relationship would also be expected because the fumigant was applied near the soil surface where the greatest amounts of organic matter occur. However, the deeper group of segments at location 1 containing EDB residues were unassociated with any variables in the regression. The discriminant analysis also incorrectly predicted the lower portion as being free of residues. EDB in the lower segments may have leached with the water flux initiated by irrigation and rainfall, or accumulated as a result of leaching. A soil core study by Zalkin et al. (3) found organic matter, moisture and clay to be significant predictive variables of EDB residues above 2.40 meters in a sandy soil. In this study, core segments between 2.40 and 12.20 meters that contained EDB levels were generally unassociated with measured soil properties in a regression analysis and the authors suggested the residues were accumulating or leaching. Again, the lack of significant residue predictors in the lower half of the hole with relatively low organic matter would suggest that the EDB was migrating. This is speculation, but it appears to fit the analysis.

Clay type at locations 1 and 2 was not significant in predicting the presence or absence of EDB. Adsorption of a nonionic compound like EDB to wet clays would not be expected (12).

The movement of volatile pesticides in fine, moist soil has been studied by McKenry and Thomason (14) who found that 1,3-Dichloropropene (1,3-D) moved no more than 45.7centimeters into a silty clay loam with 23% moisture in 20 days. For comparison, a soil coring study performed by the California State Water Resources Control Board measured 1,3-D in a fine sandy loam to a depth of 198 centimeters 45 days after application (15). The fine texture and generally high moisture levels throughout the profile at location 1 in the present study would probably make the transport of EDB from the surface to lower profile levels a lengthy process. Therefore it may be of significance that the well associated with location 1 did not have a history of detected EDB residues despite a 10 year history of EDB use on a nearby field.

In contrast to location 1, the profile at location 2 was shallower, contained more sand, and below a depth of 0.60 meters would be considered a loam. The sandy layers at the profile base would tend to accelerate movement of the pesticide associated with the bulk flow of water and perhaps would account for the lack of residues in the deeper layers of the profile and for contaminated well water nearby. Based on texture and moisture analyses, one would expect a longer time required for downward movement of EDB at location 1 compared with location 2.

In summation, EDB may tend to leach more slowly through the finer soil of location 1 than through the relatively coarse soil of location 2. This is a possible explanation for the presence of EDB in well water at location 2 and its absence in well water at location 1.

Comparisons between soil core analyses and the presence or absence of pesticides in nearby wells are speculative because of well and profile depth differences and because of a lack of information about strata within the specific groundwater region of the field and well sampled. However, correlations between residues in well water and soil properties can often be shown. Work in progress by Teso and Younglove (4), shows that a discriminant analysis performed on well sample results for DBCP in association with soil type classifications correctly predicts contamination or no contamination 75% of the time. A total of 532 wells were sampled in agricultural areas where DBCP had been used. The results were compared with soil type data on township, range, section surveyed coordinates. This suggests that well data may be related spatially to soil core data with some degree of certainty. The relationship can be used to help focus soil core research on specific areas where groundwater residues are found due to agricultural usage. Ultimately, these data in addition to laboratory results and predictions made from models will be used to support the development of measures to refine regulation of pesticide use.

Acknowledgments

The authors wish to thank the California State Water Resources Control Board (SWRCB) for support in funding this work, and Syed Ali (SWRCB) for assistance in preparation of the study. We are grateful to A. Hugh Sinclair, Stanislaus County Agriculture Commissioner, for aid in determining soil core locations and to Nolan Petz and Daniel J. Perez for use of their land.

Literature Cited

1. van Berkum, J. A. In "Soil Disinfestation"; Muller, D. ED.; Elsevier: Amsterdam, 1979; pp. 58,59, 82.
2. "Ethylene Dibromide (EDB) - Position Document 4," U.S Environmental Protection Agency, Office of Pesticide Programs, 1983.
3. Zalkin, F.; Wilkerson, M.; Oshima, R. J. "Pesticide Movement to Groundwater Volume II: Pesticide Contamination in the Soil Profile at DBCP, EDB, Simazine and Carbofuran Application Sites," California Department of Food and Agriculture, 1984.
4. Teso, R. R., Work in Progress, California Department of Food and Agriculture at the University of California, Riverside.
5. Smith, C.; Margetich; S., Fredrickson, S.; "A Survey of Well Water in Selected Counties of California for Contamination by EDB in 1983," California Department of Food and Agriculture, 1983.
6. "Climatological Data, California Section" U.S. Dept. of Commerce, National Oceanic and Atmospheric Administration, 1979-83.
7. McLaughlin, J. C.; Huntington, G. L. "Soils of Westside Stanislaus Area California"; University of California, Davis, 1968; sheets 2,3.
8. Bouyoucos, G. J. Agronomy Journal 1962, 54, 464-5.
9. Quick, J. "California Soil Testing Procedures Manual"; California Fertilizer Association, 701 12th St. Suite 110, Sacramento, CA 95814, Method S 18.0.

10. Duncan, D. W.; Oshima, R. J. "Ethylene Dibromide in Two Soil Profiles" California Department of Food and Agriculture, 1985.
11. "Residues Determination of Dibromo-chloropropane in crops, soil, water"; Shell Development Co., Biological Sciences Research Center, Modesto, CA, Method #MMS-R-272-3, 1976.
12. Goring, C. A. I. Ann. Rev. Phytopath. 1967, 5, 285-318.
13. Bailey, G. W.; White, J. L. Res. Rev. 1970, 32, 29-92.
14. McKenry, M. V.; Thomason, I. J. Hilgardia 1974, 42, 393-438.
15. "1,2-Dichloropropane (1,2-D), 1,3-Dichloropropene (1,3-D)" Toxic Substances Control Program, California State Water Resources Control Board, Special Projects Report No. 83-8sp, 1983.

RECEIVED April 1, 1986

15

Chemical and Microbial Degradation of 1,2-Dibromoethane (EDB) in Florida Ground Water, Soil, and Sludge

R. A. Weintraub, G. W. Jex, and H. A. Moye

Pesticide Research Laboratory, University of Florida, Gainesville, FL 32611

The chemical and microbial degradation of 1,2-dibromoethane (EDB) in the subsurface environment was studied in laboratory incubations of groundwater, soil suspensions and sludge suspensions. EDB has been determined to have a chemical half-life of 1.5 to 2 years in Florida groundwaters (22°C). Rate constants for degradation determined at elevated temperatures were used to obtain extrapolated values via Arrhenius kinetics. Hydrolysis is the major mode of degradation, giving ethylene glycol and bromide ion, and is not pH-dependent between pH 4 and 9. Formaldehyde, an oxidation product of ethylene glycol, was shown to be a degradation product during extended incubation at elevated temperatures. Concentrations of 1 to 2 ppm of EDB were anaerobically degraded to ethylene by either methanogenic or facultative sludge in about 60 days, whereas Florida soils examined were nearly incapable of degrading EDB under these conditions.

1,2-Dibromoethane (ethylene dibromide; EDB) is used as a lead-scavenger in gasoline, a soil fumigant nematicide, a fumigant for stored grains, as a treatment to conform with quarantine regulations for certain fruit shipments and as a means of keeping milling machinery free from insects. By late 1983, EDB received great notoriety when it was reported that trace amounts of this chemical were detected in grains and grain products in the U.S. and in groundwater in Florida, Georgia, California, South Carolina, New York, and Hawaii (1). By March of 1984, all registered agricultural uses of EDB were phased out as a result of the U.S. Environmental Protection Agency's (EPA) determination that EDB's agricultural uses presented an "imminent hazard" to the health of humans. The chemical had been shown to have a high acute toxicity in all animals tested, oral LD_{50} values ranging from 50 mg/kg in the rabbit to 420 mg/kg in female mice (2). Its mutagenicity toward bacteria (3) and carcinogenicity in rats and mice (4) have

0097-6156/86/0315-0294$06.00/0
© 1986 American Chemical Society

been demonstrated and the mechanisms of toxicity have been described in progressive detail since the 1960's (5,6).

Consistent measurable levels of EDB in groundwater have sparked concern over the persistence of that portion of chemical that does not volatilize or is not readily degraded. Its chemical properties make it fairly mobile in the subsurface environment, having a solubility in water of 4300 ppm at 30°C (7) and a vapor pressure of 11 mm Hg at 25°C (8), giving it a Henry's constant of approximately 6.3×10^{-4} atm-m^3/mol. A K_{ow} value of 58 is obtained using the appropriate predictive model (9). A K_{oc} value of 66 mL/g has been experimentally obtained in two varying soils (10). Such chemical and field characteristics place EDB on the list of pesticides with the potential to be groundwater contamination hazards by the criteria established by U.S. EPA Office of Pesticide Programs (1)

Investigations in the past have have focused residual EDB on crops and on processed or stored products (11-13). Values of the order of 5-10 days (14) and 14 years (15) have been reported as the half-life due to hydrolysis in neutral aqueous solutions at ambient temperature. The biological conversion of EDB to ethylene and bromide ion by an anaerobic soil-water culture has been reported to occur in two months (16). In another study, bacterial cultures under denitrifying conditions failed to show any potential for degrading EDB (17). Further, it was found that under methanogenic incubation conditions, EDB was transformed to a water-insoluble non-halogenated gas, which was thought to be ethylene (18,19). Recent accounts of photodegradation showed EDB was completely mineralized to HBr and CO_2 by heterogeneous photocatalysis (on TiO_2) in aqueous solution (20), while in another study, the photoreaction was reported to proceed via the conversion of EDB to bromoethanol followed by the cyclization to ethylene oxide which was hydrolyzed to ethylene glycol by a process not enhanced by light (21).

Two years after the ban on agricultural use of EDB, concentrations ranging from 0.02 to about 600 ppb have been detected in well-water samples being taken by the state of Florida's EDB monitoring program. Figure 1 gives a representation of the locations and frequency of state-controlled EDB applications and water well sites found to be contaminated with EDB as of March 1985. Estimated amounts of applied EDB in Florida total 600-700,000 L/year. Greater than one-half of the contaminated well sites have been below 0.20 ppb and most of the remaining positives have been between 0.20 and 10 ppb. Depths of the wells sampled vary widely from 3 to 300 m (22). Florida's citrus, peanut, and soybean farming areas, where most of the EDB use has been concentrated for 40 to 50 years, are located on predominantly sandy soils with relatively low organic content (i.e. <1%). The soils are mostly entisol, spodosol, alfisol and ultisol types (23). The fate of subsurface residual EDB is of great concern in light of the fact that the groundwater is the source of potable water for about 88 percent of Florida's households (nationally, about one half the population's potable water is derived from groundwater sources) (23).

This study investigates the degradation of EDB by hydrolysis in groundwater samples and by microbial activity in soil samples collected from northcentral and northwestern regions of Florida.

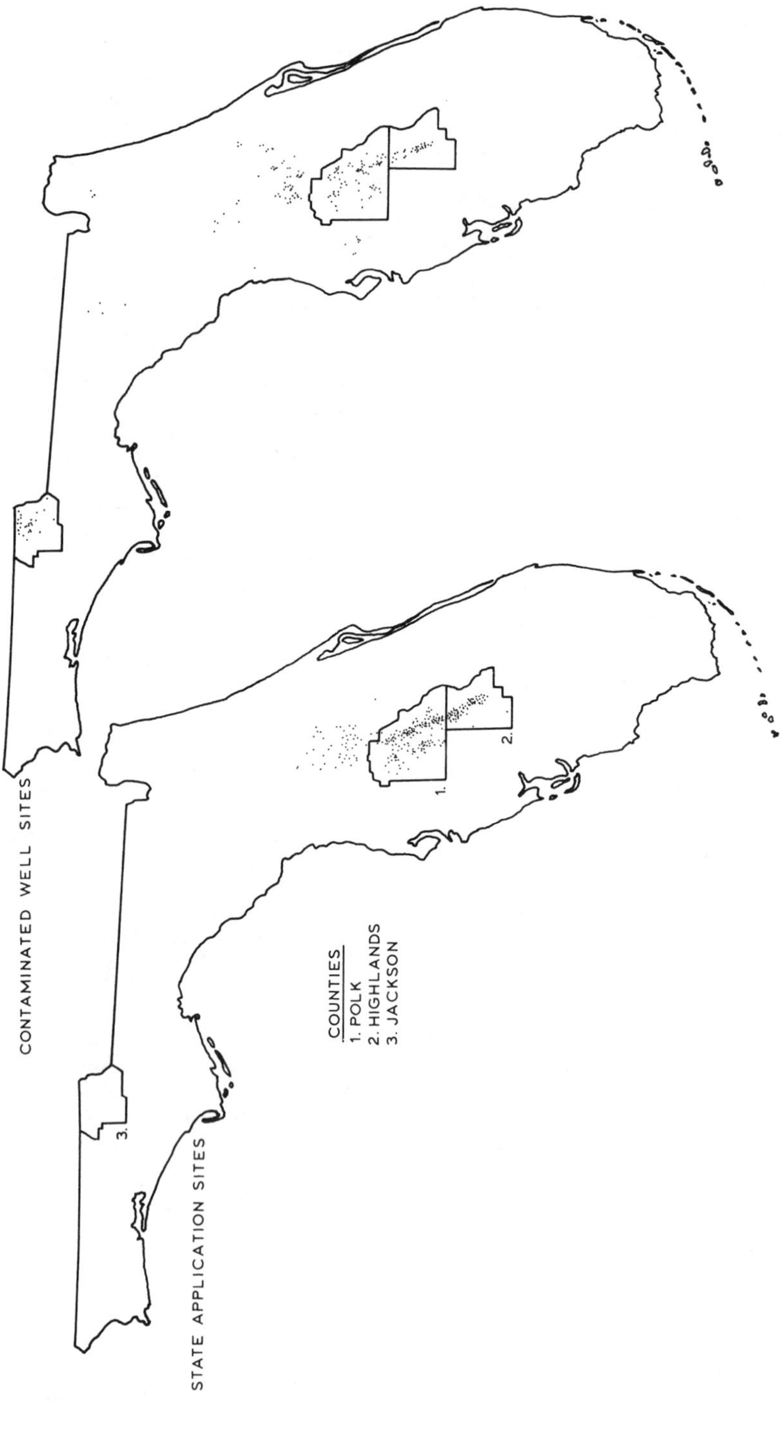

Figure 1. State-controlled EDB application sites and contaminated well sites as of March 1985.

Laboratory experiments were performed with natural samples in order to make qualified estimates of the persistence of EDB in the subsurface environment and to identify the products of degradation. The potential for biotransformation of the chemical by two different sludge preparations was also investigated.

MATERIALS AND METHODS

Groundwater Degradation Kinetics.

Groundwater obtained from shallow wells in three northcentral and northwest Florida counties (Polk (P), Highlands (H), and Jackson (J) and laboratory deionized water (DW) were fortified with EDB (EPA analytical standard, 99%+ purity) to concentrations of 10 or 100 ppb. Glass serum bottles (100 mL Wheaton, Millville, NJ) were filled with the solutions, the headspace purged with N_2 and the bottles tightly capped with Teflon-coated rubber septum seal crimp caps (Supelco, Bellefonte, PA). Before sample preparation, the serum bottles, seals, and caps were autoclaved (121°C, 15 psi, 20 minutes) and the waters were either autoclaved (same conditions) or filtered (0.20 μm filter, Millipore, Bedford, MA) to eliminate microbial activity. Sets of prepared samples in duplicate were incubated in an inverted position in a darkened water bath. Due to the relative stability of EDB at ambient temperatures observed in preliminary experiments, incubations were performed at 40, 50, 60, 70, and 80°C ± 0.5°C. Serum bottles were periodically taken out of the water bath and allowed to equilibrate to room temperature before sampling. A 10 mL syringe was used to withdraw an aliquot which was delivered to a 10 mL volumetric flask to assure a precise volume. This volume was transferred to a 30 mL screw-top test tube to which 1.0 mL of hexane was added and the tube immediately capped. It was inverted 5 times and mixed on a vortex-mixer for one minute, inverted 5 more times and allowed to stand for at least 5 minutes. The hexane layer was pipetted into a 1 mL vial for automated gas chromatographic analysis.

Degradation rate constants were obtained by linear regression least squares analysis of plots of log % EDB remaining vs time. Pseudo-first order rate constants were used to generate Arrhenius plots (log rate constant vs 1/T °K) to estimate activation energies (E_a) and to make extrapolated estimates of rate constants and half-life values at ambient temperature.

At least 2 samples in each incubation trial were fortified with ^{14}C-EDB (approximately 500 μCi/g Amersham, Arlington Heights, IL) and periodically sampled to check the integrity of the system. Total ^{14}C-activity was measured by liquid scintillation counting of 1.0 mL of sample added to 15 mL of Aqua-Sol 2 liquid scintillation cocktail (New England Nuclear, Boston, MA). Counting was done on a Searle Analytical 92 liquid scintillation counter with a Silent 700 electronic data terminal. Quenching was evaluated by comparing additions of the fortified waters or hexane extracts to additions of known amounts of ^{14}C-toluene ($4x10^6$ dpm/mL, New England Nuclear, Boston, MA). Efficiencies and precision of extraction of EDB from the waters were evaluated in a similar fashion.

Gas Chromatographic Analysis.

Hexane extracts of sample aliquots were analyzed for EDB as well as other closely related suspected metabolites (bromoethane, bromoacetic acid and bromoethanol) on a Hewlett-Packard 5840 A gas chromatograph equipped with a ^{63}Ni electron capture detector. A 2 m glass column packed with 15% polypropylene glycol on GasChrom Q 80/100 mesh was used with a 5% methane/95% Ar carrier gas at a flow of 52 mL/min. Temperatures maintained during analysis were 100°C for the column, 300°C for the detector, and 225°C for the injector port. Standard curves of detector response over the range of 20 to 1500 pg/μL of EDB were constructed and used for quantitation using linear regression least squares analysis and were prepared daily.

Soil and Sludge Degradation Experiments. Soils were sampled from sites in Jackson, Polk and Highlands counties where water from wells has been contaminated with EDB to test the EDB degrading ability of the indigenous microflora. Samples from each site were collected at depths of 1 and 3 m with auger bores. Groundwater was not reached in any of the sampling. Two sludges were used to study the EDB degrading potential of a broad spectrum of microflora. One sludge was from a primary sludge contact tank at the University of Florida containing a rich flora of facultative organisms. The second sludge was obtained from the methane digester of Dr. Paul Smith (University of Florida) and was known to contain a significant methanogenic flora.

The bottle-filling and incubation procedures were adapted from Bouwer and McCarty (17) and the medium was that of Alexander and Lurtigman (24). The sterilized medium (121°C, 15 psi, 20 min) was boiled and then flushed with N_2 while cooling. It then was fortified with EDB to a concentration of 400 ppb using an alcohol solution of the pesticide (Aldrich Chemical, Milwaukee, WI); ^{14}C-EDB (Amersham, Arlington Heights, IL) was added to the medium to give 4000-5000 dpm/mL as a tracer. $FeSO_4 \cdot 7H_2O$ (0.25 g/L) and $Na_2S \cdot 9H_2O$ (0.1 g/L) were added as reductants. The medium was maintained under N_2 until siphoned into serum bottles (100 mL) containing either 5 g soil or 10 mL of sludge. The bottles were flushed with N_2 during filling and capped with Teflon lined crimp caps. At each sampling, an unfortified soil along with three fortified and three sterilized fortified soil replicates were analyzed.

The bottles were incubated at 25°C and analyzed at 1, 5, 10 and 30 days then monthly thereafter for seven months. Three 20 mL portions from each bottle were removed for CO_2 analysis. One portion was treated with 2 mLs of concentrated H_2SO_4 and one with 1 mL of a saturated $BaCl_2$ solution. The third 20 mL portion was untreated. A 1.0 mL sample from each bottle was added to 15 mL of Aqua-sol 2 and counted by liquid scintillation. The presence of $^{14}CO_2$ is indicated by the treated samples having less ^{14}C-activity than the untreated samples.

Samples were also extracted three times (1:1) with hexane and a portion of each hexane extract was analyzed by GC, as previously described. A 1.0 mL portion of each extract and a 1.0 mL portion of the residual water was counted by liquid scintillation as described before.

Identification of Gaseous Products. Gases produced during soil and sludge incubations were analyzed by injections of headspace of the sample on a F & M Scientific 700 gas chromatograph equipped with a thermal conductivity detector in series with a Packard 894 gas proportional counter. Glass columns (2 m) packed with Chromosorb 101 (Applied Science, State College, PA) or Porapak Q (Waters Assoc., Framingham, MA) were used for determination of ethylene or CO_2, respectively. Carrier gas was N_2 at 20 mL/min and oven was at ambient temperature.

pH-Dependence of Degradation in Solutions. A series of 5 x 10^{-3} M buffer systems (25) were used to maintain the EDB fortified waters at pH values ranging from 4.0 to 9.0. The buffers included borax/succinic acid, phathalate/NaOH, borate/NaOH, borax/phosphoric acid, and carbonate/bicarbonate (analytical grade, Fisher Scientific, Fair Lawn, NJ). Incubations were conducted at 62°C and samplings and analyses were performed as described before.

Analysis for Purgeable Brominated Degradation Products. Water samples fortified with EDB to concentrations of 1 and 4 ppm and incubated at 80°C were analyzed for brominated products by bubbling helium through 25 mL of sample for 22 minutes at ambient temperature onto Tenax adsorbent. Purgeables were directed into the GC column of a Finnigan 4021 quadrupole gas chromatograph/ mass spectrometer by heating the absorbent to 120°C. A glass column packed with the same material used for EDB analysis was used under similar conditions.

Analysis of Ethylene Glycol, Formaldehyde and Bromide ion in Water. Ethylene glycol concentrations in water were determined by GC analysis of the 2-nitrophenylhydrazine derivative of formaldehyde, 2-nitrophenylhydrazone. To 2 or 4 mL aliquots of the water sample, a 20 ul aliquot of 5 x 10^{-6} M periodic acid (analytical grade, Mallinckrodt, Paris, NY) was added and the sample was then allowed to stand at room temperature in the dark for 20 hours. Excess periodic acid was precipitated by the addition of 25 uL of KNO_3 and allowed to stand for 1 hour in an ice bath. Standard aqueous solutions of ethylene glycol were assayed by the same procedure each day of analysis to construct a standard curve for quantitation. To quantitate formaldehyde in aqueous solution, either formed by the oxidation of ethylene glycol or present in solution as a degradation product or standard, 600 uL of a 2 M aqueous solution of 2-nitrophenylhydrazine (ICN Pharmaceuticals, Plainview, NY; recrystalized from hexane) was added to the reaction mixture which was then incubated in a water bath at 40°C for 50 minutes. The 2-nitrophenylhydrazone was extracted by the addition of 5 mL of hexane to the tube which was then mixed on a vortex-mixer for 1 minute. The hexane layer was pipetted to a 1 mL vial for automated GC analysis on a 2 m glass column packed with 4% OV-225 on Chromosorb Q 80/100 mesh; a 5% methane/95% Ar carrier gas at 30 mL/min, a column temperature of 173°C, a detector temperature of 300°, and an injector temperature of 225°C were employed.

Bromide ion in aqueous solutions was analyzed by Standard Methods Procedure no. 405 C (26). It was oxidized by chloroamine T to bromine which brominates phenol red (both reagents from Fisher Scientific, Fair Lawn, NJ). The brominated product was measured by absorbance of the reaction mixture at 590 nm on a Beckman DU-8 spectrophotometer. Standard aqueous solutions of bromide were assayed to construct a standard curve over a range of 50 to at least 1000 ppb.

RESULTS AND DISCUSSION

Groundwater Degradation Kinetic Studies.

All kinetic plots constructed from the data indicated that the disappearance of EDB from solution at all temperatures (40° to 70°C) in the waters tested followed simple pseudo-first-order kinetics (Figure 2). Linear regression by least squares analysis shows that correlation coefficients range from .90 for lower temperature incubation trials, to larger than .99 for higher temperature incubation trials. Kinetic results obtained in the buffered waters incubated at 63°C are shown in Table I. The rate constants observed in the different waters vary only slightly and indicate that neither acid or base-catalyzed hydrolysis is favored within the pH range examined (pH 4-9) (Figure 3). At pH 5 and 8 an increase in the rate constants of about 10% is observed, but because these portions of the plot are not consistent with a pH-dependent trend, i.e. a constant slope over a considerable pH range, these deviations can be attributed to specific contributions of the buffer type used. Such a contribution by specific buffer catalysis even at buffer concentrations of 5×10^{-3} M (as used in the present study) or less is reasonable (27).

The lack of pH-dependence shown in this study is consistent with results presented elsewhere for alkyl halides (9). This implies that for the observed degradation rate constant, k_{obs} is approximately equal to $k_{neutral}$ (or pH7) within the examined pH range.

The mechanism for the chemical reaction is assumed to be the same over the entire temperature range when a straight line is obtained in the plot of ln k vs 1/T°K for the reaction observed and the slope of the line is equal to $-E_a/RT$. Such a relationship is derived from intergration of the Arrhenius equation to Equation 1.

$$\ln k = -E_a/RT + \text{constant} \qquad (1)$$

where k is the rate constant at temperature T° (K) and R is the gas constant (1.987 cal/deg-mol). Extrapolation from the Arrhenius plot allows the calculation of the rate constant, k, at temperatures other than those tested. The half-life ($t_{1/2}$) is calculated for a first-order or pseudo-first-order rate constant by substitution into Equation 2.

$$t_{1/2} = 0.693/k \qquad (2)$$

A typical Arrhenius plot of data obtained in this study is shown in Figure 4. Extrapolations of results of Arrhenius plots for the various EDB-fortified waters to common groundwater temperatures are shown in Table II. Activation energies calculated for the hydrolysis at 22°C range from 19.6 to 24.2 Kcal/deg-mol. The pH values of the samples during the degradation experiments were in most cases about 1 pH unit higher than the recorded pH at the time of sampling at the well sites. This is presumably due to loss of dissolved CO_2. However, as the pH-dependence experiments have indicated, this pH shift should not have a significant effect on the degradation rate constants. Therefore, it was ignored in the calculations.

The differences in the rates of degradation were evident

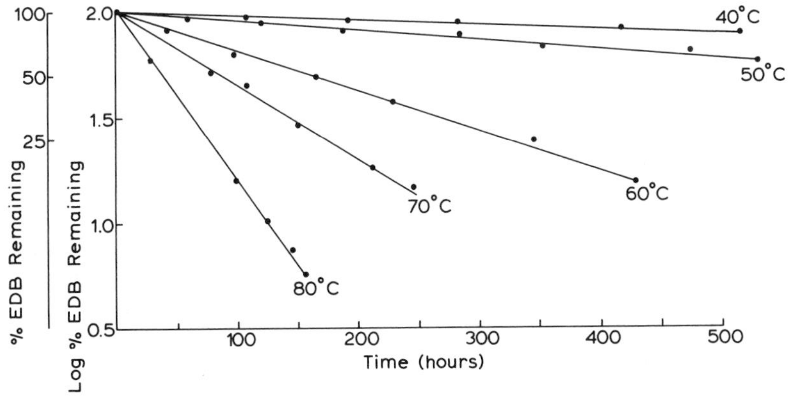

Figure 2. EDB hydrolysis in groundwaters at various temperatures.

Table I. Kinetic Data for EDB Hydrolysis in Buffered Solutions (63°C, 100ppb)

pH	*buffer	$10^3 k_{obs}$; $t_{\frac{1}{2}}$ (days) Polk	Highlands	Jackson	Deion.	**10^3k±SD (pooled data)
4.0-4.4	a	2.31 301	2.31 302	2.34 298	2.48 280	2.36±0.13
4.0-4.9	b	2.94 236	2.61 266	2.45 283	2.54 273	2.63±0.29
4.7-5.3	b	2.84 208	2.56 225	3.46 200	3.10 225	3.23±0.45
5.3-5.6	a	2.41 288	2.44 288	3.47 287	3.10 295	2.40±0.10
6.1-6.5	c	2.45 283	2.98 233	2.64 263	2.83 285	2.73±0.23
7.6-7.7	d	2.68 259	2.68 259	2.52 275	2.63 263	2.95±0.62
7.8-7.9	e	4.39 158	3.96 175	3.83 181	4.17 166	4.17±0.38
8.0-8.3	f	3.76 266	2.81 247	3.56 195	3.62 191	3.44±0.44
8.6-9.0	g	3.07 224	2.61 267	2.41 288	2.95 235	2.95±0.48

* 5×10^{-3} M buffers: a=borax/succinic acid, b=potassium phathalate, c=borax/succinic acid, d=borax/phosphate, e,f=boric acid, g=carbonate

** pooled data of all samples, n=8.

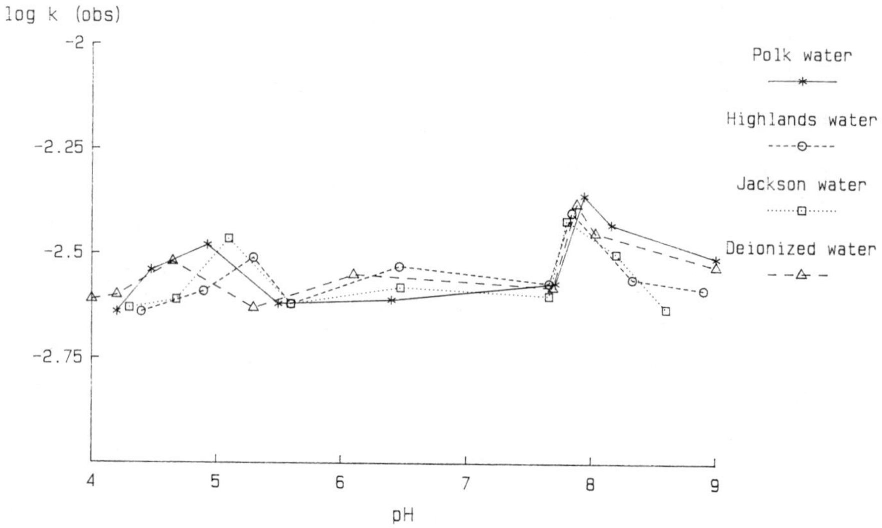

Figure 3. Profile of EDB hydrolysis rate constants vs pH in buffered solutions at 63°C.

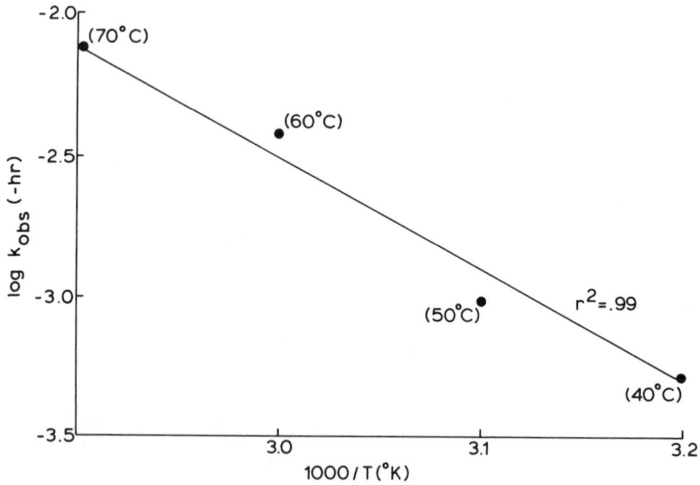

Figure 4. Typical Arrhenius plot of log k_{obs} vs 1/T for EDB hydrolysis.

(α = .05) in several groundwater samples. For example, the water from the Polk county site fortified to 10 ppb EDB had a $t_{1/2}$ about 40% lower than in the 100 ppb fortified water, and in Highlands water, the 10 ppb fortification had a $t_{1/2}$ about 60% larger than the 100 ppb fortification. The Jackson water results showed a difference between the filtered and autoclaved preparations accounting for a large variation in the 10 ppb samples making comparison with the 100 ppb samples meaningless. In contrast, the observed degradation rates in deionized water at the two concentrations were identical. These results suggest that the constituents of groundwater may have an influence on the hydrolysis of EDB.

Half-lives in the range of 1.5 to 2 years can be expected for EDB in groundwater in various regions of Florida (22°C). In more northern locations of the U.S. half-lives of about 2 to 5 years could be expected since their typical groundwater temperature is less than 15°C (Table II).

<u>Hydrolysis Degradation Product Studies.</u>
At elevated temperatures (60, 72, 80°C) the decrease in ^{14}C-activity partitioned in the hexane phase after extraction paralleled the EDB decline determined by GC as shown in Figure 5. Exhaustive extractions of the aqueous phase did not result in detection of other brominated compounds. Nor were purgeable brominated compounds detected in the aqueous phase by GC/MS.

Figure 6 shows a typical profile the concentrations of EDB, ethylene glycol, bromide ion and partitioning of ^{14}C-activity during incubations of water samples at various temperatures initially fortified to 1 to 4 ppm of EDB. During these incubations, these products accounted for 60-100% of the initial EDB. The conversion of EDB to ethylene glycol and bromide ion was essentially complete after about 7 days. However, beyond that point other products may have been forming as indicated by the changing concentrations of the observed products. This result suggests the possibility of the presence of small quantities of unidentified products. The mechanism shown in Figure 7 is proposed for the hydrolysis of EDB. Bromoethanol has not been detected as a product in incubated samples, suggesting that it is a short-lived species. Ethylene oxide has been reported to have a half-life of about 12 days in neutral solutions at 25°C (<u>9</u>).

The overall conversion of EDB to ethylene glycol is thermodynamically favored. The Gibbs free energies are -8.58 and -46.76 Kcal/mol for the neutral and base-catalyzed reactions, respectively. Thermodynamic data are not available for bromoethanol. Kinetic data for hydrolysis of the intermediate species would be helpful for further evaluation of the mechanism of EDB hydrolysis proposed here.

Prompted by a recent study which showed that EDB could be oxidized under anhydrous conditions to formaldehyde by the action of superoxide ion (<u>29</u>), we fortified water solutions with part-per-million concentrations of EDB, incubated them at elevated temperatures, analyzed them and found them to have low but consistent levels of formaldehyde, but only after all the EDB had been hydrolyzed. Studies conducted with natural and deionized water fortified to 10 ppm with ethylene glycol and incubated at 85°C showed that the amounts of formaldehyde detected varied from about 350 ppb to 2 ppm after 40 days of incubation. These findings

Table II. Extrapolated Pseudo-First-Order Kinetics Values from Arrhenius Plots for EDB Hydrolysis in Natural and Deionized Waters at Environmental Temperatures

County Water	EDB (ppb)	pH (±0.05)	10^5k (hr)	$t_{1/2}$ (days±SD)	E_a (Kcal/deg-mol)
22°C					
Polk	10	8.20	11.12	259 ± 35	19.7
	100		6.65	435 ± 47	20.9
Highlands	10	8.20	3.74	772 ± 94	24.2
	100		9.37	308 ± 66	19.2
Jackson	10	8.00	4.38	659 ±343	23.0
	100		7.83	369 ± 69	19.7
Deionized Water					
(22°C)	10	7.60	8.94	323 ± 55	19.6
	100		8.94	323 ± 55	19.6
(30°C)			25.50	130 ± 16	
(15°C)			3.57	809 ±109	
(8°C)			1.35	2136 ±461	

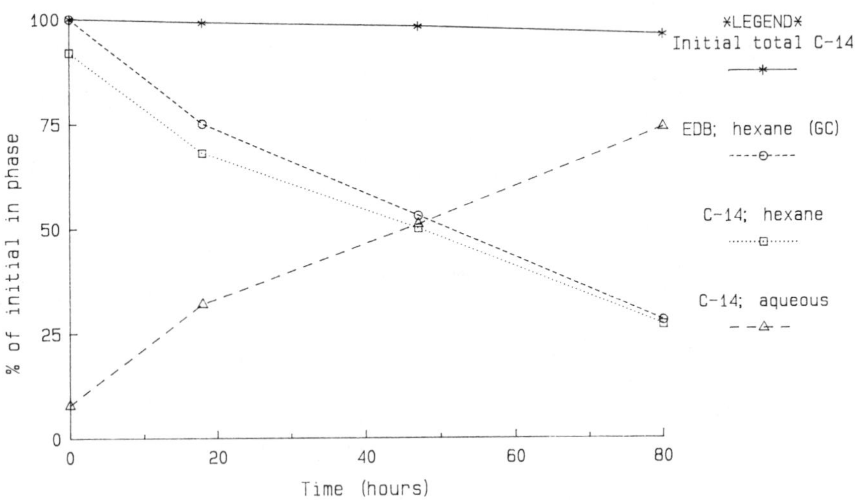

Figure 5. Partitioning of ^{14}C-EDB and products during incubation of aqueous solutions at 80°C; initial 60, 30 ppb/ 18,000, 9,000 dpm/mL.

taken together indicate that hydrolysis of EDB is required as a first step for formaldehyde production and that EDB is not converted to formaldehyde by direct oxidation as was found in the superoxide study. More work, currently underway is required to determine the factors affecting the formation of formaldehyde (i.e. oxygen concentration, other oxidants, competing reactions).

Soil and Sludge Degradation Studies.

Neither soils or sludges showed evidence of CO_2 production from EDB incubations; that is, the dpm/mL observed in H_2SO_4-treated and $BaCl_2$-treated samples were never significantly below those of the untreated samples. This result is consistent with the negative results reported by Bouwer and McCarty (18,19) for CO_2 production.

Figure 8 shows the decrease in dpm/mL found in the hexane extract of the methanogenic sludge during incubation, indicating that EDB is being degraded. Only a slight increase in dpm/mL in the residual water for both natural and sterile sludges indicates little dissolved CO_2 or other water-soluble degradation product(s) is being formed, suggesting the formation of a volatile product or products during incubation. In fact, a significant amount of gaseous product (2-4 mLs) was produced. This was observed as an increasing head-space in the bottles. Analysis by GC/gas proportional counting revealed that the headspace was a mixture of N_2, CH_4 and C_2H_4; only the C_2H_4 was radioactive. Figure 9 shows the GC peak for ethylene detected by thermal conductivity and its corresponding peak detected by gas proportional counting. Figure 10 shows the results of GC analyses for EDB of the same sludge incubations.

The results from the facultative sludge samples were strikingly similar to the methogenic sludge. In 60 days all the EDB was degraded and little radioactive material remained in the residual water after the triple hexane extraction. Similarly, the sludge produced the same gaseous products, of which only the ethylene was radioactive.

These results confirm the hypothesis of Bouwer and McCarty (19) that the water-insoluble volatile product derived from their EDB seeded culture incubations is ethylene. They observed a shorter period of time for total degradation, (14 days) probably due to the lower initial concentrations of EDB in their experiments, and reported no lag period in their incubation. Our results in Figures 8 and 10 indicate gradual decline of EDB prior to 40 days with a rapid decline thereafter. It is also noted that a significant loss of EDB in the sterile samples occurred after 40 days. This is also consistent with their results and could reflect reaction of EDB with sulfide reductant (28).

The Florida soils tested were only weakly capable of degrading EDB under anaerobic conditions. A soil preparation from the 1 m depth from Polk county showed a 40% decrease of labeled and nonlabeled EDB, after seven months. No CO_2 was produced and there was no more ^{14}C-activity in the residual water after hexane extractions than for the sterile preparations. This indicates that the degradation product or products are volatile, such as ethylene. All other soils failed to degrade EDB under similar conditions over equally long incubation periods. This implies that either appropriate organism(s) are not present or the soils contained insufficient secondary carbon sources necessary to maintain cometabolism.

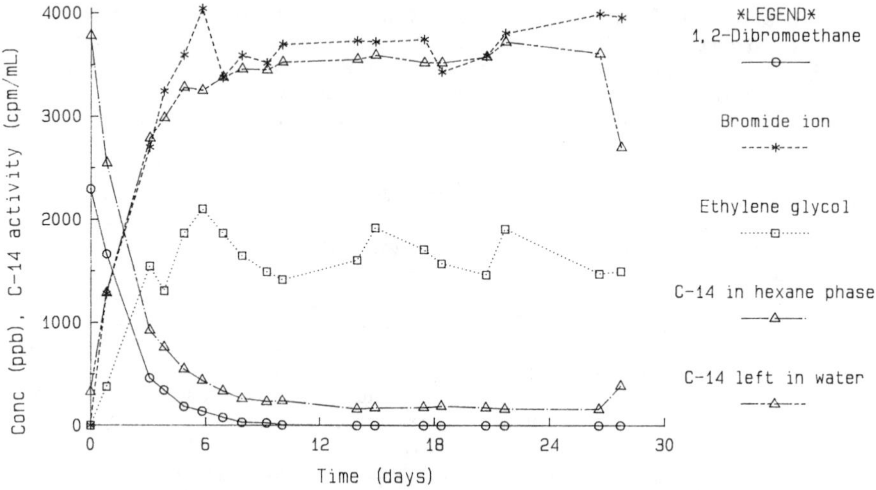

Figure 6. Products of hydrolysis of EDB during incubation of aqueous solutions at 83°C.

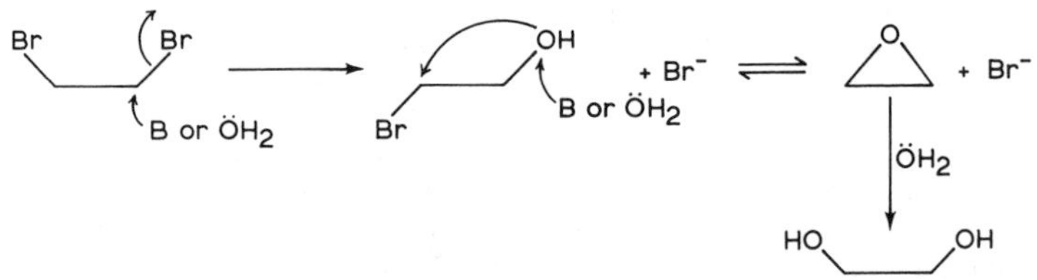

Figure 7. Proposed mechanism of hydrolysis of EDB in aqueous solutions.

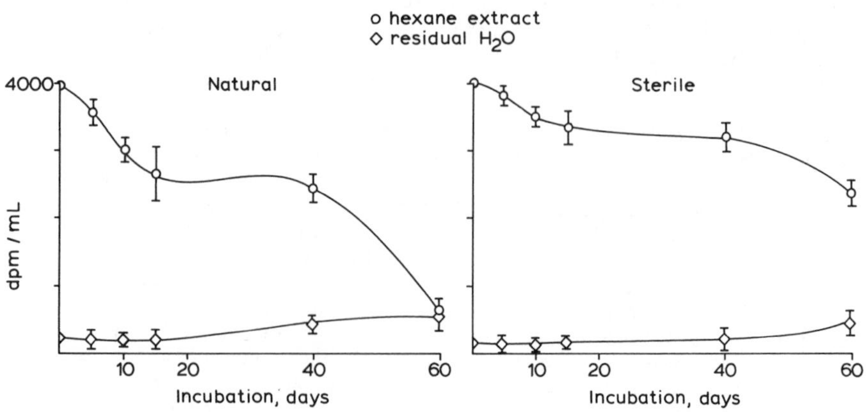

Figure 8. Degradation of EDB in methanogenic sludge monitored by extraction of ^{14}C-EDB.

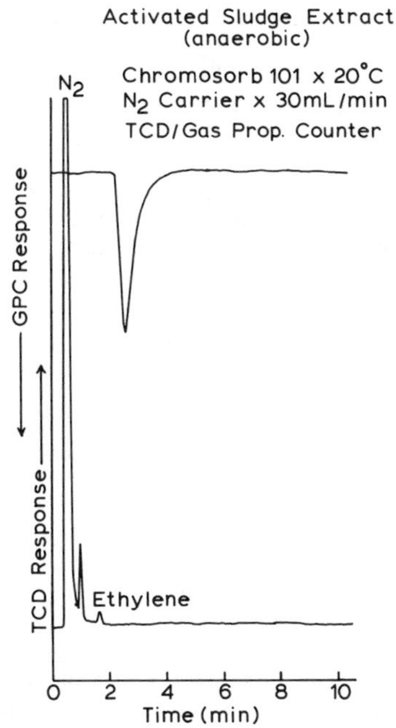

Figure 9. Detection of ^{14}C-ethylene produced during incubation of EDB in sludges.

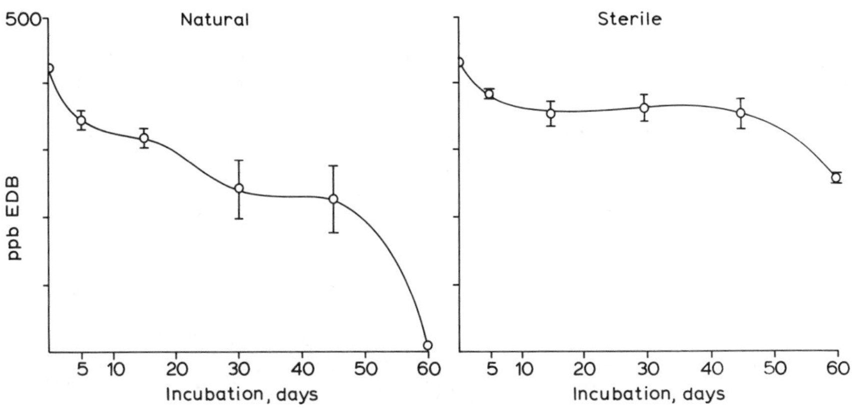

Figure 10. Degradation of EDB in methanogenic sludge monitored by GC.

Three reports, including this one, have presented evidence indicating ethylene is the major or sole product of anaerobic degradation of EDB (16,19). In contrast, Bouwer and McCarty (18) found 1,2-dichloroethane (DCE) was degraded under similar conditions to CO_2. Thus, substitution of Br for Cl appears to make the difference between cometabolism (EDB) and true metabolism (DCE).

Degradation of EDB has been observed in this laboratory in sludge preparations under aerobic conditions. The product or products, contrary to the anaerobic incubations, are not volatile, but highly water-soluble. The details of this work will be reported elsewhere.

SUMMARY

EDB was determined to have a chemical half-life of 1.5 to 2 years in the Florida groundwaters at 22°C. Whether the concentration of initial EDB present was 10 or 100 ppb made a significant difference in rate of degradation in two of the three waters tested, while in deionized water it did not, suggesting that groundwater constituents may affect degradation. Such differences due to source of the water were less evident. Hydrolysis is the major mode of degradation, giving ethylene glycol and bromide ion, and is not pH-dependent between pH 4 and 9. A reasonable mechanism in which bromoethane and ethylene oxide are intermediate products was suggested. Furthermore, ethylene glycol was oxidized to formaldehyde after extended incubation at elevated temperatures.

Two different sludge materials readily degraded EDB anaerobically, yielding ethylene as the only organic product, whereas Florida soils examined were nearly incapable of degrading EDB under these conditions or are extremely slow in doing so.

ACKNOWLEDGMENTS

We are grateful to Florida Department of Environmental Regulation for funding this and continuing work (contract EDB 005) and providing groundwater and soil samples.

We also thank Mr. F.A. Bordeaux of Paul Smith's laboratory, University of Florida, for help with methanogenic sludge and Dr. T. Phelps of D.C. White's laboratory, Florida State University, for help in performing the GC/gas proportional counting.

LITERATURE CITED

1. Cohen, S.Z; Carsel, R.F.; et al. Potential for pesticide contamination of ground water resulting from agricultural uses in "Treatment and Disposal of Pesticide Wastes." Am. Chem. Soc. Sympos. Ser. #259, Krueger, Seiber (Eds.), Wash., DC (1984).
2. Fishbein, L. Potential Industrial Carcinogens and Mutagens - Studies in Environmental Science, Elservier, Amsterdam, Vol. 4, p. 241 (1979).
3. Olson, W.A; Habermann, R.T.; et al. J. Natl. Cancer Inst., 51,1993 (1973).

4. van Bladeren, P.J.: Breimer, D.D.; et al. The role of glutathione conjugation in the mutagenicity of 1,2-dibromoethane. Biochem. Pharmacol, 29,2975 (1980).
5. Kluwe, W.M.; McNish, R.; et al. Depletion by 1,2-dibromoethane, 1,2-dibromo- 3-chloropropane, tris (2,3-dibromopropyl)-phosphate, and hexachloro-1,3,butadiene of reduced non-protein sulfhydryl groups in target and non-target organs. Biochem. Pharmacol., 2(8)499 (1981).
6. White, R.D.; et al. Characterization of the hepatic DNA damage caused by 1,2-dibromoethane using the alkaline elution technique. Carcinogen, 2(9)839 (1981).
7. Hamaker, J.W.; Kerlinger, H.O. Vapor Pressure of Pesticide in "Pesticidal Formulations Research Physical and Colloidal Chemical Aspects;" Wallenburg, J.W., Ed.; Adv. Chem. Ser. 86, Am. Chem. Soc.: Wash., DC, p. 46 (1969).
8. Gunther, F.A.; Westlake, W.E.; et al. Residue Rev., 20,45 (1968).
9. Lyman, W.J. Solubility in Water in "Handbook of Chemical Property Estimation Methods, Environmental Behavior of Organic Compounds;" Lyman, W.J.; Reehl, W.F.; et al, Eds., McGraw-Hill: New York (1982).
10. Roger, R.D.; McFarlane, J.C. Sorption of carbon tetrachloride, ethylene dibromide, and trichloroethylene on soil and clay. Environ. Monitor. and Assess., I,155 (1981).
11. Heuser, S.G. Residues in wheat and wheat products after fumigation with ethylene dibromide. J. Sci. Fd. Agric., 12,103 (1961).
12. Sinclair, W.B.; Lindgred, D.L.; et al. Recovery of ethylene dibromide residues from fumigated whole kernel and milled wheat fractions. J. Econ. Entomol., 55(6)836 (1962).
13. Beckman, H.; Crosby, D.G.; et al. The inorganic bromide content of foodstuffs due to soil treatment with fumigants. J. Fd. Sci., 32,138 (1967).
14. John, R. "Air Pollution Assessment of Ethylene Dibromide." MITRE Tech. Report, MTR7144 (1976).
15. NIOSH. Criteria for a recommended standard: Occupational exposure to ethylene dibromide. U.S. Dept. HEW, Wash., DC, DHEW Public #77-221 (1977).
16. Castro, E.E.; Besler, N.O. Biodehalogenation. Reductive dehalogenation of the biocides ethylene dibromide, 1,2-dibromo-3-chloropropane, and 2,3-dibromobutane in soil. Environ. Sci. and Technol., 2(10)779 (1968).
17. Bouwer, E.J.; McCarty, P.L. Transformations of halogenated organic compounds under denitrification conditions. J. Appl. and Environ. Micro., 45(4)1295 (1983).
18. Bouwer, E.J.; McCarty, P.L. Transformations of 1- and 2-carbon halogenated aliphatic organic compounds under methanogenic conditions. Appl. and Environ. Micro., 45(4)1286 (1983).
19. Bouwer, E.J.; McCarty, P.L. Ethylene dibromide transformations under methanogenic conditions. Appl. and Environ. Micro., in press (1985).
20. Nguyen, T.; Ollis, D.F. Complete heterogeneous photocatalyzed transformation of 1,1- and 1,2-dibromoethane to CO_2 and HBr. J. Phys. Chem., 88,3386 (1984).

21. Castro, C.E.; Besler, N.O. Photohydrolysis of ethylene dibromide. J. Agric. Fd. Chem., 33,536 (1985).
22. Atkerson, T. Groundwater Protection Task Force/Groundwater Epidem. Division, FL HRS (1985), personal communication.
23. Fernald, E.A.; Patton, D.J. (Eds.) "Water Resources Atlas of Florida." Inst. Sci. and Publ. Affairs, Fla. State University, Publ., pp 36,58 (1984).
24. Alexander, M.; Lurtigman, B.K. Effect of chemical structure on microbial degradation of substituted benzenes. J. Agric. and Fd. Chem., 14,410 (1966).
25. "Documenta Geighy-Scientific Tables," 6th ed.; Diem, K., Ed. Geigy Pharmaceuticals, p. 314 (1962).
26. "Standard Methods for the Examination of Water and Wastewater", 15th ed., APHA-AWWA-WPCF, p.261 (1980).
27. Perdue, E.M.; Wolfe, N.L. Prediction of buffer catalysis in field and laboratory studies of pollutant hydrolysis reactions. Environ. Sci. and Tech., 17(11)635 (1983).
28. Calderwood, T.S., Sawyer, D.T. Oxygenation by superoxide ion of 1,2-dibromo-1,2-diphenylethane, 2,3-dibromobutane, ethylene dibromide (EDB), and 1,2-dibromo-3-chloropropane (DBCP). J. Am. Chem. Soc., 106,7185 (1984).
29. Schwarzenbach, R.P.; Giger, W.; et al. Groundwater contamination by volatile halogenated alkanes: abiotic formation of volatile sulfur compounds under anaerobic conditions. Environ. Sci. and Tech., 19,322 (1985).

RECEIVED April 1, 1986

Movement of Selected Pesticides and Herbicides through Columns of Sandy Loam

Viorica Lopez-Avila[1], Pat Hirata[1], Susan Kraska[1], Michael Flanagan[1], John H. Taylor, Jr.[1], Stephen C. Hern[2], Sue Melancon[3], and Jim Pollard[3]

[1]Acurex Corporation, Mountain View, CA 94039
[2]U.S. Environmental Protection Agency, Las Vegas, NV 89114
[3]University of Nevada, Las Vegas, NV 89114

The mobility of chemicals through soils following landfill disposal or agricultural application is of concern whenever there is potential threat to contaminate the groundwater by leaching. This movement is evaluated by simulation modeling. Several models, including the Pesticide Root Zone Model (1) and the Seasonal Soil Compartment Model (2) are available, and are currently being evaluated by the Environmental Protection Agency-Las Vegas. The evaluation study consists of conducting a laboratory experiment in which lysimeter columns containing test chemicals at levels corresponding to agricultural application rates are irrigated with water at a constant rate for 30 days. Leachate samples collected daily from the lysimeter columns and soil cores obtained at the completion of the experiment are analyzed for the test chemicals. The observed concentrations are compared with the model predictions to establish how well the model can simulate the actual situation. Prior to the initiation of such an experiment (5 lysimeter columns 60 cm ID by 200 cm height), a small-scale experiment (3 columns 4.8 cm ID by 50 cm height) was performed to establish: breakthrough volumes of the test chemicals, whether or not the selection of the test chemicals was appropriate, and to determine if the analytical methodologies were adequate. The purpose of this paper is to summarize the results of the small-scale experiment. The six test chemicals selected for investigation were: dicamba, 2,4-D, atrazine, diazinon, pentachlorophenol, and lindane. Their physico-chemical properties are presented in Table I. Dicamba and 2,4-D were chosen for this study because of their relatively high water solubility. Consequently, they were expected to move quickly through the soil column and be the first chemicals to appear in the leachate. Despite the fact that 2,4-D was known to degrade quickly in the soil, it was felt that due to its high water solubility, 2,4-D would also leach quickly from the soil column and will probably breakthrough immediately following dicamba. The other chemicals expected to leach from the soil column were atrazine and diazinon and both were probably of intermediate mobility.

0097–6156/86/0315–0311$06.00/0
© 1986 American Chemical Society

Table I. Physico-chemical Properties of the Test Chemicals

Property	Dicamba	2,4-D	Atrazine	Diazinon	Pentachlorophenol	Lindane
Vapor pressure (mm Hg)	3.75×10^{-3} T=100°C (3)	1.05×10^{-2} (5)[a] 6×10^{-7} (1)	3×10^{-7} (7)	8.4×10^{-5} (8)	1.1×10^{-4} (9) 8.0×10^{-6} (10)	$(3.3\text{-}2.1) \times 10^{-4}$ (13) 9.4×10^{-6} (14) 1.6×10^{-4} (15)
Water solubility (mg/L)	4,500 (3)	725 (7) 520 (5) 900 (5)	70 (7)	40 (8)	14 (11)	7.52 ±0.04 (16) 5.75 to 7.40 (17-19)
Log octanol/water partition coefficient	2.41 (4)	2.74 (4)	2.68 (4)	NA	5.01 (12) 6.3 (10)	3.72 (17) 5.43 (4)
K_{oc}	0 (3)	60 (6)	170 (6)	NA	NA	80-500 (20)[b]

[a]Numbers in parentheses are references
[b]Value given is for K_d and not K_{oc}

NA -- data not available in the literature reviewed by authors

Pentachlorophenol and lindane were selected primarily because they were reported to be the least mobile chemicals; in fact, based on their octanol/water partition coefficients, it was expected that these chemicals would not leach at all. These chemicals were spiked into the soil at 16.2 ppm and were leached for 30 days with organics-free water at a rate of 55.8 mL/day. Results are presented as amount leached versus time and concentration of chemical in soil at various depths in soil column. Furthermore, mass balances are presented for each chemical. Subsequent sections present the experimental details and the results of this experiment.

Experimental

Leaching Experiment. Three polyethylene columns (4.8 cm ID by 50 cm height) were employed to investigate the mobility of dicamba, 2,4-D, atrazine, diazinon, pentachlorophenol, and lindane. Each column was packed with 1,080g of fresh soil to a depth of 40 cm (sandy loam soil from Soils Incorporated, Puyallup, Washington: pH 5.9 to 6.0; 89 percent sand; 7 percent silt; 4 percent clay; cation exchange capacity 7.5 meq/100g).

Each column was irrigated from the bottom with HPLC grade water (Baker Analyzed Reagent) and allowed to drain for 1 hour. To remove residual salt, two pore volumes (500 mL) of HPLC grade water were slowly dripped through each column. Next, 100g of soil, spiked with 1.62 mg each of dicamba, 2,4-D, atrazine, diazinon, pentachlorophenol, and lindane, to simulate an application rate of 8 lb/acre, were loaded onto each column.

The contaminated soil was packed on top of each soil column to a depth of 3.7 cm resulting in a total column depth of 43.7 cm. To this, 32g of fresh sandy loam soil were added to bring the total soil column height to 45 cm. The contaminated soil was prepared as follows: 162-μL aliquots of six stock solutions of each chemical (concentration 10 mg/mL; dicamba, 2,4-D, and pentachlorophenol stock solutions were in acetone; the rest in methanol) were composited. Acetone (100 μL) was added to the composite solution to bring the volume to 1,072 μL. The composite solution was added to 20 mL HPLC grade water which was then mixed with 100g of fresh soil. The purity of the chemicals used for this experiment was greater than 99 percent.

Chemicals were leached from the soil column by applying water, twice daily, at a rate of 55.8 mL per day, per column, for 30 days. The irrigation water (27.9 mL per application) was applied at 8 a.m. and 2 p.m., each day, for 30 days. Leachates from the small columns were captured daily and stored in the dark at 4°C until analyzed (column I leachates were analyzed within 3 days after collection; column II and III leachates were stored for almost 30 days prior to extraction). After the 30-day experiment, each column was cross sectioned into nine discrete segments (5 cm in depth). Prior to segmenting, the soil columns were frozen at -10°C for 8 hours to facilitate sectioning. All soil samples were kept frozen at -10°C until analysis was performed.

The three soil columns were kept covered with aluminum foil throughout the 30-day experiment. Analyses were performed by gas chromatography/mass spectrometry (GC/MS) following the sample extraction procedure outlined below.

Adsorption Study. The adsorption of the test compounds by the sandy loam soil (particle size less than 35 mesh) was determined by measuring the difference in the concentration of an aqueous solution of each chemical, before and after equilibration with soil. Four to five concentrations per chemical were tested and three replicate measurements were performed at each concentration. Water-to-soil ratios of 2:1 to 10:1 were employed in all experiments. The time required to reach equilibrium was determined separately for each chemical at one concentration and a specific water-to-soil ratio.

Following equilibration with soil, the aqueous solution was separated by centrifugation at 2,800 rpm and was analyzed directly by high-pressure liquid chromatography (HPLC). The experimental conditions for HPLC analysis are given in Reference 21. Attempts to analyze lindane in the aqueous phase by HPLC with ultraviolet detection, without preconcentration, were unsuccessful. Lindane was analyzed by gas chromatography with electron capture detection following preconcentration on C_{18}-reverse phase resin and elution with tetrahydrofuran. Other details can be found in Reference 21.

Soil Degradation Study. One hundred and forty-four amber jars (500 mL) were loaded with 50-g fresh sandy loam soil and divided into two groups. Seventy-two of them were adjusted at 14 percent moisture (group I). The other 72 were adjusted at 22 percent moisture (group II). The 72 samples in group I were further divided as follows: 36 of them were spiked with dicamba, 2,4-D, and pentachlorophenol at 1 μg/g; the other 36 spiked with lindane, atrazine, and diazinon at 1 μg/g. Spiking was performed after the soil was adjusted at 14 percent or 22 percent moisture. The spike was added in a concentrated solution to the soil (100 μL of a 0.5 mg/mL stock solution) and was mixed thoroughly with the soil. All samples from group II were treated similarly.

The spiked samples were incubated in an oven at 85°F, in dark, and in open system; cotton plugs were used to prevent dust from entering the flasks. The experiment was set up for 90 days. The sampling times were 0.04; 0.67; 1.67; 4; 10 or 12; 20; 32; 61 and 93 days from the initiation of the experiment. Three replicates were analyzed at each time interval for the test compounds. Soil moisture was maintained at 14 percent and 22 percent by addition of organics-free water and was verified twice a week during the experiment. Other details can be found in Reference 21.

Analytical Methodologies

Dicamba and 2,4-D Analysis. A Finnigan 1020 quadrupole mass spectrometer coupled to a Perkin-Elmer Sigma 1 gas chromatograph and an Incos 2300 data system was used for all measurements reported here. Calibration standards and sample extracts were injected manually. Compound separations were performed on a 6-ft glass column packed with Ultrabond 20M, temperature programmed from 150°C (1 min) to 240°C at 15°C/min; carrier gas He; flowrate 45 mL/min; injector temperature 220°C; transfer line temperature 250°C. The mass spectrometer operating conditions included: electron energy

70 eV; and selected ion monitoring mode for ions at m/z 188, 203, 206, 400, and 403 in a total scan time of 0.65 sec. Other details of the gas chromatographic/mass spectrometric analyses can be found elsewhere (21).

Atrazine, Diazinon, Pentachlorophenol, and Lindane Analysis. A Finnigan 4021 quadrupole mass spectrometer coupled to a Finnigan 9610 gas chromatograph and an Incos 2300 data system was used for all measurements reported here. Calibration standards and sample extracts were injected automatically by a Varian autosampler. Compound separations were performed on a fused silica capillary column 30m x 0.25 mm ID DB-5 (J&W Scientific, 0.25-μm film thickness); splitless injection at 50°C followed by temperature programming to 300°C at 15°C/min; injector temperature 260°C; transfer line temperature 280°C; carrier gas He at 10 psi pressure. The mass spectrometer operating conditions were ion source temperature 300°C; electron energy 70 eV; selected ion monitoring mode for ions at m/z 188, 200, 205, 181, 224, 304, 314, 266, and 272 in a total scan time of 0.69 sec. Other details of the gas chromatographic/mass spectrometric analyses can be found elsewhere (21-22).

Reagents and Standard Compounds. Analytical reference standards of dicamba, 2,4-D, atrazine, diazinon, pentachlorophenol, and lindane were obtained from the U.S. EPA Pesticides and Industrial Chemicals Repository (MD-8), Research Triangle Park, NC 27711. Stock solutions were prepared in pesticide grade methanol or acetone (Baker resi-analyzed) and stored at -10°C.

Stable-labeled isotopes -- dicamba-d_3, 2,4-D-d_3, atrazine-d_5, diazinon-d_{10}, and lindane-d_6 -- were synthesized by Pathfinder Laboratories, St. Louis, Missouri. Pentachlorophenol-$^{13}C_6$ was obtained from MSD-isotopes, Division of Merck Frost Canada Inc., Montreal, Canada. The deuterated compounds and the ^{13}C-labeled pentachlorophenol were spiked into every sample at known concentration and were used to quantitate the naturally abundant compounds by stable-isotope dilution procedures (23-24).

Leachate Extraction. Known volumes of leachate samples (~50 mL) were spiked with the stable-labeled isotopes at 80 μg/L and were extracted at their original pH (6 to 6.5) by vigorously shaking for 2 min, in a 250-mL separatory funnel with 50 mL methylene chloride. Extraction was performed three consecutive times, each time using fresh 50-mL aliquots of methylene chloride. The extracts were combined, moisture was removed by passing the extract through a column of anhydrous sodium sulfate, and the extract was then concentrated to 4 to 6 mL in a Kuderna-Danish evaporator. Further concentration to 1 mL was performed using nitrogen blowdown evaporation. Following extraction at pH 6 to 6.5, the aqueous phase was acidified at pH 1 and was extracted three consecutive times with 50-mL aliquots of methylene chloride. The acidic extracts were combined and were concentrated to 10 mL using a Kuderna-Danish evaporator. Prior to analysis, the acidic extract was derivatized with pentafluorobenzyl bromide according to procedures described in Reference 19.

Soil Extraction. Soil samples were thawed at room temperature for 2 hours and were extracted as follows: 50-g aliquots of the fresh sandy loam soil were slurried with 10 mL of organics-free water and spiked with known amounts of stable-labeled isotopes. The spike was added in 100 to 400-μL methanol aliquots to the wet soil and was allowed to equilibrate with the soil for 1 hour. Two separate aliquots were extracted for each sample. One aliquot was extracted at neutral pH to separate atrazine, lindane, and diazinon. The other aliquot was extracted at acidic pH to separate dicamba, 2,4-D, and pentachlorophenol.

The aliquot designated for extraction at neutral pH was extracted three times, with fresh 100-mL acetone/hexane (50/50). An ultrasonic cell disruptor, in pulsed mode at 50 percent duty cycle, was employed to enhance the contact between the extraction solvent and soil. Following each extraction, the soil was allowed to settle, the solvent decanted, and the combined supernatants from the neutral pH extraction were dried through a column of anhydrous sodium sulfate (5-cm bed height; 3-cm diameter). Concentration of the extract was performed using a Kuderna-Danish evaporator. Final volume was adjusted to 1 mL using nitrogen blowdown evaporation.

The extraction at acidic pH was performed as follows: 50-g aliquots of the soil sample were slurried with 10 mL of organics-free water and spiked with the stable-labeled isotopes. Following equilibration, the soil slurry was adjusted to pH ≤ 1 with approximately 10 mL of H_2SO_4 (1 + 1) and was extracted three times with 100 mL of acetone/hexane (50/50) using an ultrasonic cell disruptor. The extract was concentrated using a Kuderna-Danish evaporator and was split as follows: 1 mL was taken for pentachlorophenol analysis and 3 mL were taken for analysis of dicamba and 2,4-D. The extract containing dicamba and 2,4-D was derivatized with pentafluorobenzyl bromide at 60°C according to procedures given in Reference 19.

Results and Discussion

Leaching Experiment. The observed distributions of atrazine, diazinon, pentachlorophenol, and lindane in soil columns, that resulted from irrigation with water for 30 days, are presented in Figures 1 through 6. The amounts leached from each soil column daily, for 30 days, are presented in Figures 7 through 10. Dicamba and 2,4-D were not detected in any of the soil samples or their concentrations in soil were only slightly above the method detection limit; therefore, no distributions profiles for these chemicals in soil are given. Diazinon and pentachlorophenol were not detected in any of the leachates; consequently, the leachate profiles for these two chemicals are not presented.

Table II is a summary of the data generated for the six chemicals. The various parameters in Table II are defined below:

- Zone position (Z_p) is the soil column section containing the highest concentration of the test chemical
- Zone dispersion (Z_d) is the number of column sections containing the chemical divided by the concentration of the core containing the highest concentration
- Depth (d) is the distance, in cm, measured from the top of the column at which the chemical reached the maximum concentration

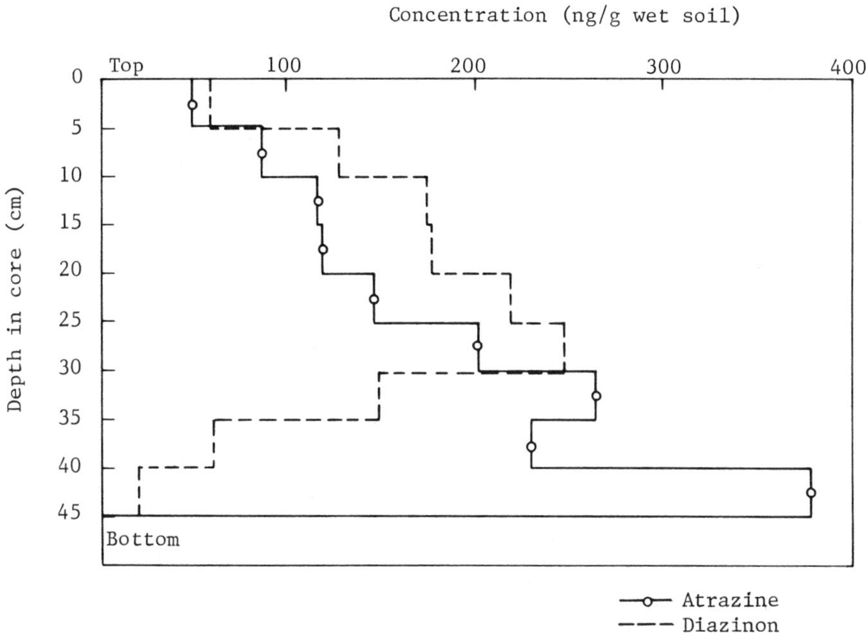

Figure 1. Distribution of atrazine and diazinon in soil following the 30-day test (column I).

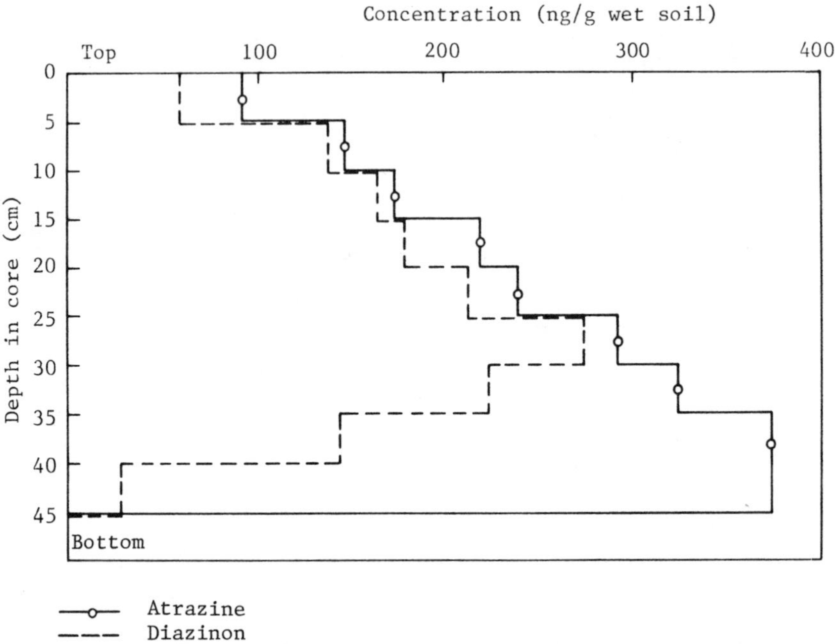

Figure 2. Distribution of atrazine and diazinon in soil following the 30-day test (column II).

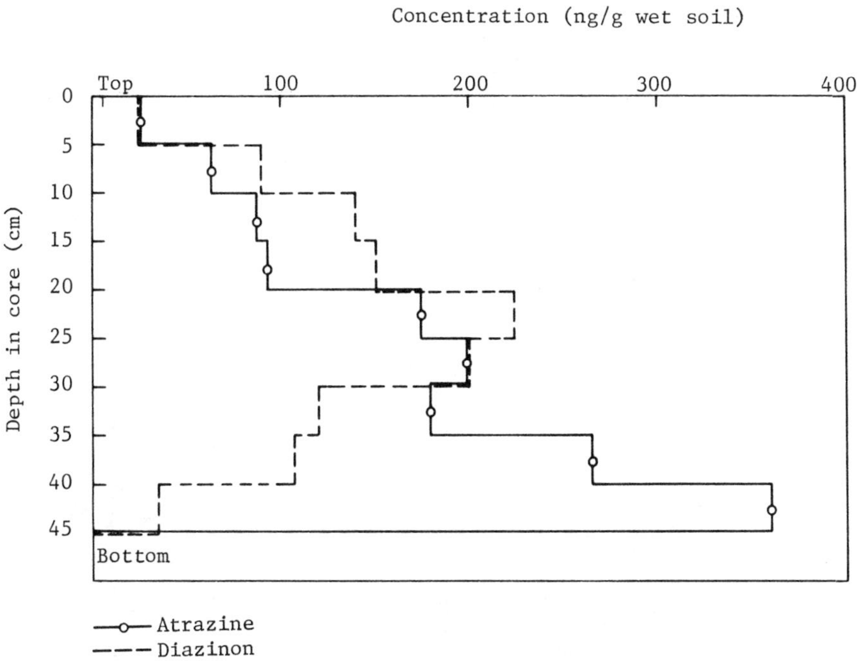

Figure 3. Distribution of atrazine and diazinon in soil following the 30-day test (column III).

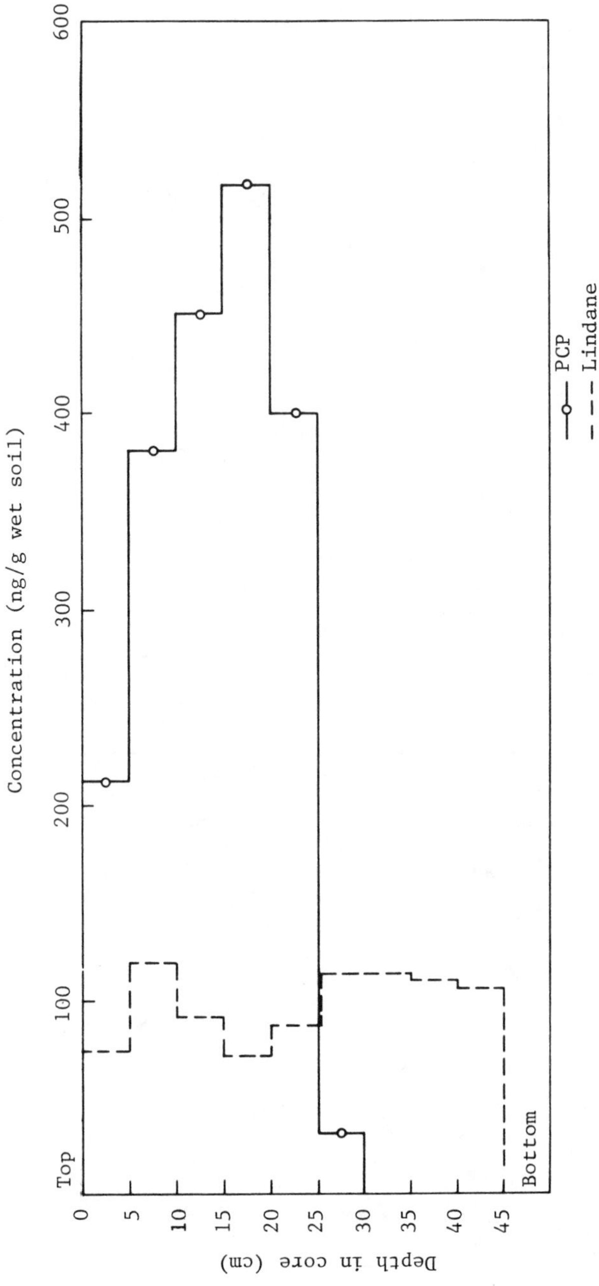

Figure 4. Distribution of lindane and pentachlorophenol in soil following the 30-day test (column I).

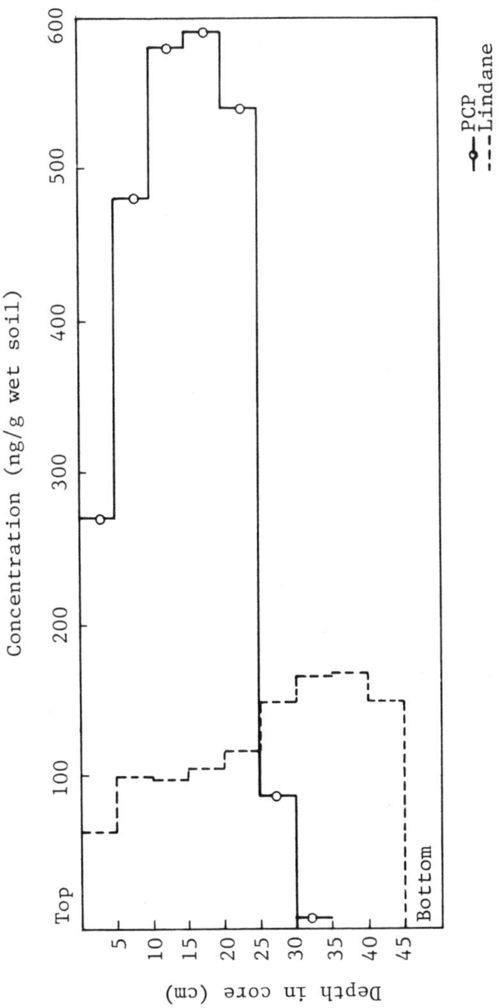

Figure 5. Distribution of lindane and pentachlorophenol in soil following the 30-day test (column II).

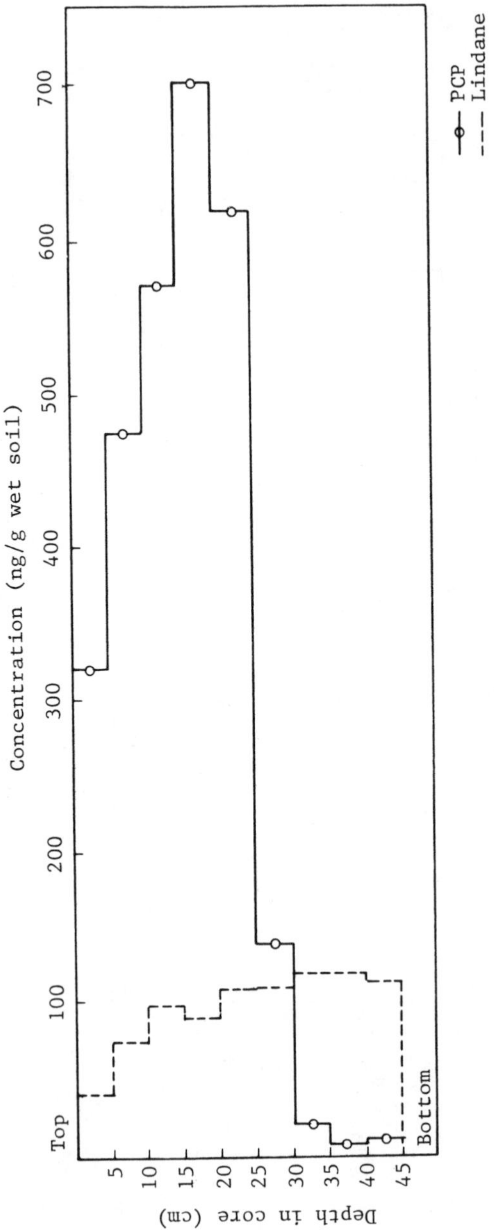

Figure 6. Distribution of lindane and pentachlorophenol in soil following the 30-day test (column III).

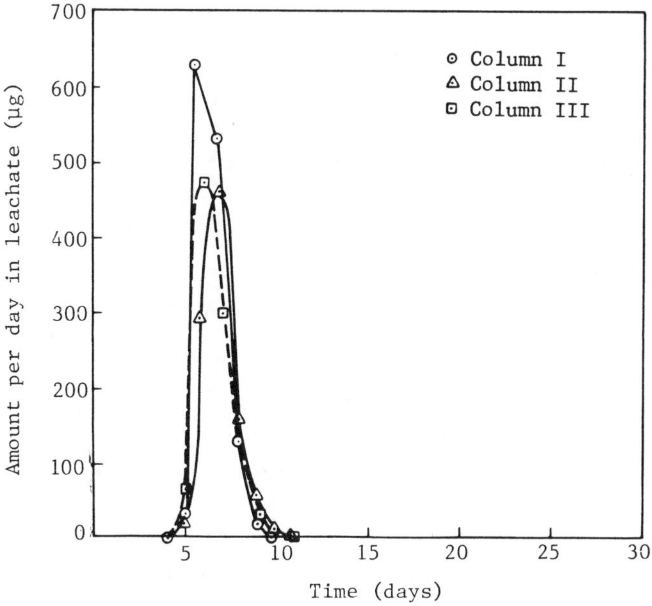

Figure 7. Amount of dicamba (µg) leached from columns I, II, III during the 30-day test.

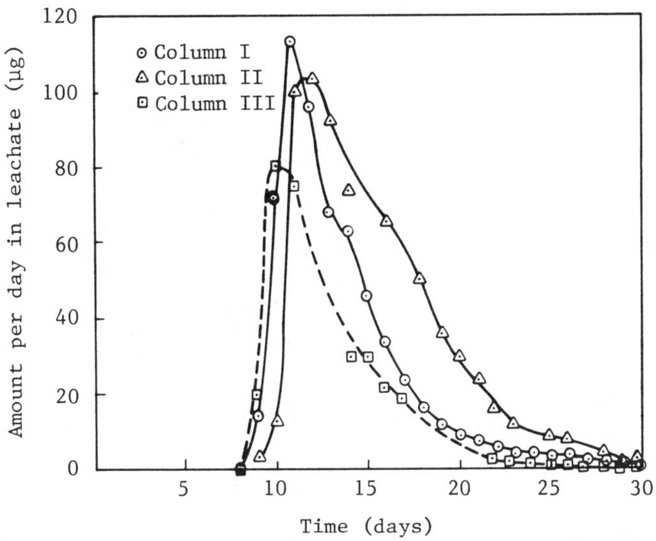

Figure 8. Amount of 2,4-D (µg) leached from columns I, II, III during the 30-day test.

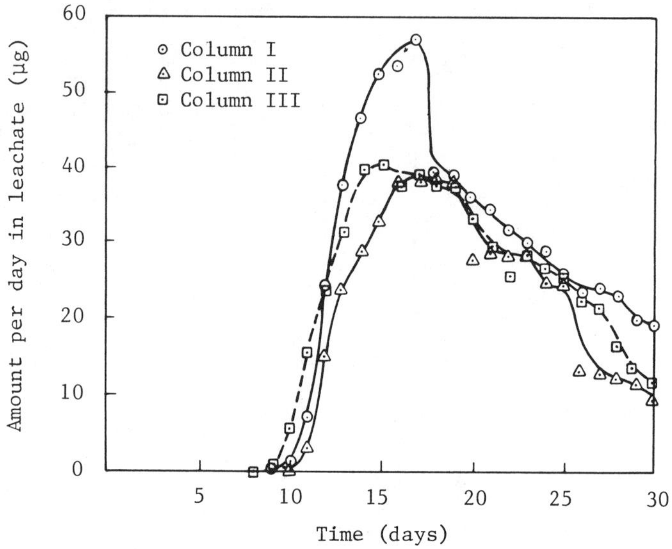

Figure 9. Amount of atrazine (μg) leached from columns I, II, III during the 30-day test.

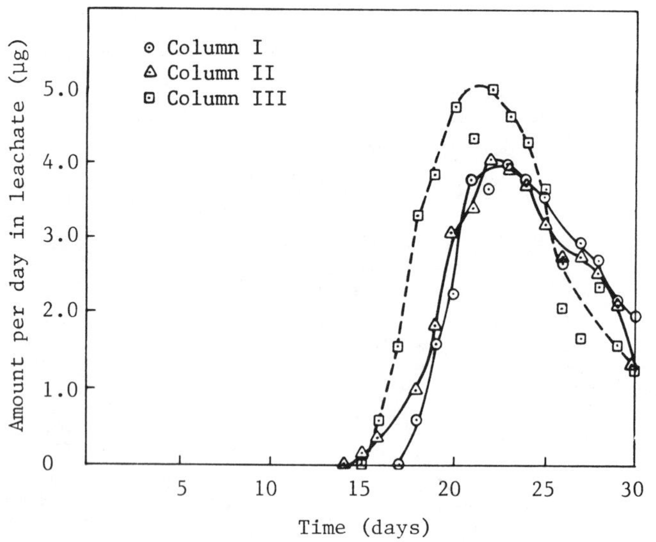

Figure 10. Amount of lindane (μg) leached from columns I, II, III during the 30-day test.

- K_{oc} is the organic carbon normalized adsorption coefficient of the test chemical. The K_{oc} values presented in Table II were determined in our laboratory by measuring the difference in the concentration of an aqueous solution of the chemical before and after equilibration with soil, for a defined time interval.

Table II. Correlation Between the Movement of Test Chemicals Through Sandy Loam and K_{oc}

Compound	Zone Position (Z_p)	Depth d (cm)	Zone Dispersion (Z_d)	$d \cdot Z_d$ [b]	K_{oc} [c]
Dicamba	a	a	a	a	93
2,4-D	a	a	a	a	108
Atrazine	1	40-45	0.23;0.24;0.25	10.2	380
Diazinon	1	20-30	0.37;0.33;0.40	9.2	1,700
Pentachlorophenol	1	15-20	0.117;0.119;0.119	2.1	2,500
Lindane	4-5	25-45	0.756;0.526;0.744	23.6	2,700

[a]Chemical leached completely from the soil column (see Table III for mass balance)
[b]$d.Z_d$ is a movement constant. It is used to describe how fast the chemical moved through the soil column.
[c]Organic carbon content of the sandy loam soil used in this study is 1,290 ±185 mg/kg

Table III is a summary of mass balances performed for each chemical and degradation half-lives at 85°F, in open system, and using sandy loam soil of 14 percent and 22 percent moisture.

Table III. Mass Balance and Degradation Half-lives for the Test Chemicals

Compound	Mass Balance[a,b] (Percent Recovered)	Half-life ($t_{1/2}$) for Degradation in Soil[c] (Days)
Dicamba	62-86	73-77
2,4-D	39-47	d
Atrazine	48-53	73-78
Diazinon	9.2-13	17-22
Pentachlorophenol	18-25	24-38
Lindane	9.5-13	21

[a]Values given represent range values for the three columns, and they refer to the total amount of chemical that was recovered from leachates and soil cores.
[b]Loss through volatilization from the soil column was insignificant for diazinon, pentachlorophenol, and lindane.
[c]$t_{1/2}$ was determined at 85°F, in open system, and at 14 percent and 22 percent moisture.
[d]$t_{1/2}$ is a function of concentration ($t_{1/2}$ is 73 hours at 1 µg/g and 213 hours at 10 µg/g).

The following conclusions can be drawn from the data presented in Figures 1 through 10 and Tables II and III.

- Dicamba was the first chemical that appeared in the leachate. It was detected in the leachates collected at day 4 from each of the three columns and reached maximum concentration at day 6 or 7 of the experiment (Figure 7). The amounts recovered in leachates account for 86 percent, 62 percent, and 62 percent from the amounts spiked in columns I, II, and III, respectively. Lower recoveries for columns II and III might be due to the fact that leachates from columns II and III were stored for almost 30 days prior to extraction, during which time compound degradation might have occurred. Dicamba was not detected in any of the soil samples or the levels found were close to the method detection limit (1 to 2 ng/g).
- 2,4-D was the second chemical to leach from the soil columns. The data shown in Figure 8 indicate that 2,4-D was present in the leachates collected at day 9 of the experiment, its concentration reached a maximum at day 10, 11, or 12 of the experiment, and continued to leach from the soil columns until day 30 of the experiment. The amounts of 2,4-D recovered in leachates account for 39 percent and 47 percent from the amounts spiked in columns I and II, respectively. 2,4-D was detected in the soil samples at concentrations of 10 to 20 ng/g which account for less than 0.5 percent of the amount recovered in leachate.
- Atrazine was the third chemical to leach from the soil columns ($d \cdot Z_d$ value is 10.2). The data shown in Figure 9 indicate that atrazine was detected in leachates collected at day 9 of the experiment; concentration reached a maximum at days 17, 19, and 15, of the experiment for columns I, II, and III, respectively, and continued to leach until day 30 of the experiment. The total amounts of atrazine recovered in leachates account for 75 percent, 59 percent, and 74 percent from the total amounts of atrazine recovered from columns I, II, and III, respectively. Atrazine was found in soil at levels ranging from 28 to 406 ng/g. The soil profiles given in Figures 1 through 3 show that atrazine moved through the soil column and reached a maximum at a depth of 40 to 45 cm. The total amounts recovered from each of the three columns represent 53 percent, 51 percent and 48 percent of the amount spiked.
- Diazinon was not detected in any of the leachates. The soil profiles (Figures 1 through 3) indicate that diazinon moved through the column, since the highest concentration was found in the soil cores at 25 to 30 cm in columns I and II, and at 20 to 25 cm in column III ($d \cdot Z_d$ value is 9.2). The amounts recovered from soil account only for 10 percent, 13 percent, and 9.2 percent. These recoveries are not unexpected if one takes into consideration the fact that the half-life for degradation of diazinon in soil at 22 percent moisture was approximately 17 days.
- Pentachlorophenol was not detected in any of the leachates. The soil profiles (Figures 4 through 6) indicate that pentachlorophenol moved through the soil column, and reached

highest concentration at depths of 15 to 20 cm ($d \cdot Z_d$ value is 2.1). The amounts recovered from soil account for 18 percent, 23 percent, and 25 percent for columns I, II, and III, respectively. Likewise in the case of diazinon, there seems to be a correlation between the degradation half-life in soil and the amount recovered at the end of the 30-day experiment.

- Lindane was the fourth chemical to leach from the soil columns. The data shown in Figure 10 indicates that lindane was detected in the leachates collected from the three columns at day 18, 15, and 16, respectively; concentration reached a maximum at day 23, 22, and 22 of the experiment for columns I, II, and III, respectively; and continued to leach from the soil column until the day 30 of the experiment. The amounts recovered in leachate account for 22 percent, 18 percent, and 27 percent, respectively. Lindane was found in soil at levels ranging from 39 to 171 ng/g (Figures 4 through 6). It is interesting to note that, in the case of lindane, the Z_p is 4 to 5. The fact that lindane was detected in the leachate indicates that indeed lindane moved through the soil column ($d \cdot Z_d$ value is 23.6) and this is also confirmed by the fact that all soil cores showed detectable levels of lindane. The mechanism by which leaching of lindane took place remains yet to be explained. The total amounts recovered from each of the three columns represent 9.5 percent, 13 percent and 10 percent.

In summary, it can be concluded that the observed distributions of the test chemicals in the three soil columns that resulted from irrigation with water, at constant rate, for 30 days are in excellent agreement. Furthermore, the amounts leached from the soil columns and the order in which the chemicals were leached are also in good agreement. There seems to be a good correlation between the K_{oc} value and the movement constant ($d \cdot Z_d$) for three of the chemicals (atrazine, diazinon, pentachlorophenol) which suggests that the movement through the soil column can be predicted from adsorption data. Surprisingly, this is not the case for lindane. A large-scale experiment is now underway and the results of this study may help to clarify the movement of lindane through soils.

Literature Cited

1. Carsel, R. F.; Smith, C. N.; Mulkey, L.A.; Dean J. D.; and P. Jowise "Users Manual for the Pesticide Root Zone Model (PRZM)," EPA-600/3-84-109, 1984.
2. Bonazountas, M.; I. Wagner; and B. Goodwin "Evaluation of Seasonal Soil/Ground Water Pollutant Pathways," Arthur D. Little, Inc., Final Report, prepared for U.S. EPA, Monitoring and Data Support Division, EPA Contract No. 68-01-5949/9.
3. Worthing, C. R. "The Pesticide Manual -- A World Compendium," 6th Ed., British Crop Protection Council, 1979.
4. Dao, T. H.; T. L. Lavy; J. Dragun Residue Rev. 1983, 87, 91.
5. Hee, S. S. Q.; Sutherland, R. G. "The Phenoxyalkanoic Herbicides, Volume I, Chemistry, Analysis, and Environmental Pollution," CRC Press, Inc., Boca Raton, Florida, 1981.
6. "Test Protocols for Environmental Fate and Movement of Toxicants," Proceedings of a Symposium Association of Official

Analytical Chemists, 94th Annual Meeting, October 21-22, 1980, Washington D.C., p. 106.
7. Bailey, G. W.; White, J. L. Residue Rev. 1970, 32, 29.
8. Gunther, Residue Rev. 1971, 36, 147.
9. Bevenue, A.; Beckman, H. Residue Rev. 1967, 19, 83-134.
10. Leo, A.; Hansch, C; Elkins, D. Chem. Rev. 1971, 7, 52-616.
11. Verschueren, K. Handbook of Environmental Data on Organic Compounds. Van Nostrand/Reinhold, New York. 1977, p 659.
12. Drahonovsky, J.; Vachek, Z. Col. Czech. Chem. Commun. 1971, 36(10), 3431-3440.
13. Demozay, D.; Marechal, G. "Physical and Chemical Properties in Lindane: Monograph of an Insecticide," E. Ulmann 1972, pp. 15-21. K. Schiller, Freiburg im Breisgau.
14. Benchmark, 1975. Draft of Preliminary Summaries of Literature Surveys of Benchmark Pesticides, George Washington University Medical Center, October, 30, 1975.
15. Masterton, W. L.; Lee, T. P. Environ. Sci. & Technol. 1973, 6(10), 919-921.
16. Kurihara, N.; Uchida, M.; Fujita, T.; Nakajima, M. Pestic. Biochem. Physiol. 1973, 2(4), 383-390.
17. Biggar, W. J.; Riggs, R. L. Hilgardia 1974, 42(10), 383-391.
18. Bhavagary, H. M.; Jayaram M. Bull, Grain Technol. 1974, 12(2), 95-99.
19. Lee, H. B.; Chau, A. S. Y. Assoc. Off. Anal. Chem. 1983, 66, 1023-1028.
20. Wolfe, N. L.; Zepp, R. G.; Baughman, G. R.; Fincher, R. C.; Gordon, J. A. "Chemical and Photochemical Transformation of Selected Pesticides in Aquatic Environment," EPA 600/3-76-067.
21. Lopez-Avila, V.; Hirata, P.; Kraska, S.; Flanagan, M.; and Taylor, J. H., Jr. "Analysis of Water and Soil Samples from Lysimeter Columns," Acurex Final Report prepared for EPA-Las Vegas, Environmental Assessment Division, EPA Contract No. 68-03-3100, 1984.
22. Lopez-Avila, V.; Hirata, P.; Kraska, S.; Flanagan, M.; Taylor, Jr., J. H.; Hern, S. C.; manuscript accepted for publication to Anal. Chem., 1985.
23. Colby, B. N.; Rosecrance, A. E.; Colby, M. E. Anal. Chem. 1981, 53, 1907-1911.
24. "Method 1625 Revision B -- Semivolatile Organic Compounds by Isotope Dilution GC/MS"; Environmental Protection Agency, Federal Register 1984, 49, 184-198.

RECEIVED April 1, 1986

MODELING AND MODEL VALIDATION

17

Principles of Modeling Pesticide Movement in the Unsaturated Zone

R. J. Wagenet

Department of Agronomy, Cornell University, Ithaca, NY 14853

The basic principles of modeling the physical, chemical and biological processes that determine pesticide fate in unsaturated soil are reviewed. The mathematical approaches taken to integrate diffusion, convection, sorption, degradation and volatilization are presented. Deterministic and stochastic models formulated to describe these processes in a soil-water pesticide system are contrasted and evaluated. The use of pesticide models for research or management purposes dictates the degree of resolution with thich these processes are modeled. The ability of each type of modeling approach to consider pesticide fate in spatially variable field systems is discussed.

The movement of water and chemicals through the unsaturated zone has been studied for approximately the last one hundred thirty years. Since the early 1950s, these studies have intensified and utilized increasing levels of mathematical sophistication and, recently, computer technology, as tools to integrate and summarize the physical, chemical and biological processes that determine the dynamics of water and solutes in the system. Early efforts focused on inorganic chemicals, such as chemical fertilizers or salts found in irrigation water or as natural constituents of the soil profile. Not only were these chemicals of immediate interest to agriculturalists during the 1950s and 1960s, it was also true that the use of organic chemicals was not very widespread until relatively recently. The result of the research activity on inorganic salts was the accumulation of a substantial body of experimental information and theoretical conceptualizations concerning the physics of water movement in homogeneous soil profiles and the resulting displacement and chemical reactions of such solutes as chloride, nitrate, sulfate and the common accompanying cations. Most of this information has been developed under carefully controlled laboratory soil column conditions. Confirmation that these field scale physical and chemical

0097–6156/86/0315–0330$06.00/0
© 1986 American Chemical Society

processes operate in the same manner has been attempted during the last ten years, with mixed results.

These results have served as the beginning point for description of pesticide movement through the unsaturated zone. Yet, as we investigate the field regime more closely and actually measure the displacement of both inorganic and organic solutes under field conditions, two points are becoming increasingly clear. First, there are many cases where current state-of-the-art modeling approaches do not provide accurate description of the spatial and temporal distribution of pesticides or more common inorganic salts within the unsaturated zone. The presumably basic principles we have identified in laboratory studies apparently do not always operate similarly under field conditions. Second, based upon these observations, it is worthy to consider the reformulation of the basic approach taken to predict water and solute movement under field conditions. Strictly mechanistic, deterministic models, although currently the best tool we have, are probably not the type of tool we should be considering as we look twenty years ahead in the development of pesticide models.

The principles of modeling pesticide fate in unsaturated field regimes are currently being re-examined in the light of such concerns, and new mdoels are under development that offer the promise of more accurate predictions under field conditions. Yet, we must make decisions today regarding pesticide registration, use and management, and the best tools at our disposal must be used in the process. It is therefore important that we appreciate the basic principles currently used in modeling pesticides, understand the structure and organization of contemporary modeling approaches, and recognize the limitations of these approaches that are being increasingly demonstrated as we learn more of the field regime. These models must be used in the short run, but in a wise and cautious manner consistent with their limits. Examination of these limits in terms of currently recognized basic principles will illustrate why future models will probably bear little resemblance to contemporary approaches.

Modeling Approaches

There are different types of models intended to serve different purposes. Both the models and the purposes are often confused. Although all the models attempt to include description of important basic processes, the degree of resolution used to represent each process determines the purpose for which the model can be used. Understanding these different types of models is an important first step in evaluating the usefulness of models developed from basic principles.

There are at least two major criteria that can be used to classify models (1). One criterion is the manner in which basic processes are considered, i.e., whether they are assumed to be deterministic or stochastic. All pesticide models currently in the scientific literature or in use are deterministic. That is, they presume that the soil-water-pesticide system operates such that the occurrence of a given set of events leads to a uniquely-definable outcome. Such models can only simulate the system's response to a single set of assumed conditions, and whether these predictions are accurate depends upon the nature and extent of the variability of physical, chemical and biological processes within the system. The uncertainties inherent in estimating the rate or magnitude of processes in the field is ignored in these formulations. The alternative is the stochastic approach which considers the uncertainty of the system. The system processes, and the outcome of those processes, are characterized in statistical terms, such as the mean, variance or other statistical moments. Predictions are not made with respect to a particular coordinate position in the soil, but in terms of areas or volumes. There are no pesticide models of this type currently being used, although this could well change in the future, as discussed below.

A second way of classifying models is based upon the intended use. Three types can be distinguished: research, management/educational, and screening models. Examples of each exist as pesticide models, and all are deterministic in form. Research models (2-5) are developed as tools to aid in the testing of hypotheses and the exposure of areas of incomplete understanding. These models represent basic processes in fundamental and mechanistic terms, and, represent our most complete understanding of basic principles. These are usually complex models that demand a large amount of input characterization data, are based upon numerical differencing methods, and are generally not used by anyone other than the model developer. Management or educational models are substantially simplified conceptualizations of the natural system, intended to provide qualitative guidance concerning pesticide fate as a function of major soil and chemical properties, and management practices. These models (6-8) are less mechanistic than research models, are flexible in the types of situations they can consider, are not intended to provide quantitative predictions of pesticide concentrations or fluxes in the unsaturated zone, and are intended only to evaluate the relative, approximate behavior or chemicals without demanding a large amount of input characterization data. Screening models, such as developed by Jury et al. (9), are developed for a third purpose: the quantitative comparison of pesticide behavior under a very limited, yet carefully and comprehensively described, set of conditions. Such models retain a description of basic mechanisms as a foundation, but are not intended to be used under field conditions. They are exact analytical solutions for a single well-defined case, and because of the assumptions needed to make the mathematics tractable, are not useful in predicting site-specific behavior. Their strength lies in relative comparisons of pesticides.

Basic Processes Affecting Pesticide Fate in the Unsaturated Zone

All contemporary modeling approaches to predicting pesticide fate in the unsaturated zone are simply the logical representation of our understanding of the interacting physical, chemical and biological processes. This understanding has evolved from laboratory studies of the basic physics and chemistry of solute transport, combined with information on pesticide-soil interaction, degradation and volatilization. The result of such dependence on laboratory studies is that a conceptual one-dimensional soil column of homogeneous properties is used to formulate these models. Less complicated management or screening models have generally evolved from these approaches. A quick review of the basic processes included in these models will provide the framework for evaluating our appreciation of basic principles.

Transport. The mechanisms responsible for transport are considered to be both physical (convection or mass flow) and chemical (diffusion). When considered simultaneously, these processes have been summarized in the convective-dispersive, or miscible displacement, equation. For a non-interacting solute (such as chloride) under steady state water flow conditions in a homogeneous soil, this equation can be written as (10):

$$\frac{\partial c}{\partial t} = D \frac{\partial^2 c}{\partial z^2} - v \frac{\partial c}{\partial z} \qquad (1)$$

where c = solute concentration (m/L^3), D = apparent diffusion coefficient (L^2/T), incorporating both chemical diffusion and hydrodynamic dispersion, v = pore water velocity (L/T), defined as the ratio of the water flux to the volume water content, and z and t are depth (L) and time (T), respectively. The lower case m represents mass of solute. When a pesticide is to be described, we must consider chemical-soil interaction, degradation and volatilization. The first two processes are often included in Equation 1 as:

$$\frac{\partial c}{\partial t} + \frac{\partial s}{\partial t} = D \frac{\partial^2 c}{\partial z^2} - v \frac{\partial c}{\partial z} - \phi \qquad (2)$$

where s = adsorbed concentration (m/M), ϕ = degradation (chemical or biological) and the other terms have been defined. The upper case M denotes mass of soil. If plant uptake is present, it can be included as a second term analagous to ϕ. The volatilization process can be included in several ways. Plant uptake, volatilization and functional relationships used for s and ϕ are outlined below.

The analytical solution to Equation 2 for a range of boundary conditions is a model of pesticide fate that has been used under a variety of laboratory situations to study the basic principles of soil-water-pesticide interaction. It is in fact limited to such laboratory cases, as steady state water flow is an assumption used in deriving the equation. As a modeling approach it is useful in those research studies in which careful control of water and solute fluxes can be used to study degradation and adsorption. For example, Zhong et al. (11) present a study of aldicarb in which the adsorption and degradation of aldicarb, aldicarb sulfone and aldicarb-sulfoxide were simultaneously determined from laboratory soil column effluent data. The solution to a set of equations of the form of Equation 2 was used. A number of similar studies for other chemicals could be cited that have provided useful basic information on pesticide behavior in soil (4,12,13). Yet, these equations are not useful in the field unless re-formulated to describe transient water and solute fluxes rather than steady ones. Early models of pesticide fate based upon Equation 2 (14) were constrained by such assumptions, but were used to describe field cases because no other more appropriate tools were available.

The transient field regime requires that Equation 2 be derived to reflect the depth and time dependence of water and solute. Equation 2 then becomes (10).

$$\frac{\partial(\theta c)}{\partial t} + \frac{\partial(\rho s)}{\partial t} = \frac{\partial}{\partial z}\left[\theta D(\theta,q)\frac{\partial c}{\partial z} - qc\right] - \phi(z,t) \qquad (3)$$

where $\theta = \theta(z,t)$ = volumetric soil-water content (L^3/L^3), ρ = soil bulk density (M/L^3), q = water flux (L/T) and D is now dependent on θ and q. Pesticide degradation is also dependent on z and t, as it will vary according to the environmental factors that vary with depth and time. Equation 3, when solved by numerical methods, forms the core of models used to describe pesticide fate in the unsaturated zone. It requires knowledge of the water content and water flux changes with depth and time. This information is most accurately provided by calculating water movement with the numerical solution of a second non-linear differential equation of the form

$$\frac{\partial\theta}{\partial t} = \frac{\partial}{\partial z}\left[K(\theta)\frac{\partial H}{\partial z}\right] - U(z,t) \qquad (4)$$

where $K(\theta)$ = hydraulic conductivity (L/T), dependent upon water content, U(z,t) = plant uptake of water (1/T), H = hydraulic head (L), for unsaturated soils defined as H = h+z, where h is the soil-water matric potential (L), and the other terms have been defined. Coupled solutions of Equations 3 and 4 have been used to describe water and inorganic salt movement in a variety of cases (15-17) and have recently been extended to pesticides (5,8). A key point to recognize in such models is that any uncertainty in the K-θ-h relationship, or in the value of U, will be manifested in an uncertainty in calculated water contents or water fluxes (calculated as the water content change over time within a depth interval). These uncertain values of θ and q used in Equation 3 to predict pesticide fate will impose an uncertainty in the prediction of pesticide concentrations with depth and time. That is, the answers obtained using Equations 3 and 4 are only as good as the information provided on K-θ-h, and the relationship between D, θ and q.

Mediating Processes

Sorption. The interaction of a pesticide with the soil solids or organic matter, termed sorption, retards the pesticide movement through the unsaturated zone. This interaction is being increasingly associated with the soil organic matter (18), with the extent of the interaction dependent upon the type of organic materials and the molecular characteristics of the pesticide (19). Linear and Freundlich isotherms have been used to describe sorption, where

$$s = K_D c \qquad (5)$$

and

$$s = K_F c^N;\ N \leq 1 \tag{6}$$

with s and c defined in Equation 2, and K_D and K_F the appropriate distribution coefficients. The values of N, an empirical constant, and K_D and K_F depend upon soil and chemical properties, and are usually obtained from laboratory batch-type equilibrium studies, although a recent paper (11) has obtained them from soil column flow experiments. Reviews of such methods and the data obtained from them are available (20,21).

Equations 5 and 6 are often used in Equations 2 and 3 for s. Knowledge of K_D or K_F and N thereby provide the ability (in a modeling context) to retard pesticide movement through the soil in a manner consistent with the sorption characteristics of the particular soil and chemical of concern. Often the values of K_D, K_F or N are unknown, which has led to estimation methods based upon easily available information.

The first step in the estimation process is the definition of a normalized sorption coefficient, K_{OC}, defined as

$$K_{OC} = (K_D \text{ or } K_F)/f \tag{7}$$

where f = decimal fraction of organic carbon in the soil on a weight basis. This normalization has been found to produce K_{OC} values that are essentially independent of soil type. Values of K_{OC} can be estimated if only the melting point and aqueous solubility of the pesticide are known from (22):

$$\log K_{OC} = -0.921 \log X_{sol} - 0.00953\ (MP-25) - 0.405 \tag{8a}$$

and

$$X_{sol} = \left[\frac{(c_{sol}/MW) \times 10^{-3}}{55.56}\right] \tag{8b}$$

where c_{sol} = pesticide aqueous solubility (g/ml), MW = pesticide molecular weight (g/mole), and MP = pesticide melting point (°C), which is set equal to 25°C for pesticides that are liquids at temperatures less than or equal to 25°C. Equations 8a and 8b allow the estimation of K_D or K_F if the soil organic carbon content is known, as long as organic carbon content is neither very high nor low (23).

Additional efforts to measure K_{OC}, K_D or K_F as well as more accurate methods to estimate them are needed. Almost all pesticide models, whether research, management or screening use one of these three parameters to represent pesticide-soil/organic matter interaction. Although more complicated expressions of sorption phenomena may better represent fundamental relationships, such resolution is not needed in most modeling approaches, given the level of resolution with which other processes are considered (water flow, plant uptake, degradation) and the sensitivity of the soil-water system to classical sorption phenomena. In fact, future modeling approaches may reformulate the sorption process to a large degree, considering part of the chemical to be sorbed to a stationary soil/organic phase, and part of the chemical sorbed to a soluble, mobile organic fraction that is capable of moving through only a fraction of the total pore space. Such reformulation is currently being suggested as the result of field observations of pesticide movement.

Degradation. Pesticide loss to both microbiological and chemical transformational processes is collectively termed degradation. In the root zone, degradation proceeds by microbiological processes that are faster than the chemnical ones. However, there is little biological activity below the root zone, and degradation is therefore accomplished at a much slower rate in the deeper unsaturated zone, as well as in groundwater.

A number of experimental studies have established that both microbial and chemical degradation can be approximately described by first-order kinetics (24). Most pesticide models employ such an approach. As with linear sorption, this relatively naive representation of a fundamentally more complicated process is a simplifying assumption to make mathematical solutions possible and data requirements reasonable. Implicit in the assumption is the belief that the accuracy of simulation of pesticide fate is more dependent upon other factors than a very precise representation of the degradation process. These factors include spatial and temporal variability of the degradation process itself as affected by water, temperature, substrate, and pH, and variability in the transport of pesticide through the soil profile. There is little information to substantiate this assumption, although some field experiments on water and solute movement (discussed below) indicate it to be reasonable at this point in model development.

First order kinetics of degradation are defined by

$$\phi = \frac{dc}{dt} = -kc \qquad (9)$$

where and c are defined in Equation 2 and k = first-order rate coefficient (1/T). Equation 9 can be integrated from $c(0) = c_o$ to $c(t)$ and rearranged in terms of t to give the half-life, $t_{0.5}$, as the time required for c to equal $c_o/2$. Rao and Davidson (20) have compiled the values of both k and $t_{0.5}$ for a number of pesticides. Values measured in both the field and laboratory exhibited coefficients of variation generally less than 100%. This represents a relatively narrow range considering the diverse conditions in which they were determined. Additionally, it was found that laboratory measured values of k were generally smaller than those measured under field conditions. This consequence of the multiple degradation pathways operating in the field indicates that the use of laboratory-derived values in models tends to over-estimate pesticide persistence under field conditions.

The present approach by the modeler is to estimate k from laboratory studies, assuming that these studies approximate the degradation process under field conditions. Recognizing the probability that degradation rates are both spatially and temporally variable, deterministic research and management models should both be executed with a range of k values to represent the influence upon pesticide fate of the field variation of degradation processes. Yet, sensitivity analysis of models or comparison of such predictions with field data on this basis is almost non-existent. Development of functional relationships between k and the environmental variables cited above would be very useful, and would provide a feedback mechanism whereby predicted water contents or temperatures could be used to estimate appropriate k values. Additionally, while some progress has been made in estimating sorption coefficients using physical-chemical properties of the pesticides, similar developments in relating microbial or chemical degradation rates to chemical structure are not well developed, and would be quite useful. In summary, first-order kinetics is used because it is an appropriate representation of degradation in the context of model structure and also because not enough is known of degradation under field conditions to utilize more complicated kinetics.

Volatilization. The volatilization flux of pesticide is usually determined by first considering its aqueous solubility and sorption. Excess pesticide beyond that which will dissolve in soil water and be sorbed by the soil is considered available for diffusion across the soil surface and into the atmosphere. Most models that consider volatilization therefore require as input the pesticide aqueous solubility and the saturated vapor density. One method of partitioning between the liquid and vapor phases is (9).

$$c_g = K_H c \tag{10}$$

where c_g = gas concentration of pesticide (m/L^3), K_H = Henry's constant and c has been defined above. The value of K_H is calculated from

$$K_H = c_g^* / c_{sol} \tag{11}$$

where c_g^* = saturated vapor density and c_{sol} is defined in Equation 8b. These values are usually included as manufacturer's data, and are assumed constant with temperature and changes in soil solution composition.

Once values of $c_g(z,t)$ are estimated, the volatilization flux is calculated from a diffusion equation that considers the gradient in c_g and a partially water-filled soil porosity. Such flux calculations are difficult due to the near infinite gradient in gas concentration from soil to atmosphere. Often the volatilization flux is calculated using Equations 10-11 and considering only the very shallowest upper layer of the profile. Volatilization models are not included in most pesticide management models, and in only a few pesticide research models.

Plant uptake. Pesticide uptake by plants has not been considered in most modeling efforts. This is primarily due to an almost total lack of quantitative experimental information available to the modeler, and the presumption that the absolute mass of pesticide absorbed by the plant is small compared to the mass remaining in the system. Due to these considerations, modelers have apparently assumed that any inaccuracy in simulation of pesticide fate that results from not considering plant uptake is within the "noise" of inaccuracies produced by other assumptions about the physical, chemical, and biological processes operating in the system. While this assumption is unproven for pesticide absorption, it clearly cannot be accepted for water absorption by the plant (the U(z,t) term in Equation 4). Plant extraction of water greatly influences water flux, which affects pesticide displacement. The validity of almost all assumptions related to plant extraction of water and pesticide are relatively unproven and will certainly need to be investigated before more effort is expended incorporating further detail into pesticide models relative to degradation, sorption or volatilization.

Validity of Current Modeling Approaches

All the above information on the principles of pesticide modeling have a common origin. Whether we consider the basic water flow (Equation 4) and solute transport (Equations 2-3) equations or the approaches taken to describe sorption, degradation, volatilization and plant uptake, they all derive primarily from laboratory experience. Any credibility of the supposed "basic principles" related to describing pesticide fate that has evolved from these studies remains almost completely unsubstantiated in the field. As more field data are collected, there is in fact more doubt cast upon the use under field conditions of the classical equations and formulations of basic processes. The net result is that although we understand how to model well-defined soil columns, we understand very little regarding the magnitude, intensity, variation and interaction of the processes under field conditions. A few examples will illustrate the point, and although these examples do not specifically treat pesticides (because pesticides are not much studied in the field), they are entirely appropriate for the general case of solute movement, of which pesticides constitute a subset.

Much of the present concern about the credibility of model predictions focuses upon issues related to spatial variability of water and solute movement. The first large-scale experiment (25) that measured on a field basis the K-θ-h relationship used in Equation 4 and the values of both D and v (Equation 2) demonstrated quite clearly that all these values were spatially variable. Subsequent studies by a number of individuals have confirmed over a wide range of soil types that a single field is characterized by values of K, D and v that each vary by orders of magnitude for a given water content. Such physical realities, presumably due to micro-scale variation in the geometry of the porous media, infer that we are being extremely naive in constructing deterministic models of the type formulated from Equation 2-4. Whether the model is intended for research or management purposes, the credibility of its predictions are cast into doubt by the realization that whatever unique set of conditions are used in a particular execution of the model, these conditions and the resulting predictions represent only one possible scenario of many that could be occurring simultaneously in the field. The assumption that the model is describing the field case is therefore unrealistic. The usefulness of the model to predict pesticide concentration on a very limited or a very large-scale basis is questionable.

The formulation of a pesticide model upon the basis of Equation 2-4 has also been cast into doubt by field studies that demonstrate that many field soils do not meet the underlying assumptions used in developing those equations. The assumption that a single pore size distribution prevails in a field soil, through which water and solutes move uniformly over the entire cross-sectional area, is not often met. Many soils contain cracks, wormholes, preferred pathways or relatively immobile porosity resulting from specific soil forming or aggregation processes. These soils are not characterized by water flow through the entire soil pore space, but transmit large quantities of water during wet periods through a relatively few large pores. This "artificial" or non-matrix transport of water and solute, often termed "short-circuiting", can result in substantial pesticide movement to deep soil depths in very short time periods. Such displacement is not described by the models constructed from Equations 2-4, and again the predictions of such models will be quite misleading in such cases.

Several studies currently in progress cast further doubt upon the existing deterministic research and management models. These studies are demonstrating that sorption coefficients are spatially variable, and that even strongly sorbed pesticides can be simultaneously found at the soil surface and at deeper depths in the soil profile. This indicates that some pesticide may be sorbed to a mobile, soluble organic fraction and displaced in the same manner as a non-interacting solute. The balance of the pesticide is sorbed and retained near the soil surface as would be predicted with Equations 5-6. Other studies are demonstrating that in some sites the non-uniformity of pesticide application greatly influences the resulting variability of pesticide distribution in the soil profile. None of these processes is included in the "basic principles" discussed above, yet may play an important role in determining the accuracy and reliability of any modeling predictions.

Future Use and Development of Pesticide Models

The above issues raise serious questions about the manner in which current pesticide models should be used, the reliability of their predictions, and the direction of future pesticide modeling efforts.

Our current understanding of the basic principles that determine pesticide fate in the field is incomplete, yet we must make decisions now considering pesticide regulation and management. Current pesticide models used by regulators and academics represent the best tools we have to estimate pesticide fate as a function of soil, climate and management factors. Yet, we have every indication that their predictions are not universally reliable, and almost no proof of their credibility in the field. The question is whether we can feel confortable about the predictions produced by these models, or whether we should abstain from their use as predictive tools until their credibility is better established. A healthy and continuing intellectual argument is in progress on this issue, and will probably persist for some time. During this debate, the use of existing models for regulatory and management purposes will continue, and will result in some good decisions, and probably some mistakes.

Several points are clear. First, no pesticide model exists that has been proven to estimate consistently and accurately the spatial and temporal distribution of pesticide concentrations in the unsaturated zone. This is true regardless of the resolution used to represent basic principles in the model, and whether the model falls into the research or management category. Second, it follows that current models should be used only to compare the relative, not absolute, behavior of pesticides in field soils. Third, the first two points indicate that our approach to modeling pesticide fate in unsaturated field soils must change if we are to develop a new generation of pesticide models that do not suffer from the limitations of the current models.

It appears that a stochastic, rather than deterministic, approach should be considered when modeling water and chemical movement in the unsaturated zone. This will represent no small change in our conceptualization of basic principles of pesticide modeling. The resulting models will almost certainly not represent basic processes in fundamental mechanistic terms, but will instead represent the soil-water-pesticide system in statistical terms. Predictions of such models should be probabilistic, and should include the confidence limits of the prediction. Given the heterogeneous, variable nature of soil systems, this approach may offer the best opportunity to predict pesticide fate in terms commensurate with our ability to quantify soil processes. Several such models that can serve as beginning points have been recently reported (26,27), and development work is continuing. Until such time that their reliability is established, we should be quite cautious in our use and interpretation of contemporary pesticide models.

Literature Cited

1. Addiscott, T.M.; Wagenet, R.J. J. Soil Sci. 1985, 36, 411-424.
2. Leistra, M. Plant Soil 1978, 49,569-580.
3. Leistra, M. Soil Sci. 1979, 128,303-311.
4. Lindstrom, F.T.; Boersma, L.; Gardiner, H. Soil Sci. 1968, 105,107-113.
5. Hutson, J.L.; Wagenet, R.J. "LEACHM: A model for simulating the leaching and chemistry of solutes in the plant root zone"; New York State Agric. Exp. Stn.; Ithaca, N.Y., 1985, (in press).
6. Rao, P.S.C.; Davidson, J.M.; Hammond, L.C. In "Residual Management by Land Disposal"; U.S. Environ. Prot. Ag. EPA-600/9-76-015, 1976; pp 235-242.
7. Steenhuis, T.; Pacenka, S.; Hughes, H.; Gross, M. "Mathematical model summary", Dep. of Agric. Engineering, Cornell Univ., Ithaca, NY, 1983; 12pp.
8. Carsel, R.F.; Smith, D.N.; Lorber, M.N. "Users manual for the Pesticide Root Zone Model (PRZM), Release 1"; U.S. Environ. Prot. Ag. EPA-600/3-84-109, 1984; 216 pp.
9. Jury, W.A.; Spencer, W.F.; Farmer, W.J. J. Env. Quality. 1983, 12,558-564.
10. Wagenet, R.J. In "Chemical Mobility and Reactivity in Soil Systems"; Nelson, D.W.; Elrick, D.E.; Tanji, K.K., Eds.; Spec. Publ. No. 11, Am. Soc. Agronomy, Madison, WI., 1983; pp. 123-140.
11. Zhong, W.-Z.; Wagenet, R.J.; Lemley, A.T. In chapter 4 of this book.

12. vanGenuchten, M.Th.; Davidson, J.M.; Wierenga, P.J. Soil Sci. Soc. Amer. Proc. 1974, 38,29-35.
13. Selim, H.M.; Davidson, J.M.; Rao, P.S.C. Soil Sci. Soc. Amer. J. 1977, 41,3-10.
14. Enfield, C.G.; Carsel, R.F. In "Test Protocols for Environmental Fate and Movement of Toxicants"; Assoc. Official Anal. Chem.: Washington, D.C. 1981; pp. 233-250.
15. Childs, S.W.; Hanks, R.J. Soil Sci. Soc. Amer. Proc. 1975, 39,617-622.
16. Davidson, J.M.; Graetz, D.A.; Rao, P.S.C.; Selim, H.M. "Simulation of Nitrogen Movement, Transformation and Plant Uptake in the Plant Root Zone", U.S. Environ. Prot. Ag. EPA-600/3-78-029, 1978; 106 pp.
17. Tillotson, W.R.; Wagenet, R.J. Soil Sci. 1982, 133, 133-143.
18. Karickhoff, S.W. In "Proc. of the Symposium on Processes Involving Contaminants and Sediments"; Baker, R.A., Ed.; Ann Arbor Sci. Publ. Co., Ann Arbor, MI, Vol. 2, 1980; pp. 193-205.
19. Helling, C.S.; Dragun, J. In "Test Protocols for Environmental Fate and Movement of Toxicants"; Assoc. Official Anal. Chem.: Washington, D.C., 1981; pp. 43-88.
20. Rao, P.S.C.; Davidson, J.M. In "Environmental Impact of Nonpoint Source Pollution"; Overcash, M.R.; Davidson, J.M., Eds.; Ann Arbor Sci. Publ. Co., Ann Arbor, MI, 1980; pp. 23-67.
21. Green, R.E.; Davidson, J.M.; Biggar, J.W. In "Agrochemicals in Soils"; Banin, A.; Kafkaffi, U., Eds.; Pergammon Press, New York, 1980; pp. 73-82.
22. Karickhoff, S.W. Chemosphere 1981, 10,833-846.
23. Hamaker, J.W.; Thompson, J.M. In "Organic Chemicals in the Environment"; Goring, C.A.I.; Hamaker, J.W., Eds., Marcel Dekker, New York, 1972; pp. 49-143.
24. Goring, C.A.I.; Laskowski, D.A.; Hamaker, J.W.; Meikle, R.W. In "Environmental Dynamics of Pesticides"; Hague, R.; Freed, V.H., Eds,; Plenum Press, New York, 1975; pp. 135-172.
25. Nielsen, D.R.; Biggar, J.W.; Erh, K.T. Hilgardia 1973, 42, 215-259.
26. Jury, W.A. Water Resour. Res. 1982, 18, 363-368.
27. Dagan, G.; Bresler, E. Soil Sci. Soc. Amer. J. 1979, 43, 461-467.

RECEIVED April 1, 1986

18

A Method for the Assessment of Ground Water Contamination Potential

Using a Pesticide Root Zone Model (PRZM) for the Unsaturated Zone

M. N. Lorber and Carolyn K. Offutt

Office of Pesticide Programs (TS-769C), U.S. Environmental Protection Agency, Washington, DC 20460

The PRZM model was used to evaluate the potential of aldicarb to leach through soil and contaminate ground water in three use sites: tobacco grown on a sandy loam soil in North Carolina and potatoes grown on a sandy loam and a loamy sand in Wisconsin. Calibration with the use of field data on these sites allowed field values of aldicarb decay rate and partition coefficient to be estimated. Long term simulations then permitted evaluation of the effect of soil type, date of application, and irrigation on the leaching potential of aldicarb in these use sites. Results showed little to no potential for leaching on the sandy loam soils, but significant potential on the loamy sand, with between 1 and 19% of applied to leach below 2 meters, and six-month plume solution concentrations as high as 103 ppb at two meters. Applying later in the season reduced leaching by about one-half, and irrigation increased leaching 3-5 times.

The findings in 1979 of 1,2-dibromochloropropane in ground water in five states and aldicarb in ground water in Long Island, New York, began an intense effort on the part of the Office of Pesticide Programs in the U.S. Environmental Protection Agency (EPA) to evaluate the potential for pesticides used in agriculture to leach through soils and cause contamination of ground water (1). A critical component of this effort is the use of mathematical models, which can predict the fate and transport of pesticides. An early model used was called PESTANS, the Pesticide Analytical Solution model (2), which employed a steady state solution to the analytical equation describing solute transport in a homogenous soil. However, the transient aspect of the wetting and drying cycle in the root zone, and the inhomogenous character of soil necessitated a model which would consider variations in weather patterns and allow for temporal and vertical reassignment of model parameters. The Pesticide Root Zone Model (3,4) was developed at

This chapter not subject to U.S. copyright.
Published 1986, American Chemical Society

the US EPA Environmental Research Laboratory in Athens, Georgia, to meet this need.

The ability to predict the transport of pesticides leaching through soil allows for an evaluation exercise known as an "exposure assessment". Strictly defined, an exposure assessment would result in an estimation of the concentrations of pesticide in drinking water. However, PRZM and related models such as PESTANS only predict the mass flux and the concentration in the "vadose", or unsaturated zone, which includes the root zone, and some defined distance below the root zone. Although the science and mathematics of predicting transport in the saturated zone exist, and aldicarb transport in the saturated zone has been modeled (5,6), there are still several practical difficulties in modeling this process. Quantifying the dispersion process and estimating the rate of pesticide decay in the saturated zone are the two main problems in parameter estimation. As well, there are no adequate field data bases available to test a model which is capable of predicting the movement of pesticide from the point of application to and through an aquifer.

Although also a complex phenomena, movement of water and solute in the unsaturated zone has been studied more extensively. Several field site data bases, including the three used to calibrate PRZM in this paper (7-9), are becoming available to use in conjunction with unsaturated models. Quantification of pesticide movement in the unsaturated zone can be accomplished effectively with an adequate number of soil cores taken at regular intervals following pesticide application (10). In contrast, wells penetrating the aquifer either above or below the pesticide plume, or well samples taken before or after the plume has passed the well will result in false negative findings.

A major advantage to models such as PRZM or PESTANS is that they are transportable: they can simulate a variety of situations with simple changes in weather input and parameters. More importantly, however, is the fact that in most situations, 90% or more of applied pesticide would have runoff, volatilized, been taken up by the plant, or otherwise decayed before any of it leaches below the root zone. It makes sense, therefore, to develop the capability to predict the fate of pesticides in the root zone, and hence determine the potential for pesticides to contaminate ground water.

The purpose of this paper is to present an assessment exercise of a leaching pesticide using the PRZM model. The assessment begins with a calibration of PRZM for the pesticide aldicarb applied to tobacco in North Carolina and potatoes in Wisconsin. Following these calibrations, long term simulations are performed using these same calibration scenarios. Examination of key PRZM output indicates the "potential" for aldicarb to contaminate ground water in the scenarios modeled.

Description of PRZM

As the name of the model implies, PRZM models the unsaturated zone, which includes the root zone and a user-specified depth below the root zone within the "vadose" zone. The simulation uses a daily time step, and mass balances of water and pesticide are maintained in "zones" of finite depth, usually 5 cm. Model parameters can vary as a function of both space and time. Complete details of model theory, including equations, sensitivity analysis, and other applications are presented elsewhere (3,4).

The water balance algorithm is based on the Soil Conservation Service Curve Number approach (11), which estimates daily runoff as a function of the antecedent moisture condition (wetness of the soil profile prior to a storm) and a curve number determined from field conditions (soil type, crop type, etc). Simply put, rainfall is partitioned into runoff and infiltration - that which doesn't run off, infiltrates. Following a storm, the soil drains by gravity to field capacity in one day. A "slow drainage" option allows the accumulation of water above field capacity, which then drains over the next several days. Between storms, water is extracted from the root zone via evapotranspiration, which can be determined from daily pan evaporation, if it is available, or from an empirical equation based on average daily air temperature. The parameters for the soil water model include: runoff curve numbers, soil field capacity and wilting point, soil bulk density, crop planting, maturity, and harvest dates, root depth, and crop surface area coverage.

Mass balance equations of pesticide fate and transport are developed for the surface and subsurface zones in PRZM. In the surface zone, avenues of loss include soluble loss in runoff, percolation to the next zone, sorbed loss in erosion, and decay in both phases. In the subsurface zones, losses include plant uptake and percolation in the soluble phase, and decay in both phases. A backward difference, implicit numerical scheme is used to solve the partial differential solute transport equations, with a time step of one day and a spatial increment specified by the user. The important assumptions for the pesticide model are: instantaneous, linear, reversible adsorption described by an adsorption partition coefficient, K_d, and first-order decay described by an overall decay rate, k. Parameters for the pesticide model include: universal soil loss equation parameters (if erosion loss is to be modeled), pesticide application information (rate, date, and method of application), K_d, k, and a dispersion coefficient.

Calibration of PRZM

The appropriate means to "test" a model is dependent on the biases of the model tester and the purposes of his exercise. Words such as calibration, validation, and verification have been used to describe a model testing procedure. For this study, PRZM was calibrated to three field sites to determine appropriate parameters for longer term simulations. These long term simulations employed the same parameters as the calibration simulations, and their purpose was to examine trends in pesticide leaching as expressed by PRZM output.

PRZM was calibrated to field data on the pesticide aldicarb applied to two potato sites in Wisconsin and one tobacco site in North Carolina (7-9). Field data required for the calibration exercise include: soil physical parameters, crop cultural information, aldicarb rate and date of application, depth of incorporation, on-site weather data including daily rainfall and average air temperature (or pan evaporation if available), and field observations of pesticide leaching over time. Specifically, these observations were soil cores taken at several dates following application and measured for total aldicarb residues. Aldicarb degrades rapidly in soil to aldicarb sulfoxide and aldicarb sulfone. In this paper, aldicarb refers to the sum of parent aldicarb and degradates sulfoxide and sulfone. The physical parameters for the three sites are given in Table I.

The purpose of the calibration exercise was to determine appropriate parameters which would result in a best-fit match between model simulations and the soil core field observations. The first step in the procedure was to assign values to field-measured, physically-based, parameters which include all parameters for the water balance and crop development portions of the model. The second step was to calibrate chemically and biologically based parameters which are not easily measured in the field. These include the adsorption partition coefficient, K_d, and the first-order rate of decay, k, of aldicarb. Using reasonable ranges of K_d and k as defined by the literature, a trial-and-error method was used until model predictions matched field observations.

In Table II are the values of the aldicarb parameters which were calculated from field data. The calculated K_d was determined assuming a K_{oc} for aldicarb of 36 (1), a ratio of organic matter to organic carbon of 1.7, and field data for soil organic matter for the different soil layers. The decay rate was assumed uniform within the soil profile. The range of "observed" decay rates was determined from the field observations made on different dates. They were calculated by estimating mass of aldicarb remaining on the observation date (based on soil concentrations and soil bulk density), and then applying the first-order equation of pesticide decay. In estimating the decay rate from field observations, the important assumption is made that the residues measured represent the fate of all aldicarb applied, i.e., that no aldicarb was lost in runoff, leached below the depth of sampling, or was taken up by the crop.

Table III summarizes the several calibration scenarios, which are now described.

Site 1 - North Carolina tobacco. This site is located in Hertford County, North Carolina. The soil is classified as a sandy loam, with 0.85% organic matter in the top 0.3 m of soil. The mechanical analysis for the top 3 meters is as follows: 55-69% sand, 18-38% silt, and 7-16% clay. Aldicarb was incorporated to a depth of 10 cm at a rate of 3.36 kg/ha at tobacco transplanting on May 13, 1983. Pretreatment samples taken prior to aldicarb application insured that the soil was free of aldicarb residues. Daily rainfall and pan evaporation for this site were obtained from a nearby weather station. Soil cores were obtained by bucket auger to a depth of 3 meters, separated into increments of 0-0.3 m, 0.3-0.6 m, 0.6-1.2 m, 1.2-1.8 m, 1.8-2.4 m, and 2.4-3.0 m. The plot was divided into four subplots, and four samples were taken per subplot and composited at each date of observation, resulting in a total of four samples per date per core depth. These four samples were

Table I. Summary of PRZM soil and crop parameters for calibration and exposure assessment exercises for North Carolina and Wisconsin (parameters developed from data in 7-9).

Parameter Description	North Carolina	Wisconsin Cameron	Wisconsin Hancock	
- Soil Classification	sandy loam	sandy loam	loamy sand	
- Runoff Curve Number Assumptions	row crops straight row good condition "B" soil group residue after harvest	row crops straight row good condition "B" soil group residue after harvest	row crops straight row good condition "A" soil group residue after harvest	
- Water Holding Capacities, cm^3/cm^3				
field capacity				
0-30 cm	0.18	0.38	0.11	
30-300 cm	0.24	0.20	0.08	
wilting point				
0-30 cm	0.06	0.20	0.05	
30-300 cm	0.10	0.08	0.03	
initial soil water				
0-30 cm	0.16	0.20	0.10	
30-300 cm	0.24	0.20	0.08	
- Bulk Density, gm/cm^3				
0-30 cm	1.47	1.46	1.48	
30-60 cm	1.34	1.54	1.54	
60-300 cm	1.56	1.54	1.54	
- Zone Depths, cm				
core	300	300	300	
root	45	30	30	
evaporation	20	15	15	
storage	5	5	5	
- Erosion	not considered	not considered	not considered	
- Crop Grown	tobacco	potatoes	potatoes	
- Year	1983	1982	1982	1983
- Crop Development				
emergence	May 20	June 4	May 28	June 21
maturity	Aug 1	July 15	July 15	Aug 1
harvest	Nov 1	Nov 1	Nov 1	Nov 1
- Rain interception, cm	0.20	0.15	0.15	

Table II. Summary of PRZM calibration parameters for aldicarb use in North Carolina and Wisconsin (calculated values for K_d and k from field site data are given in parentheses besides calibrated values).

Parameter Description	North Carolina	Wisconsin Cameron	Wisconsin Hancock
Adsorption Partition Coefficient, K_d, ml/gm			
0 - 15 cm	0.50 (0.18)	1.00 (0.22)	1.00 (0.16)
15 - 30 cm	0.15 (0.18)	0.50 (0.22)	0.50 (0.16)
30 - 60 cm	0.01 (0.04)	0.05 (0.04)	0.05 (0.01)
60 - 300 cm	0.01 (0.01)	0.01 (0.01)	0.01 (0.01)
First-Order Decay Rate, k, day^{-1}			
planting, 0-300 cm	0.016 (0.018-0.026)	0.010 (0.011)	0.010 (0.013-0.019)
emergence, 0-300 cm	-	0.015 (0.016)	0.015 (0.014-0.049)
Equivalent half-life, days			
planting, 0-300 cm	43 (27-39)	69 (63)	69 (36-53)
emergence, 0-300 cm	-	46 (43)	46 (14-49)
Date of Application			
1982 planting	-	May 15	May 19
emergence	-	June 4	May 28
1983 planting	May 13	-	May 10
emergence	-	-	June 21
Rate of Application, kg/ha 1982 & 1983			
planting	3.36	3.36	3.36
emergence	2.24	2.24	2.24
Depth of Incorporation, cm	10	10	10

Table III. Summary of PRZM calibration runs.

Run #	Location	Crop	Soil	Date/Rate	Irr*
1	North Carolina	tobacco	sandy loam	5/13/83; 3.36	N
2	Cameron, Ws.	potato	sandy loam	5/15/82; 3.36	M
3	Cameron, Ws.	potato	sandy loam	6/4/82; 2.24	M
4	Hancock, Ws.	potato	loamy sand	5/19/82; 3.36	M
5	Hancock, Ws.	potato	loamy sand	5/28/82; 2.24	M
6	Hancock, Ws.	potato	loamy sand	5/19/82; 3.36	H
7	Hancock, Ws.	potato	loamy sand	5/28/82; 2.24	H
8	Hancock, Ws.	potato	loamy sand	5/10/83; 3.36	M
9	Hancock, Ws.	potato	loamy sand	6/21/83; 2.24	M

* Irrigation regimes practiced: N = no irrigation; M = medium irrigation schedule; H = high irrigation schedule

averaged to represent "observed" results. Observations were taken on June 6, July 14, Sep. 15, and Nov. 30, and these are represented by the dashed lines shown in Figure 1. Further details on this field site are given in Jones et al. (7). The simulated results are shown by the solid line curves in Figure 1. The observations are matched against predictions for the Nov. 30 date in tabular form in Figure 1. Table IV summarizes the simulated water balance, and Table V summarizes the aldicarb fate and transport for this North Carolina calibration site.

Site 2 - Cameron, Wisconsin potatoes. This field site is located on a commercial potato farm in northwestern Wisconsin in Cameron. The soil is an Onamia sandy loam, with 1.03% organic matter in the top 0.3 m of soil. The mechanical analysis for the top three meters of soil is as follows: 83-99% sand, 0-13% silt, and 1-8% clay. The experimental field was divided in half, with one-half receiving 3.36 kg/ha aldicarb incorporated at potato planting on May 15, 1982, and the other half receiving 2.24 kg/ha at emergence on May 28. Soil cores were taken prior to application to insure that the profile was free of aldicarb. The procedure in North Carolina for post-treatment sampling including subplots, composites, and soil coring was also followed in Wisconsin. In addition, 11.7 cm of irrigation was applied between May 15 and Dec. 8 to meet the evapotranspirative demands of the crop when rainfall was insufficient. This irrigation was directly input as part of the rainfall record and as such, was treated as rainfall by PRZM. Daily totals of rainfall were recorded on-site (8). Evapotranspiration was estimated on-site using Penman's equation between June and September (8), and was directly input to PRZM. Otherwise, daily average air temperature was used to estimate evapotranspirative demand. Model predictions are compared with observations on June 8 and Dec. 8. Samples were also taken on July 13 and Sep 22, but only the first and last were chosen for illustration in this study. Additional details on this field site are given in Wyman et al (8). Figure 2 shows the planting and emergence simulations vs. observations, including tabular summaries for the Dec. 8 date. Table IV summarizes the water balance for the Cameron calibration, and Table V summarizes aldicarb fate and transport.

Site 3 - Hancock, Wisconsin potatoes. Aldicarb was applied in 1982 and 1983 at the University of Wisconsin Experimental Farm at Hancock in central Wisconsin. The soil at this site is a Plainfield loamy sand. The organic matter in the top 0.3 m in this soil is 0.77%. The mechanical analysis for the three meter core depth sampled was: 89-97% sand, 0-6% silt, and 1-9% clay. Like the Cameron site, there was a planting (May 19, 3.36 kg/ha) and an emergence (June 4, 2.24 kg/ha) application of aldicarb in 1982, and two similar applications in 1983 (May 10, 3.36 kg/ha; and June 21, 2.4 kg/ha). There were two irrigation regimes practiced in 1982, and one in 1983. One irrigation scheme was similar to the irrigation in Cameron in that it sought to meet evapotranspirative demand. This was called the "medium schedule". The second irrigation scheme, which was only practiced in 1982, was termed the "heavy schedule" in that 60% more water than estimated to meet the evapotranspirative demand was applied with identical timing as the medium schedule. In summary, there were a total of six separate scenarios at Hancock: four in 1982 which included permutations of application date (and rate) and irrigation strategy, and two in 1983 having different application dates. Daily rainfall was obtained on-site for 1982 and 1983. Actual evapotranspiration was estimated on-site for 1982 with Penman's equation and directly input to PRZM, while pan evaporation was available for 1983. The methodology for soil coring was similar to the Cameron and North Carolina sites. Additional details on the Hancock site can be found in Wyman et al (8,9). Figures 3-5 summarize the Hancock calibration results, and Tables IV and V summarize the water balance and aldicarb fate and transport for the Hancock scenarios.

Discussion of Calibration Results

Table IV summarizes the water balance results of all the calibration scenarios. A trend that can be seen from this data is that evapotranspiration demand in Wisconsin is consistantly around 43 cm for the summer months between May and September, regardless of irrigation added or soil type. Recharge was higher for the loamy sand soil in Hancock, 35 cm, than for the sandy loam in Cameron, 26 cm. Recharge increased to 53 cm when 19 cm of extra irrigation water was added to the sand soil, indicating that this extra irrigation was not needed by the crop. Evapotranspiration in North Carolina

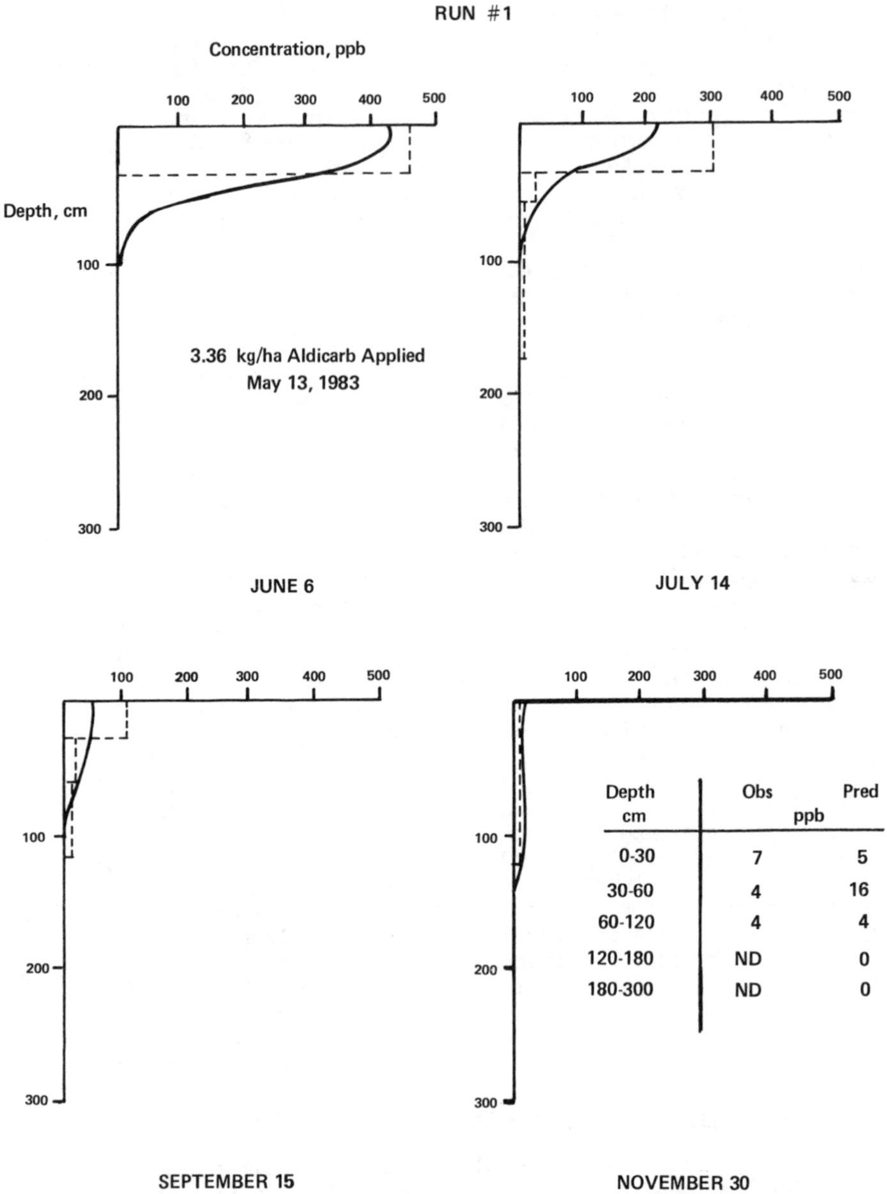

Figure 1. Calibration results of PRZM, scenario 1 (see Table 3): concentration-depth profiles for predicted (smooth curves) vs. observed (dashed lines) aldicarb applied to tobacco in North Carolina (observed data from 7).

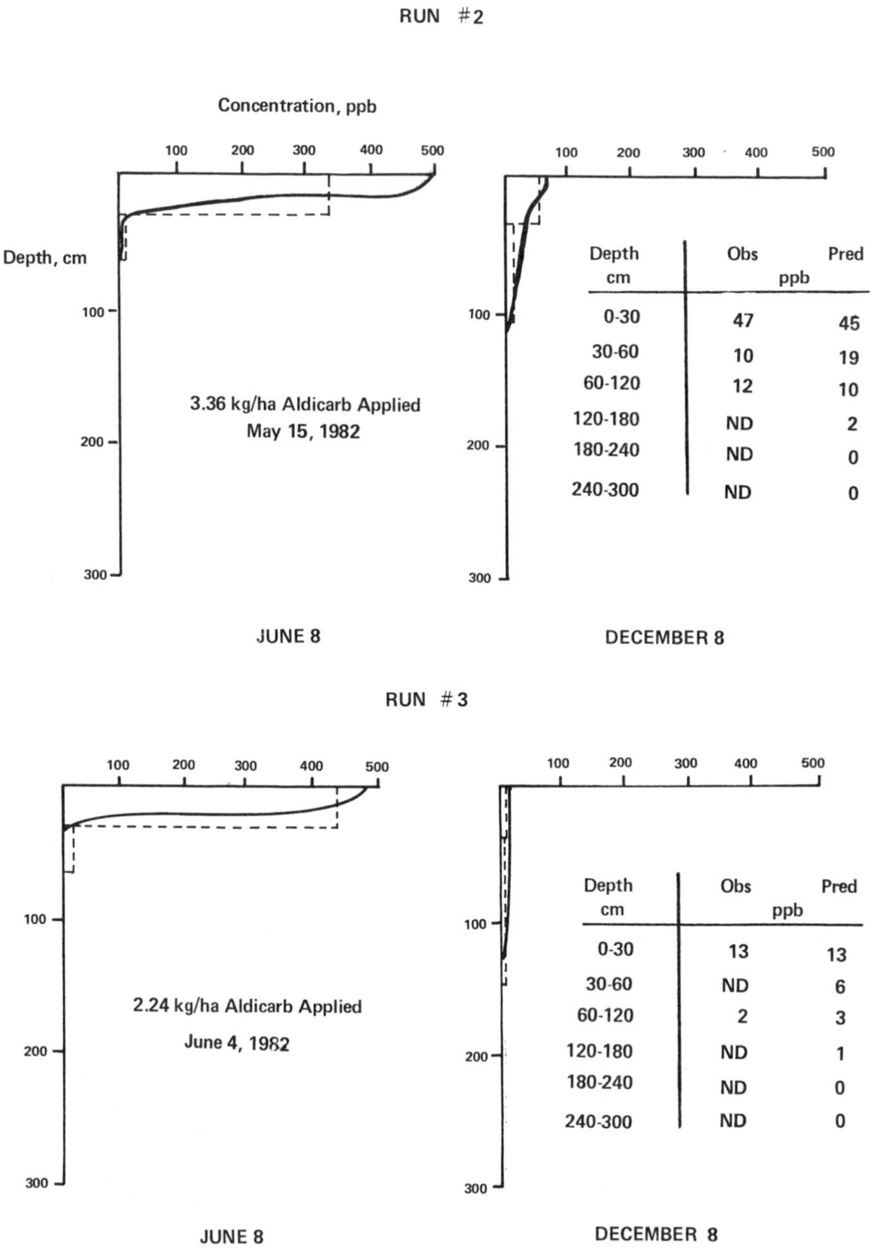

Run #2, December 8:

Depth cm	Obs ppb	Pred ppb
0-30	47	45
30-60	10	19
60-120	12	10
120-180	ND	2
180-240	ND	0
240-300	ND	0

Run #3, December 8:

Depth cm	Obs ppb	Pred ppb
0-30	13	13
30-60	ND	6
60-120	2	3
120-180	ND	1
180-240	ND	0
240-300	ND	0

Figure 2. Calibration results of PRZM, scenarios 2 & 3 (see Table 3): concentration-depth profiles for predicted (smooth curves) vs. observed (dashed lines) aldicarb applied to potatoes in Cameron, Wisconsin (observed data from 8).

was higher than in Wisconsin, at 50 cm for the period of testing between May and November. Recharge was only a small fraction of rainfall during this period, equalling 14 cm.

Table IV. Summary of water balance results for calibration scenarios (all quantities in cm/yr).

Description	Calibration Run #				
	1	2,3	4,5	6,7	8,9
Dates	May-Nov 1983	May-Nov 1982	May-Sep 1982	May-Sep 1982	May-Sep 1983
Precipitation	71.4	86.4	44.7	43.4	59.3
Irrigation	0.0	11.7	31.7	50.8	24.4
Runoff	6.9	2.9	0.9	1.0	3.6
Evapotranspiration	50.4	44.6	41.4	41.9	44.4
Recharge	14.0	25.6	35.1	52.5	37.0
Snow Storage	0.0	24.8	0.0	0.0	0.0
Change in Soil Storage	+0.1	+0.2	-1.0	-1.2	-1.3

The calibrated half-life for aldicarb is longer than the half-life which was calculated based on field data (Table II). This occurs because runoff loss of aldicarb as well as leaching below the depth of sampling are not accounted for in field-calculated half-lives, which are calculated based only on aldicarb remaining at each sampling date. An additional possible avenue of loss is plant uptake of aldicarb. However, the total amount of uptake was not estimated in the field, nor was it simulated in PRZM. As such, it can be considered that plant uptake loss was "lumped" in the calibrated (and calculated) half-lives.

Table V summarizes the fate and transport of aldicarb in the calibration scenarios. In North Carolina, the simulations predict that 4.2% of applied aldicarb was lost via runoff and none leached below the depth of sampling. The runoff result cannot be verified since field data of runoff were not taken. However, since the soil was a loam soil, some water runoff would be expected, and a fraction of the soluble aldicarb present in top zone on the date of runoff would also run off. As a result, the calibrated half-life, 43 days, is higher than the range calculated, 27-39 days.

In the Cameron, Wisconsin simulations, aldicarb did not leach below the depth of sampling for the planting or emergence appli-

Table V. Summary of fate and transport results for aldicarb calibration scenarios (all results expressed in percent of applied).

Description	Calibration Run # 1	2	3	4	5	6	7	8	9
Applied	100.0	100.0	100.0	100.0	100.0	100.0	100.0	100.0	100.0
Decay	91.9	73.5	92.9	73.8	83.9	67.9	78.5	75.9	77.7
Runoff	4.2	2.1	1.3	0.3	0.4	0.2	0.3	0.0	0.0
Leached Below 3 meters	0.0	0.0	0.0	1.5	1.3	15.2	10.9	2.1	1.3
Remained in profile	3.9	24.4	5.8	24.4	14.4	16.7	10.3	22.0	21.0

cation. However, a small percentage of applied was found to runoff for both application dates, 2.1 and 1.3%. As a result, the calibrated half-life, 69 and 46 days for planting and emergence applications, respectively, were only slightly higher than the calculated half-lives, 65 and 43 days.

In Hancock, where there was a loamy sand soil, leaching below the depth of sampling was simulated with all scenarios, particularly the scenarios of high irrigations. Approximately 15 and 11% of applied aldicarb leached below 3 meters in these high irrigation scenarios (runs # 6 & 7). The field-calculated half-life for these intense irrigation scenarios is in the neighborhood of 35 days; the calibrated half-life was 69 days. As noted earlier, the extra irrigation did not add to the consumptive use of water. Therefore, the field calculated half-life was misleading for these scenarios of intense irrigation. Between 1.3 and 2.1% of applied aldicarb was simulated to leach below 3 meters for the scenarios of medium irrigation, again resulting in higher calibrated half-lives than were calculated.

One other important observation that is evident from both field observations and simulations is that the half-life decreases with later application dates. This occurs because the aldicarb is applied during warmer weather and is hence subject to a healthier environment for microbes, which can degrade it more quickly. As seen in Table V, the amount of leaching in Hancock decreased from planting applications, 1.5, 15.2, and 2.1% of applied (runs 4,6,8), to emergence applications, 1.3, 10.9, and 1.3 (runs 5,7,9). A more appropriate simulation approach might be to make the rate of decay a function of air or soil temperature, or to simulate a step-wise change in rate of decay as a function of time. Nonetheless, this more rapid rate of decay for later applications results in less leaching for later applications.

In order to force the simulated profiles to match the observed

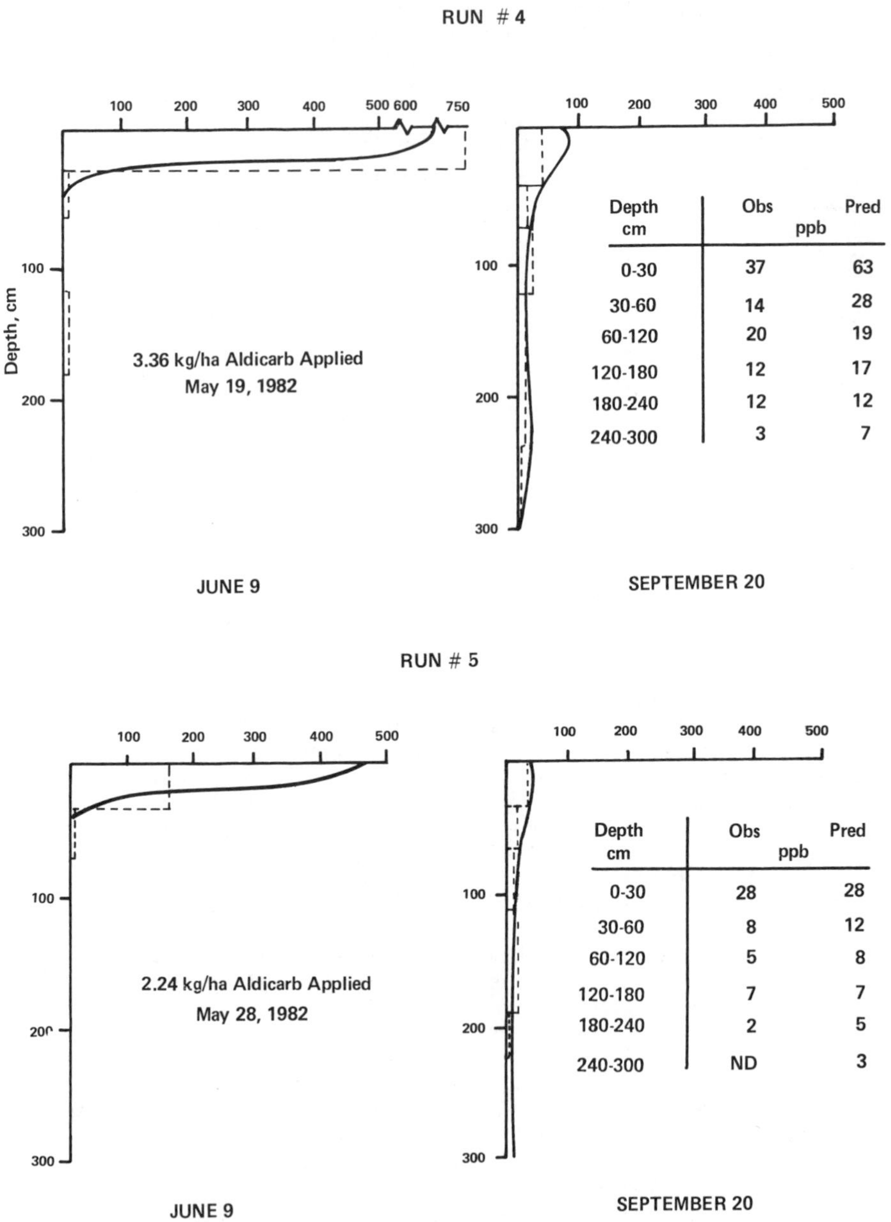

Figure 3. Calibration results of PRZM, scenarios 4 & 5 (see Table 3): concentration-depth profiles for predicted (smooth curves) vs. observed (dashed lines) aldicarb applied to potatoes in Hancock, Wisconsin (observed data from 8).

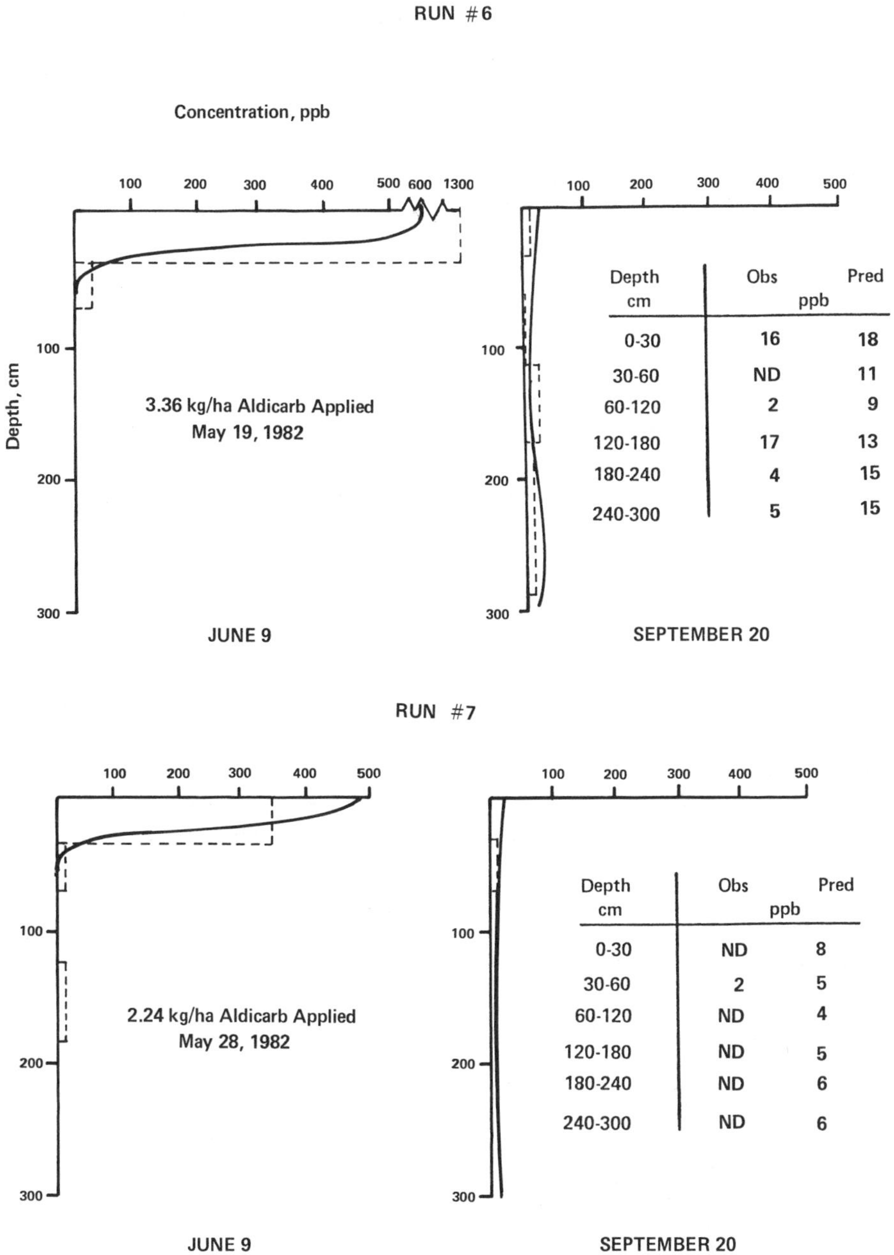

Figure 4. Calibration results of PRZM, scenarios 6 & 7 (see Table 3): concentration-depth profiles for predicted (smooth curves) vs. observed (dashed lines) aldicarb applied to potatoes in Hancock, Wisconsin (observed data from 8).

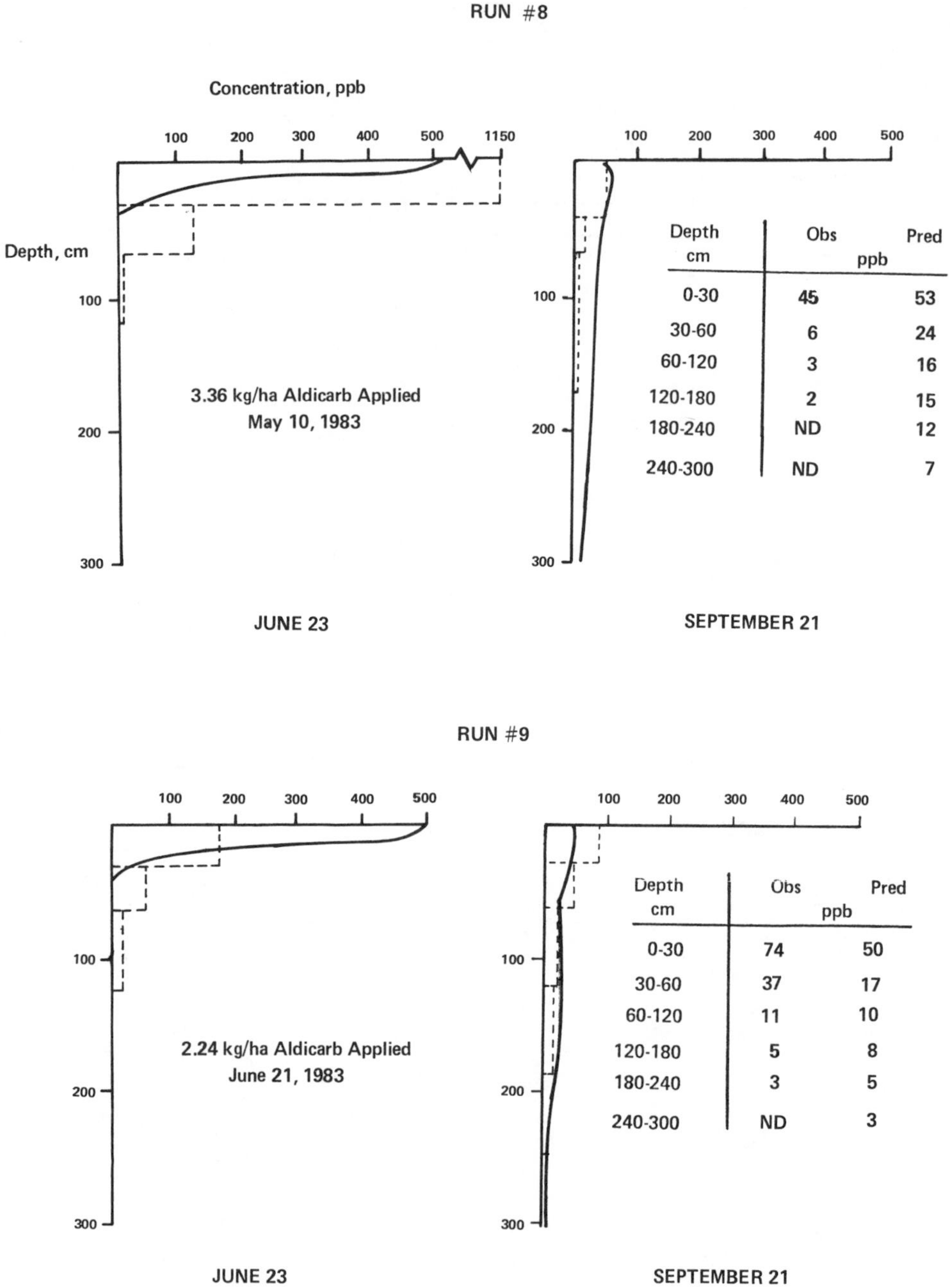

Figure 5. Calibration results of PRZM, scenarios 8 & 9 (see Table 3): concentration-depth profiles for predicted (smooth curves) vs. observed (dashed lines) aldicarb applied to potatoes in Hancock, Wisconsin (observed data from 9).

profiles, a higher adsorption partition coefficient, K_d, was required for the top zone than would be calculated based on aldicarb K_{oc} and soil organic matter. In North Carolina, the K_d required for the top 15 cm was 0.50, and in Wisconsin, was 1.00 for the top 15 cm and 0.50 for the 15-30 cm layer. These K_d are the highest assumed for aldicarb as compared to other published modeling efforts on aldicarb, which range in K_d from 0.0-0.3 (3,5,12,13) The reasons for this discrepancy are not known. Several possibilities exist which include:

1) PRZM may be overestimating the amount of water to percolate from the top zones to lower zones. This would occur if the water holding capacities measured in the field and assigned in PRZM are too low, resulting in less evapotranspiration and more percolation. This could also occur with inacurrate estimations of runoff - if more runoff occurs than is being simulated, then percolation will be overestimated.

2) It may be connected to the granular formulation of aldicarb. The outer protective layer dissolves upon contact with moisture, releasing the active ingredient, aldicarb. Furthermore, granules are incorporated, leading to the possibility of pockets of high granular concentration. If this is the case, a wetting event might not dissolve the protective coating of all the granules, leaving some partially dissolved granules. However, in several aldicarb test plots, high chloride concentrations (where chloride was applied as a tracer) also remain near the surface. This would seem to indicate that the granular formulation is not the cause of high surface aldicarb concentrations (14).

3) It is possible that evaporation demand at the soil surface results in water translocating upward from shallow depths, <15 cm, keeping aldicarb near the soil surface. Since PRZM does not simulate the upward movement of water, the high K_d assigned for aldicarb near the surface is an artificial way to maintain high surface concentrations.

4) The assumption made in PRZM and similar models is that equilibrium between sorbed and soluble phases occurs instantly. However, it has been shown that while adsorption occurs rapidly, there is a time delay associated with desorption. Incoming rainfall typically infiltrates a dry soil surface rapidly, not allowing the time necessary for equilibrium to be reestablished. This tendency for rapid infiltration has been called "macropore" flow, which can be visualized as the flow of water through preferential flow paths, created by inhomogeneous soil structure, wormholes, and so on. When this occurs, not all the pesticide "sees" the water flowing by during the initial period in a storm. Only a portion of actual infiltration carries soluble pesticide available through the equilibrium adsorption/desorption process. Therefore, a higher K_d may be required to limit the transport of pesticides residues below the soil surface. Another problem which may occur as a result of macropore flow (one not associated with the necessity for a high K_d) is that soluble pesticide picked up by macropore flow can be rapidly transported deep into the profile, resulting in a skewed distribution of pesticide concentration. This has also been seen in the field.

The dilemma of requiring a higher K_d to keep a pesticide near the surface in leaching simulations also occurred with simulations of DBCP in Hawaii (15). In that case, K_ds of 4 and 17 were required to match the observed pattern of DBCP remaining near the surface over time. The reported K_d of DBCP is in the neighborhood of 0.13-2.10 (1).

Below 30 cm, the calibrated and calculated K_d for aldicarb are similar. At that depth, phenomena such as undissolved granules, upwardly moving evaporation water, and macropore flow are much less likely to occur. Therefore, the calibration of aldicarb K_d is more straightforward and a function of soil organic matter.

As stated earlier, the means by which to test a model are dependent on the biases of the model tester and the purposes of his exercise. The purpose of the calibration in this study is to set up the exposure assessment simulations, which will determine the potential for aldicarb to contaminate the ground water in the scenarios modeled. As will be seen shortly, this potential is represented by model results at a point deep in the unsaturated zone. As such, it becomes imperative to accurately portray aldicarb fate in the unsaturated zone. Since these field studies showed that high concentrations of aldicarb were maintained near the soil surface, parameters were adjusted to portray that behavior. If anything, this exercise has uncovered a discrepancy between model theory and reality, and a future direction in PRZM development and/or field testing might be to test the theories proposed. Nonetheless, the calibration in this study is valid since the purpose is to duplicate reality, given the limitations of the model.

Figures 1-5 show the final calibrated results. Despite some differences, it can be seen that the important considerations of mass balance and trends in concentration-depth profiles are well met in the calibration exercise.

Leaching Assessment of Aldicarb

The purpose of the leaching assessment was to determine trends in the potential for aldicarb to leach to ground water in the scenarios modeled. The "potential" to contaminate ground water in the context of this exercise is indicated by predictions of "significant" concentrations (and/or mass) of aldicarb to leach to a "low depth" in the unsaturated zone. These predicted concentrations will not, in most cases, be equivalent to concentrations in the ground water, since the aldicarb will continue to decay as it traverses vertically to the water table and then horizontally to a well where it could be extracted. However, the rate of decay will be slower once the aldicarb leaves the biologically active root zone. Its major mode of decay in the saturated zone is hydroloysis, and the aldicarb hydrolysis half-life is estimated to be 10-650 weeks (1), depending on conditions. Therefore, the predicted concentrations at two meters in these long-term simulations represent an upper bound on what might be expected in the ground water.

A measure of "significance" for aldicarb is given by the "Health Advisory Level" (HAL) for aldicarb. "Health Advisories" suggest concentrations of a contaminant in drinking water at which adverse health effects would not be anticipated, with a margin of safety, for 1-day, 10-day and longer-term exposure periods (from a few months to 1-2 years)." (16). The HAL for aldicarb is 10 ppb (17). Wisconsin adopted this level of 10 ppb in legislation concerning aldicarb (18), while New York state opted for a more stringent level of 7 ppb (19).

The appropriate "low depth" in the unsaturated zone is not as well defined, and is arbitrarily chosen in this exercise to be two meters. PRZM was calibrated to a depth of two meters in North Carolina and three meters in Wisconsin. Two meters is felt to be a generally valid depth for PRZM assessments considering such factors as the unlikelihood of water table intrusion above that level, and sufficient depth such that conclusions of "potential" to contaminate ground water are meaningful. As well, the assumption is made that the soil column drains from saturation to field capacity in one day in PRZM. This assumption loses its validity as the depth of the soil column increases, i.e., the velocity of water increases as the depth of flow increases. For example, it is reasonable to assume that drainage below two meters will occur in one day, but less reasonable to assume that drainage will be complete below ten meters in one day. (note: a slow drainage option is available in PRZM, which would allow for column drainage in several days, but it was not used in the simulations of this paper).

The scenarios chosen for the assessment were essentially identical to the calibration exercise. All the parameters, including soil, crop, and most importantly, aldicarb K_d and k, were unchanged in the leaching assessment. The rate of aldicarb application was set at 2.24 kg/ha for all planting dates so that concentration comparisons between scenarios were meaningful. Five years of weather record were generated using a statistical model of weather generation developed by Richardson (20). The weather record for the North Carolina scenario was statistically similar to that of Raleigh, and the weather for Wisconsin was statistically similar to Madison. Pan evaporation data were not statistically generated, so evapotranspiration was estimated as a function of daily average temperatures. Aldicarb was applied at planting to a North Carolina tobacco crop on a sandy loam, at planting to a Wisconsin potato crop on a sandy loam, and at planting and emergence to a Wisconsin potato crop on a loamy sand. Irrigation was also evaluated for its effect on leaching on the loamy sand, leading to four scenarios on the loamy sand (planting and emergence applications; irrigation and no irrigation). The irrigation scenario was designed from the aldicarb field trials used for calibration. In these trials, approximately 25 cm/yr of irrigation was added for the "medium" schedule, with daily rates ranging from 0.5 to 2.5 cm. For the assessment, an average of 24 cm/yr of irrigation was applied during June, July, August, and September, at daily rates of 2.5 cm only when rain had not fallen for at least four days. A summary of the leaching assessment scenarios are given on Table VI.

Discussion of Leaching Assessment Results

A summary of the water balance results is given in Table VII. The statistically generated annual average rainfall, 108 cm/yr, for

Table VI. Summary of aldicarb exposure assessment PRZM simulations.

Run #	Location	Crop	Soil	Date/Rate	Irr*
1	Raleigh, N. Carolina	tobacco	sandy loam	May 1; 2.24	N
2	Madison, Ws.	potato	sandy loam	May 15; 2.24	N
3	Madison, Ws.	potato	loamy sand	May 15; 2.24	N
4	Madison, Ws.	potato	loamy sand	June 10; 2.24	N
5	Madison, Ws.	potato	loamy sand	May 15; 2.24	M
6	Madison, Ws.	potato	loamy sand	June 10; 2.24	M

* Irrigation schedule: N = no irrigation; M = medium irrigation; 24 cm/yr average

Raleigh, North Carolina, was similar to the historical average for Raleigh, 117 cm/yr (21). An equal amount of water recharged as did evapotranspire, 52 cm/yr (48% of precipitation). These water balance results differ from the calibration water balance, which showed only 14 cm/yr recharge (20% of precipitation) for a simulation between the months of May and November. The major reason for this is that 42 cm/yr of precipitation typically falls between December and April (21) and most of this would percolate rather than evapotranspire in the cool winter months in North Carolina. Another possible reason has to do with the use of pan evaporation in the estimation of evapotranspiration. For all calibration scenarios, pan evaporation or actual on-site evapotranspiration was supplied and used to estimate evapotranspiration, while an empirical formula used air temperature to estimate evapotranspiration for the leaching assessment exercises. It has been speculated that this empirical formula will underestimate evapotranspiration. Testing with PRZM showed this to be true; however, only for hot summer months typical of the Southeast (22).

The generated rainfall in Wisconsin, 80 cm/yr, was similar to the historical average for Madison, Wisconsin, 76 cm/yr (21). Recharge equalled 35% (28 cm/yr) of incoming precipitation for the sandy loam, while evapotranspiration equalled 58% (46 cm/yr) of precipitation. On the other hand, recharge comprised a larger proportion of incoming rainfall in the loamy sand soil. In the 5-year simulations, recharge and evapotranspiration were 56% (45 cm/yr) and 43% (34 cm/yr) of incoming water, respectively. The 24 cm of annual irrigation on the loamy sand increased consumptive use of water by the potato crop by 6 cm and recharge by 17 cm. This may have been an overestimation of recharge because the 2.5 cm/day irrigation amount may have been too high and caused recharge itself. Still, irrigation will have a tendency to increase recharge in general, not so much on the day of irrigation, but two or three days later, when a rainfall does not encounter a dry soil profile, but rather recharges through a wet profile.

Table VII. Summary of water balance for leaching assessment simulations (all units in cm/yr average).

	Leaching Assessment Run #			
Description	1	2	3,4	5,6
Precipitation	108.0	80.0	80.0	80.0
Irrigation	0.0	0.0	0.0	24.0
Runoff	2.0	6.0	1.0	2.0
Evapotranspiration	54.0	46.0	34.0	40.0
Recharge	52.0	28.0	45.0	62.0

Table VIII summarizes aldicarb fate and transport results for the 5-year leaching assessments, including the percent of applied aldicarb predicted to leach below 2 meters and soluble concentrations at 2 meters. These soluble concentrations are described as "plume" and "average". The plume concentration has been arbitrarily defined as an average of a continuous six-month period of high aldicarb concentration at the 2-meter depth, which occurs before the next application of aldicarb. This recognizes that aldicarb does, in fact, move with a plume and the plume will pass the 2-meter mark for approximately a 6 month period. For a May application, the plume usually reached the 2-meter mark by June. The "average" concentration is simply the 12-month average soluble concentration, taking into account non-plume months of low or absent concentrations of aldicarb.

As shown in Table VIII, there was essentially no potential for aldicarb to leach in the sandy loam soil modeled in Wisconsin, and very little potential to leach in the North Carolina sandy loam. The small amount of aldicarb which leached in North Carolina, 0.5% of applied, can be attributed to high water recharge, 52 cm/yr vs. 29 cm/yr in Wisconsin. One can conclude from this exercise that aldicarb will most likely not pose a leaching threat in North Carolina and Wisconsin soils similar to the sandy loam soils modeled in this study.

On the other hand, aldicarb shows the potential to leach on all scenarios of the loamy sand in Wisconsin. Between 1.4 to 18.9% of applied leached below 2 meters, the plume concentration ranged from 11 to 103 ppb, and the average concentration ranged from 8 to 64 ppb.

Table VIII. Leaching assessment results for aldicarb including: percent of applied to leach below two meters, average annual soil water concentrations at two meters, and plume concentrations at two meters*.

Description	Leaching Assessment Run # 1	2	3	4	5	6
Percent of applied to leach below two meters	0.5	0.0003	5.4	1.4	18.9	8.6
Plume soil water concentrations at two meters, ppb.	3.0	<0.1	40	11	103	53
Average soil water concentrations at two meters, ppb.	2.0	<0.1	31	8	64	33

* plume concentration = average concentration for consecutive six-month period following application
average concentration = average annual concentration

Irrigation significantly increased aldicarb leaching, by as much as 3-5 times. There are two reasons for this. One is related to the increase in recharge water discussed earlier. Intuitively, an increase in the quantity of water recharge will result in an increase in aldicarb recharge. Secondly, the irrigation event itself, while not necessarily a leaching event, will still cause aldicarb to move lower within the unsaturated zone as incoming irrigation water fills the profile.

The date of application also affected leaching, with later applications at emergence resulting in a reduction in leaching. This is due primarily to more rapid decay rate at the later application date. It is also possible that one or more storms occurred between the earlier and the later application date, resulting in more leaching for the earlier date. In any case, it is clear from this exercise that applying aldicarb at the date of potato emergence will reduce leaching of aldicarb by approximately one-half as compared to planting applications.

Summary and Conclusions

The PRZM model was calibrated to aldicarb used on three use sites comprising a total of nine distinct scenarios, which are summarized in Table III. One use site was tobacco on a sandy loam soil in North Carolina, one on potato on a sandy loam soil in Wisconsin, and one on a loamy sand soil in Wisconsin. Tables I and II show the parameters and assumptions used in the PRZM simulations and Figures 1-5 show the results of the calibration expressed as concentration-depth profiles comparing predicted and observed concentrations of aldicarb. Tables IV and V summarize the water balance and fate and transport results for these calibration scenarios, respectively. As can be seen in Figures 1-5, the PRZM model accurately portrayed the field behavior of aldicarb for the important considerations of mass balance and trends in concentration depth profiles.

The calibrated rates of decay in the unsaturated zone were equivalent to half-lives of 6 weeks in North Carolina, 10 weeks for planting applications in Wisconsin, and 6.5 weeks for emergence applications in Wisconsin. These calibrated half-lives are slightly longer than the half-lives that are calculated from soil core data alone. The half-life calculated from soil core data alone can be unrealistically short if other avenues of transport, including runoff and leaching below the depth of sampling, are not accounted for. Modeling offers the opportunity to estimate these quantities. Therefore, the "lumped" calibrated half-life is a more accurate estimate of all other avenues of pesticide dissipation, including plant uptake, volatilization, and degradation processes of photolysis, hydrolysis, and microbial decay.

The calibrated adsorption partition coefficient was 0.50 in the top 15-cm zone for North Carolina, and 1.00 in the top 15-cm zone in Wisconsin. These are higher than would be calculated based upon the k_{OC} of aldicarb and the soil organic matter of these field sites, and higher than has been used in published aldicarb modeling exercises. However, the observed data of these and other aldicarb field studies indicate that aldicarb, in fact, does stay near the surface more than would be surmised based on a calculated partition coefficient. The reasons for this are unclear, but four possible explanations were offered: 1) Top soil water percolation is overestimated due either to an underestimation of water holding capacity of the soil or an underestimation of surface runoff. 2) The granular formulation of aldicarb requires a wetting event to dissolve the protective outer cover, releasing the active ingredient. Since aldicarb is incorporated, the potential exists for localized pockets of granules to exist, and wetting events may not completely dissolve all the outer layers of the aldicarb granules immediately. 3) Atmospheric evaporation demand may result in upward movement of soil water near the soil surface, bringing with it leached aldicarb back to the surface. 4) The assumption of instantaneous equilibrium between sorbed and soluble phases is inaccurate: desorption is slower than adsorption. "Macropore" flow has been defined as rapid and deep infiltration of rainfall at the onset of a storm as the rain follows preferential flowpaths in the soil. Hence, not all the aldicarb in the surface zones will "see" the water as it flows by, and desorb according to the equilibrium assumption. It is difficult to determine which, if any, of these possibilities explain the maintainance of high aldicarb concentrations near the soil surface. Nonetheless, the high assigned partition coefficient is the way in which the model artificially duplicates this behavior.

Long term simulations were performed using identical scenarios as the calibration scenarios. The purpose was to assess trends in aldicarb leaching by examining key PRZM outputs of aldicarb mass and concentration at 2 meters below the soil surface. The results indicate that there was little to no potential for aldicarb to leach in the sandy loam soils modeled in North Carolina and Wisconsin.

On the other hand, a significant potential was indicated for leaching of aldicarb in the loamy sand modeled in Wisconsin. The percent of application to leach below 2 meters ranged from 1.4-18.9. The plume concentration (defined as an average concentratrion for a continuous 6-month period of high aldicarb concentration) ranged from 11-103 ppb, and the average annual solution concentration of aldicarb at two meters ranged from 8-64 ppb. With a Health Advisory Level of 10 ppb for aldicarb, clearly the potential for ground water contamination exists for this and similar soil types.

Irrigation increased the amount of recharge water, if not adding to recharge on the day of irrigation, than at the next rainfall, when the rain encountered a wet rather than a dry profile. Irrigation also tended to move aldicarb lower in the profile. As a result, 3-5 times as much aldicarb leached below two meters with irrigation. Applications at potato emergence decreased the amount to leach by about one-half as compared to earlier applications at planting. The reasons for this are two-fold: avoidance of some spring rain for later applications, and a more rapid rate of decay for later applications.

The appropriate direction to take at this point to complete the leaching assessment for aldicarb would be to determine the extent of the loamy sand soils in Wisconsin (and elsewhere), and define the hydrogeology of these areas in detail. Based on information such as depth to ground water, rate of ground water movement, pH of ground water, and other factors affecting transport and persistence in ground water, one can draw further conclusions as to the potential of aldicarb to contaminate the ground water in these areas.

Literature Cited

1. Cohen, S.Z.; Creeger, S.M.; Carsel, R.F.; Enfield, C.G. In "Treatment and Disposal of Pesticide Wastes"; Krueger, R.F.; Seiber, J.N., Eds.; ACS SYMPOSIUM SERIES No. 259, American Chemical Society: Washington, D.C., 1984; pp. 297-325.
2. Enfield, C.G.; Carsel, R.F.; Cohen, S.Z.; Phan, T.; Walters, D.M. Ground Water 1982, 20(6), 711-22.

3. Carsel, R.F.; Mulkey, L.A.; Lorber, M.N.; Baskin, L.B. Ecological Modeling 1985 (in press).
4. Carsel, R.F.; Smith, C.N.; Mulkey, L.A.; Dean, J.D.; Jowise, P. "Users Manual for the Pesticide Root Zone Model (PRZM). Release I." EPA Research Report EPA-600/3-84-109, 1984.
5. Intera Environmental Consultants, Inc. "Mathematical Simulation of Aldicarb Behavior on Long Island: Unsaturated Flow and Ground-Water Transport"; 1980.
6. Anderson, M.P. "Field Validation of Groundwater Models", 189th American Chemical Society National Meeting, PEST 57, Florida, 1985.
7. Jones, R.L.; Hansen, J.L.; Romine, R.R.; Marquardt, T.E. "Movement of Aldicarb and Aldoxycarb Residues in Soil". To be submitted for publication, 1985.
8. Wyman, J.A.; Jensen, J.S.; Curwen, D.; Jones, R.L.; Marquardt, T.E. Env. Tox. and Chem. 1985, 4, 641-651.
9. Wyman, J.A.; Medina, J.; Curwen, D.; Hansen, J.S.; Jones, R.L. "Movement of Aldicarb and Aldoxycarb Residues in Soil". To be submitted for publication, 1985.
10. Cohen, S.Z.; Eiden, C.; Lorber, M.N. "Ground-Water Monitoring of Pesticides in the USA", 189th American Chemical Society National Meeting, PEST 34, Florida, 1985.
11. "SCS National Engineering Handbook," U.S. Department of Agriculture, Soil Conservation Service, 1971.
12. Jones, R.L.; Back, R.C. Env. Tox. and Chem. 1984, 3, 9-20.
13. Jones, R.L.; "Movement and Degradation of Aldicarb Residues in Soil and Ground Water" presented at the Society of Environmental Toxicology and Chemistry Conference on Multidisciplinary Approaches to Environmental Problems, Crystal City, Va, 1983.
14. Jones, R.L., personal communication.
15. Green, R.E.; Liu, C.C.K.; Tamraker, N. "Modeling Pesticide Movement in the Unsaturated Zone in Hawaii Soils Under Agricultural Use", 189th American Chemical Society National Meeting, PEST 54, Florida, 1985.
16. "Health Advisory Program. Part 7," Office of Drinking Water, U.S. EPA.
17. Cotruvo, J.A.; Director, Criteria and Standards Division, Office of Drinking Water, U.S. EPA, letter to Hank Weiss, Dept. of Health and Social Service, Madison, Wisconsin, dated May 30, 1980.
18. "Proposed Rules Relating to Special Restrictions on the Use in Wisconsin of Pesticides Containing Aldicarb," State of Wisconsin, Department of Agriculture, Trade and and Consumer Protection, proposed S. Ag. 29.17, Wis. Adm. Code, submitted October 1982.
19. Mahfouz, H.Z.; Moran, D.; Harris, D. American Journal of Public Health 1982, 72, 1391-95.
20. Richardson, C.W. Water Resources Research 1981, 17, 182-90.
21. "Climates of the States," National Oceanic and Atmospheric Administration, 1978.
22. Carsel, R.F., personal communication.

RECEIVED April 1, 1986

19

Modeling Pesticide Movement in the Unsaturated Zone of Hawaiian Soils under Agricultural Use

R. E. Green[1], C. C. K. Liu[2], and N. Tamrakar[2]

[1]Department of Agronomy and Soil Science, University of Hawaii, Honolulu, HI 96822
[2]Department of Civil Engineering, University of Hawaii, Honolulu, HI 96822

Two instances of pesticide leaching to groundwater have been examined by modeling approaches which incorporated appropriate pesticide inputs to the soil, measured pesticide sorption coefficients and water balance calculations based on actual rainfall and estimated evapotranspiration and runoff. First, an analytical model was used to examine movement of DBCP which was accidentally spilled near a deep well on Oahu. The model features the convective-dispersive solute transport equation with an upper boundary condition which approximates volatile losses over time from a soil layer near the surface. The second study examined DBCP movement under pineapple culture. The analysis combined a two-dimensional analytical model, to compute the time varying distribution and dissipation of DBCP in the plow layer, and a one-dimensional numerical model to simulate both water and DBCP movement in a layered profile below the surface layer. Both modeling efforts were only moderately successful in predicting pesticide concentration profiles, as evidenced by a comparison of computed and measured DBCP concentrations with depth, but the results were useful in assessing the impact of key processes and in identifying data requirements for future modeling efforts. Model calculations and laboratory experiments both showed that sorption coefficients for long-term DBCP residues were much larger than was indicated by conventional batch equilibration measurements. Thus, successful prediction of the leaching and volatilization of DBCP residues requires a better understanding of sorption mechanisms.

The detection of DBCP and other toxic organics in Hawaii well waters has led to modeling studies with the objective of analyzing the interaction of chemicals and soils in relation to their properties and the dynamic processes involved in pesticide movement to groundwater. It was anticipated that these studies would (a) explain

0097-6156/86/0315-0366$06.00/0
© 1986 American Chemical Society

why DBCP moved to groundwater when it was not previously considered a likely contaminant, (b) assist in the assessment of the impact of present DBCP resides on future water contamination, and (c) provide a foundation for modeling chemical movement to groundwater in the context of Hawaii's unique hydrogeology and agricultural chemical practices. This paper summarizes and integrates the results of modeling studies and field monitoring and laboratory measurements which have or will be reported in detail elsewhere (1-3). It is hoped that the apparent inadequacies of the models used in these studies will encourage the development of improved mathematical descriptions of actual processes in the field. Only then can dynamic modeling of pesticide behavior in soils become a useful part of risk assessment procedures which will assist in regulating pesticide use in Hawaii in the context of groundwater quality requirements.

Background

Until recent years concerns about contamination of water by pesticides in Hawaii focused primarily on surface waters (4,5). Assessments of potential contamination of groundwater were limited to chemicals used in sugarcane culture. While nitrates from fertilizers were known to reach groundwater in irrigated sugarcane areas, the likelihood of herbicides reaching basal waters was considered remote (6). In retrospect, the failure of researchers to examine the leachability of volatile nematicides used in pineapple culture was short-sighted, but nematicide residues in soils were considered insignificant in relation to the amounts applied.

The detection of widespread DBCP contamination of wells in California in 1979 led to the monitoring of Hawaii wells for toxic organics in 1980. Contamination of basal waters by DBCP and EDB was first detected in a well supplying potable water near Kunia village on Oahu; subsequently DBCP was detected in several wells on Oahu and in one well on Maui island. A program of soil boring was initiated in 1980, first in the vicinity of the Kunia well (1,7), and later in 1983, in seven pineapple fields on Oahu (8) and three pineapple fields on Maui (2). The availability of measured DBCP concentration profiles at the Kunia spill site and on pineapple fields on Maui encouraged the modeling efforts which are summarized in this paper.

Modeling Strategy

The sequence of models used in these studies constituted a progression from a simple analytical model of the convection-dispersion type with fixed parameters, associated with the assumption of a semi-infinite homogenous profile, to a convection-dispersion numerical model, which incorporated dynamic water and solute movement through a multilayered profile. In brief, the sequence was as follows: (a) one-dimensional analytical model with an upper boundary condition that allowed first-order decrease in concentration over time (Kunia site); (b) two-dimensional numerical model to calculate the approximate short-term vertical concentration profile in the plow layer for shank-injected DBCP; the results were used to specify initial conditions in a subsequent one-dimenstional numerical model (Maui field sites); (c) simple one-dimensional analytical model with an upper boundary condition which provides for DBCP vapor loss to the atmosphere over time; this model was used to estimate an appropriate value of K_d for the plow layer of the Maui fields by calibration of the model with measured field data; (d) a dual-model approach in which the analytical model in (c) above was used to calculate the concentration-time relationship at the bottom of the zone of DBCP application; this output was used in defining the upper boundary condition for the one-dimensional analytical model of (a) above, allowing the use of different K_d values in the zone of application and below the plow layer; (e) one-dimensional numerical model incorporating both water and solute movement through multiple layers, with K_d adjusted with depth.

All of the models used in these studies are based upon the convection-dispersion equation for solute transport through porous media and thus are constrained by the inherent limitations of this mathematical representation of actual processes. These limitations, analyzed in some detail in a number of recent papers (9,10,11,12,13), are real for many field conditions. On the other hand, alternative approaches (e.g. stochastic transfer models) are still in an early state of development for solute transport applications. Consequently, we have initiated our modeling efforts with the traditional transport equations. Hopefully, improved approaches will be developed in the near future.

Each of the models identified in the above sequence is described briefly below with only the most important features being noted. Detailed descriptions are given in the referenced papers.

<u>One-Dimensional Analytical Model With 1st-Order Loss At Upper Boundary</u>. The unique initial and boundary conditions associated with the Kunia pesticide spill (1,7) required that the surface layer, containing the concentrated pesticide, lose DBCP rapidly by volatilization. Thus an upper boundary was established such that

$$C(z,t) = C_o e^{\gamma t} \qquad z = 0,\ t>0 \qquad (1)$$

where C is the solute concentration in solution $[ML^{-3}]$, C_o is the concentration when downward movement starts, z and t are the space and time variables, respectively, and γ is an empirical source decay coefficient $[T^{-1}]$. Loss of DBCP was assumed to occur by volatilization only. The convection-dispersion transport equation was used to describe DBCP movement in the soil solution, viz.

$$(\rho K_d + \theta)(\partial C/\partial t) = \theta D(\partial^2 C/\partial z^2) - q\,\partial C/\partial z - k'C \qquad (2)$$

in which ρ is the soil bulk density, K_d is the sorption distribution coefficient, θ is the soil water content, D is the diffusion-dispersion coefficient, q is the Darcy water flux, and k' is the first-order rate coefficient, with $k'C = kC_T$, where $k' = k(\rho K_d + \theta)$ and kC_T denotes the first-order decay of the total concentration of pesticide residual in the soil, i.e. $kC_T = k(\rho K_D + \theta)C$. In order to use the simple analytical model, constant values of θ , D and q are used. The adequacy of this simplifying approach depends mainly on the assumed value of q which indicates an average water flux over a long period.

The system equation along with boundary and initial conditions were solved analytically using a Laplace integral transform (1); a computer program by Cleary (14) was modified for use in this study.

Two-Dimensional Numerical Model. Application of DBCP by shank injection in a pineapple bed (Maui fields) provides a line source of pesticide at a depth of about 30 cm which cannot be modeled in a one-dimensional mode without first evaluating the impact of initial concentration distribution. The geometry of the system is indicated by the cross-sectional diagram of two pineapple beds in Figure 1. The section of Figure 1 delineated by AA-BB represents a typical section. The objective was to model the movement of DBCP from an assumed source zone (8 cm wide x 10 cm deep section centered at 30 cm depth) to determine the approximate time required to justify the assumption of a uniform lateral distribution of DBCP in the plow layer for subsequent one-dimensional modeling. An appropriate model and numerical solution for this case was given by Hemwell (15) and the details of the present application are presented elsewhere (3). In brief, the two-dimensional equation solved is

$$\partial C/\partial t = D'(\partial^2 C/\partial x^2 + \partial^2 C/\partial z^2) - kC \tag{3}$$

in which $D' = D/(\rho K_d + \theta + \underline{a}K_H)$ and $\underline{a}$ is the air-filled porosity, K_H is Henry's Law constant [L^3 solution/L^3 gas] and the other variables are as defined previously. The boundary condition at the soil surface (z=0) is defined by

$$-D'(\partial C/\partial z) = -(D_G^{air}/R_G d)\ C \tag{4}$$

in which D_G^{air} is the vapor-air diffusion coefficient, d is the thickness of the stagnant air layer, and $R_G = (\rho K_D + \theta + \underline{a}\ K_H)/K_H$. The initial concentration of pesticide in the zone of application was calculated from the amount applied (with assumed 10% application loss), giving 3.54×10^5 ppb for Maui Pineapple Company Field 234 and 2.96×10^5 ppb for Field 210.

One-Dimensional Analytical Model With Diffusive Vapor Loss At Upper Boundary. This model was developed by Jury et al. (16) to provide a computational method for classifying organic chemicals for their relative susceptibility to different loss pathways (volatilization, leaching and degradation). Although the basic equation is essentially the same as Equation 2, in contrast to Equation 2 it includes transport in both the vapor and liquid phases. An effective diffusion coefficient, D_E, is defined such that it includes both the vapor component, $K_H D_G$, and liquid component, D_L, in the following manner:

$$D_E = (K_H D_G + D_L)/R_L$$

where $R_L = \rho K_D + \theta + \underline{a}\ K_H$. The initially applied chemical was assumed (based on 2-D calculations) to be uniformly distributed in the top layer of soil of depth L, giving an initial condition of $C(z,0) = C_o$ for $0 < z \leq L$ and $C(z,0) = 0$ for $z > L$.

Although this simple analytical model could not be expected to describe DBCP movement through a soil profile which varied several fold in organic carbon content (and hence in K_d) over a vertical distance of even a few meters, it was considered adequate to compute

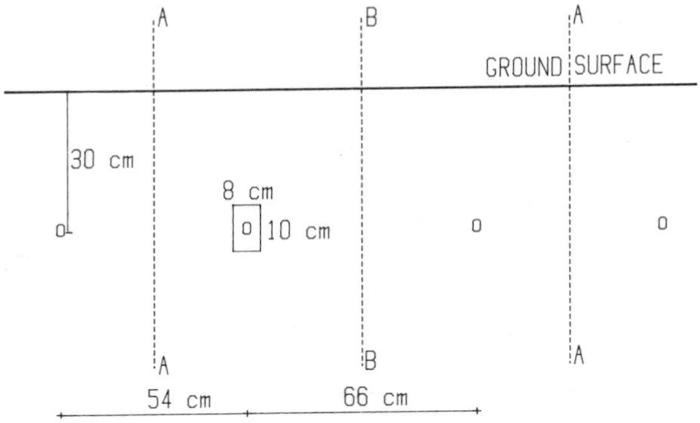

Figure 1. Cross-Sectional Diagram of Shank-Injected DBCP in Two Pineapple Beds.

chemical movement in the tilled surface layer, as was done with the two-dimensional model. The model was used in a calibration mode to determine an apparent K_d value for the surface layer. Using the estimate of K_d obtained by calibration of the model with measured concentration at a depth of 30 cm, the model was used to calculate the total concentration of DBCP at z = 36 cm over time. The results of these calculations were later used in the dual-model approach described below.

Dual Analytical Model Approach. This simulation involved the sequential use of two one-dimensional analytical models for the purpose of incorporatiang different parameters and boundary conditions for the upper and lower soil zones. First the model of Jury et al. (16) was applied to the surface layer as described above. The computed concentrations of DBCP at 36 cm over time were then fitted to a first order equation so that Equation 1 could be used as input for the model of Liu et al. (1), which was applied to the lower layer.

One-Dimensional Numerical Model. The equations are in finite difference form and are based on the model described by Selim and Iskandar (17). The water flow equation was solved in terms of soil water pressure head, h, rather than water content, θ, giving

$$Cap(h)\ \partial h/\partial t = \partial/\partial z\ (K\partial h/\partial z) + \partial K/\partial z \qquad (5)$$

in which K is the hydraulic conductivity (a function of h) and Cap(h) = dθ/dh is the water capacity term. The solution, stability conditions and other details are given elsewhere (3). The numerical solution to the water movement model was tested against the experimental data of Warrick et al. (18).

The accompanying solute transport equation is the finite difference equivalent of Equation 2. The upper boundary condition is represented by a flux equation,

$$-\theta D\ \partial C/\partial z + vC = -(D_G^{air}/d)\ K_H C \qquad (6).$$

The model was formulated with three or four layers in an attempt to represent changes in K_d with depth. Details of the solution and a listing of subroutines are given by Tamrakar (3).

Chemical and Hydraulic Inputs

Kunia. To determine initial conditions for the Kunia spill it was necessary to estimate the area of the spill and calculate the approximate depth of pesticide penetration. A simple mass balance analysis indicated that the soil solution was saturated with DBCP to a depth of 5 cm soon after the spill. We estimated volatilization loss of DBCP in a three-day period prior to the first rainfall with a first-order decay coefficient of 1.26 day^{-1} (19). The calculated initial solution concentration in the soil solution at the first rainfall was 3480 ppb. Clearly, this quantity could be in error, so that subsequent calculated concentrations are suspect.

An average water flux for the Kunia site for the 3-year period after the spill was calculated from a simple water balance, using local estimates of runoff and evaporation, during and between each rainfall event to get infiltration (1). This highly aggregated soil was considered to have an average effective water content of 0.15 cm^3/cm^3 contributing to solute transport.

Maui. DBCP was applied by shank injection as illustrated in Figure 1 at 75.6 kg/ha in Field 234 and 46.9 kg/ha in Field 210. When a 10% application loss was assumed, the initial concentration in the small rectangular zone of application shown in Figure 1 was 3.54×10^5 ppb for Field 234 and 2.96×10^5 ppb for Field 210. The two-dimensional model was calibrated with field data for Field 234 (two weeks post-application) and then used to calculate the average DBCP concentration in the plow layer for subsequent specification of initial conditions for the one-dimensional models.

The water recharge for the Maui fields was calculated with a simple hydrologic balance for each field using rainfall data from nearby rain gauges. Both fields were non-irrigated. A rainfall-runoff relationship was developed for each field using the USDA-Soil Conservation Service "curve number" approach. The validity of this approach for Hawaii conditions was demonstrated by Cooley and Lane (20). Potential evapotranspiration from pineapple fields was assumed to be one-third of pan evaporation throughout the study period, based on results of Ekern (21), but actual ET was adjusted for the water input for each month (3).

Water balance calculations for each month resulted in cumulative infiltration amounts of 29.4 cm for Field 234 (6-month period) and 541 cm for Field 210 (3.5 years). For the analytical models, average water fluxes were calculated from the cumulative infiltration and elapsed time. For the one-dimensional numerical model the rainfall for each rainy day was assumed to be distributed uniformly throughout the day. The stability condition for the water flow submodel requires that the time increment be small during the period of infiltration, resulting in large computation times for each rainfall event. To reduce computation costs for the 3.5 year simulation of Field 210 the rainfall for each month was lumped as a single rainfall event, occuring for a period of one or two days, depending on the total rainfall in the month. A sensitivity analysis indicated this approximation was justified (3). Another feature of the program which reduces computational time is a limitation on when downward movement of DBCP can occur: only during the period of infiltration and significant redistribution. At other times the pesticide is assumed to remain immobile, except in the top layer in which DBCP volatilization occurs.

Parameter Estimation

A major limitation of solute transport simulation (even when the model represents field processes well) is the difficulty of obtaining good estimates of key parameters. Some aspects of this problem will be considered in the Discussion section later. Fundamental parameters identified in Equations 1, 3, 4, and 6 which require specification are: K_d, D, k, D_G^{air} K_H and ρ . Most other parameters are derived from these, with the exception of the parameters required to define hydraulic conductivity and water content as functions of soil water pressure head, for the water flow equation in the numerical model. The estimation of the fundamental parameters will be discussed briefly; details are given in references (1) and (3).

Sorption Distribution Coefficient. Measurements of DBCP adsorption on soils from the Kunia site at solution concentrations ranging from about 0.25 μg/ml to 290 μg/ml indicated that a linear isotherm described sorption reasonably well; when sorption data were fitted with the Freundlich equation, $S = K_f C_e^N$, the values of N on soils from three depths were 0.92, 0.76 and 0.95 (1). Subsequently, sorption was determined at only one initial concentration, 0.35 μg/ml, and the equation $S = K_d C$ was assumed to apply. Sorption was measured by the batch-suspension method (22) for most samples; for deep saprolite samples having very low sorption a flow-equilibration technique (22) provided improved sensitivity and precision.

In addition to the conventional sorption distribution coefficients which were based on the addition of a solution of known DBCP concentration to soil, desorption coefficients were measured by desorbing DBCP residues into water with a 3-hour equilibration period (2). When distribution coefficients determined by these two methods were found to be very different, estimates of K_d were also obtained by calibrating the one-dimensional analytical model to field data. The various estimates of K_d are compared in the Results section.

Solute Diffusion-Dispersion Coefficient. In these studies D was assumed to be constant and not a function of flow velocity. For low flow velocities this assumption seems reasonably valid (23). For the natural rainfall conditons of these studies the flow velocities were generally quite low. In lieu of actual measured D values on the two Maui soils, an estimate of D obtained on a somewhat similar soil on Oahu by Khan (24) was used in this study. An average value of D = 0.6 cm^2/hr was measured in the field with a steady water flux of 10 cm/day.

First-Order Degradation Rate Coefficient. DBCP is highly resistant to biodegradation and is also stable against hydrolysis. We have no measurements of DBCP degradation in Hawaii soils, but Williams et al. (2) present evidence that DBCP is debrominated in the root zone of pineapple. Assuming 15 percent DBCP being debrominated within two weeks, the decay rate was calculated to be 4.8×10^{-4} hr^{-1} (3), which is similar to the value 4.2×10^{-4} hr^{-1} obtained for DBCP in soil by Castro and Belser (25). Degradation losses of DBCP are probably sufficiently small relative to volatilization losses to be considered insignificant in the present modeling context.

Vapor and Liquid Diffusion Coefficients. The values of 247.1 cm^2/hr for D_G^{air} and 6.184×10^{-5} m^2/day for D_L^{water} were calculated by methods utilizing chemical properties (3). Other approximations suggested by Jury et al. were used to obtain the effective diffusion coefficient, D_E, the vapor phase diffusion coefficient, D_G, and the liquid phase diffusion coefficient, D_L, for soils. Using the equation $D_{EF} = D_G K_H + D_L$ for the Maui soils with 59% porosity and 30% water, a value of $D_{EF} = 0.195$ cm^2/hr was obtained (3). The calculated vapor phase diffusion coefficient, D_G, for these same conditons was 11.5 cm^2/hr. Experimental measurements on a similar Hawaii soil by Pringle et al. (26) gave a DBCP liquid-vapor diffusion coefficient of 9.5 cm^2/hr by the method of Jury et al. (27), suggesting that the estimation methods used to obtain diffusion coefficients are reasonably valid.

Henry's Law Constant. The calculation of K_H' from vapor pressure, P, and solubility, S, at a given temperature, as suggested by Mackay and Wolkoff (28), resulted in K_H' = 0.3558 atm - L/mole at 20°C. For use as a partition coefficient K_H is best defined as a ratio of concentration in the vapor phase to concentration in the liquid phase, $K_H = C_V/C_L$. The value obtained for DBCP was K_H = 0.0148 (3) with dimensions of [L^3 liquid/L^3 vapor].

Other Parameters Defining Soil Properties. In lieu of measured data on the Maui soils, the hydraulic properties and bulk density for the Wahiawa soil on Oahu were used in numerical simulations (3). Experimental results (29) for the soil depths of 20, 60 and 90 cm were used in simulations of a three layer system with a tilled layer of 45 cm depth, a subsoil layer 45 to 90 cm, and a substratum layer at depths exceeding 90 cm. Bulk density values corresponding to these three layers were 1.25, 1.25 and 1.4 g/cm^3.

DBCP Residues in Field Soils

Kunia. Background information on the Kunia site and core sampling were reported by Mink (7). The spill area has an elevation of 260 m. The soil is in the Kunia series, of the Ustoxic Humitropept subgroup; kaolin clays dominate the mineral fraction. A three-meter soil profile is underlain by weathered basalt (saprolite) to a depth of 46 meters and unweathered basalt continues to the basal water table. At each sampling location, sixteen cores 45 cm long were taken from each coring hole to a total depth of about 16 meters (52 feet).

Maui. Data from two of three Maui Pineapple Company fields which were sampled for DBCP movement were selected for the modeling study. Both fields are located on East Maui at elevations below 185 m, and the soils, Pauwela clay for Field 210 and Hamakuapoko clay for Field 234, are members of the taxonomic group Tropohumults. These soils are highly weathered, with low-activity kaolin and oxide minerals. Field 210 had received only one application of DBCP 3.5 years prior to the deep borings which were obtained at two locations in the field. Field 234 had received six pre-plant applications of DBCP since 1958, the last being six months prior to soil borings taken at five locations in the field. In addition to the soil borings (taken to a maximum depth of 19 m) samples were also taken by hand two weeks after application from a 2-meter deep trench which had been dug with a backhoe. Other details of sampling and sample preparation are given by Williams et al (2).

Analysis of DBCP in soil samples was by co-distillation with benzene and analysis of the benzene by electron-capture gas chromatography.

Results

While we have used various models, both analytical and numerical, in this preliminary modeling effort, the principal focus of the analysis of results here is on the way in which the one-dimensional models reflect the impact of sorption changes with depth in the soil profile. Thus we observed the performance of the models using measured K_d values from various soil depths, and also used the analytical model of Jury et al. (16) to derive K_d values for the surface layer by calibration of the model with measured field concentrations.

DBCP Sorption on Soil and Saprolite. For the Kunia site, sorption K_d values were measured on samples from several depths in Boreholes 2 and 3 as well as from three shallow depths near the well. The borehole sorption data, obtained by flow equilibration, are given in Table I. Batch equilibration gave results which were about 20% higher for samples with the highest sorption, but the batch method had inadequate precision when sorption was low, as for samples 2-1, 2-3 and 3-2. The precision of the flow-equilibration method for DBCP is limited only by the GC analysis. Zero values in Table I indicate sorption K_d < 0.01 ml/gm. Of the several borehole samples, only sample 3-1 showed substantial sorption of DBCP. In both boreholes there was little sorption below 1 meter. The diminished sorption of DBCP with increasing depth appears to be related principally to a decrease in organic carbon with depth.

Table I. DBCP Sorption by Flow-Equilibration on Soil from Two Boreholes at the Kunia Site, Oahu, Hawaii

Borehole Sample	Depth (meter)	Organic Carbon (%)	K_d (ml/gm)
2-1	0.5	2.3	0.01
2-3	2.3	0.6	0.02
2-7	5.9	0.3	0
3-1	0.5	4.1	1.94
3-2	1.4	1.3	0.06
3-3	2.3	0.8	0
3-5	4.1	0.4	0
3-7	5.9	0.2	0
3-9	7.8	0.2	0

Sorption K_d values for Maui Field 234 are given in Table II. These data are selected from sorption values measured by Williams et al (2) to show the large difference in K_d values obtained by the conventional batch equilibration method and by desorption of residual DBCP in field samples. Desorption K_d values are clearly several times larger than those obtained by batch equilibration. Although it might be expected that desorption equilibration times much longer that those used in these tests (3 hours) might have resulted in smaller differences, recent measurements using 24-hour sorption and desorption equilibrations also indicated a several-fold difference in K_d between sorption of DBCP from recently added DBCP solution and desorption of DBCP added to dry soil a few weeks earlier (30).

Table II. Comparison of DBCP K_d Values from Conventional Batch Equilibration and Desorption of Field Residues (Maui Pineapple Co. Field 234)

Soil Core No.	Depth (m)	K_d (ml/gm) Sorption	Desorption
B	0.3	0.60	18.0
B	1.2	0	3.4
B	3.0	0	3.8
E	0.3	0.46	24.3

The large effective desorption K_d for DBCP suggests that DBCP residues in the surface soil are much less mobile than is indicated by the conventional sorption measurement.

One-Dimensional Simulation of DBCP Movement at Kunia. DBCP distribution three years after the pesticide spill was simulated by the one-dimensional analytical model with exponential decay source term at the surface (1); predicted concentrations are shown in Figure 2. Measured concentrations from Boreholes 2, 3 and 5 are also shown in Figure 2 for comparison with simulated results. The three measured DBCP concentration profiles are quite variable, both in shape and in magnitude of concentrations. The reason for the variation in measured profiles is not known, but may be due to differences in the amount of DBCP which entered the soil at each location and variation in soil properties between borehole sites. There appears to be a correlation between the sorption values in Table 1 for Boreholes 2 and 3 and the retention of DBCP near the surface at these two locations, ie. high sorption in the surface soil at site 3 resulted in high retention of DBCP, in contrast to site 2.

The simulated concentration profile in Figure 2 was obtained using a weighted sorption coefficient (K_d = 0.15) for the 0 to 7 meter zone of Borehole 3. Use of a weighted value is an undesirable simplification, especially when the sorption coefficient changes so rapidly with depth (Table I). However, we were constrained by the analytical model to use a single sorption coefficient for the entire profile. The simulated DBCP concentration profile has the shape of Borehole 5 data and the concentration range of Borehole 3 data.

Simulated DBCP Movement in Plow Layer -- Two Dimensional Numerical Model. Of the several input parameters required for Equations 3 and 4 (Hemwell model) the most difficult to specify with confidence are K_d and D. Our objective in using the 2-D model was to calculate initial DBCP concentration in the upper layer in order to specify initial conditions for the one-dimensional models used to predict deep penetration of DBCP. For this reason the 2-D model was calibrated with field data by adjusting K_d and D within reasonable limits. Combinations of parameters used and the results are shown in Table III. The use of K_d = 4.10 was based on a calibration of the Jury et al. analytical model with field surface layer data six months after application as described earlier. This value is obviously too high for recently applied DBCP at the extremely high concentration existing soon after application. The combination of K_d = 0.1 ml/gm and D = 1.05 cm^2/hr gave results which matched the field measured average concentration at 30 cm two weeks after application. The concentration profile for Field 234, based on the average computed concentration at each depth, is shown in Figure 3. Tamrakar et al. (3) show also the plot for an uncovered surface.

Table III. Comparison of 2-D Model and Field Data, Maui Pineapple Field 234, Two Weeks after Application

K_d (ml/gm)	D (cm^2/hr)	DBCP Loss in 2 Weeks Simulated (%)	DBCP Loss in 2 Weeks Measured (%)	Average DBCP Conc. in Soil at 30 cm. Simulated (ppb)	Average DBCP Conc. in Soil at 30 cm. Measured (ppb)
4.10	0.20	0	75	32200	3408
0.10	0.20	20	75	12940	3408
0.10	1.05	76	75	3580	3408

Dual Analytical 1-D Model. A comparison of calculated and measured DBCP concentration profiles for Field 210 is shown in Figure 4. The K_d parameter for the surface 36 cm used in this simulation was the calibration value obtained with the Jury et al. model for Field 234 surface soil, K_d = 4.1. The two fields were thought to be sufficiently similar to use the same value; measured K_d values were about 0.6 ml/gm for surface soil in both fields (2). For soil below 36 cm, K_d = 0.56 ml/gm was used, even though this value was initially thought to be more appropriate for the surface layer than for deeper layers (Table II). The comparison between predicted and measured data in Figure 4 does not encourage the use of analytical models to predict residual DBCP concentrations in the soil profile, even with accomodation for parameter changes between the upper and lower layers. The model predicts leaching of DBCP from the top layer in a broad zone, with a peak at about 400 cm; no such peak is evident in the measured results. The limitations of the analytical model encouraged development of the numerical model which would allow more detailed specification of soil properties with depth.

Numerical 1-D Model Results. Simulation of DBCP movement was accomplished with the DBCP application and hydrologic inputs for both Maui fields. The measured DBCP distributions shown in the following figures represent average values for five sets of core samples in Field 234 and two sets in Field 210. Although the concentrations between sites within a given field varied up to several fold, especially at the shallowest depths (0.3 and 1.2 m), the shape of the concentraion profile was consistent between sampling sites.

Initial concentrations in the top 36 cm were obtained from the 2-D simulations discussed earlier. The absence of significant rainfall during the two weeks after DBCP application allowed this approach. One-dimensional numerical simulations indicated little difference in computed profiles whether initial distribution of DBCP

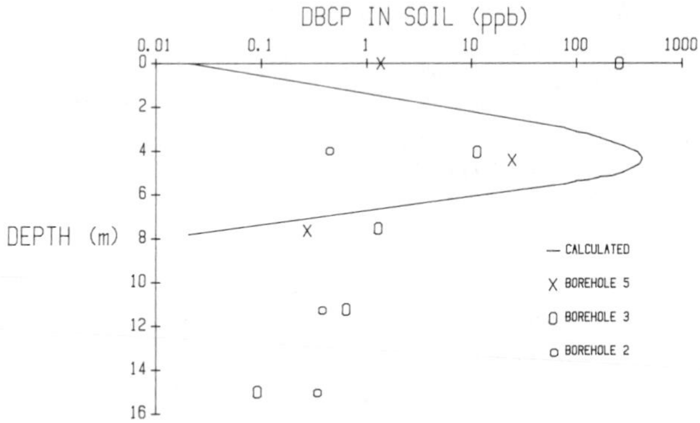

Figure 2. Simulated (Curve) and Measured (Data Points) DBCP Concentrations in Soil at the Kunia, Oahu Site.

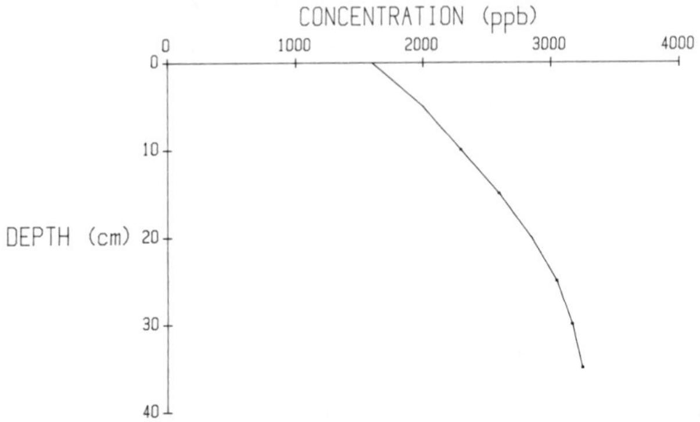

Figure 3. Simulated Average Lateral DBCP Concentrations (Dry Soil Basis) in Maui Pineapple Field 234 Two Weeks After Application; Surface 43% Covered; 2-D Numerical Model.

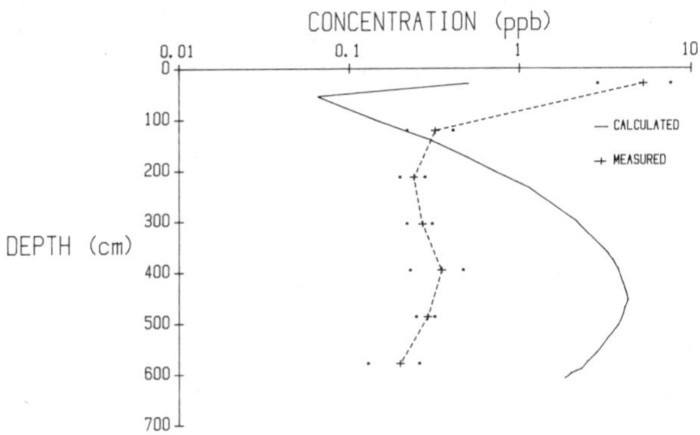

Figure 4. Simulated (Dual-Model Approach) and Measured DBCP Concentrations (Dry Soil Basis) 3.5 years after Application, Maui Pineapple Field 210. Average deviation is indicated for mean of two boreholes.

was as shown in Figure 3 or uniformly distributed in the top 36 cm (3). For Field 234, DBCP residues existed in the profile prior to the DBCP application considered in the simulation. The pre-application concentrations at depths of 0.3, 1, 1.5 and 2.1 meters were 22, 6, 13, and 11 ppb. There were no residues in Field 210 prior to the application considered in the simulation.

Initially, numerical simulations were done for both fields with three soil layers having depth intervals of 0 to 45 cm, 45 to 90 cm, and greater than 90 cm. Input parameters included K_d values of 4.1, 4.1 and 0.12 ml/gm and k (decay rate) values of 1.6×10^{-4}, 1.4×10^{-4} and 6×10^{-5} day^{-1} for the three layers, respectively. Other parameter values were defined earlier in the paper.

Simulated results for Field 234 are shown in Figure 5. The simulated concentrations to a depth of about 100 cm are reasonably close to measured values, partly as a result of using a K_d value in the upper profile which had been obtained by calibration of the 1-D analytical model with Field 234 data. The secondary simulated peak at 200 cm is apparently due to the pre-application DBCP residue, which was not actually as mobile as the simulation suggests.

Use of the same sorption coefficients for Field 210 as were used in the Field 234 simulation resulted in concentrations in the surface layer which were about an order of magnitude too low and concentrations at greater depths which were an order too high. Thus it was not possible to use the K_d obtained from Field 234 for Field 210. Subsequently, the 1-D analytical model was used to determine by calibration an appropriate K_d value for Field 210; the result was $K_d = 17.0$. Thus for the numerical simulation, $K_d = 17$ was used for 0 to 90 cm and $K_d = 0.12$ was assigned for depths greater than 90 cm. The result of simulation with these values is shown in Figure 6. The effect of the abrupt change in K_d at 90 cm is evident in the simulated result. An additional simulation was done using four layers with K_d values of 17, 12 and 1.8 for the first three layers (each 45 cm thick) and 0.12 for the underlying layer. The resulting simulated concentration profile, shown in Figure 7, indicates a considerable improvement in the performance of the model when the sorption variation with depth is adequatedly represented by the model. It is of interest to note that the measured K_d values for Field 210 were 0.64, 0.11 and 0.074 ml/gm for 0.3, 1.2 and 2.1 meters, respectively (2). Thus the K_d values required to give a reasonable simulation are much larger than those measured by the conventional batch equilibration.

Discussion and Conclusions

The modeling exercises reported here cannot be considered rigorous tests of the suitability of the models examined for two principal reasons: (a) key input parameters were not adequately quantified, and (b) measured DBCP concentration profiles varied considerably within a given field. On the otherhand, some general observations as to the apparent suitability of the various models are justified.

The simple analytical model is inadequate principally because it cannot accomodate changes in soil properties with depth. Measured sorption coefficients decreased strikingly with depth at both the Kunia and Maui sites. The attempt to incorporate some change in sorption with depth by using the dual analytical model did not result

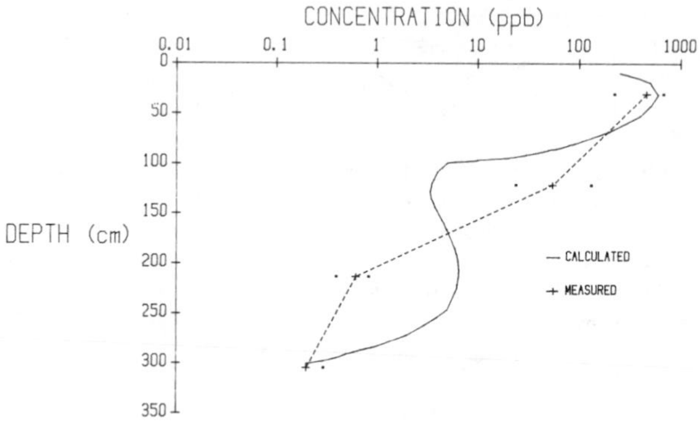

Figure 5. Numerical Simulation (Three Soil Layers) of DBCP Profile Six Months after Application, Maui Pineapple Field 234. Average deviation is indicated for mean of five boreholes.

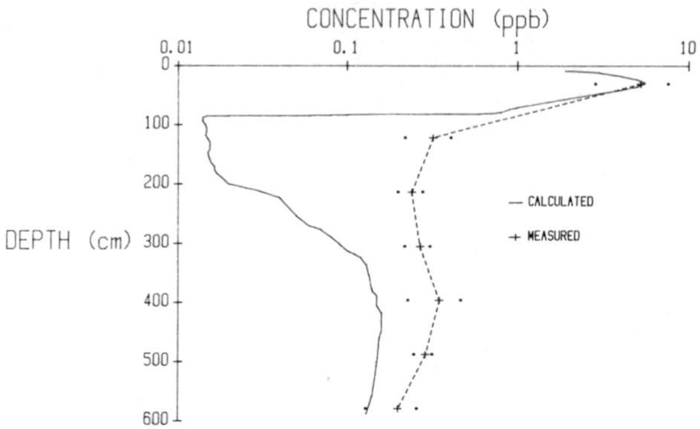

Figure 6. Numerical Simulation (Three Soil Layers) of DBCP Profile 3.5 Years after Application, Maui Pineapple Field 210. Average deviation is indicated for mean of two boreholes.

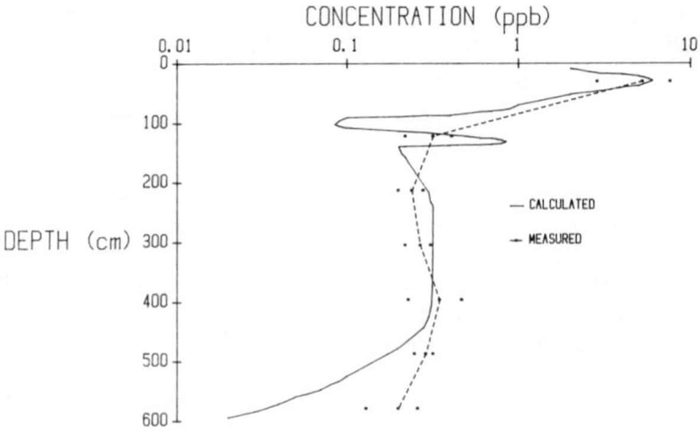

Figure 7. Numerical Simulation (Four Soil Layers) of DBCP Profile 3.5 Years after Application, Maui Pineapple Field 210.

in a satisfactory prediction of field behavior. The numerical model, on the other hand, provides for changes in soil properties with depth and also incorporates a more realistic description of water flow. Considering the increased availability of high speed computers and the low cost of computation relative to other aspects of groundwater contamination analyses, we conclude that numerical modeling is by far the most appropriate.

How Useful Is a Measured Sorption Coefficient? The numerical model provided reasonably good predictions of field DBCP leaching behavior only when the K_d values used in the top layer were obtained by calibration. Sorption coefficients measured by conventional batch equilibration were much too low to describe the leaching of residual DBCP over a period of months or years. It is significant that the K_d value obtained for Field 234 by calibration of the analytical model (K_d = 4.1) was greater than the sorption coefficient measured by batch equilibration (K_d = 0.6), but less than the value obtained by desorption of residual DBCP in field samples (K_d = 22). Similar results were obtained for Maui Pineapple Company Field 210, with batch equilibration giving K_d = 0.64 and model calibration giving K_d = 17. These results have implications for a) analysis of mechanisms of pesticide transport in soils and b) the procedures used to measure sorption coefficients for modeling purposes.

First, some assumptions inherent in the use of the convection-dispersion equation are suspect for most field soils. It is now abundantly clear that an average water flow velocity does not generally represent adequately the contribution of convection to solute transport. Preferential flow through the largest pore sequences contributes to a large apparent dispersion, with small quantities of solute being carried large distances and a major portion of the solute in small micropores being retained near the soil surface because of "flow by-passing". A number of experimental and modeling studies have demonstrated this phenomenon (e.g. 31, 32, 33, 34, 35, 36, 37); one of these studies (34) was conducted on a Hawaii Oxisol with structural properties similar to those of the Maui soils of our study. Recent analytical reviews (9, 10, 11, 12) have examined the considerable body of evidence and suggest that prediction of solute transport in field soils will require modeling approaches which incorporate some description of the wide range of pore velocities, whether by stochastic or deterministic means.

Another common simplifying assumption is that of sorption equilibrium, which is inherent in the use of K_d in Equation 2. During a rainfall or irrigation event, and subsequently during water redistribution, the dynamics of water movement has a great impact on the rate of solute diffusion out of aggregate micropores and consequently on the extent of sorption equilibrium both in micropores and macropores. Thus preferential water flow, solute diffusion and sorption kinetics are linked processes which are essentially ignored when we use the conventional convection- dispersion equation. Progress has been made in modeling this complex system (36) but application to field situations is limited by the difficulty of specifying key parameters.

The results of our studies suggest that equilibrium sorption coefficients measured by conventional batch equilibration may be useful for modeling the short-term movement of pesticides but will not adequately represent sorption of residues. Considering the possible interaction of preferential water flow through large pores and sorption-desorption dynamics, one might expect that the apparently inflated K_d values obtained by model calibration were the result of pesticide in micropores being by-passed in the leaching process. But in fact, K_d values measured by desorption of DBCP residues were also much larger than K_d values measured by the conventional sorption procedure, indicating that the model-derived K_d values were a true representation of actual sorption behavior in the field. We conclude that it may be appropriate to incorporate a sorption-desorption hysteresis submodel in the numerical model of pesticide transport. The sorption-desorption relationship is likely to be time dependent in most cases and thus may be difficult to specify apriori for a given system.

Acknowledgment

The work described in this paper was funded in part by the U.S. Environmental Protection Agency through the Pesticide Hazard Assessment Program, Pacific Biomedical Research Center, University of Hawaii. The contents do not necessarily reflect the views of the Agency and no official endorsement should be inferred.

Literature Cited

1. Liu, C.C.K.; Green, R.E.; Lee, C.C.; Williams, M.K. Project Completion Report, Pesticide Hazard Assessment Project, University of Hawaii, 1983. 51pp.
2. Williams, A.E.H.; Williams, D.D.F.; Green, R.E.; Liu, C.C.K., unpublished report of Maui Pineapple Company and University of Hawaii.
3. Tamrakar, N.K.; MS. Thesis, University of Hawaii, Honolulu, 1984.
4. Green, R.E.; Goswami, K.P.; Mukhtar, M.; Young, H.Y. J. Environ. Qual. 1977, 6, 145-150.
5. Bevenue, A.; Hylin, J.W.; Kawano, Y.; Kellys, T.W. Pest. Monit. J. 1972, 6, 56-64.
6. Green, R.E.; Young, R.H.F. Hawaiian Sugar Tech. 1970 Reports, 1971, 88-96.
7. Mink, J.F., unpublished report for Del Monte Corporation, 1981.
8. Wong, L., unpublished report of Pesticide Branch, Department of Agriculture, State of Hawaii.
9. Davidson, J.M.; Rao, P.S.C.; Nkedi-Kizza, P. In "Chemical Mobility and Reactivity in Soil Systems"; Nelson, D.W., Ed.; Soil Science Society of America: Madison, 1983; pp. 35-47.
10. Jury, W.A. In "Chemical Mobility and Reactivity in Soil Systems"; Nelson, D.W. Ed.; Soil Science Society of America: Madison, 1983; pp. 49-64.
11. Nielson, D.R.; Wierenga, P.J.; Bigger, J.W. In "Chemical Mobility and Reactivity in Soil Systems"; Nelson, D.W. Ed.; Soil Science Society of America: Madison, 1983; pp. 65-78.
12. Green, R.E. In "Prediction of Pesticide Behavior in the Environment"; EPA-600/9-84-026, 1984, 42-71.
13. Addiscott, T.M.,; Wagenet, R.J.; Submitted to J. Soil Science, 1985.

14. Cleary, R.W. Water Resources Series, School of Engineering/Applied Science, Princeton Univ., 1977.
15. Hemwell, J.B. Soil Sci. 1959, 88, 184-190.
16. Jury, W.A.; Spencer, W.F.; Farmer, W.J. J. Environ. Qual. 1983, 12, 558-564.
17. Selim, H.M.; Iskandar, I.K. In "Modeling Wastewater Renovation"; Iskandar, I.K., Ed.; John Wiley and Sons: New York, 1981; pp. 478-507.
18. Warrick, A.W.; Biggar, J.W.; Nielsen, D.R. Water Resour. Res. 1971, 7, 1216-1225.
19. Saltzman, S.; Kliger, L. J. Environ. Sci. Health 1979, B14(4), 353-366.
20. Colley, K.R.; Lane, L.J. J. Soil Water Conserv. 1980, 35, 137-141.
21. Ekern, P.C. Plant Physiology 1966, 40, 736-739.
22. Green, R.E.; Davidson, J.M.; Biggar, J.W. In "Agrochemicals in Soil"; Branin, A.; Kafkafi, U., Eds.; Pergamon Press: New York, 1980, pp 73-80.
23. Smiles, D.E.; Philip, J.R. Soil Sci. Soc. Amer. Proc. 1978, 42, 537-544.
24. Khan, M.A. Ph.D. Thesis, University of Hawaii, Honolulu, 1979, p. 35-41.
25. Castro, C.E.; Belser, N.O. Environ. Sci. and Tech. 1968, 2, 779-783.
26. Pringle, K.W.; Liu, C.C.K.; Green, R.E. DBCP Volatilization from Soil and Water. U. Hawaii Water Resour. Res. Center Tech Rept. 157, 1984, 95 pp.
27. Jury, W.A.; Grover, R.; Spencer, W.F.; Farmer, W.J. Soil Sci. Soc. Am. J. 1980, 44, 445-450.
28. Mackay, D.; Wolkoff, A.W. Environ. Sci. Tech. 1973, 7, 611-614.
29. Green, R.E.; Ahuja, L.R.; Chong, S.K.; Lau, L.S. Water Conduction in Hawaii Oxic Soils, Tech. Rep. 143, Univ. Hawaii Water Resour. Res. Center, 1982.
30. Buxton, D.S.; Green R.E. 1985 unpublished data.
31. Biggar, J.W.; Nielsen, D.R. Soil Sci. Soc. Amer. Proc. 1962, 26, 125-128.
32. Green, R.E.; Rao, P.S.C.; Corey, J.C. Proc. Joint Symp. Fund. Transport Phenomena in Porous Media, Guelph, Ont. 1972, 2, 732-752.
33. Kissel, D.E.; Ritchie, J.T.; Barnett, E. J. Environ. Qual. 1974, 3, 401-404.
34. Rao, P.S.C.; Green, R.E.; Balasubramanian, V; Kanehiro, Y. J. Environ. Qual. 1974, 3, 197-202.
35. Quisenberry, V.L.; Phillips, R.E. Soil Sci. Soc. Am. J. 1978, 43, 675-679.
36. van Genuchten, M.Th.; Wierenga, P.J. Soil Sci. Soc. Am. J. 1976, 40, 473-480.
37. Amoozegar-Fard, A.; Nielsen, D.R.; Warrick, A.W. Soil Sci. Soc. Am. J. 1982, 46, 3-9.

RECEIVED April 1, 1986

20

Evaluation of Pesticide Transport Screening Models under Field Conditions

William A. Jury, Hesham Elabd, L. Denise Clendening, and Margaret Resketo

Department of Soil and Environmental Sciences, University of California, Riverside, CA 92521

Two field experiments investigating the mobility of pesticides leached by sprinkler irrigation are discussed. The first experiment applied a pulse of chloride and napropamide, a moderately adsorbed herbicide, to a 0.6 ha field which was subsequently leached with 25 cm water low in chloride over a two-week period. Soil core samples were taken at 36 locations to 300 cm and were analyzed in 10 cm increments for chloride and pesticide. Distribution coefficient K_D measurements were made at each sampling location by batch equilibrium and column flow through methods. K_D values obtained by either method had coefficients of variation of approximately 30% between replicates. Correlation between K_D measurements by the two methods was not significant, even though their field average values and variances were similar. Approximately 80% of the pesticide recovered in the cores was found within 20 cm of the soil surface, which was consistent with the predicted behavior based on the measured K_D values. However the remaining 20% of the pesticide was located randomly between 20 and 180 cm, suggesting that this portion of the applied chemical moved without adsorbing.

A second experiment was conducted on 14 plots on the same site. The soil surface on these plots was sprayed with a mixture containing four chemicals: bromide, bromacil, napropamide and prometryn, which were leached during two weeks of sprinkler irrigation. Field-wide leaching behavior was similar to that of the earlier experiment, with the strongly adsorbed chemicals napropamide and prometryn showing split concentration peaks in each core with most of the chemical located near the soil surface, and the remainder randomly located between the surface and depths approaching those reached by the mobile chemical (bromide). It was concluded that existing chemical

0097-6156/86/0315-0384$06.00/0
© 1986 American Chemical Society

transport models after calibration with laboratory K_D values are capable of describing the location of the adsorbed portion of the chemical but are unable to predict the location of the portion which bypasses the adsorption sites. The predicted relative mobility of the four chemicals was consistent with field observations, suggesting that environmental screening tests may still be useful in classifying pesticides.

Widespread occurrence of pesticide chemicals in groundwater and surface water supplies has prompted an increased research effort to develop improved methods for estimating pesticide mobility. Although the most comprehensive method for estimating mobility is to conduct transport experiments under natural conditions, this procedure is clearly not feasible to carry out on the large number of existing chemicals and new chemicals introduced each year. For this reason, pesticide transport or environmental screening models have been developed to aid in pesticide mobility classification.

Pesticide simulation models are not new. Considerable success has been achieved over the last 15 years in simulating outflow concentrations from laboratory columns using a convection-dispersion chemical transport model coupled with a representation of pesticide adsorption onto mineral or organic solid phases (1-6). The most comprehensive adsorption models include a non-linear adsorption isotherm and rate-limited diffusive transport from solution to adsorbed phases (4,6). However, these models contain many unknown parameters and are difficult to calibrate using field data. Most proposals for mobility classification, therefore, are based on simple linear adsorption isotherms containing a single parameter, the distribution coefficient K_D (cm^3/g), to represent the adsorption affinity of the chemical (7,8). Because the parameter K_D is soil specific, it has also been simplified by defining an organic carbon coefficient K_{oc} which is equal to the distribution coefficient K_D divided by the organic carbon fraction f_{oc}. The K_{oc} formulation has been shown to reduce the coefficient of variation of a given chemical's adsorption affinity between different soils compared to K_D (7,8).

These mobility indices or equivalent representations such as thin-layer chromatography retention time (9) play a significant role in environmental screening classifications of relative mobility (10). Environmental screening offers one means whereby large numbers of chemicals may be grouped into a small number of categories displaying similar environmental behavior to reduce the experimental requirements (10).

Although the laboratory experimental evidence is considerable, there have been to date no comprehensive field-scale studies of pesticide movement under natural conditions. In fact, until 10 years ago there were virtually no studies of water or mobile chemical transport in the natural field environment.

Information obtained from field-scale experiments conducted during the last decade on mobile water-tracer chemicals is not encouraging. Extensive vertical and lateral variability of water and chemical transport and retention parameters has created difficulties in testing laboratory scale models (11). Furthermore, a growing body of evidence suggests that the consensus laboratory scale model for solute movement, the convection-dispersion equation, may not correctly describe field-scale mobile chemical movement near the soil surface (12-14).

The limited success that laboratory-scale models have had in simulating mobile chemical movement experiments on the field scale led us to set two priorities for our first field experiments with organic pesticides. Rather than seeking to calibrate and validate an actual simulation model, we focused on two more restricted objectives. The first experiment was designed to characterize the variability of the distribution coefficient of a single pesticide over a large field and also to compare its values measured by two different methods, batch equilibrium and flow through methods. A two-week leaching experiment followed by a field-wide sampling provided the first comprehensive information about the extent of lateral and vertical variabilty of pesticide movement into the soil. The second experiment focused on a test of the predictive capability of K_{oc} or K_D values for determining the relative behavior of a group of pesticides subjected to leaching in field plots. These experiments are presented below.

Materials and Methods

Napropamide field study. Both experiments were conducted on the inner 0.64 hectare of a loamy sand (mixed, thermic, typic xeropsamments) located in Etiwanda, California. In the first experiment a concentrated pulse of water containing 15 meq/l KCl was added just after 50 ppm of napropamide, a commerically available herbicide, was sprayed over the entire field. The chemicals were leached into the soil during two weeks of daily sprinkler irrigation until 25 cm net applied water had entered the soil. At this time a hydraulic sampling truck was brought onto the field and 36 soil cores were taken on a 6 x 6 grid to a depth of 300 cm. These cores were subdivided into 10 cm increments and analyzed for napropamide and chloride.

Thirty-six undisturbed soil columns were taken on a 6 x 6 sampling grid immediately adjacent to the field core locations, and were brought to the laboratory. The columns were leached at 2 cm/d until steady state was reached, at which time a pulse of KCl and napropamide was added to the inlet end. Effluent breakthrough curves for each chemical were fitted to the convection-dispersion equation by the method of moments (11). The effective retardation factor R, which may be calculated from the ratio of the chloride V_{Cl} and napropamide V_{nap} velocity parameters obtained by fitting the convection-dispersion equation, is equal to

$$R = V_{cl}/V_{nap} = 1 + \rho_b K_D/\Theta \qquad (1)$$

where ρ_b is dry bulk density (g/cm^3) and Θ is volumetric water content (2). Equation 1 was used to calculate K_D.

Each hole left in the field where the 36 undisturbed columns were taken was sampled and the soil was brought to the laboratory for a batch equilibrium determination of K_D, estimated as the linear slope of the plot of adsorbed versus dissolved napropamide concentration at equilibrium. Further details of the experiments are given in Elabd (15).

Plot experiments. The second experiment on the same field used 14 2m x 2m plots randomly located across the field. On each of these plots, a concentrated spray containing potassium bromide, bromacil, napropamide and prometryn were added. Half of the plots equilibrated for 72 hours after the spray application while covered to prevent evaporation and exposure to sunlight. After the 72-hour period, the remaining plots were sprayed and the entire field was leached daily by sprinkler irrigation. The plots were covered after each irrigation was completed. Eight of the plots were sampled after 10 cm of applied water had been added to the field and the remaining six plots were sampled after 18 cm of applied water had been added. A single core profile was taken to 300 cm from the center of the plot and was analyzed for the concentration of each chemical in 10 cm increments. Further details of this experiment are given in Clendening (16).

Results

Soil and chemical properties. Table I gives various soil adsorption properties of the four chemicals used in the two experiments. The organic carbon partition coefficients were obtained from a literature survey, except for the K_{oc} values for napropamide, which were calculated from the K_D values measured by batch equilibrium and measured organic carbon fraction values at each location. Also given in Table I are the bulk density, water content and organic carbon fraction measurements for the field and their variability among the 36 replicates (15). The retardation factor values R given in the table are calculated from Equation 1. The R value defines the predicted retardation of a linearly adsorbed pesticide reaching equilibrium. It may be viewed as an index of pesticide mobility relative to a free-water tracer such as bromide or chloride (10). Thus, when a bromide pulse moves 10 cm, the predicted movement of an adsorbed chemical in the same solution is 10/R cm. In the absence of direct field observations, the information in Table I represents the consensus laboratory prediction of relative mobility for the chemicals in our study.

Spatial variability of adsorption. Table II shows the mean, standard deviation, and coefficient of variation of various pesticide adsorption parameters measured in the study of Elabd (15), including correlation coefficients between different properties measured at the same location. Two significant features are evident. First, although the batch equilibrium and soil column flow through methods for determining K_D gave approximately the same field average over the 36 replicates, (2.01 and 1.91 cm^3/gm respectively), the correlation between individual measurements at the same location was negligible. Second, division of K_D by organic carbon fraction did not reduce the coefficient of variability for either the batch or flow through method measurements, suggesting that on our sandy soil both mineral and organic adsorption sites were significant. For either adsorption index, the field wide coefficient of variability is large ($\approx$ 30%), but the field average value could still be estimated relatively accurately from a few samples.

Table I. Various Adsorption Parameters for the Chemicals Used in the Field Experiments

	K_{oc} (cm^3g^{-1})	Number of soils or replicates	K_D* (cm^3g^{-1})	R+	Reference Number
Bromide	0		0	1	
Bromacil	41 ± 30%	7 soils	0.25	2.2	(17)
	34 ± 25%	6 soils	0.20	2.0	(18)
	72 ± 102%	2 soils	0.43	3.1	(8)
Napropamide	335 ± 30%	36 replicates	2.01	10.6	(15)
Prometryn	810	1	4.86	24.1	(19)
	518 ± 17%	102	2.40	12.4	(7)
	614 ± 99%	38	3.68	18.5	(8)

f_{oc} = .006 ± 31% (15)

ρ_b = 1.38 ± 6% (15)

Θ = 0.29 ± 5% (15)

*$K_D = K_{oc} \cdot f_{oc}$

+$R = 1 + \rho_b K_D/\Theta$

Table II. Means, Standard Deviations, Coefficients of Variation and Cross Correlations (N=36) Between Various Adsorption Parameters

Parameter	Mean	Standard Deviation	Coefficient of Variation
K_D batch	2.01 (ml/g)	0.63	31
K_D column	1.91 (ml/g)	0.49	26
K_{oc} batch	357.7 (ml/g)	142.6	39
K_{oc} column	338.9 (ml/g)	84.6	25
f_{oc} 0-20 cm	.0058	.0013	23
ρ_b	1.38 (g/cm^3)	.08	6
Θ	.29	.01	4

X	Y	r
K_D batch	K_D column	- .21
K_D batch	f_{oc} 0-10	.18
K_D batch	f_{oc} 0-20	- .06
K_D batch	ρ_b	- .28
K_D column	f_{oc} 0-10	.38
K_D column	f_{oc} 0-20	.57
K_D column	ρ_b	- .28

Napropamide field experiment. Figure 1 presents napropamide concentrations as a function of depth measured in 18 of the field cores which were taken at the end of the two-week leaching period. A significant amount of deep penetration is present in all cores, suggesting that a portion of the applied chemical at each location is moving downward without adsorbing to solid surfaces. Each of the replicate soil core concentrations at a given depth were averaged to produce a field-mean concentration-depth distribution for both napropamide and for chloride, which was added in a pulse at the same time as napropamide (Figure 2). This average curve shows most of the napropamide concentrated in the top 20 cm with the remainder dispersed between 20 and 180 cm. The chloride, which acts as a water tracer, is found to depths as great as 300 cm.

Also shown in Figure 2 is the predicted napropamide distribution, calculated by assuming that it follows the same water flow pathways as chloride, but is retarded by a constant factor R = 10.6 (Table I). This curve is constructed by plotting R • Cl(Z) versus Z/R where Cl(Z) is the chloride concentration at depth Z in the figure. Since both the chloride and napropamide curves are normalized to have unit area under the concentration-depth curve, the predicted curve plotted in this manner has the same scale.

Plot experiments. Figures 3 and 4 show field average profiles for bromacil, bromide, prometryn and napropamide for each of the two replicated plot experiments. These profiles are scaled so as to give unit area underneath the curves for each chemical. The profiles of both strongly adsorbed pesticides, napropamide and prometryn, show substantial retention in the surface layers, and deep penetration of a portion of the applied chemical, which is consistent with the behavior of napropamide in the earlier experiment. Similarly, these dispersed profiles occurred in all replicates of the plot studies. Furthermore, no significant differences were observed between the plots which equilibrated for 72 hours and those which were leached immediately after application.

Discussion

Spatial variability of adsorption. The field average concentration curves for napropamide and chloride (Figure 4) clearly illustrate the influence of both water flow variability and adsorption variability on the chemical displacement process. The chloride curve may be regarded as a representation of the spatial variability of water flow across the field, resulting in a large dispersion of the pulse which was initially added with a small quantity of irrigation water. However, the napropamide distribution cannot be explained solely on the basis of water flow variability. The field-wide napropamide curve is characterized by significant detainment within the top 20 cm and a broad displacement between 20 and 180 cm. Since the maximum depth reached by the chloride was 300 cm (15), then the maximum depth which would be reached by napropamide if it were completely adsorbed is 300/10.6 = 28 cm, assuming a constant partition coefficient R = 10.6. Although the K_D values measured by the two methods were variable, the observed 30% coefficient of variation even when superimposed on the variable water flow could not explain the extremely deep penetration of part of the chemical. One is left

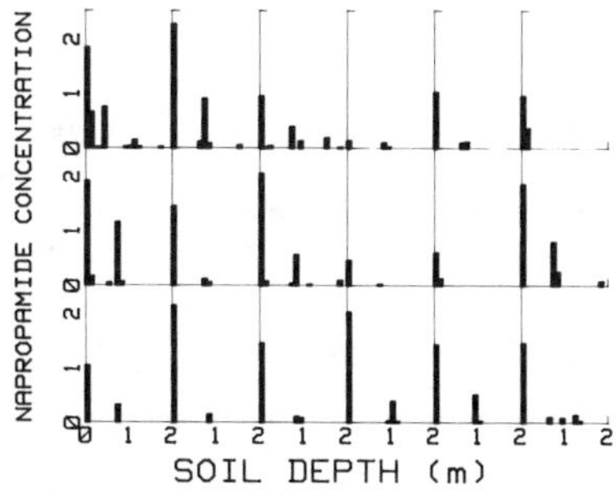

Figure 1. Napropamide concentration versus soil depth found in 18 of the soil cores in the experiment of Elabd (15).

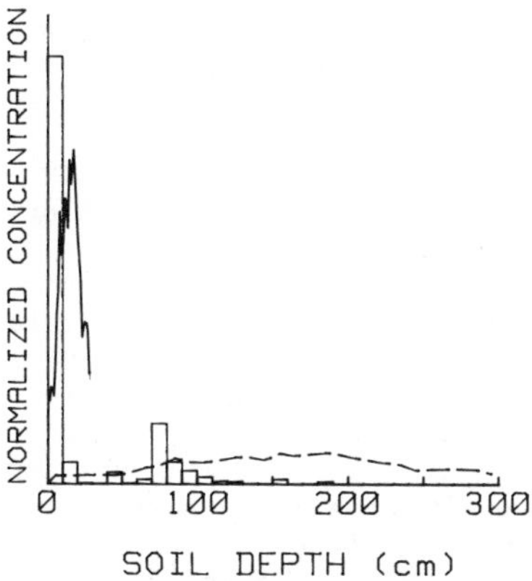

Figure 2. Field average pesticide curve (rectangles) together with field average chloride curve (dashed line) and predicted pesticide curve (dotted line) assuming constant K_D = 2.01 (ml/g) measured in laboratory and variable water flow.

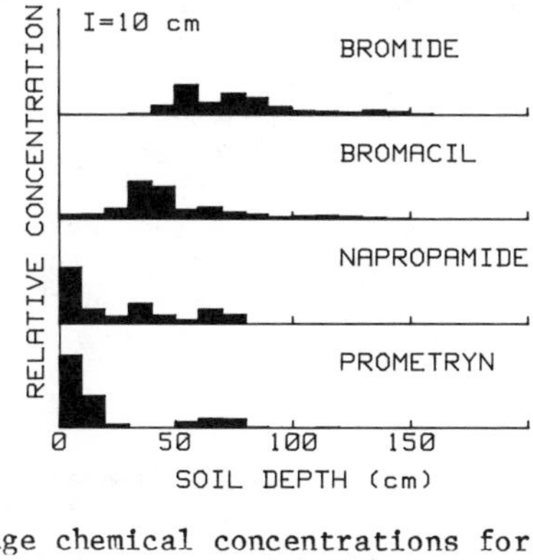

Figure 3. Average chemical concentrations for all replicates of field plot experiment sampled after 10 cm applied water.

with the conclusion that the portions of napropamide which reached depths significantly greater than 30 cm were not adsorbed.

The lack of correlation between the flow through method and the batch equilibrium method of determining K_D is quite surprising, particularly since the means and variances obtained with the two methods were similar. Since at a given site these experiments were performed on immediately adjacent samples, one would not expect significant variations in organic carbon or mineral adsorption site distribution. Furthermore, since the methods gave the same average values, differences between methods cannot be attributed to rate-limited adsorption by loss of adsorption sites caused by soil structure in the soil column method. Clearly, further research is needed on the relationship between flow through experiments and batch experiments for measuring chemical adsorption coefficients.

Plot experiments. The two field plot studies summarized in Figures 3 and 4 offer evidence that the deep penetration of napropamide observed in the earlier experiment (Figure 1) was not an artifact. Both napropamide and prometryn concentration profiles showed the same partial penetration observed earlier with most of the applied chemical pulse located near the surface and the remainder found sporadically distributed at greater depths. The variability of the bromide leaching is similar to that of the chloride profile observed in the earlier experiment, suggesting that the water flow regime was similar in both experiments. The bromacil profile, although maintaining a continuous structure, was somewhat retarded with respect to the bromide profile.

A major goal of the second experiment was to determine the extent to which laboratory characterizations of relative mobility were useful in describing relative field behavior. The consensus laboratory information in Table I predicts that the relative order of penetration of the four chemicals should be bromide: bromacil: napropamide: prometryn:. Figures 5 and 6 present graphs of the cumulative probability of being found above depth Z as a function of Z for each of the four chemicals in the two experiments, obtained by integrating the concentration profiles in Figures 3 and 4 with depth. These curves substantiate the laboratory predictions for the relative movement of the chemicals in the 10 cm sampling in Figure 5, showing that at any given depth the relative retardation of the chemicals was in the proper order. The 18 cm sampling in Figure 6 showed a greater retardation of napropamide than prometryn in the top 20 cm, although this was reversed below that depth. However, as shown in Table I, the literature values of K_{oc} for prometryn are quite variable. In fact, one of the studies reported in (7) had a K_{oc} value of 310 which is smaller than the napropamide value measured for our field. It is possible that these chemicals should be classified as similar in mobility based on their experimental evidence.

Concluding Remarks

The field experiments report above offer a pessimistic forecast for the future of pesticide simulation models. Adsorbed chemicals in all replicates of two experiments on our field showed significant bypass which is evidence that adsorption is not retarding a fraction

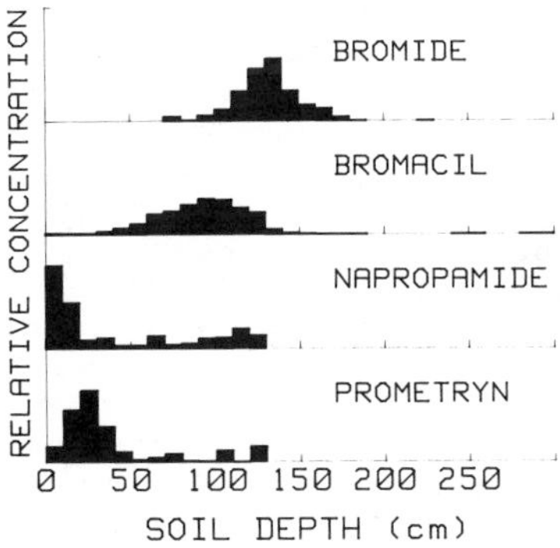

Figure 4. Average chemical concentrations for all replicates of field plot experiment sampled after 18 cm applied water.

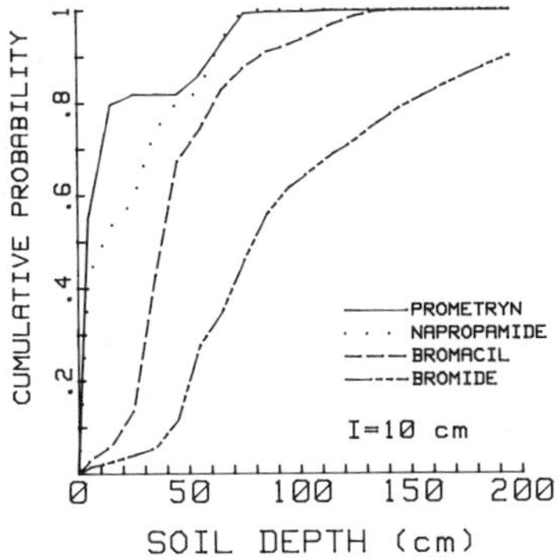

Figure 5. Cumulative probability curve for plot averages calculated from Figure 3.

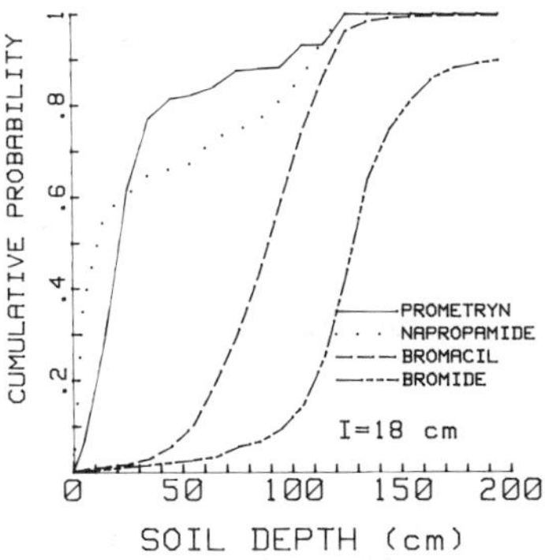

Figure 6. Cumulative probability curve for plot averages calculated from Figure 4.

of the chemical added to the soil surface. This phenomenon was present even when the plots equilibrated for 72 hours prior to leaching. No existing or proposed pesticide transport model is capable of describing this deep penetration.

However, the limited test of the screening model mobility prediction is more encouraging. Given that a certain amount of variability will occur under field conditions, the laboratory K_{oc} values did present a picture of the relative order of leaching of the four chemicals in the two plot experiments. If this result is obtained when the field experiment is repeated on large numbers of chemicals in different soils, it may eventuallly allow future comprehensive field research studies to focus on representatives of different mobility categories rather than requiring prohibitive numbers of experiments to be conducted on each chemcial.

The research also points out a number of areas for future study. The lack of correlation between the two methods of measuring K_D is surprising and should be investigated for other chemicals. The partial deep leaching of both prometryn and napropamide on our field site may indicate that an overlooked mechanism of adsorbed chemical transport is operating, which could involve either a mobile organic complex or conceiveably attachment to fine colloidal suspended particles. These hypotheses should be investigated not only on different field sites but also under controlled laboratory conditions. Our research group at Riverside is planning such experiments in the future.

Acknowledgment

The authors would like to thank the Southern California Edison Company and the US-Israel BARD Fund for financial assistance in this project.

Literature Cited

1. Davidson, J. M.; Chang, R. K. Soil Sci. Soc. Amer. Proc. 1972, 36, 251-261.
2. van Genuchten, M. Th.; Davidson, J. M.; Wierenga, P. J. Soil Sci. Soc. Amer. Proc. 1974, 38, 29-34.
3. van Genuchten, M. Th.; Wierenga, P. J. Soil Sci. Soc. Amer. J. 1977, 41, 278-285.
4. van Genuchten, M. Th.; Wierenga, P. J. Soil Sci. Soc. Amer. J. 1976, 40, 473-480.
5. van Genuchten, M. Th.; Wierenga, P. J. Soil Sci. Soc. Amer. J. 1977, 41, 272-8.
6. Lindstrom, F. T.; Haque, R.; Freed, V. H.; Boersma, L. Env. Sci. Tech. 1967, 1, 561-5.
7. Hamaker, J. W.; Thompson, T. M. In "Organic Chemicals in the Soil Environment"; Goring, I.; Hamaker, J. S., Eds.; Marcel Dekker, New York, 1972.
8. Rao, P. S. C.; Davidson, J. M. In "Environmental Impact of Nonpoint Source Pollution"; Overcash, M. R., Ed.; Ann Arbor Science, Michigan, 1980.
9. Helling, C. S. Soil Sci. Soc. Amer. Proc. 1971, 35, 737-743.
10. Jury, W. A.; Spencer, W. F.; Farmer, W. J. J. Env. Qual. 1983, 12, 558-564.

11. Jury, W. A.; Sposito, G. Soil Sci. Soc. Amer. J. 1985. Nov.-Dec., (In press).
12. Dagan, G.; Bresler, E. Soil. Sci. Soc. Amer. J. 1979, 43, 461-6.
13. Jury, W. A.; Stolzy, L. H.; Shouse, Peter. Water Resources Res. 1982, 18, 369-374.
14. Jury, W. A.; Elabd, H. Proc. of NWWA/USEPA Conference on Characterization and Monitoring of the Vadose Zone. 1983.
15. Elabd, H. Ph.D. Thesis, University of California, Riverside, 1984.
16. Clendening, L. Denise. Master's Thesis, University of California, Riverside, 1985.
17. Corwin, D. Ph.D. Thesis, University of California, Riverside, 1985.
18. Gerstl, Z.; Saltzmann, S.; Kliger, L.; Yaron, B. Irrig. Sci. 1981, 2, 155-166.
19. Kenaga, E. E. Etox. and Env. Safety. 1980, 4, 26-38.

RECEIVED January 21, 1986

21

Field Validation of Ground Water Models

Mary P. Anderson

Department of Geology and Geophysics, University of Wisconsin–Madison, Madison, WI 53706

Rigorous field validation or calibration of a groundwater model is often impossible because of uncertainties in input parameters. Nevertheless, an attempt should always be made to demonstrate that a model is capable of predicting concentrations measured in the field. The biggest uncertainty in model input often lies in quantifying the source function. The amount and concentration of contaminants entering the groundwater system depend not only on the amount infiltrating the land surface but also on uptake and chemical reactions occurring in the unsaturated zone. Ideally, a groundwater model should be linked to a model of the unsaturated zone. Other uncertainties typically arise owing to a lack of information on the degradation characteristics of the contaminant in the subsurface and the heterogeneous nature of the porous material that constitutes the aquifer. Heterogeneities influence the configuration of the velocity field and may be critical to proper quantification of contaminant movement but there is no consensus among researchers on the proper way to simulate dispersion of contaminants in groundwater. Moreover, modelers must confront the fact that most groundwater models are designed to simulate two-dimensional flow fields when in reality many groundwater contamination problems are three-dimensional. Whereas, groundwater flow problems generally can be simplified to two dimensions, it appears doubtful that three-dimensional contaminant plumes can be adequately represented by two-dimensional models. The expense of collecting sufficient three-dimensional field data to validate a three-dimensional model will be prohibitive for most studies. These points are demonstrated through discussion of a case example involving the application of a two-dimensional contaminant transport model to simulate the movement of aldicarb in groundwater in Wisconsin.

0097–6156/86/0315–0396$06.00/0
© 1986 American Chemical Society

Predicting the future is not an easy task. Based on engineering and geologic judgement, it is certainly possible to make predictions about the nature and extent of the migration of chemicals in groundwater. However, it is often preferable to use a mathematical model in order to remove some of the subjectivity inherent in making predictions. A mathematical model consists of a set of equations that have been demonstrated to mimic the effects of various physical processes operative in the real world. However, the derivation and solution of the equations always require a number of assumptions that simplify the way in which processes occur in the real world. For this reason, models are never completely accurate and cannot be expected to predict the future with certainty. The degree of confidence that can be placed in modeling predictions depends on: 1. how well the real world situation conforms to the assumptions imposed by the model; 2. the certainty with which various input parameters are known from field data.

For the purposes of this paper, groundwater models can be classified into two general classes: flow models and contaminant transport models. Among groundwater hydrologists, attention is currently focused on contaminant transport models. However, the configuration of the groundwater flow field is necessary input to these models and thus it is essential to have information on the head distribution. A groundwater flow model is often used as an aid in conceptualizing the flow field and defining the head distribution. The velocity distribution, which is used directly in a contaminant transport model, is calculated from the head distribution.

Calibration of models refers to the process by which the values of certain parameters are adjusted by trail and error until the model yields results which approximate a set of field data. When calibrating a groundwater flow model, the objective is to match the observed head distribution. When calibrating a contaminant transport model, attempts are made to reproduce the measured concentration distribution of a given chemical constituent. The process of calibration has also been called history matching (1). Loosely speaking, field validation of models is synonymous with calibration. Strictly speaking, field validation refers to a model prediction made several years into the future, which is later verified in the field. Under this strict definition, no groundwater contaminant transport model has been field validated to date.

In this paper, procedures for calibrating flow models and contaminant transport models are outlined and some of the difficulties frequently encountered during calibration are discussed. An example of a contaminant transport model applied to a problem involving aldicarb migration in groundwater in Wisconsin is also presented.

Dimensionality

Real world problems are three-dimensional in nature but three-dimensional models are seldom used in practice because it is rare to have adequate three-dimensional field data. Moreover, three-dimensional models are cumbersome to use. Therefore, a key concern in selecting a mathematical model is whether the problem at hand can be simplified to one or two dimensions.

Most flow problems can be readily simplified to two-dimensions and most of the standard methods for treating groundwater supply problems are built around two-dimensional analyses. It is less easy to justify the use of two-dimensional analyses for contaminant transport problems. However, because three-dimensional contaminant transport models are particularly unwieldy, most readily available transport models and most reported applications are two-dimensional in nature.

Flow Models

The two-dimensional governing equation used in most simulations of groundwater flow is:

$$\frac{\partial}{\partial x}\left(T_x\frac{\partial h}{\partial x}\right) + \frac{\partial}{\partial y}\left(T_y\frac{\partial h}{\partial y}\right) = S\,\frac{\partial h}{\partial t} - W \quad (1)$$

where h is head; T_x and T_y are components of the transmissivity tensor; S is storage coefficient and W is the recharge rate.

The calibration process for flow models ideally consists of two steps--a steady state calibration phase and a transient calibration phase, sometimes called model verification (2). During the steady state calibration phase, the transmissivity distribution and the recharge rate are adjusted within a pre-determined reasonable range until the steady state heads observed in the field are matched. Boundary conditions may also be adjusted. Ideally, the parameters determined during the steady state calibration phase are verified during a transient calibration phase. During transient calibration, the storage coefficient is adjusted and minor adjustments may also be made in the transmissivity distribution. However, transient data sets such as drawdown data from well pumping, or a record of the decline of water levels during a drought, are seldom available and it is common to skip the verification phase. A good discussion of the steps involved in calibration of a flow model can be found in (3).

Transmissivity and storage coefficient are standard aquifer parameters that can be estimated from geologic data. Recharge rate is one of the most difficult parameters to estimate with confidence and it is standard practice to let recharge equal a fraction of the average annual precipitation. The ratio of recharge to precipitation will vary with the geographic location of the study site. Groundwater recharge in Wisconsin is roughly estimated to be one-third of precipitation, or about 10 in/yr (254 mm/yr). However, it is likely that actual average annual recharge varies within the state from close to zero in parts of eastern Wisconsin, where there are nearly impervious soils, to perhaps as much as 15 in/yr (381 mm/yr) in the central and northern portions of the state, where there are sandy glacial deposits at the surface.

Ideally the entire subsurface should be treated in a single model. The governing equation for groundwater flow can be generalized to include the unsaturated zone and a model based on this governing equation allows the recharge process to be simulated directly within the model (4-6). However, these types of models are unwieldy and usually are avoided for practical applications. Other investigators (7) advocated the use of linked models in which a one-dimensional unsaturated column model is used to calculate amounts of recharge arriving at the water table. This approach has recently been applied to a problem involving pesticide movement in the subsurface (8).

The output of a flow model consists of the head distribution in time and space. Darcy's Law is used to convert the head distribution to a velocity distribution suitable for input to a contaminant transport model. In a two-dimensional application, Darcy's Law is used to compute two sets of velocity components:

$$v_x = \frac{-K_x}{n}\frac{\partial h}{\partial x}$$
$$v_y = \frac{-K_y}{n}\frac{\partial h}{\partial y} \qquad (2)$$

where v_x and v_y are the components of the average linear velocity; K_x and K_y are components of the hydraulic conductivity tensor and n is effective porosity. Hydraulic conductivity is related to transmissivity as follows:

$$T_x = K_x b$$
$$T_y = K_y b \qquad (3)$$

where b is the saturated thickness of the aquifer. The effective porosity is a measure of the interconnected void space and generally ranges between 0.15 and 0.35. For sandy materials, effective porosity can be taken equal to specific yield. A compilation of representative values for groundwater flow parameters for use in modeling can be found in (9).

Contaminant Transport Models

The two-dimensional governing equation used in most contaminant transport applications is:

$$\frac{\partial}{\partial x}\left(D_{11}\frac{\partial c}{\partial x} + D_{12}\frac{\partial c}{\partial y}\right) + \frac{\partial}{\partial y}\left(D_{21}\frac{\partial c}{\partial x} + D_{22}\frac{\partial c}{\partial y}\right) - \frac{\partial}{\partial x}(cv_x)$$
$$- \frac{\partial}{\partial y}(cv_y) = R\frac{\partial c}{\partial t} + \lambda cR - \frac{c'W}{bn} \qquad (4)$$

$$D_{11} = D_L \cos^2\theta + D_T \sin^2\theta; \; D_{22} = D_L \sin^2\theta + D_T \cos^2\theta;$$

$$D_{12} = D_{21} = (D_L - D_T)\sin\theta\cos\theta$$

where c is concentration and c' is the source concentration; D_L is the longitudinal dispersion coefficient and D_T is the transverse dispersion coefficient; θ is the angle of rotation between the local and global coordinate systems and R is the retardation factor. When sufficient field data are available, calibration consists of attempts to reproduce the configuration of an observed plume of contaminated water. Unknown parameters subject to adjustment during calibration include the dispersion parameters known as longitudinal and transverse dispersivity (10). Dispersivities are related to the dispersion coefficients as follows:

$$D_L = a_L \left| v \right| + D^*$$

$$D_T = a_T \left| v \right| + D^* \qquad (5)$$

$$v = \sqrt{v_x^2 + v_y^2}$$

where a_L and a_T are the dispersivities and D^* is the coefficient of molecular diffusion. Representative values for parameters used in contaminant transport models can be found in (9).

A key consideration during model calibration is the dimensionality of the problem. If a two-dimensional model is used to simulate a three-dimensional plume, some discretion must be used in selecting concentration data against which to calibrate the model. For example, in a two-dimensional areal modeling application it is generally assumed that the contaminant is uniformly distributed throughout the entire saturated thickness of the aquifer. If the plume is stratified or does not penetrate the full thickness of the aquifer, concentration data obtained from vertically nested wells should be averaged before comparing field data to model results. Another stratagem may be necessary if there are not enough field data to calculate reliable vertically averaged concentrations.

Examples of model calibration are cited in (11) and (12) and an example of an attempt to field validate a model is discussed in (13). An example of a two-dimensional model application for a case in which the field data were insuffecent for calculating reliable vertical averages and were also insufficient for defining the areal extent of the plume, is presented below.

Case Study

Introduction. A problem involving the movement of aldicarb in groundwater in the central sand plain area of Wisconsin (Figure 1) will be presented to illustrate the difficulties involved in model calibration. Aldicarb is a systemic pesticide manufactured by Union Carbide under the trade name Temik. In Wisconsin, Temik is applied in potato furrows during planting to control a variety of insects, mites and nematodes. The field study which supplied the field data

Figure 1. Location map of the Central Sand Plain in Wisconsin.

for this simulation was part of an investigation of aldicarb contamination of groundwater beneath several agricultural fields in Wisconsin (14-17). The field data reported in this paper were taken from these sources.

The computer code used in this problem solves the advection-dispersion equation in two dimensions using a random walk technique (18). The code also contains a two-dimensional flow model component that interfaces with the random walk model. The grid shown in Figure 2 was adapted from preliminary modeling simulations (16) and was used for all the simulations reported here. Aldicarb was applied in 1979 and 1980, to the portion of the field shaded in Figure 2, at a rate of 3 lbs/acre of active ingredient of Temik. Aldicarb was applied to the entire shaded area in 1979, but to only the western half of the shaded area in 1980. Aldicarb was detected in four observation well sets, C2, C7, C4 and C9, located beneath and downgradient of the aldicarb-treated portion of the field. The locations of the well sets are shown in Figure 2. A well set (or well nest) consists of two or more piezometers finished at different depths below the water table. Each piezometer was constructed with a 3 foot (0.91 m) well screen. A bundle piezometer similar to the type described in (19) was located near the conventional nested piezometers at site 4. The bundle piezometer (or multilevel sampler) consisted of 9 sampling ports, each open to approximately 6 inches of the aquifer. The distance between sampling ports was about 1.5 feet (0.46 m).

Flow Modeling. The flow component of the random walk model was used to produce the head distribution shown in Figure 3a. The hydraulic conductivity of the aquifer was set equal to 200 ft/day (61 m/day). The saturated thickness of the aquifer is equal to the elevation of the water table above the impermeable bedrock; the water table elevation is adjusted automatically during the iteration process used to solve the flow equation.

Boundary conditions used in the simulation consisted of specified heads along the eastern and western sides of the modeled area and no flow conditions along the northern and southern edges. Under these boundary conditions water is supplied to the aquifer as a result of an imposed head gradient. Hence, it was not necessary to supply water to the aquifer via recharge and the recharge rate was set equal to zero for the purposes of creating the flow field shown in Figure 3a. However, setting recharge equal to zero in the flow model causes an inconsistency between the flow component and the random walk component of the model. When solving the random walk component, it is necessary to specify a loading rate, defined to be the volumetric recharge rate to the water table times the concentration of the leachate. During calibration efforts described below, the loading rate was estimated to be 0.0548 lb/acre (6.14 kg/km^2) of aldicarb. Ideally, one should specify a non zero recharge rate during those time periods when leaching is assumed to occur and solve for the transient head distribution during and after leaching episodes. Sensitivity testing of the flow model assuming a recharge rate of 10 in/yr (254 mm/yr) during leaching episodes, demonstrated that transient head distributions are not appreciably

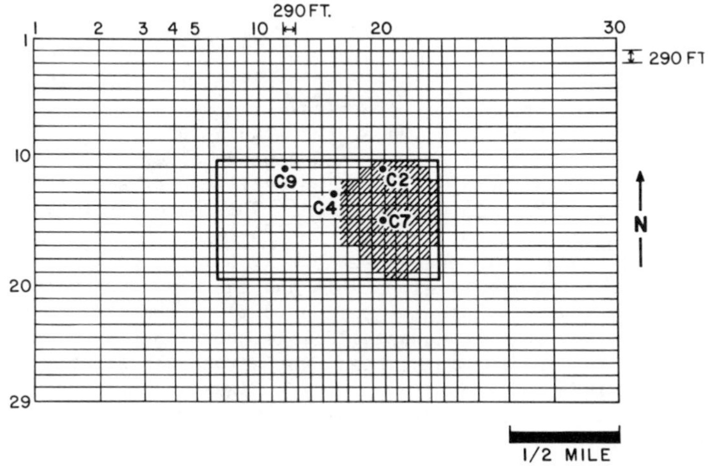

Figure 2. Grid used in the modeling simulation. Locations of wells used in the model calibration are also shown. The shaded area designates the portion of the field treated with aldicarb in 1979-80.

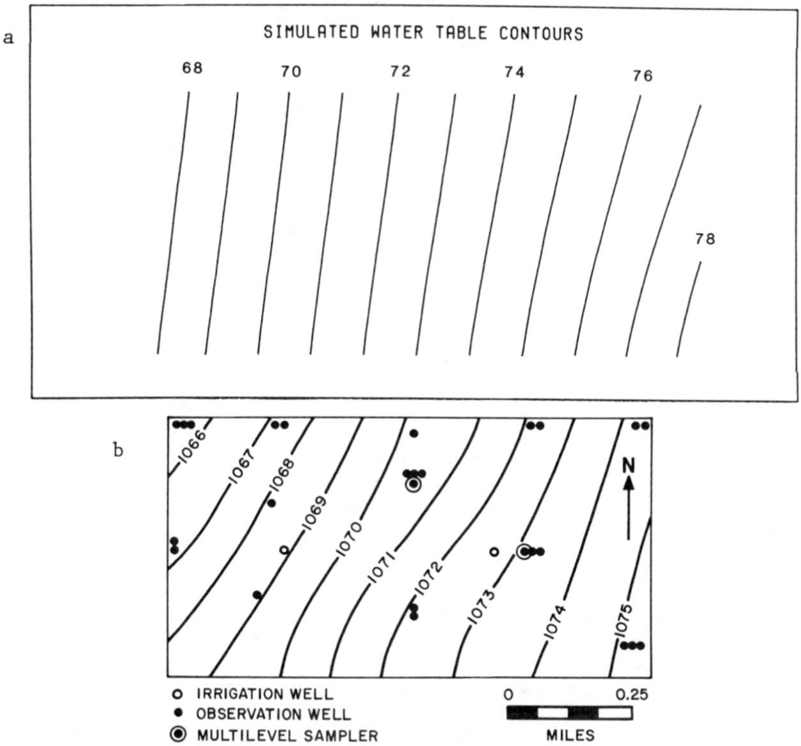

Figure 3. (a) Simulated water table contours. Datum is 1000 ft above sea level. Magnitudes of simulated water levels are the same as those measured in May 1981.
(b) Water table contours based on heads measured in the field during August 1981 (adapted from 15). Datum is sea level.

different from the steady state distribution shown in Figure 3a. Assuming a steady state flow field greatly simplifies model calculations. Hence, the steady state head distribution shown in Figure 3a was assumed for the purposes of calculating velocities which were held constant for the duration of the simulation.

The heads along the eastern and western sides of the modeled area were adjusted during calibration until the steady state flow field shown in Figure 3a was obtained. This head distribution resembles the flow fields reported in (15-16). The magnitudes of the heads in this field change seasonally but the configuration of the water table is relatively stable and the head drop across the field is relatively constant (10 ft or 3 m). The flow pattern measured in the field in August 1981, is shown in Figure 3b. The pattern of flow and the total head drop across the field is roughly the same as the simulated flow field in Figure 3a although the absolute magnitudes of the heads are different. The magnitudes of heads shown in Figure 3a are the same as the heads measured in May 1981 (15-16).

The steady state head distribution shown in Figure 3a was used to calculate the velocity distribution, using a hydraulic conductivity of 200 ft/day (61 m/day) and an effective porosity of 0.30. Velocities ranged from 0.8 to 1.2 ft/day (0.24 - 0.37 m/day) but were around 1.1 ft/day (0.34 m/day) beneath most of the field. Groundwater flows in a westerly or northwesterly direction, as can be inferred from the groundwater potentials shown in Figure 3.

Contaminant Transport Modeling. A major difficulty in the calibration of any two-dimensional contaminant transport model is relating the two-dimensional simulated plume to the real three-dimensional plume. A model based on Equation 4 can simulate two dimensions in cross section or areal view. An areal view was selected for the problem considered here. Use of a two-dimensional areal view model implies that the contaminant is uniformly spread out through the entire saturated thickness of the aquifer. However, in the field the aldicarb plume is only around 10 feet (3 m) thick while the aquifer is around 70 feet (21 m) thick. Moreover, the concentration data were collected from wells having 3 ft (0.91 m) well screens and hence are representative of only a small fraction of the total aquifer thickness. It was decided to calibrate the model to concentrations representative of the center of the plume vertically. That is, the model was calibrated to maximum measured concentrations in each well nest. As a result, the loading rate to the model is inflated over probable field values. The model assumes the load to the model is distributed over the full aquifer thickness, when in the field the zone of maximum concentration is probably no more than 3 feet thick. Therefore, the probable loading rate in the field is roughly 3/70 or 4% of that used to calibrate the model.

The calibration procedure consisted of attempts to reproduce the transient response at each of the four well sets, to the pulse of aldicarb moving through the aquifer. The unknown parameters subject to calibration include the retardation factor, dispersion parameters, loading rate and half life--in short, all the parameters in Equation 4. Of these parameters, the retardation factor is known with the most certainty.

Retardation Factor. Reported distribution coefficients for aldicarb range from 0.16 - 0.073 ml/gm (20), or are effectively equal to zero, suggesting that aldicarb moves at the same speed as the groundwater. In order to eliminate the distribution coefficient as a variable subject to sensitivity analysis, it was set equal to zero for all simulations. A distribution coefficient of zero is equal to a retardation factor of one.

Dispersion Parameters. The dispersion parameters (longitudinal and transverse dispersivity) are unknown but sensitivity analyses proved that at the scale of this model, results were insensitive to variations in these parameters. The longitudinal dispersivity was varied from zero to 70 ft (21 m) and the transverse dispersivity was varied from zero to 7 ft (2.1 m), with no appreciable effect on the simulation results. Dispersivity values of 70 ft and 7 ft were selected for the remaining simulations.

Loading Rate. Trial and error simulations assuming no degradation suggested that a loading rate of 0.0548 lb/acre (6.14 kg/km^2) during 200 day leaching episodes in 1979 and 1980 provided a reasonable fit to the early concentration data in wells C2 and C7 (Figure 4). The actual loading rate in the field is estimated to be 4% of the simulated value or 0.0022 lb/acre (0.246 kg/km^2), which is less that 0.1% of the application rate at the surface. Wells sets C2 and C7 are located directly beneath the treated portion of the field where the effects of degradation can be expected to be minimal at early times. Therefore, during the early phases of the calibration, emphasis was placed on fitting the model to the early data at well sites C2 and C7. Figure 4 shows the results of one of these early simulations in which it was assumed that aldicarb does not degrade below the water table. Actual measured concentrations at each well site are also shown for comparison. It is apparent that concentrations are too high at later times at well sites C2 and C7 and also too high at the downgradient well sites C4 and C9, suggesting that degradation does occur below the water table.

Half Life. Union Carbide scientists have suggested that the half life of aldicarb in groundwater in Wisconsin is around 900 days (21). Figure 5 shows the results of simulations in which half life was set equal to either 900 days or 1800 days. Field concentrations are also shown. Both simulations provide reasonable fits to the observed concentrations. The exception to this is at well site C9, where a large sudden increase in aldicarb was detected in the field but not reproduced by the model when the half life was set equal to 900 days. With half life equal to 1800 days, the model predicts a concentration peak closer to that observed in well 9C, although the simulated peak is offset. The timing of the initial arrival of aldicarb is also slightly offset in both runs of the model, suggesting that the simulated velocity field is not accurate between the treated field and well site C9.

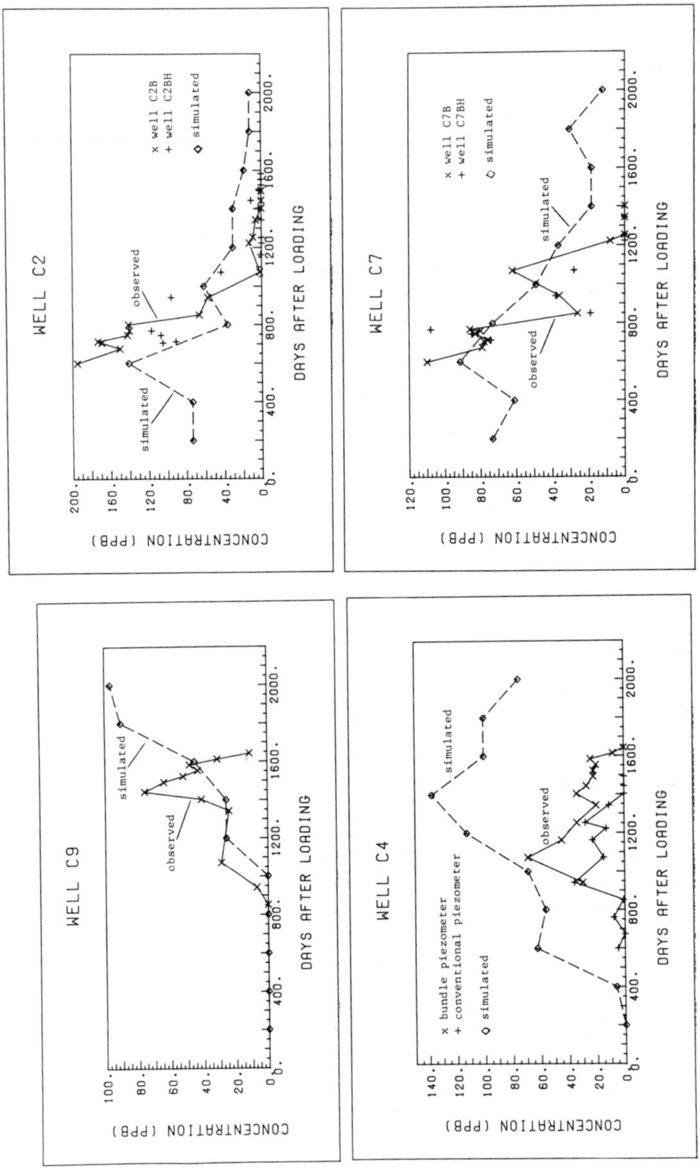

Figure 4. Comparison of predicted vs. observed concentrations assuming no degradation of aldicarb.

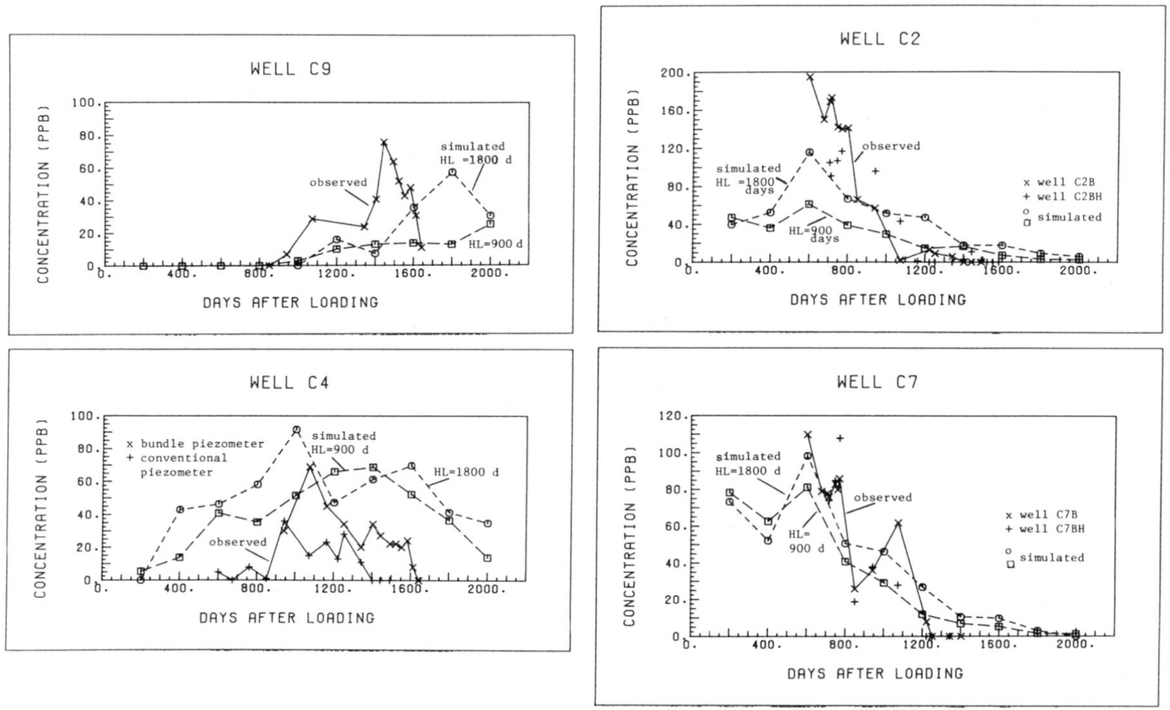

Figure 5. Comparison of predicted vs. observed concentrations assuming that half life of aldicarb equals either 900 or 1800 days.

Discussion of Results. Although there are discrepancies between field observations and model predictions for both choices of half life shown in Figure 5, the model does provide a reasonable fit to the field data. The discrepancies which stand out are the long tails on the simulated breakthrough curves for well C4 and the offset of the sharp peak measured in the field in well C9 at 1400 days. In addition, the model underestimates the concentrations at early times at well site C2.

There are many factors which undoubtedly contribute to these discrepancies. Among them are:

1. A steady state flow field was assumed when the heads do change in response to fluctuations in recharge rate. Hence, the head distribution used in the model is an approximate one.
2. The velocity field is not exactly reproduced. The model assumes constant hydraulic conductivity and effective porosity and uses an approximate head distribution to compute the velocity field.
3. A two-dimensional model was used to represent a three-dimensional plume.
4. Concentrations calculated by the model are averages for the 290 ft by 290 ft cells shown in Figure 2. The field measured concentrations are essentially point values.
5. A constant leaching rate and a 200 day leaching period were assumed. The leaching rate probably varies both spacially and temporally. The random walk model used in this example introduced the contaminant randomly in space by means of an algorithm that involves a random number generator. Consequently, different runs of the model give somewhat different results. The curves shown in Figures 4-6 are representative of the general nature of the solutions. Other runs of the model using the same parameters gave similar but slightly different results.
6. Degradation rate and retardation factor may vary spacially within the field.

Another key question is related to the uniqueness of the calibration. Specifically, is it possible that another set of parameters might also yield a good fit to the field data? For example, it has been suggested that the half life of aldicarb in groundwater in Wisconsin could be as low as 475 days (22). Figure 6 shows the results of a simulation in which the loading rate was doubled to 0.1096 lb/acre (12.28 kg/km^2) and the half life was reduced to 450 days. This simulation also provides a good fit to the data for well site C4 but the fit is less satisfactory for the other three well sites.

A question that is frequently posed is: "How many wells are necessary to ensure that enough field data can be collected to calibrate a groundwater model?" Given information from one well, "calibration" is relatively easy. That is, it is easy to find a set of parameters for which the model results will be in agreement with the data set from one well. No one, of course, would put much faith in this sort of calibration. The calibration task becomes more and more difficult as information from other wells are added to the data base. As before, the model can be readily calibrated to one well, but then other wells go out of calibration. As more and more wells are added to the data base, the calibration objective is

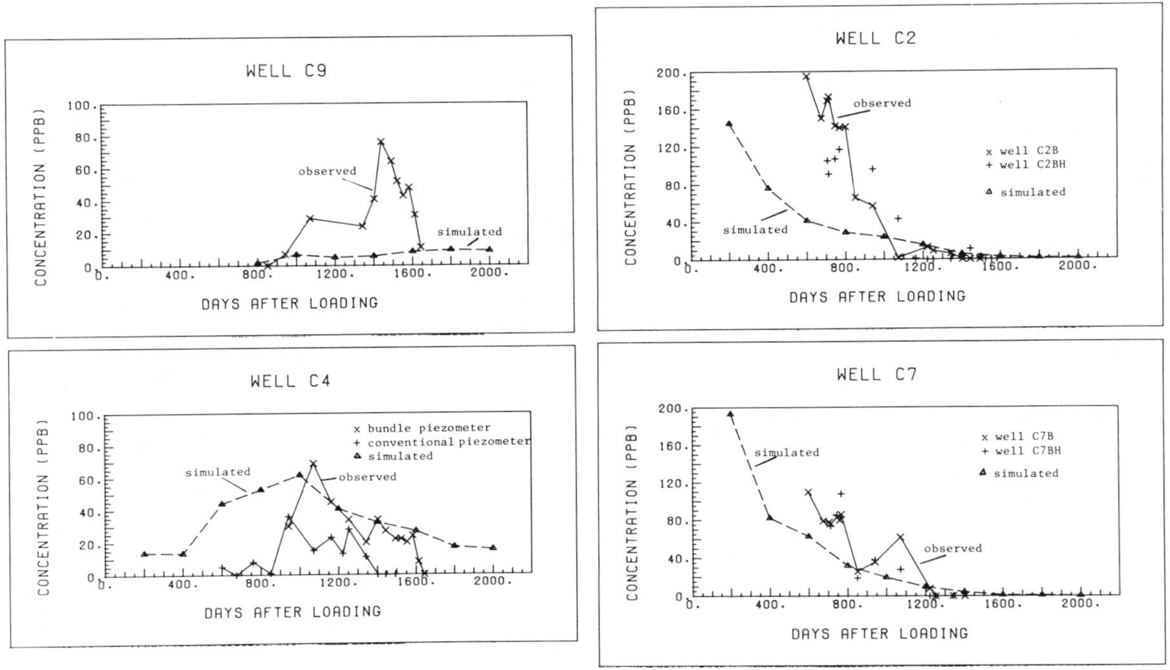

Figure 6. Comparison of predicted vs. observed concentrations assuming a half life of 450 days and a loading rate double that used in Figures 4 and 5.

to find a realistic set of parameters that will allow most of the wells to be approximately in calibration.

In the present example, a value of half life between 900 and 1800 days and a simulated loading rate of 0.0548 lb/acre (6.14 kg/km^2), which corresponds to an estimated field loading rate of 0.0022 lb/acre (0.246 kg/km^2), give a relatively good overall fit to the available field data. Hence, one is tempted to conclude that at this site the half life is between 900 and 1800 days and that less than 0.1% of the aldicarb applied to the land surface in 1979-80 reached the water table. However, in citing these conclusions one must bear in mind that every model is an approximation of reality and that this calibration is based on data from four well sites. Whether or not four well sites are enough to achieve a good calibration is a subjective judgement.

After a model calibration is accepted as valid, the next step generally is to use the model for predictive purposes. In the present application, one might want to predict the areal extent of the peak concentrations within the plume. Figure 7 shows the spacial concentration distributions at three different times as computed by the model with half life equal to 900 days.

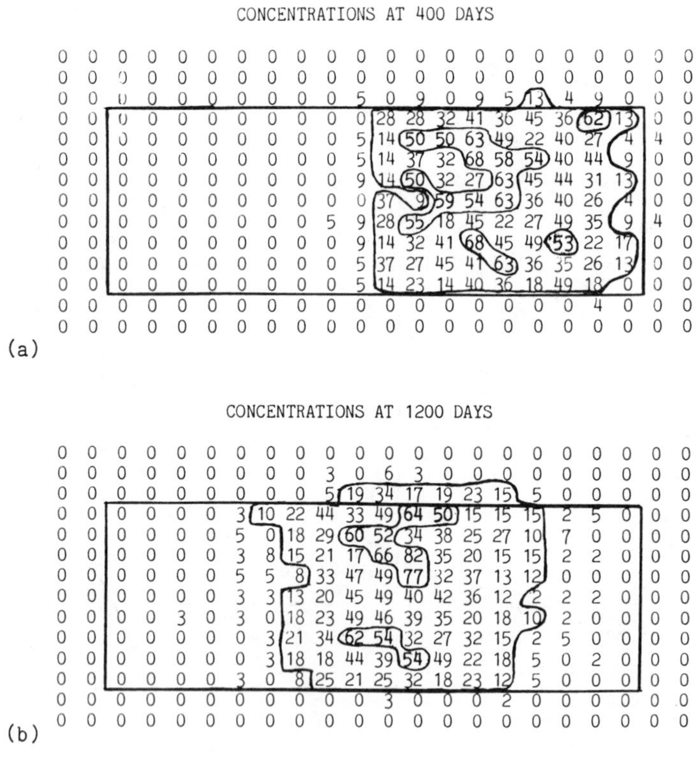

Figure 7(a,b). Predicted areal distribution of the plume using a half life of 900 days. The numbers represent aldicarb concentrations in ppb; contour lines for 10 and 50 ppb are shown.

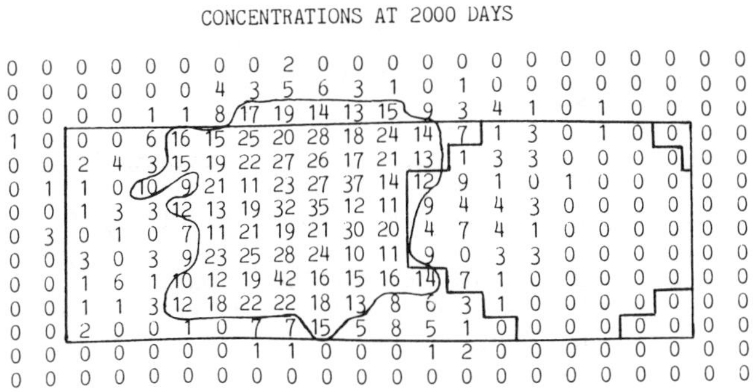

(c)

Figure 7(c). Predicted areal distribution of the plume using a half life of 900 days. The numbers represent aldicarb concentrations in ppb; contour lines for 10 and 50 ppb are shown. The aldicarb-treated portion of the field is shown here.

Acknowledgments

Rich Manser and Frank Jones collected the field data used in this modeling exercise. Rich Manser also tested the modeling procedure during preliminary computer runs. Aldicarb analyses were performed by laboratories at the Univ. of Wisc.-Stevens Point and the Univ. of Wisc.-Madison. Quality control checks on spiked samples were performed by Union Carbide. Funds for the computer simulation were provided by a grant from the ARCO Foundation.

Literature Cited

1. Mercer, J.W.; Faust, C. "Ground-Water Modeling"; National Water Well Assoc.: Worthington, Ohio, 1981, p. 60.

2. Wang, H.F.; Anderson, M.P. "Introduction to Groundwater Modeling: Finite Difference and Finite Element Methods"; W.H. Freeman and Co.: San Francisco, 1982; p. 237.

3. Karanjac, J.; Altunkaynak,M.; Ovul, G. Ground Water 1977, 15(5), 348-57.

4. Freeze, R.A. Water Resour. Res. 1971, 7(2), 347-66.

5. Frind, E.O.; Verge, M.J. Water Resour. Res. 1978, 14(5), 844-56.

6. Winter, T.C. Water Resour. Res. 1983, 19(5), 1203-18.

7. Pikul, M.F.; Street, R.L.; Remson, I. Water Resour. Res. 1974, 10(2), 295-302.

8. Jones, R.: "Movement and Degradation of Aldicarb Residues in Soil and Groundwater"; presented at the Envir. Toxicology and Chemistry Conf. on Multidisciplinary Approaches to Envir. Problems, Crystal City, VA, 1983.

9. Mercer, J.W.: Thomas, S.D., Ross, B. "Parameters and Variables Appearing in Repository Siting Models"; NUREG/CR-3066, Nuclear Regulatory Commission: Washington, D.C., 1982, p. 244.

10. Anderson, M.P. In "Groundwater Contamination"; Bredehoeft, J.D., Ed.; National Academy Press: Washington, D.C., 1984, pp. 37-45.

11. Anderson, M.P. CRC Critical Reviews in Environmental Control 1979, 9(2), 97-156.

12. "Groundwater Contamination from Hazardous Wastes," Princeton Water Resources Program, Prentice-Hall: New Jersey, 1984, p. 163.

13. Lewis, B.D; Goldstein, F.J. "Evaluation of a Predictive Ground Water Solute Transport Model at the Idaho National Engineering Laboratory, Idaho"; U.S. Geol. Surv. Water Resour. Investig. 82-25, U.S. Geol. Survey: Washington, D.C., 1982, p. 71.

14. Rothschild, E.R.; Manser, R.J.; Anderson, M.P. Ground Water 1982, 20(4), 437-45.

15. Chesters, G., Anderson, M.P., Shaw, B.H., Harkin, J.M., M. Meyer, E. Rothschild, Manser, R.J. "Aldicarb in Groundwater"; Wisc. Water Resources Center: Madison, Wisc., 1982, 38 pp.

16. Manser, R.J. M.S. Thesis, Univ. of Wisc.-Madison, 1983.

17. Jones, F. Prospectus for Ph.D. Research, Univ. of Wisc.-Madison, 1984.

18. Prickett, T.A.; Naymik, T.G.; Lonnquist C.G. "A Random-Walk Solute Transport Model for Selected Groundwater Quality Evaluations" Bull. 65 Ill. State Water Survey: Champaign, Ill., 1981, p. 103.

19. Pickens, J.F.; Cherry, J.A.; Grisak, G.E.; Merritt, W.F.; Risto, B.A. Ground Water 1978, 16(5), 322-27.

20. Enfield, C.G., Carsel, R.F., Cohen, S.Z., Phan T., Walters, D.M. Ground Water 1982, 20(6), 711-22.

21. Hansen, J.L.; Spiegel, M.H. Envir. Toxicology and Chemistry 1983, 2, 147-153.

22. Harkin, J.M.; Jones, F.A.; Falhulla, R; Dzantor, E.K.; O'Neill, E.J.; Kroll, D.G.; Chesters, G. "Pesticides in Groundwater beneath the Central Sand Plain of Wisconsin"; WIS WRC 84-01, Wisc. Water Resour. Center, Madison, Wisc., 1984, p. 46.

RECEIVED November 4, 1985

RISK ASSESSMENT AND ITS TOXICOLOGICAL SIGNIFICANCE

22

Reproductive and Developmental Toxicity Risk Assessment

Jerry M. Smith

Rohm and Haas Co., Independence Mall West, Philadelphia, PA 19105

Risk assessment of reproductive and developmental toxicity requires 1) hazard identification, 2) dose-effect, dose-response evaluation, 3) exposure assessment, and 4) risk characterization. Reproductive hazards may involve either male or female, and range from decreased libido to failure of the mother to properly nurse the infant. Developmental hazards affect the conceptus and range from death to the formation of viable but societally dependent offspring. Subchronic, chronic, multigeneration and teratogenic studies provide the majority of data used for hazard identification and dose-effect, dose-response evaluation. Genetic, pharmacokinetic, metabolism, and specially designed investigative studies provide data on the mode of action, target site, delivered dose, primary versus secondary effects, etc. required for characterization of risk. With good characterization of reproductive and developmental risk and exposure assessments, sound risk assessments can be made.

Due to vast improvements in food supply and nutrition, reproductive success has improved markedly in many areas of the world. This has been accomplished largely through advances in agricultural practices and the use of pesticides. However, the thalidomide and Minamata tragedies, the more recent dibromochloropropane (DBCP) event, and the hysteria of Love Canal demonstrate the potential, real or perceived, of a reproductive or developmental calamity that might occur with the contamination of ground water by a potent reproductive or developmental pesticide toxicant. Therefore, the developed world must be alert to potential reproductive and developmental hazards and associated risk that may accompany the use of pesticides.

While the potential for a reproductive or developmental calamity through contamination of underground or surface water is real, it need not happen and will not happen with prudent pesticide use and surveillance. We have the tools, the awareness,

0097–6156/86/0315–0414$06.00/0
© 1986 American Chemical Society

and the responsibility to assure that it does not happen. Essential for continued appropriate use of pesticides is good sound scientific risk assessment including an assessment of risk associated with contamination of ground water.

Assessment of reproductive and development risks associated with the use of pesticides, whether for mixer-loader or consumer, follows the same general guidelines used for any toxicity risk assessment. The National Research Council (NRC) (1) has recently published a monograph on risk assessment and management and their procedure is recommended for reproductive and developmental risk assessment with some slight additions and modifications.

The NRC document calls for hazard identification, dose-response assessment, exposure assessment, and risk characterization. In an effort to place descriptive experimental toxicity results in a clearer perspective and place more emphasis on evaluation, this outline deviates slightly from the NRC document and calls for hazard evaluation, hazard extrapolation, exposure assessment and risk characterization. In addition, a few comments on risk acceptability are given. Exposure assessments have been adequately discussed elsewhere in this symposium and will be discussed here only as they relate to hazard identification, evaluation, extrapolation and risk characterization.

Laboratory studies provide data for assessment of reproductive and developmental toxicity, and laboratory and environmental studies provide data for assessment of potential exposure. These data have demonstrated that for some pesticides and pesticide candidates, a potential may exist for exposure, and reproductive or developmental toxicity. This paper does not present an assessment of the potential risk for these pesticides but examines the essential steps and crucial elements of data necessary for appropriate assessment of hazard and risk.

Unfortunately, the information and data available to assessors of hazards and risks are not always ideal and evaluations must be made on available data. Even more unfortunately, those responsible for risk assessment rely too frequently on secondary and tertiary data sources for their evaluations. This may lead to misinterpretation and over interpretations of original observations.

Finally, it must be remembered that when pesticides are properly evaluated and used, the benefits of pesticides greatly outweigh the risks associated with their use. The greater potential for disaster, as is being demonstrated in certain underdeveloped areas of the world, is famine.

Definitions and Terminology

The EPA (2) has recently produced a good draft document for assessment of developmental toxicants and to be consistent the same or equivalent definitions and terminology are used.

Reproductive toxicity. Adverse effects on either parent that may interfere with copulation, conception, gestation, parturition or maturation of the conceptus.

Developmental toxicity. Adverse effects on the developing organism occurring from the time of conception to the time of sexual maturation.

Embryotoxicity and fetotoxicity. Any toxic effect on the conceptus as a result of prenatal exposure; the distinguishing feature between the terms is the period during which the insult occurred.

Altered growth. A significant alteration in fetal or neonatal organ or body weight.

Functional teratology. Alterations or delays in functional competence of the organism or organ system following exposure to an agent during critical periods of development either pre- or postnatally.

Malformations. A permanent structural deviation which generally is incompatible with or severely detrimental to normal postnatal survival or development.

Variations. A divergence beyond the usual range of structural constitution which may not have as severe an effect on survival or health as a malformation.

Hazard Identification

General toxic hazards, reproductive hazards, and developmental hazards must be identified and considered separately, yet the interrelationships between them must be clearly understood. This is particularly difficult when examining the effects of pesticides upon the female since general toxic effects may indirectly affect the ability of the female to conceive, carry the conceptus, or care for her offspring. To identify the potential of pesticides for reproductive or developmental hazards, the toxicologist and regulatory agencies rely primarily on multigeneration and teratogenic studies which normally are adequate. Different protocols and their strengths, weaknesses, limitations, etc. have been subject of numerous publications (3) (4) (5) and as such are not reviewed here. Basically, multigeneration studies examine the effects of exposure of both males and females to pesticides from before mating through weaning over at least two generations and teratogenic studies examine the effects of exposure of the pregnant animal and its conceptuses to the pesticides during the period of major development of the conceptus.

Reproductive Hazards (Effects). Multigeneration studies not only detect reproductive hazards, but in many incidences they provide data for identification of the reproductive segment that failed. Reproductive failure or reproductive hazards can be classified as impaired mating, impaired conception, impaired gestation, impaired parturition or impaired nursing and care for the neonate or young. Impaired mating or conception can be the result of adverse effects upon the male, female, or both. Impaired gestation, parturition, nursing and peri/post natal care are the results of adverse effects upon the female. (Note: Some adverse effects on the conceptus or offspring may adversely affect gestation, nursing or peri/post natal care.) The effects may be direct or indirect.

Developmental Hazards (Effects). Over the last few years our knowledge and concerns for developmental hazards have increased. We are no longer concerned only with classical teratogenic effects, i.e. structural anomalies induced during organogenesis, but we also recognize the potential for embryo/fetal toxicity and functional teratogenicity. Consequently, the toxicologist now considers structural teratogenic hazards as a subset of developmental hazards. Therefore, not only are "teratogenic" studies required to identify developmental hazards, but studies are required where conceptuses are exposed during major organogenesis and the development of the organism is followed through sexual maturity. Multigeneration studies fulfill the requirement.

Developmental hazards may be expressed as death of conceptus, live malformed offspring, developmentally delayed/runted offspring, and functionally impaired offspring. Death and altered development may be the results of a poor environment presented by the dam or a direct effect upon organogenesis and development of the conceptus/offspring.

Parental Hazards (Effects). For assessment of reproductive and developmental risk, parental hazards, both paternal and maternal, must be identified and evaluated. Parental hazards can be expressed as altered nutritional state, functional impairment, and systemic toxicity. Because of possible indirect affects, knowledge and evaluation of non-reproductive/non-developmental toxicity studies are useful. This information is available by examination of subchronic and chronic toxicity studies.

Hazard Evaluation

Hazard evaluation is used here to mean examination and evaluation of the hazards (adverse effects) observed in toxicity studies, including reproductive, teratogenic, and general toxicity studies plus other pertinent data. Because of the complexities of the reproductive system as well as embryogenesis and maturation of the conceptus, hazard evaluations must go beyond the mere determination of statistical effects, no observed effect levels (NOELS), relationships between dose and response, dose and effects, and the determination of margins of safety. Most importantly, for assessment of risk of reproductive and development toxicity the assessor must determine if the reproductive system of the parents or development of the conceptus/offspring are uniquely sensitive to the pesticide. While it is possible to determine a number of variables for examination, the complexities of the interactions of reproduction preclude a practical set of an all inclusive list for examination. Therefore, hazard evaluations can be guided most effectively by asking a series of questions of the observed results and professionally comparing the answers to known biological facts, data, theories, etc. Some of the most important questions are listed below. Not all questions will have answers, however, the experienced trained investigator should be able to provide appropriate assessment of the data and statements of significance (toxicological, biological or statistical).

Effect(s). What were the types of effects observed? Did they involve a parent, both parents, or the conceptus/offspring? Were they reproductive or developmental? What system(s) were involved? Were the effects the results of direct or indirect toxicity?

Species. What species were affected? Were all species affected in the same way? Were all species affected to the same degree?

Dose-effect/Dose-response. What were the dose-effect and dose-response relationships? (Note: Dose-effects means that the type of effect will differ with different doses; dose-response means that the degree of response or number responding will differ with different doses.) With increasing doses, were the observed effects biologically (toxicologically) different, similar, related, or an extension only in degree. How should lesions be tabulated to express increased effects with increased doses? Does the development of a particular lesion preclude the development of a different lesion. For example, a heart malformation is different from a malformation of the brain, an anomaly of the long bones of the fore limb may be similar to an anomaly of the hind limb, a severe malformation of the CNS may result in death precluding observation of a lesser severe CNS lesion, etc. What were the no observed effect levels; the minimal effect levels? What were the slopes and characterisitics of the dose response curves?

Route(s) of Exposure. What route(s) of exposure were used? What were the duration and characteristics of the exposure? Were the route and duration of the exposure appropriate for evaluation of the expected exposure of humans?

Experimental Design(s). What are the weaknesses and strengths of the experimental designs? What is the unit of analysis? (The individual conceptus or offspring, or the litter.) Should the results be weighted for statistical analysis? What are the appropriate, inappropriate methods of analysis? Should each experiment be given the same weight of evidence?

Relative Sensitivity. What are the relationships between, parental and maternal toxicity, and reproductive effects? What are the relationships between maternal toxicity and developmental toxicity? Was the reproductive system uniquely sensitive to the pesticide? Was the developing organism uniquely sensitive to the pesticide?

Hazard Extrapolation.

Hazard extrapolation is used here to mean extrapolation both from the observed effect levels to non-tested levels within the tested species and from tested species to non-tested species. Hazard extrapolation beyond the observed should be attempted only after hazard identification, and a full and complete evaluation of all relevant available toxicity data has been completed. The reproductive/developmental toxicologist must be very specific in stating his/her certainties and uncertainties about extrapolation beyond the observed. Again, a series of questions can greatly aid the regulator and the investigator in arriving at judgements of the significance of the observed data for extrapolation to the unobserved. Listed below is a series of questions helpful to the regulator and the investigator for extrapolation beyond the observed.

Test Model(s). What was the appropriateness of the test model(s) used for reproductive and developmental hazards associated with pesticide contamination of underground (drinking) water. Common sense tells us that extrapolation of results from a subchronic or chronic drinking water study is more appropriate than extrapolation from some fraction of the LD50 dose given intraperitoneally for 5 days. Likewise, chronic feeding is more appropriate than gavage dosing which is more appropriate than dermal exposure, etc. The same concepts apply to reproductive and developmental hazard assessments.

Delivered Dose. What was the dose delivered to the target tissue(s)? Variations and differences between species, individual animals and even units within an animal are common when one compares response observed to various exposure regimens. However, examination of delivered dose of toxicant to target tissues frequently reduces the variations and differences. Kimmel and Young (8) have recently published a paper demonstrating improved prediction of teratogenic outcome of rats to sodium salicylate when blood levels were used. The more complete the knowledge and understanding of the absorption, distribution, metabolism, and excretion within the studied species and the species of concern, the greater the confidence the investigator has in extrapolating data from one species to another or from an observed dose within a species to a non-tested dose in the same species.

Structural Activity Relationships. Are there any structural activity relationships (SAR) between the pesticide under consideration and pesticides that have a history of safe uses? Structural activity relationships must be used, not misused or abused. To be used properly, SAR evaluations must take into consideration not only chemical structure class and toxic categories, but subtle chemical structural differences as well as all known biological activity and metabolism of the chemical structure class.

Mathematical Models. There are no biological appropriate mathematical models for evaluation of reproductive or developmental hazards. The best that can be done is to mathematically characterize the dose-response curve in the observable range and estimate the threshold levels. The EPA (2) has rejected the use of mathematical modeling for estimating developmental toxicity responses below the applied dose range.

Statistics. What are the statistical limits of the experimental designs and what are the limits to the statistics that can be applied to reproductive studies and evaluations? Statistical evaluations of reproductive and developmental studies present numerous problems including from whether the litter or the individual conceptus is the experimental unit for evaluation, to the effects of weighting of body weights due to litter size or 0 and 100% responses within a litter (Gaylor (9) and Nelson and Holson (10)). While it is important in all toxicological studies to consult a statistician before conducting an experiment, reproductive and developmental studies are more demanding for proper statistical consultation both before and after the conducting of a study(ies).

Uncertainties. In any hazard or risk assessment it is extremely important that uncertainties be expressed. Frequently the uncertainties are expressed only in statistical terms. It is important that the uncertainties also be expressed in biological terms. Statistical modeling or modeling statistics has become very popular in health risk assessments, particularly for "low dose" extrapolations. As previously stated, there are no mathematical models with a biological basis for reproduction or developmental hazard evaluation.

Gaylor (11) and Crump (12) have pointed out some of the deficiencies of using no-observed-effect-levels (NOELS's) and minimal effect levels (MEL's) in risk management. Basically, they point out that NOEL's and MEL's are as much a function of experimental dose selections as the true dose-response nature of the test animal to the toxicants. Both suggest that a statistical acceptable mathematical model be applied to the data in the observed range and that a consistent end point such as a 0.1 or 0.01 i.e. the 10% or 1% effect level be estimated. Examination of data from well conducted toxicological studies, including reproductive and teratogenic studies suggest that an estimated 10% extra risk corresponds well to minimal observed effect levels and an estimated 1% extra risk corresponds well to a threshold or no-observed-effect level. Also, consistent end points improve comparison between effects, studying and in general aid evaluation. Estimation of confident limits for risk and doses about such estimated points help define confidences in a particular study, but add little to certainties or uncertainties for extrapolations to low doses or to other species. Therefore, most likely effect dose estimates should be used.

Unfortunately, certainties and uncertainties cannot be expressed with mathematical precision and should not be attempted. Confidences and uncertainties must be expressed as expert opinions with statements of support. Experts can usually agree that above a stated "high" exposure a given response is likely and that below a stated "low" exposure the response is unlikely. However, between the "low" and "high" exposures there is an area of uncertainties where experts will disagree about the likelihood of a given response to a particular exposure.

Risk Characterization

Risk characterization is an expert, integrated, summarization of hazard identification, evaluation, and extrapolation, plus a characterization of the population(s) at risk and likely exposures. It is the conclusion of the scientific endeavor to assess risk. It must be precise, inclusive and stated in terms that can be understood and used appropriately for societal judgements on the acceptability of a risk.

Risk Acceptability

As all materials are toxic to some organisms under some exposure conditions, there is no "safe material" in the purest sense, there are only "safe handling" or "acceptable handling" procedures, or risk management techniques. This is particularly true for pesticides.

In this sense, safety factors, margin of safety, low dose extrapolations, etc. are not risk assessment tools, they are risk management tools. Thus while it has been noted in scientific experiments and observations that animal to animal variations are seldom greater than an order of magnitude and sensitivity from one species to another is usually within a factor of ten, the 1/100 safety factor for establishing acceptable daily intake (ADI) is not a risk assessment decision, it is a risk managerial decision with scientific input.

For a controlling, regulatory, perspective I would like to paraphrase Marshall Johnson (13). It can be concluded that a pesticide would need to be (regulated) controlled as a reproductive or developmental hazard only if the reproductive system or the embryo/fetus are uniquely susceptible to the pesticide. Pesticides that are coeffective reproductive or developmental toxins, that is, adversely affecting the reproductive system or the developing conceptus only at dose levels that adversely affect the adult, would not necessarily be controlled as reproductive or developmental hazards. Thus only non-coeffective reproductive or developmental toxins, which are those affecting reproduction or the developing conceptus below the dose needed to adversely affect the adult would require control.

Conclusion

In conclusion, regulators must make decisions and investigators must express opinions based on available data and cannot wait for a second or third recycling through testing and evaluations. If hazard identification and preliminary exposure estimates suggest the potential of reproductive or developmental hazard and the potential for exposure, both must conduct a hazard assessment and hazard extrapolation utilizing all available information and best science. And, if the assessment demonstrates that reproduction or development is uniquely sensitive, ie, that the pesticide may affect reproduction or the developing conceptus below the dose needed to adversely affect the adult, then with the best estimate of exposure, they must perform a risk assessment. The risk assessment must be a scientific endeavor while the decisions on risk acceptability and risk management should be left to societal judgement.

Literature Cited

1. "Risk Assessment in the Federal Government: Managing the Process," National Research Council, National Academy Press, Washington, D.C., 1983.
2. "Proposed Guidelines for the Health Assessment of Suspect Developmental Toxicants," Federal Register 49 CFR, November 23, 1984, pp 46325-46331.
3. "U.S. Environmental Protection Agency. 1982. Health effects test guidelines, Chapter II. Specific organ/tissue toxicity-teratogenicity." Office of Toxic Substances. Available from: NTIS, Springfield, VA. PB82-232984.
4. Kimmel, C. A.; Holson, J. F.; Hogue C. J.; Carlo, G. L.; "Reliability of experimental studies for predicting hazards to human development." NCTR Technical Report for Experiment No. 6015. 1984; NCTR, Jefferson, Arkansas.
5. "Guidelines for reproduction and teratology of drugs." Bureau of Drugs. Food and Drug Administration, 1966.
6. Food and Drug Administration. Advisory Committee on Protocols for Safety Evaluations. "Panel on Reproduction Report on Reproduction Studies in the Safety Evaluation of Food Additives and Pesticide Residues." Toxicol. Appl. Pharmacol. 1970, 16, 264-296.
7. Collins, T. F. X. "Multigeneration Reproduction Studies"; In "Handbook of Teratology"; Wilson, J. G.; Fraser, F. C., Ed., Plenum Press, New York, N.Y., 1978; Chap 7.
8. Kimmel, C. A.; Young, J. F. Fundam. Appl. Toxicol. 1983, 3, 250-5.
9. Gaylor, D. W. "Methods and Concepts of Biometrics Applied to Teratology"; In "Handbook of Teratology 4 Research Procedures and Data Analysis": Wilson, J. G.; Fraser, F. C., Eds.; Plenum Press, New York, N. Y. 1978; Chap. 14.
10. Nelson, C. J.; Holson, J. F. J. Environ. Pathol. Toxicol. 1978, 2, 187-99.
11. Gaylor, D. W. J. Toxicol. Environ. Health. 1983, 11, 329-36.
12. Crump, K. S. Fundam. Appl. Toxicol . 1984, 4, 854-71. 13. Johnson, E. M. Ann. Rev. Pharmacol. Toxicol. 1981, 21, 417-29.
13. Johnson, E. M. Ann. Rev. Pharmacol. Toxicol. 1981, 21, 417-29.

RECEIVED November 4, 1985

23

The Toxicological and Epidemiological Effects of Pesticide Contamination in California Ground Water

Peter E. Berteau and David P. Spath

California State Department of Health Services, Berkeley, CA 94704

The contamination of California groundwater by pesticides is a problem which is becoming more evident as monitoring programs are being implemented. In a recent study, 54 pesticides were detected in 2,893 wells in the State. Among the most pervasive toxic pesticide chemicals found have been 1,2-dibromo-3-chloropropane (DBCP), ethylene dibromide (EDB) and 1,2-dichloropropane (1,2-D). DBCP alone has resulted in the closure of about 1,000 drinking water wells. Toxicological evidence from animal studies indicate that these substances are carcinogenic and in one epidemiological study there was an excess of stomach cancer in census tracts with high DBCP content. An acutely toxic chemical, aldicarb, has also been detected in a number of wells in the State. Adverse reproductive effects and neurotoxicity occur with several of these pesticides.

Because of the importance of agriculture to the California economy, widespread use of pesticides has occurred during the past four decades. While this use has led to concerns about poisonings of farmworkers, questions of residues in food and the establishment of safe levels, it has only recently been recognized that hazards from the use of pesticides may occur from sources other than occupation or food supply. Now it is emerging that the contamination of groundwater in California by pesticides and industrial chemicals is a problem of previously unforeseen proportions. In this connection, it must be recognized that about 43% of the state of California or about 10 million people rely on groundwater for their water supplies.

In Table I an idea of the scope of the problem of pesticide contamination in groundwater is given by listing some of the pesticide chemicals which are being most commonly encountered. It is only a partial list but represents the most commonly occurring chemicals, in particular those arising from agricultural use, and the ones which are considered in this section.

0097–6156/86/0315–0423$06.00/0
© 1986 American Chemical Society

Table I. Major Pesticides Encountered in California Groundwater

Soil Fumigants (Nematocides)
1,2-Dibromo-3-chloropropane (DBCP)
Ethylene dibromide (EDB)
1,2-Dichloropropane (1,2-D)
Insecticides
Aldicarb (Temik)
Herbicides
Atrazine
Simazine
Fungicides
Pentachlorophenol (PCP)
Dithiocarbamates (e.g., Maneb, Ziram, Thiram)

In 1984 the California State Water Resources Control Board (SWRCB) reported that during the past 12 years pesticide contaminants have been found in 2,893 wells (1, 2). Although 54 different pesticides have been detected in these wells about 85% of the contamination has been caused by 1,2-dibromo-3-chloropropane (DBCP), a soil fumigant that was banned seven years ago but that is still being detected in increasing amounts in wells.

Figure 1 shows a map of California giving the locations of the most abundantly occurring pesticides which are being found in groundwater. Of course, the contamination is not limited to agricultural chemicals such as pesticides; industrial chemicals have contributed almost equally to overall groundwater contamination, however, it is pollution by pesticides which we are concerned with here.

Most of this contamination arises from the use of pesticides in agriculture but some pesticide contamination of groundwater has arisen in the same manner as from industrial chemicals, namely from careless use and poor housekeeping in industrial plants. The contamination by pentachlorophenol(PCP) is thought to arise from the use of this chemical for treating poles in manufacturing plants.

The presence of pesticides and other chemicals in groundwater has destroyed the long held belief that groundwater is immune to chemical pollution because soil would filter out or break down the chemicals. Although the contamination has probably been present for years, laboratory techniques and equipment to detect low levels of the complex chemicals have only been developed in recent years. State programs specifically designed to protect water from toxic substances began less than 10 years ago.

Types of Pesticides which Contaminate Groundwater

By far the most ubiquitous pollutants which are contaminating groundwater are the soil fumigants, particularly DBCP, but also ethylene dibromide (EDB) and 1,2-dichloropropane (1,2-D). Organophosphorus insecticides have not been found to any appreciable extent in groundwater probably because they are usually readily

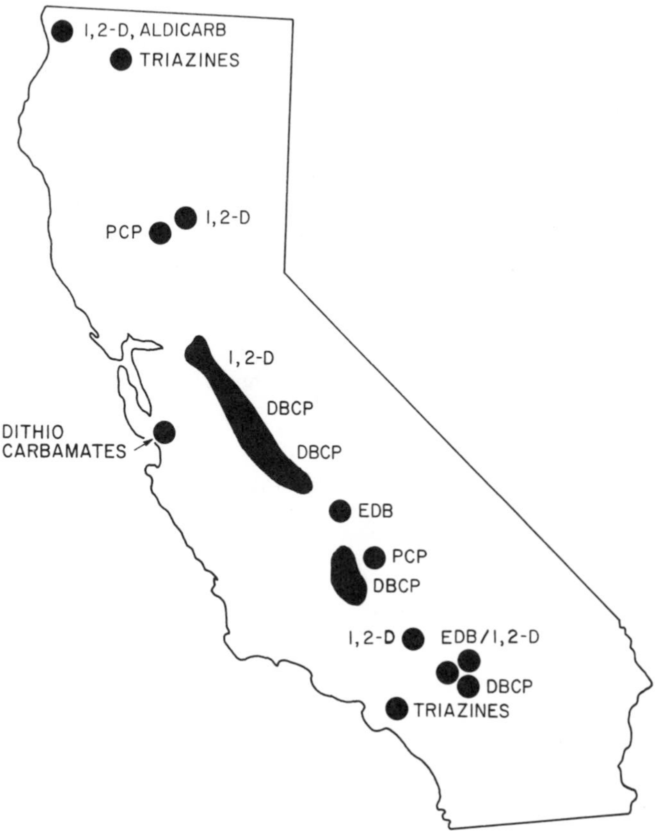

Figure 1. Pesticide Contamination - California Groundwaters.

broken down into phosphoric acid dialkyl esters and lose their toxic character. A related class of compounds, the carbamates, which owe their toxic actions to inhibition of cholinesterase, are widely used in California. One of these in particular is being found in groundwater in many parts of the country. This is the highly toxic oxime carbamate aldicarb (Temik) which, in California, has been found along with its sulfoxide in the extreme Northern part of the state close to the Oregon border.

Herbicides have also been found in water although their occurrence is more often seen in surface waters arising from direct run-off from agricultural use as in rice growing. Certain of the triazine herbicides have been found in wells in Northern California and in the Los Angeles Basin. Fungicides have also been found in groundwater. The occurrence of the wood preservative pentachlorophenol (PCP) has already been mentioned. More recently the metal containing dithiocarbamates have been encountered. Many of these are the heavy metal salts of ethylene-bis-dithiocarbamic acid.

Pesticide chemicals are related to specific toxic effects. With the soil fumigants, reproductive effects and carcinogenicity are the main concern. With the carbamates, such as aldicarb, we are concerned with an acute toxic effect on the peripheral nervous system. With pentachlorophenol high acute toxicity and the potential for carcinogenicity are the main concerns.

Soil Fumigants

As already mentioned, three soil fumigants, shown below, along with their chemical structures, are presenting probably the greatest problem of groundwater chemical contamination by pesticides in California.

Table II. Chemical Structures of Soil Fumigants

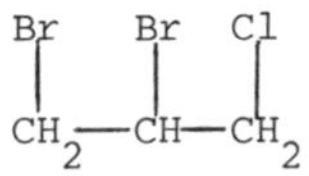

1,2-Dibromo-3-chloropropane (DBCP)

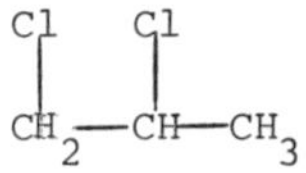

1,2-Dichloropropane (1,2-D)

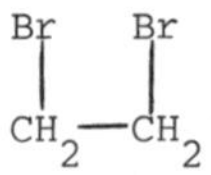

Ethylene dibromide (EDB)

These three compounds are halogenated as are most of the industrial groundwater pollutants. They are well known as being hepatotoxic and have been demonstrated to cause cancer in one or more species of animal. Two of these, DBCP and EDB, are brominated compounds and these are the more potent animal carcinogens, 1,2-D being only weakly carcinogenic. Most soil fumigants are injected into the soil up to a foot in depth and some of the chemical is carried downward into water-bearing strata (aquifers) by irrigation and rainwater. DBCP has been the most pervasive of these soil fumigants because of its past extensive use starting in the late 1950's. Nearly one million pounds per year were being used by the time DBCP was banned in 1977. Interestingly, more of this chemical is showing up now than when it first began to be detected in wells in 1979. Millions of dollars have already been spent to drill water-wells into deeper, untainted aquifers in Fresno county in the Central Valley where most of the DBCP contamination has occurred. However, as of November, 1984, 1,473 wells had been rendered unuseable for drinking or cooking because they exceeded the action level of 1 ppb established by the state of California (3). (An action level is a concentration at which the California Department of Health Services recommends, that for health protection, the water not be used for drinking or cooking). If DBCP has a half life in soils of more than 100 years, as some of the literature indicates, (4) its persistence is going to be a problem.

The toxicological effects of DBCP were investigated in rats in the late 1950's not long after it became commercially available (5). At that time, the testing requirements for registration of pesticides were very much more limited than they are today and adverse findings were often disregarded. The study included standard skin and eye irritation studies as well as repeated vapor inhalation studies on rats, rabbits and guinea pigs. The inhalation exposure concentrations were 0 (control), 5, 10, 20 and 40 ppm for the rat. Exposure times were 7 hours a day, 5 days a week for a total of 50 exposures (10 weeks).

The most significant finding was that at the lowest level, 5 ppm, there was a specific histological alteration in the testes in the male rat. It was reported that testicular weight was reduced to 50% of normal in half the rats although this reduction was not statistically significant because of the large internal variation within the group. Changes were also reported in kidney, liver and bronchioles and weight gain was significantly reduced. At higher concentrations the effect in the testes of rats was particularly severe, resulting in atrophy, degenerative changes, reduction of spermatogensis and the development of abnormal sperm. Less extensive studies were done with rabbits and guinea pigs but decreased testicular weight was reported in both these species at the only concentration administered, 12 ppm. A 90-day feeding study was also conducted with rats at dose levels ranging from 5 to 1,350 ppm. No mention was made of any testicular effect in this study and 150 ppm was judged to be a no-effect level. It is not known if the testes were examined.

In general this study, reported in 1961, is well conducted and the data are still useful for the specific purposes of risk assessment today. The effect on the testes at the lowest inhalation exposure is mentioned several times and the authors recommended an exposure level below 1 ppm if prolonged exposure is likely. No effort was made to repeat the exposure study at lower levels in order to determine a no-effect level as would be required today. More important, however, the study was ignored for about 15 years until the appearance of the well publicized report of sterility in male workers at a DBCP manufacturing plant in Lathrop, California (6).

Since that time, extensive toxicological and epidemiological studies have been conducted with DBCP. These include three studies on humans. One was a study on plant workers. Of 25 non-vasectomized men, in those who had been exposed to DBCP for three years, sperm counts were drastically reduced in 11 men (< 1 million/ml) and nine of these had no detectable sperm cells. Of those who had only a short duration of exposure, sperm counts were normal (> 40 million/ml) (7). In another study, an increase in sperm cells containing two Y chromosomes was noted (8). The third human study was an epidemiological study involving subjects exposed to well-water containing DBCP and will be discussed later.

Since the 1961 study, DBCP has been tested in standard long term rat and mouse bioassays. These studies are reviewed in the monograph on DBCP issued by the International Agency for Research on Cancer (9). In the studies conducted for the National Cancer Institute, DBCP was highly carcinogenic at the lowest doses administered (about 10 mg/kg/day for mice and 15 mg/kg/day for rats). Tumors in both species resulted in early mortality. Tumors in general were squamous cell carcinomas of the forestomach which metastasized into the abdominal viscera and lungs. Testicular atrophy was also noted in the rats in these studies. A high incidence of adenocarcinomas of the mammary gland was seen in female rats exposed to both the low and high dose levels (10).

In studies conducted at the same laboratory but sponsored by the Dow Chemical Company (11), DBCP was incorporated into the diet of rats and mice. Dietary levels were adjusted so that both species received 0 (control), 0.3, 1.0 and 3.0 mg/kg/day. There were 60 rats per group which received the treated diet for 2 years and 50 mice per group which received the diet for 1 1/2 years. In the rats a tumorigenic effect was seen in the kidneys and non-glandular region of the stomach with the greatest incidence of neoplasia being observed in the high dose group of rats, with fewer tumors at the mid-dose level and only rare lesions at the low-dose level. In mice, a dose related increase in the incidence of stomach nodules was noted in all treated animals sacrificed at termination. Histological examination of control and high dose groups revealed a high incidence of treatment related neoplasia (squamous cell papillomas and carcinomas) in the non-glandular region of the stomach.

In an initiation-promotion skin bioassay conducted at New York University Medical Center (12), 69.0 mg of DBCP was applied to the skin of female mice, followed by repeated application of phorbol myristate acetate three times weekly for 499 days. The most significant finding from this study was the occurrence of lung papillomas and squamous cell carcinomas of the forestomach indicating that the occurrence of stomach tumors in the oral studies may be systemic and not merely related to an irritant effect at the site of application.

Based upon these data accumulated from animal studies, the Environmental Protection Agency's Carcinogen Assessment Group (CAG) performed a risk assessment for lifetime exposure to DBCP in drinking water (13). At a concentration of 0.005 ppb the risk of excess cancer cases would be 1.1 in 1 million; at 0.05 ppb, 11 in 1 million and at 1.0 ppb, 220 in 1 million. Based on the available evidence on the carcinogenicity, sensitivity of analytical methods, impact on water delivery capabilities, and additional factors the California Department of Health Services adopted an action level of 1 ppb for contamination of domestic drinking water by DBCP (3). This policy recommends that contamination of water supplies at this level or above indicates that the water should not be used.

A study conducted by the California Department of Health Services is the only known actual epidemiological study that has been conducted on people exposed to water contaminated with pesticides (14). Population groups in an area of Central California, Fresno county, where wells had been contaminated with DBCP were studied for cancer mortality. The various census tracts under study were divided into low (0.0-0.05 ppb), medium (0.05-1 ppb) and high (> 1 ppb) levels of DBCP in the drinking water. The cohorts were examined for those cancers seen in animals which are associated with ingestion of DBCP, namely cancers of the stomach, esophagus, liver, kidney and female breast. Incidences of lymphoid leukemia were also included because of anecdotal reports of this kind of cancer in the exposed population. The results of the studies revealed a statistical trend ($p < 0.005$) in male stomach cancer deaths with increasing DBCP concentrations in drinking water by census tract. A weaker trend was seen with kidney cancer in females ($p < 0.01$) and lymphoid leukemia in males ($p < 0.01$).

The results of this epidemiological study, must be regarded with caution. One plausible explanation is that DBCP does truly increase the rate of stomach cancer and other cancers. However, other variables may be responsible. The census tracts containing high levels of DBCP were more rural, more Hispanic and of lower socioeconomic status than were the low DBCP tracts. Low socioeconomic status has been associated with gastrointestinal cancers but not lymphoid leukemia. Mexican-American ethnicity may be a risk factor in stomach cancers. Other common risk factors such as smoking and alcohol consumption should be considered. In addition, persons living in rural areas may have more exposure to other pesticides some of which are carcinogenic.

Another important soil fumigant is ethylene dibromide (EDB) and the use of this chemical was suspended in several counties in Central California when contaminated domestic wells were encountered in the San Joaquin Valley. Later it was banned nationally by EPA as a soil fumigant. Levels in California water-wells ranged from 70 to 4800 ppt (15, 18). Studies in animals indicate that is has a similar toxicological profile to DBCP. The California State Department of Health Services has recommended that persons with EDB-contaminated wells switch to other drinking water supplies. The available information on the toxicology of EDB has been extensively summarized both by the International Agency for Research on Cancer (16) and more recently by the Occupational Safety and Health Administration (OSHA) as a notice of proposed rule making for an occupational workplace exposure criterion (17) and by also the California Department of Health Services as a basis of recommending an ambient air standard (18). In both mice and rats given EDB by gavage, doses as low as 60 mg/kg/day for mice and 40 mg/kg/day for rats resulted in squamous cell carcinomas of the forestomach which began to appear early in the studies. The female rats also had significant increases in hepatocellular carcinomas. Both sexes of mice had statisically significant increases in alveolar/bronchiolar adenomas and there were significant increases in hemangiosarcomas in the male rats. Similar incidences of cancer were seen with rats and mice in inhalation studies. Testicular degeneration was also seen in the male animals in these studies. EDB was also shown to be mutagenic in several *in vivo* and *in vitro* systems. No epidemiological studies are reported on persons exposed to drinking water containing EDB. In one published epidemiological report involving 161 workers occupationally exposed to EDB it was concluded that the study can neither rule out nor establish EDB as a human carcinogen because of the small size of the population studied.

Risk assessments conducted on the basis of animal studies estimate a lifetime excess cancer risk from inhalation exposure to EDB to range from 0.06 to 67 per 1000 for 0.1 ppm and 160 to 1000 per 1000 for 20 ppm depending upon the risk assessment models used (17, 18).

One substitute for DBCP is 1,2-dichloropropane (1,2-D). This compound has been found in several wells in the extreme Northern part of California and also in an area between Fresno and Bakersfield. In particular, 1,2-D has been found in areas where the water table is shallow. Usually the concentrations are around 0.1 to 5 ppb but levels as high as 30 ppb have been encountered (19). Studies sponsored by the National Toxicology Program have indicated possible carcinogenic activity in laboratory animals (20). The manufacturer of 1,2-D has voluntarily withdrawn D-D, the formulation that contains 1,2-D, from sale in California.

Insecticides

The only primarily insecticidal chemical arising from agricultural use which has been encountered in California groundwater is aldicarb. Levels of 6 to 26 ppb have been reported in some wells in Northern counties of the state (21). The presence of aldicarb is a problem in many other parts of the country and its fate in groundwater is discussed in the section of this symposium dealing with field monitoring. The chemical structure of aldicarb is shown below.

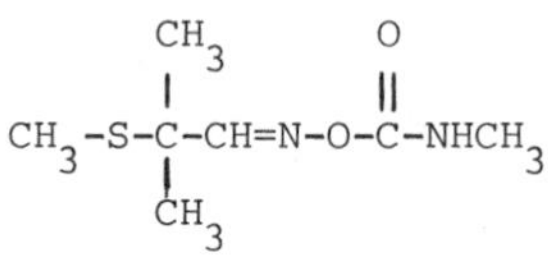

Aldicarb

The toxicology of this compound has been summarized by the National Academy of Sciences/National Research Council (22) and by the World Health Organization Expert Group on Pesticide Residues (23). The main concern about this compound is its very high acute toxicity. It is a potent cholinesterase inhibitor with an LD50 to the rat of less than 1 mg/kg. However, the compound has not been reported to display any adverse long term effects at sub-lethal doses. It is not carcinogenic, teratogenic, nor does it display any adverse effects on reproduction in a three-generation feeding study. Based on studies conducted with rats and dogs, the World Health Organization established an acceptable daily intake (ADI) of 0.005 mg/kg for aldicarb (23).

The National Academy of Science/National Research Council (NAS/NRC) gives a suggested no adverse response level (SNARL) for aldicarb in drinking water of 7 ppb (22). The USA Environmental Protection Agency (EPA) has recommended a health advisory level of 10 ppb. The California Department of Health Services has adopted EPA's limit as an action level (24).

Herbicides

The triazine herbicides, atrazine and simazine have recently been encountered in groundwater in Northern California and in the Los Angeles Basin at levels ranging from 0.4 to 2.0 ppb (25). The chemical structures of these two compounds are shown below:

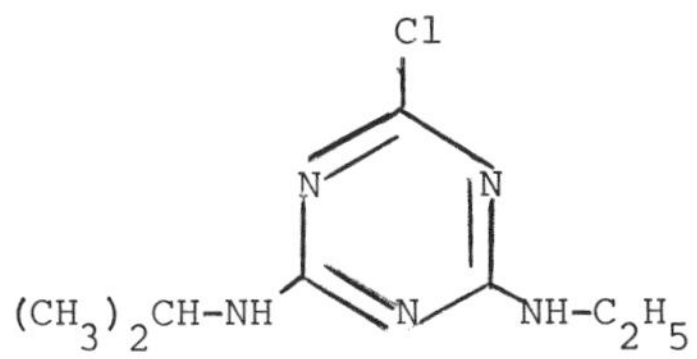

Atrazine

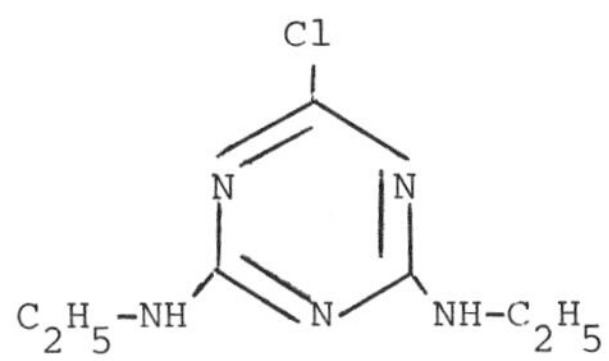

Simazine

The toxicology of these compounds has been summarized by the National Academy of Sciences/National Research Council (26). Atrazine is slightly toxic when given in single oral doses to rats and simazine is practically non-toxic under the same conditions. The compounds have been submitted to tests for carcinogenicity to mice at the highest tolerated doses and were determined not to be tumorigenic under the conditions of the tests. Simazine at 50 and 100 ppb in the rat had no adverse effects on reproduction of rats or their offspring over three generations. The NAS/NRC recommends an acceptable daily intake of 0.0215 mg/kg and 0.215 mg/kg for atrazine and simazine respectively. Recommended levels in drinking water are 7.5 ppb for atrazine and 75.25 ppb for simazine.

Fungicides

The occurrence of the wood preservative pentachlorophenol (PCP), in groundwater is thought to arise from the manufacturing process involving treatment of poles. The chemical structure is shown below

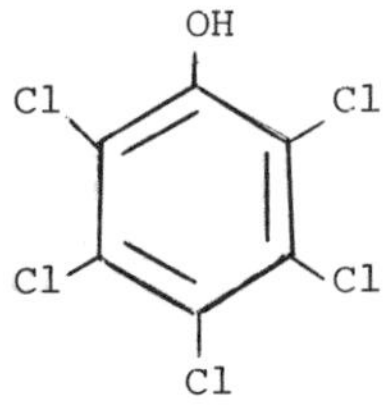

Pentachlorophenol (PCP)

PCP is a highly toxic compound producing a variety of toxic effects, in particular histopathological changes in the liver. The compound contains impurities of chlorinated dibenzodioxins, but recently manufactured batches do not have detectable levels of the highly toxic 2,3,7,8-tetrachlorodibenzo-p-dioxin. For chemically pure PCP a no effect level of 3 mg/kg/day was estimated in the rat. The NAS/NRC recommend an ADI of 0.003 mg/kg and a suggested no adverse effect level in drinking water of 21 ppb (27).

Another class of fungicides recently encountered in groundwater are the ethylene-bis-dithiocarbamates (EBDC's) (28). These compounds are often salts of metals and the names and chemical structures of some of them are shown:

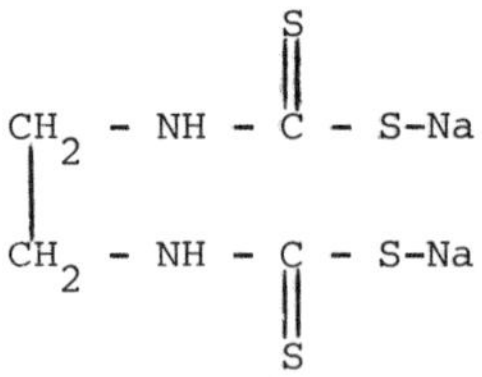

Nabam

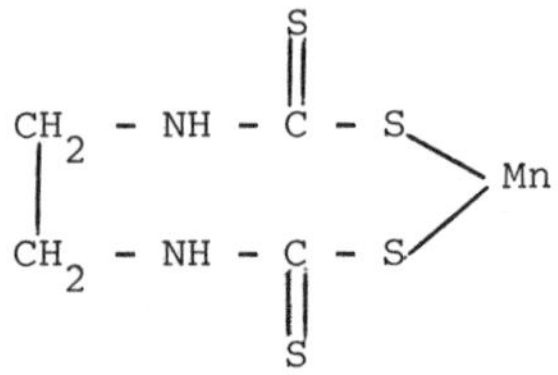

Maneb

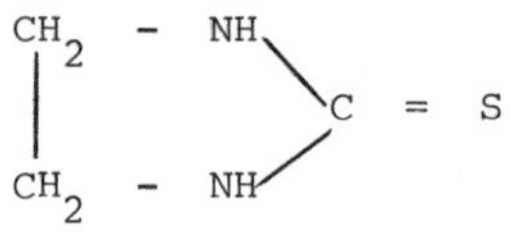

Ethylenethiourea

The main toxicological concern about this class of compounds is their breakdown to the goitrogen ethylenethiourea (ETU), a potent animal carcinogen (29). Testing for ETU in groundwater is presently being conducted.

Conclusions

An attempt has been made to give an overview of the significant toxicological and epidemiological effects that have been reported from those pesticides which are being encountered in groundwater in California. Although a relatively large number of pesticides have been reported, only a few are occurring to such an extent that their effects warrant consideration from a toxicological point of view. The fumigants DBCP and EDB comprise by far the major part of the problem in California and it is probably the occurrence of these two compounds that represents the greatest concern to human health. Both are potent reproductive toxins and the evidence for their being carcinogenic is overwhelming. The epidemiological study conducted in Fresno county in census tracts containing appreciable levels of DBCP is suggestive but more data need to be generated before it can be confirmed or refuted that DBCP in well-water has been responsible for increased human cancer deaths.

Literature Cited

1. Groundwater Contamination by Pesticides. California State Water Resources Control Board, Publication No. 83-45 sp, 1983.

2. Cohen, D.B., and Bowers, G.W. Water Quality and Pesticides: A California Risk Assessment Program (Vol. 1), State Water Resources Control Board, Toxic Substances Control Program, Sacramento, CA, November, 1984.

3. Gaston, J.M. to Williams, M. Letter from Sanitary Engineering Branch, California Department of Health Services to Special Pesticides Review Division, U.S. Environmental Protection Agency, May 31, 1979.

4. Burlinson, N.E.; Lee, L.A.; Rosenblatt, D.H. Environ. Sci. Technol. 1982, 16, 627-632.

5. Torkelson, T.R.; Sadek, S.E.; Rowe, V.K.; Kodama, J.F.; Anderson, H.H.; Loquvam, C.S.; Hine, C.H. Toxicol. Appl. Pharmacol. 1961, 3, 545-559.

6. Emergency Temporary Standard for Occupational Exposure to 1,2-Dibromo-3-chloropropane (DBCP); Hearing. Federal Register 1977, 42 (175), 45536-45543.

7. Whorton, D.; Krauss, R.M.; Marshall, S.; Milby, T.H. Lancet 1977, ii, 1259-1261.

8. Kapp, R.W.; Picciano, D.J.; Jacobson, C.B. Mutat. Res. 1979, 64, 47-51.

9. Evaluation of the Carcinogenic Risk of Chemicals to Humans, Vol. 20. Some Halogenated Hydrocarbons. International Agency for Research on Cancer, Lyon, France, 1979, pp. 83-96.

10. Bioassay of Dibromochloropropane for Possible Carcinogenicity (Carcinogenesis Technical Report Series, No. 28), National Cancer Institute, 78-828, 1978.

11. Dibromochloropropane (DBCP): Final position document. Special Pesticides Review Division, Office of Pesticide Programs, U.S. Environmental Protection Agency, September, 1978.

12. Van Duuren, B.L.; Goldschmidt, B.M.; Loewengart, G.; Smith, A.C.; Melchionne, S.; Selman, I.; Roth, D. J. Natl. Cancer Inst. 1979, 63, 1433-1439.

13. Carcinogen Assessment Group (E.P.A.) review of risk assessment for DBCP in drinking water in a memo to Jeff Kempter, Special Pesticides Review Division, U.S. Environmental Protection Agency, June, 1979.

14. Jackson, R.J.; Greene, C.J.; Thomas, J.T.; Murphy, E.L.; Kaldor, J. Literature Review on the Toxicological Aspects of DBCP and An Epidemiological Comparison of Patterns of DBCP Drinking Water Contamination with Mortality Rates from Selected Cancers in Fresno County, 1970-1979. Unpublished report to the California Department of Food and Agriculture from the California Department of Health Services, June 1, 1982.

15. Smith, C.; Margetich, S.; Fredrickson, A.S. A Survey of Well-Water in Selected Counties of California for Contamination by EDB in 1983. Unpublished report from the California Department of Food and Agriculture, HS-1123, September 12, 1983.

16. Evaluation of the Carcinogenic Risk of Chemicals to Humans. Vol. 15. Some Fumigants, the Herbicides 2,4-D and 2,4,5-T, Chlorinated Dibenzodioxins and Miscellaneous Industrial Chemicals. International Agency for Research on Cancer, Lyon, France, 1977, pp. 195-209.

17. Occupational Exposure to Ethylene Dibromide; Notice of Proposed Rulemaking. Federal Register 1983, 48 (196) 45956-46003.

18. Health Effects of Ethylene Dibromide (EDB). Unpublished report prepared by the California Department of Health Services, April 15, 1984, 25 pp.

19. Organic Quality of Groundwater in the Fruitvale Area, Kern County. Unpublished report prepared by the Sanitary Engineering Branch, California Department of Health Services, January, 1984.

20. NTP Technical Report on the Carcinogenesis Bioassay of 1,2-Dichloropropane (propylene dichloride) in F3441N rats and B6C3F1 mice. National Toxicology Program. Draft Report, February 28, 1983.

21. Klant, R.R. Unpublished information from the California Regional Water Quality Control Board, North Coast Region, June 17, 1983.

22. "Drinking Water and Health" Vol. 5. Toxicity of Selected Contaminants, Aldicarb. National Academy Press, 1983, pp. 10-12.

23. Pesticide Residues in Food, Evaluations 1982. Report of the Joint Meeting of the FAO Panel of Experts on Pesticide Residues in Food and the Environment and the WHO Expert Group on Pesticide Residues, Rome, 1983, pp. 7-15.

24. Spath, D. to Rogers, P. Aldicarb "Action Level". Internal Memorandum from Sanitary Engineering Branch, California Department of Health Services, July 25, 1983.

25. Sturm, G., personal communication. Sanitary Engineering Branch, California Department of Health Services, September, 1984.

26. "Drinking Water and Health" Vol. 4. Organic Solutes, Triazines. National Academy of Sciences, 1977, pp. 533-537.

27. "Drinking Water and Health" Vol. 4. Organic Solutes, Pentachlorophenol. National Academy of Sciences, 1977, pp. 750-753.

28. Bowen, C., personal communication, Sanitary Engineering Branch, California Department of Health Services, February, 1985.

29. "Drinking Water and Health" Vol. 4. Organic Solutes, Pesticides: Fungicides, Dithiocarbamates. National Academy of Sciences, 1977, pp. 650-657.

RECEIVED March 25, 1986

24

Safety Evaluation of Pesticides in Ground Water

D. D. Sumner and J. T. Stevens

Agricultural Division, CIBA-GEIGY Corporation, Greensboro, NC 27419

As the sensitivity of analytical methods increases, more instances of pesticide detection in groundwater will occur. In order to avoid a series of crisis situations, it is necessary that standardized procedures for the identification of safe levels of chemicals in groundwater be established. The same basic principles used to establish safe levels in food, can be used to establish safe levels in groundwater. Some methods used to establish acceptable levels in groundwater are discussed. The setting of acceptable levels at zero or limits of detection, should be avoided since practical considerations preclude these approaches.

The presence of chemicals in groundwater has become a prominent and emotional issue. In the past few years, few issues have produced the emotional responses that accompany groundwater contamination. This response seems intensified when the chemical which contaminates the water is a pesticide [1].

Many scientific and regulatory bodies have considered the question of groundwater and chemicals, and an even greater number are currently considering this issue. Federal agencies under several statues [2, 3, 4, 5] and several state regulatory or legislative bodies [6, 7, 8, 9] are actively investigating groundwater quality and standards. Industry as well as environmental groups are active in directing attention to groundwater.

Although there is uniformity in the interest in groundwater, considerable differences exist among the respective groups in their approaches to setting acceptable limits for chemicals in groundwater. Some groups want no acceptable levels for any chemical in groundwater, while other groups want to set standards which ensure safety of the water [6, 7, 10, 11]. The latter are typified by the Maximum Contaminant Levels (MCL's) set by the EPA Office of Drinking Water (ODW), which are legally enforceable standards.

0097-6156/86/0315-0436$06.00/0
© 1986 American Chemical Society

As discussed within this symposium, it is not possible to keep all chemicals out of groundwater. As limits of detection are reduced more chemicals will be detected. Clearly there are many different chemicals in our potable water. Some are intentionally introduced, such as chlorine or fluoride salts; some are natural products; some are unintentionally present, such as pathogens, industrial chemicals, pharmaceuticals, and pesticides; and some are synthesized _in situ_, such as chloroform. Although clear documentation of adverse effects is not common, epidemiological implications of water quality effects have been reported [12]. The present paper will address questions concerning the safety of pesticides in groundwater from the perspective of human health.

Several factors contribute to the attention currently given to pesticides in groundwater. One of the most important factors is the progress which has been made in analytical methodology. In the '60's, methods were available to measure materials in the parts per million (ppm) range. In the '80's, it is not uncommon to have methods which can detect some materials in the parts per trillion (ppt) range. This constitutes a one million-fold increase in sensitivity.

An additional factor contributing to the proliferation of reports on pesticide contamination of groundwater is ease of pesticide analysis. Pesticides are a predefined analytical target. Federal regulations [2] require the development of defined analytical methods which make identification easier than many other materials, especially natural products. Finally, the current interest in pesticides in groundwater probably focuses efforts which may otherwise have been directed to other chemicals.

The reaction to the presence of pesticides in groundwater is also influenced by a multiplicity of factors. In addition to the potential toxicity of the materials or their potential environmental effects, emotional factors seem to contribute to the negative response.

For example, simazine is involved in a groundwater issue. Simazine has an LD_{50} of approximately 5,000 mg/kg [13]. This is less acutely toxic than table salt, which has an LD_{50} of approximately 4,000 mg/kg [14]. Simazine can be fed in the diets of rodents for their lifetime at 3,000 ppm without any remarkable effect [15] or administered daily at 215 mg/kg/day in an National Cancer Institute (NCI) bioassay without oncogenic effect [16].

Simazine has received considerable notoriety based upon the finding of levels equal to or less than 3.5 ppb in six out of 217 wells evaluated in a study in California [17]. These six wells are suspected of being point source contaminations [18]. Based upon the toxicity data and its use in algae and weed control in ponds, simazine has a potable water tolerance of 10 ppb. It has a Suggested No Adverse Response Level [SNARL] established by the National Academy of Sciences of 1,500 ppb. However, because of the notoriety associated with the findings in the six wells, there is a move to restrict the product from use on sand and loamy sand soils [19].

Given the emotionalism aroused by pesticides found in groundwater, it is difficult to propose an acceptable solution to the dilemma. Furthermore, considering the complexities of the issues involved, it would be presumptuous to try to address all issues and possible resolutions in a single paper. Instead our discussion will focus on some of the approaches that have been taken or are being taken to consider the vital question of establishing safe/acceptable limits of pesticide levels in groundwater.

The importance of setting acceptable limits for pesticides in groundwater cannot be underestimated. With the sensitivity of current methods and the expansion of monitoring efforts, instances of pesticide detection in water are inevitable. Each of these instances will precipitate a crisis if the analytical values cannot be placed in perspective with regard to relative hazard.

In the absence of acceptable limits, instances of detection could be expected to result in overreaction to trivial events. If false alarms occur frequently, the consumer may become complacent and hazardous situations could be ignored.

The establishment of acceptable limits is critical from a legal/public relations perspective as well as from a safety perspective. Setting acceptable limits for chemicals in water will reduce the number of crisis situations and afford opportunities for orderly decisions concerning water quality and safety. Hence, it is appropriate to give consideration to some of the contemporary approaches for the establishment of these safe limits.

The National Academy of Sciences (NAS) addressed the groundwater issue in response to the Safe Drinking Water Act of 1974 [4]. The NAS examination [20] dealt with a broad spectrum of possible materials including microbial, inorganic, organic, and radionucleotide contaminants; several pesticides included.

The NAS used toxicology data to establish an Acceptable Daily Intake (ADI) for each of the materials considered. The ADI was established using the No-Observed-Effect level (NOEL) from animal studies and a series of Safety Factors ranging from 10 to 1,000 depending upon the duration of animal toxicity studies and the nature of the toxic effects. For oncogens the NAS recommended consideration of the Multi-Stage model of carcinogenesis. If the data available were not sufficient for use of the Multi-Stage Model, then the One-Hit Model was suggested.

The NAS used the ADI to establish an acceptable maximum limit designated the Suggested No Adverse Response Level (SNARL). The NAS assumed 20% of the ADI was available for water contribution (reserving 80% for other sources). Calculation of the SNARL was based upon a 70 kg person drinking two liters per day.

NAS SNARL = NOEL (mg/kg/day) SF X 20% 2L/day X 70 kg

The procedures utilized by the NAS in establishing Safety Factors and evaluating the animal data are essentially equivalent to the current standards [21]. The suggested use of the Multi-Stage Model is further compatible with later recommendations [22, 23]. The NAS limited its consideration only to published data, reasoning the published data has undergone peer review. However, in the case of pesticides most of the data resides in corporate and EPA files and not in the literature. The restricted database thus is a weakness and a limiting factor associated with the NAS approach. Several sources [24, 25] have recommended that all available data be used in safety evaluation. Although not all of these data will have been published, the data will have undergone regulatory review.

Fortunately, other groups have considered an expanded database. The National Agricultural Chemical Association (NACA) established a committee to look into groundwater issues in 1983. That committee made recommendations similar to but not identical with the NAS procedures. The NACA committee recommended using the exisiting database within EPA. The database required for pesticides is very extensive (Table I) and provides considerable information upon which to develop a safety evaluation.

Table I. A Partial List of Pesticide Data Requirements

Standard Acute Studies	Kinetics of Environmental Degradation
90-Day Studies in Rats	Mobility in the Environment
1-Year Studies in Dogs	Potential for Bioaccumulation
Teratology in 2 Species	Avian Acute Toxicity
Lifetime Feeding Studies in 2 Species	Avian Reproductive Effects
Multigeneration Reproduction Studies	Avian Acute Toxicity
Metabolism in Plants, Animals, and Soil	Aquatic Chronic and Reproduction Effects
Residue Determinations in Food and Nonfood	Marine Effects
Methods to Determine Residues	Aquatic Vegetation
	Toxicity on Earthworms, Bees, and Other Beneficial Invertebrates

In fact, pesticides are some of the most thoroughly studied chemicals in our environment. Typical data needed for a food tolerance include: acute, subchronic, reproduction, and chronic, as well as oncogenicity, mutagenicity, and environmental studies [26]. Pesticides are unique in that not only studies from the mammalian health perspective are completed but also environmental investigations [27, 28].

It was the opinion of the NACA committee [10] that the existing ADI established for all products with food tolerances is adequate to establish maximum safe levels in groundwater. The ADI established by EPA is based upon the database in the Agency.

The development of an ADI is essentially the same in the NAS procedures, the EPA Food Tolerance procedures, and the NACA proposal for groundwater. An ADI is determined by dividing the NOEL in the most sensitive species by a suitable Safety Factor (SF). Safety Factors for subchronic or repeat administration are usually 1,000; for chronic or lifetime studies, 100 is used. Species conversions can be based upon mg/kg, ppm in the food, or body surface area conversion [29]. Currently, non-oncogenic effects are considered on an mg/kg basis without attempts to correct for species differences. Risk assessment procedures for oncogenic risk employed by the EPA are based upon surface area extrapolations in an attempt to relate to man [30].

The establishment of a maximum acceptable concentration of any chemical in potable water should not be done in isolation, but should consider all sources of exposure. Intake of pesticides can result from worker exposure, ingestion, and water exposure. Work-related exposure is time and space limited and is restricted to limited populations which are evaluated in separate studies and are not as all encompassing as food and water exposure.

Ingestion of food is examined using Theoretical Maximum Residue Contribution (TMRC) approach.

The TMRC is based upon tolerances established for raw agricultural commodities. It assumes all crops are treated and that all residues occur at the maximum level seen in use situations. Tolerances are usually set at the limit of analytical detection when residues are not expected to occur. For instance, fruit trees sprayed during dormancy with a product that degrades before fruit are formed; hence such tolerances in the fruit are meaningless. Therefore, TMRC greatly exaggerates pesticide intake.

Market basket surveys have shown actual residues to be less than the expected TMRC values. For most products the dietary contributions are less than 10% of the TMRC. Typical values from the market basket survey are shown in Table II [31, 32].

Since the TMRC for food residues so vastly exaggerated pesticide intake, the NACA committee did not sponsor this approach in the setting of pesticide limits in water. On the other hand, they did not want to ignore the dietary contribution to pesticide intake.

The dietary contribution was evidently resolved by adopting a different model for consumption. The NACA committee recommended use of a 10 kg child drinking one liter of water per day; the TMRC was ignored. This model provides an additional 3.5

TABLE II

Relationship of "Theoretical Intake" to "Actual Intake" of Seven Pesticides in the Diet*

Pesticide	Theoretical Intake mg/day	Actual Intake mg/day[2]	Relationship of Actual Intake to Theoretical Intake	Acceptable Daily Intake mg/day[3]	Relationship of ADI to Actual Intake
DDT	6.79	0.029	230:1	0.300	10:1
dicofol	4.45	0.006	740:1	1.500	250:1
dieldrin	0.23	0.006	40:1	0.006	1:1
lindane	9.21	0.003	>1,000:1	0.750	250:1
malathion	12.56	0.008	>1,000:1	1.20	150:1
parathion	1:18	<1.001	>1,000:1	0.300	>300:1
carbaryl	9:50	0.031	300:1	0.600	20:1

Calculated from USA tolerance and 9th decile consumption figures minimum figures since small consumption commodities are excluded. USA tolerances used for calculations were those in effect during 1964-1970. Tolerances for DDT and dieldrin on certain commodities were reduced during 1968-69; however, the higher tolerances were used in the calculations.

[2]Dietary Intake of Pesticide Chemicals in the United States (III), R. E. Duggan and P. E. Corneliussen, Pesticide Monitoring Journal, 5, No. 4, March, 1972.

[3]Pesticide Residues in Foods - Report of the 1971 Joint FAO/WHO Meeting, FAO Agricultural Studies, Number 88. Based on a 60 kilogram individual.

*From FAO/WHO Food Standards Program, 1974. [31]

Safety Factor over the 70 kg person drinking two liters. The original Safety Factor determinations by the NAS [33] included consideration of susceptible population subgroups so children in some respects had already been considered. The NACA committee suggested the term Health Advisory (HA) for the maximum acceptable pesticide levels.

HACA HA = NOEL (mg/kg/day) SF $\div$ 1L/day X 10 kg

Likewise, ODW at EPA had used the 10 kg child as a model in setting MCL's. By their adoption of the model using a 10 kg person drinking one liter per day [34], it can be deduced that the NACA committee compensated for dietary contribution by adopting 28.5% of the ADI for water, which affords a considerable margin of safety considering the market basket survey data (Table 2).

For oncogenic materials the NACA committee recommended a calculated risk of one in a million using the Multi-Stage model. The NACA committee further indicated that Health Advisories are not intended to encourage practices which will result in health advisory levels [10].

Recently the State of Wisconsin has adopted groundwater legislation which introduces several interesting aspects. The Wisconsin bill [8] establishes a basis for numerical standards based upon toxicology data and avoids "limits of detection." Regulation based upon limits of detection may seem attractive on the surface, but toxicology is independent of analytical technique. At best, "limit of detection" provides little relevance to biology; at worst, it rewards sloppy or inept analytical development.

The Wisconsin bill recognizes the possibility of future developments. It recognizes the existence of present and future federal standards including the EPA-established ADI values. It incorporates provisions to accept technical advancements in toxicology or analytical chemistry.

The Wisconsin bill introduces an interesting concept called the Preventive Action Limit (PAL). Conceptually, this can be envisioned as some major portion of the enforcement standard which could trigger investigations into sources for groundwater contamination, i.e., improper disposal, point sources, etc. In practice it is an additional safety factor since the state regulatory response is nearly the same whether a standard or PAL is exceeded.

Recent rulemaking in New York has also introduced another concept [35]. The state has proposed that for ordinary toxicity, the water standard shall utilize calculations based on a 10 kg child drinking one liter of water. But for oncogenicity, a 60 kg person drinking two liters per day is used. This is logical since oncogenicity studies are based upon lifetime feeding studies, thus covering the adult period.

In addition, New York has provided a classification mechanism for groundwater so that aquifers which are unusable for natural reasons will not need as much protection as those which are useful.

It would, of course, be a substantial omission not to mention EPA. Considerable activity is in progress, but this has been previously discussed by Dr. Stara [36] within this symposium.

Although it would be presumptuous to attempt to define a program to resolve all questions concerning pesticides in groundwater, it is possible to recommend certain elements which would appear to be necessary factors in any resolution. Some of the most important aspects seem to be:

- In order to avoid a crisis every time a pesticide is detected, maximum acceptable levels should be set before samples are analyzed.

- Safety evaluations should include use of the entire data base. Data in the published literature and that in EPA files should all be considered. To do otherwise, short changes the public and the industry.

- Acceptable levels should be based wherever possible on toxicology data, not limits of detection. Toxicity is independent of analytical technique.

- Acceptable levels should consider environmental as well as mammalian health effects.

- More public education to provide information concerning the basis for setting containment levels and to counteract "emotionalism".

Literature Cited

1. Knudson, T., Groundwater Chemicals Cancer Linked, Des Moines, Sunday Register, June 16, 1985.
2. Federal Insecticide Fungicides, Rodenticide Act
3. Toxic Substances Control Act.
4. Clean Water Act.
5. Resources Conservation and Recovery Act.
6. Kim, N. K. and D. W. Stone, "Organic Chemicals and Drinking Water". New York State Department of Health, April 1981.
7. State of Florida Statutes, Chapter 487, 1983.
8. State of Wisconsin, Act 410, 1983.
9. State of California, Pending Legislation, Jones AB2133, 1985.
10. "Groundwater Position Paper," National Agricultural Chemicals Association: Washington, D.C., 1984.
11. "Groundwater Management by Use Classification," Chemical Manufacturers Association: Washington, D.C., 1983.
12. Davidson, I. W. F., et. al. Chloroform: A Review of its Metabolism Teratogenic, Mutagenic and Carcinogenic Potential. Drug and Chem. Toxicol. 1982, 5, 1-87.
13. "Pesticide Reference Standards and Supplemental Data," Office of Research and Development, US - EPA Research Triangle Park, NC, 1973.
14. Loomis, T. A. "Essentials of Toxicology" 2nd Ed.; Lea and Febiger: Philadelphia, 1974.
15. CIBA-GEIGY Corporation, Internal Communications.
16. Innes, J. R. M., et. al. Bioassay of Pesticides and Industrial Chemicals for Tumorigenicity in Mice. A Preliminary Note. J. Nat. Cancer Inst. 1969, 42, 1101-1114.

17. Weaver, D. J., et. al. "Pesticide Movement to Groundwater, Volume I Survey of Groundwater Basins for DBCP, EDA, Simazine and Carbofuran"; State of California, Department of Food and Agriculture": Sacramento, CA, 1983.
18. Roux, P., personal communication.
19. "Partially Closed Meeting of FIFRA Science Advisory Panel," Federal Register, 50, 25783, EPA, 1985.
20. "Drinking Water and Health Report of the Safe Drinking Water Committee," National Academy of Sciences - National Research Council, Washington, 1977.
21. Hayes, A. W. "Principles and Methods of Toxicology"; Raven Press: New York, 1982.
22. Crump, K. S.; Guess, H. A. "Drinking Water and Cancer: Review of Recent Findings and Assessment of Risks." Science Research Systems, Inc.: Ruston, LA., CEQ Contract No. EQ10AC018.
23. Crump, K. S., et. al. Confidence Intervals and Tests of Hypothesis Concerning Dose Response Relations Inferred from Animal Carcinogenicity Data, Biometrics 1977, 33, 437.
24. Mantel, N. and W. R. Bryan. "Safety" Testing of Carcinogenic Agents. J. Nat'l. Cancer Inst. 1961, 27, 455-470.
25. "Chemical Carcinogens: Review of the Science and its Associated Principles", Office of Science and Technology Policy, Fed. Reg. 49, 21595-21661, 1984.
26. "Hazard Evaluation Human and Domestic Animals" Pesticide Assessment Guidelines Subdivision F, EPA, PB, 82-153916, NTIS 1982.
27. "Chemistry; Environmental Fate" Pesticide Assessment Guidelines Subdivision N, EPA, 83-153973, NTIS, October, 1982.
28. "Hazard Evaluation: Wildlife and Aquatic Organisms PB" Pesticide Assessment Guidelines Subdivision E, EPA, 83-153908 NTIS, October, 1982.
29. Casarett, L. F.; Doull, J. "Toxicology the Basic Science of Poisons" 2nd Ed.; Macmillan: New York, 1975.
30. "Report of the Safe Drinking Water Committee," National Academy of Sciences, National Research Council Drinking Water and Health Volume 3, Washington, 1980.
31. "Relationship Between Tolerances and Actual Daily Intake of Pesticides" FAO/WHO Food Standards Programme, World Health Organization, Geneva, 1974.
32. Frawley, J. and R. Duggan. "Techniques for Deriving Realistic Estimates of Pesticide Intakes from Advances in Pesticide Science", Part 3.; H. Geissbuehler Ed.; Pergamon Press: Oxford, 1979.
33. Lehman, A. J.; Fitzhugh, O. G. 100-Fold Margin of Safety Assoc. Food Drug 088. Q. Bull. 1954, 18, 33-35.
34. Lappenbusch, L.; S. Moskowitz Proc. AMA/EPA Symp. on Drinking Water and Human Health, Washington, DC, 1984.
35. "Proposed Changes to Subdivision 701", State of New York, 1985.
36. Stara, G. "Evaluation of Pesticides in Groundwater"; Division of Pesticide Chemistry; American Chemical Society: Washington, D.C., 1985.

RECEIVED March 25, 1986

25

Risk Assessment Approaches for Ground Water Contamination by Pesticides and Other Organic Substances

J. F. Stara, J. Patterson, and M. L. Dourson

Environmental Criteria and Assessment Office, Office of Research and Development, U.S. Environmental Protection Agency, Cincinnati, OH 45268

The Environmental Criteria and Assessment Office in Cincinnati (ECAO-Cin) of the U.S. Environmental Protection Agency (EPA) has been preparing health risk assessment documentation and developing methods useful for assessment of health hazards for single chemical chronic exposure for several years (1,2). In addition, ECAO-Cin has been instrumental in developing new and improving existing methods for health risk assessments of toxicants and mixtures of toxic chemicals (3-7). These methodologies are used to derive "acceptable intakes" for systemic toxicants or "risk specific intakes" for carcinogens. Sufficient data are needed in order to develop a satisfactory program for assessment of the human health risks associated with ground water contamination by pesticides and other organic and inorganic pollutants. These data should include the types and concentrations of pollutants present, the potential and extent of exposure of the population at risk, and the toxicity data base on the individual chemicals or their mixtures.

In the United States, both surface waters and ground waters are used for drinking water supplies. The use of ground water as a source of drinking water has been steadily increasing with about half of the U.S. population currently relying on ground water for its source of drinking water. Municipal water systems supply much of the need, but ground water from individually-owned wells represents a major drinking water resource in many rural areas. Reliance on ground water appears to vary with geographic location. For example, the midwest and western states tend to use a greater proportion of ground water than do the eastern states (8). Some of these states with high ground water usage (e.g., Nebraska, Kansas, Oklahoma) are also characterized by high agricultural activity, with the expected increased possibility of contaminating ground water sources with pesticides and fertilizers.

This chapter not subject to U.S. copyright.
Published 1986, American Chemical Society

Ground water contamination by environmental pollutants is a growing concern in the United States today. Ground water may be contaminated by a number of sources, such as land disposal of hazardous wastes, leaking underground storage tanks, land spreading of sludge, and Superfund sites, as well as the use of fertilizers and pesticides in agriculture.

Three classes of pollutants (conventional, nonconventional and toxic) were established by the Clean Water Act of 1977 and are listed in Table I. Those that have been traditionally controlled by waste water treatment are considered conventional pollutants. The toxic pollutants are those identified in the 1976 Natural Resource Defense Council (NRDC) Consent Decree. Pollutants not otherwise designated are classified as nonconventional (9). All three types have been detected in both ground and surface waters.

Table I. Pollutant Classes

CONVENTIONAL POLLUTANTS	
Biochemical Oxygen Demand	pH
Suspended Solids	Fecal Coliform
Oil and Grease	
NONCONVENTIONAL POLLUTANTS	
Chemical Oxygen Demand	Nitrogen
Ammonia	Phosphorus
Total Organic Carbon	Pesticides
Sulfides	
TOXIC POLLUTANTS	
Heavy Metals	Halogenated Aliphatics
Pesticides	Phthalate Esters
PCBs	Nitrosamines
Phenol & Cresols	Polycyclic Aromatic Hydrocarbons
Ethers	

Source: CEQ, 1982.

Table II lists the highest concentrations of selected nonconventional and toxic organic pollutants which have been measured in drinking water wells and consequently reflect possible ground water contamination. For comparison purposes, this Table also lists the highest measured surface water concentrations. Generally, the levels of organics in surface water are much lower than in individual wells. One exception is the chlorinated compounds that are related to the chlorination of surface water. Since these values represent the highest concentrations measured, they may be related to specific site contamination instead of representing a general trend.

Table II. Toxic Organic Chemicals Found in Drinking Water Wells (With Corresponding Surface Water Concentrations)

Chemical	Highest Drinking Water Well Concentration Reported (ppb)	Highest Surface Water Concentration Reported (ppb)
Trichloroethylene (TCE)	27,300	160
Toluene	6,400	6.2
1,1,1-Trichloroethane	5,440	5.1
Methylene chloride	3,000	13
Tetrachloroethylene	1,500	21
Chloroform	490	700
Carbon tetrachloride	400	30
Benzene	330	4.4
1,2-Dichloroethylene	323	9.8
1,2-Dichloroethane	250	4.8
Xylene	300	24
1,1-Dichloroethylene	280	0.5
Dibromochloromethane	55	317
Vinyl chloride	50	9.8
Chloromethane	44	12
Bromoform	20	280
1,1-Dichloroethane	7	0.2
Parathion	4.6	0.2

Adapted from 10.

The potential contamination of drinking water sources by pesticides is a primary concern of the U.S. EPA. The Office of Pesticide Programs and the Office of Drinking Water of the U.S. EPA are now in the preliminary stages of a national pesticide survey that will identify pesticides which are contaminating ground water supplies as well as assess the extent of the contamination. Pesticides have been detected in the groundwater of numerous states, often as a result of agricultural use. Some of the major pesticides that have been identified include alachlor, aldicarb, atrazine, bromacil, carbofuran, DBCP, DCPA, 1,2-dichloropropane, Dinoseb, EDB, Oxamyl and Simazine. These pesticides have been detected at levels ranging from 0.02-700 ppb (11).

The clean-up of contaminated ground water poses problems that are different from contaminated surface waters, particularly because ground water is not easily accessible. Due to the typical slow movement of pollutants in the aquifer and the relatively low degree of dispersion, concentrations of contaminants can remain high and detection can be difficult. In the case of pesticides, some of the factors that may contribute to ground water contamination include physical and chemical properties of the pesticide, application methods used for their application, and characteristics of the soil and site. Once an aquifer is contaminated, its restoration as a usable drinking water supply is extremely difficult or expensive, or both. However, with an increasing reliance on ground water it is important to identify and characterize the health risk from all pollutants, including pesticides, in drinking water that is derived from these sources. Development of improved control measures and clean-up techniques are also necessary.

Establishing Criteria and Standards

Under the Clean Water Act, the EPA prepares criteria documents to be used in developing water quality standards. The documents contain the latest scientific information on the human health and environmental effects of individual pollutants or a class of pollutants. The criteria are based on scientific data and, at times, on scientific judgment. Criteria are established as either "safe" levels for chemicals where the toxicity is presumed to have a threshold, or as incremental risk levels for presumably non-threshold chemicals such as carcinogens. These are usually estimated in regards to lifetime exposure. Ambient water quality standards are then set using these criteria and taking into consideration modifying factors, which may include societal factors, economic and technical considerations such as best available technology, natural background levels, and formal risk-to-benefit assessments. However, by necessity some of these considerations cannot involve precise values; an example is the determination of the cost of a case of cancer to society. Balancing the criteria with modifying factors to establish standards is a part of the risk management process.

Similar criteria documents are also prepared under the Safe Drinking Water Act. These documents specify recommended maximum contaminant levels (RMCLs) as nonenforceable health goals for chronic exposure and 1-day and 10-day health advisory levels. Maximum contaminant levels (MCLs), the enforceable standards, are set as close to the RMCL as is feasible considering the best available technology or treatment techniques and costs (12).

The U.S. EPA is currently developing RMCLs and MCLs for more than two dozen organic chemicals, including many pesticides, that are being considered for control under the National Revised Primary Drinking Water Regulations (NRPDWR). Table III lists some of these organics and includes RMCLs for the six pesticides that were included in the National Interim Primary Drinking Water Regulations (NIPDWR). These pesticides are those that have either been detected in drinking water, are registered for use in or around drinking water sources, or are used in a way that may result in their entering drinking water supplies.

Risk Assessment for Single Chemical Chronic Exposure and Chemical Mixtures

In 1980, EPA published 64 Ambient Water Quality Criteria Documents (AWQCDs) covering 128 priority pollutants. Of these 64 documents, 19 were on pesticides such as dieldrin, chlordane and DDT. Criteria

Table III. Organic Chemicals in Drinking Water Considered for the NRPDWR[ab]

	NIPDWR[c] RMCL (mg/l)	NRPDWR
Endrin	0.0002	Alachlor
Lindane	0.004	Aldicarb
Methoxychlor	0.1	Carbofuran
Toxaphene	0.005	Chlordane
2,4-D	0.1	Chlorobenzene
2,4,5-TP	0.01	cis- and trans-1,2-Dichloroethylene
		Dibromochloroprane (DBCP)
		Dichlorobenzenes
		Dioxin
		EDB
		Endothall
		Epichlorohydrin
		Ethyl benzene
		Heptachlor
		Hexachlorobenzene
		1,2-Dichloropropane
		PCBs
		Pentachlorophenol
		Styrene
		Toluene
		Xylene

[a] Source: Adapted from 12
[b] National Revised Primary Drinking Water Regulations
[c] National Interim Primary Drinking Water Regulations, 40 CFR 141.12

were established in nine of these pesticide documents on the basis of the pollutant's potential carcinogenicity, and in nine documents criteria were established based on toxicity. Chlorinated Benzenes and Chlorinated Phenols were determined to pose both a carcinogenic risk and toxicological risk. In addition, some of the chemicals exhibit taste and odor (organoleptic) effects. Table IV lists these 19 pesticides and EPA's criteria levels.

Table IV. Ambient Water Quality Criteria (Human Health)[a]

Priority Pollutant	Criterion Based on:		
	Toxicity	Carcinogenicity	Organoleptic
Acenaphthene			0.02 mg/l
Aldrin		0.74 ng/l	
Arsenic		22 ng/l	
Chlordane		4.6 ng/l	
Chlorinated Benzenes[b]	X	X	X
Chlorinated Phenols[b]	X	X	
DDT		0.24 ng/l	
Dichlorobenzenes	400 ug/l		
2,4-Dichlorophenol	3.09 mg/l		0.03 ug/l
Dichloropropene	87 ug/l		
Dieldrin		0.71 ng/l	
Endosulfan	74 ug/l		
Endrin	1.0 ug/l		
Heptachlor		2.0 ng/l	
Hexachlorocyclohexane[b]		X	
Hexachlorocyclopentadiene	206 ug/l		1.0 ug/l
Naphthalene		No criterion	
Pentachlorophenol	1.01 mg/l		30 ug/l
Toxaphene		5.1 ng/l	

[a] As of 3/82
[b] These documents covered a class or group of chemicals, resulting in more than one criterion.

The health assessment chapters of these documents contain the available dose-response data from animal experiments and human epidemiology studies for the chemical or class of chemicals of concern. By assessing the risks associated with various doses, acceptable daily intakes (ADIs - for systemic toxicants) or risk-specific doses (for carcinogens) were derived. These levels were divided by appropriate exposure assumptions (e.g., estimated average water consumption) to derive a criterion.

The criteria documents, and the risk assessment methodologies used in the development of criteria, address the need for protective standards. "Safe" levels are needed to ensure that populations exposed to these chemicals over a lifetime will not face potential health hazards. There is also a need to be able to predict or estimate the human health risk in a particular setting. The passage of Superfund reflected a growing concern with serious localized contamination situations at hazardous waste sites. A primary concern at these sites is the potential for migration of toxic chemicals at the site into ground and surface waters. Once this has occurred, options for controlling exposure are severely compromised. The difficulty in predicting the human health risk from exposure to specific mixtures of chemicals on a site-specific basis has become evident. For this reason, the methodologies developed to establish ambient water quality criteria and drinking water criteria were adapted by the ECAO staff to estimate risks from specific chemicals or chemical mixtures at such sites (3,4). For example, one adaptation is to characterize incremental exposures as a function of site-specific factors. This approach adds additional complexity to the risk assessment process, but should result in reduced uncertainty.

Health risk assessments consider estimates of exposure to a toxic chemical and the health hazards associated with that chemical. In agreement with the National Academy of Sciences and other scientific reports (13,14), EPA assumes that carcinogenesis is a nonthreshold phenomenon, whereas other toxic effects exhibit thresholds (i.e., doses below which no adverse effects will occur). Therefore, one must determine the potential carcinogenicity or systemic toxicity of the chemical and then proceed, utilizing one of two parallel methodologies that have been designed to address nonthreshold or threshold effects. These assessments include the basic toxicologic concept of dose-response relationships. For carcinogens, only the incremental risks associated with a pollutant level in a specific environmental medium are considered. For systemic toxicants the actual exposure is compared with levels that do not present a human health hazard. The next two sections briefly describe the methodologies used by EPA to derive ADIs and risk-specific intakes. Dieldrin and aldicarb have been used as examples to illustrate the quantitative approaches.

Carcinogens (Nonthreshold Effects). After a compound has been determined to have the potential to cause cancer in humans the relationship between risk and exposure is estimated. Two types of data are used for quantitative estimates: human studies where excess cancer risk has been associated with exposure to the agent, and animal bioassays. If human epidemiologic data are available with sufficiently valid exposure information on the compound, the data are analyzed by appropriate statistical procedures that assume a linear dose-response relationship. If the epidemiologic data show no significant carcinogenic effect when positive animal evidence is available, an upper limit of the cancer incidence is calculated, assuming that the true incidence is just below the level of detection in the epidemiologic studies. Cancer risk assessment for low exposure levels is based on estimates of the cancer potency, i.e., the slope of the dose-response curve in the low dose region. The estimated human potency is derived directly when adequate epidemiologic data are available. When animal studies must be used, the human potency estimate is calculated using the linearized multistage model fitted to the animal data (15,1). First, the upper 95% confidence limit ($q_1^*{}_{(A)}$) of the linear coefficient is determined. Then, $q_1^*{}_{(A)}$ is adjusted for exposure duration and species differences to give the estimated human potency or $q_1^*{}_{(H)}$ in $(mg/kg/day)^{-1}$ using Equation 1:

$$q_1^*\,(H) = \frac{q_1^*\,(A)\;(70)/W_A)^{1/3}}{(le/Le)\;(Le/L)^3} \qquad (1)$$

where:

$q_1^*(A)$ = animal potency $(mg/kg/day)^{-1}$
70 = assumed human weight, kg
W_A = animal weight, kg
l_e = length of exposure
L_e = length of experiment or observation period
L = lifespan of the animal.

The cube root of the ratio of body weights is used to adjust for species differences on the assumption that metabolic rate is proportional to body surface area, which is proportional to the 2/3 power of body weight. The factor l_e/L_e adjusts the actual dose to a daily dose averaged over the length of the experiment. The third factor, $(L_e/L)^3$, is used to estimate risk from lifetime exposure when the animal experiment is for only partial lifetime. This adjustment is necessary to allow for positive responses that would have occurred had sufficient time been allowed for the tumors to develop (1).

After the human potency has been calculated, the risk-specific dose (RSD, in mg/day) associated with an upper limit estimate of the excess lifetime cancer risk (e.g., 10^{-5} or 1 in 100,000 people) is determined in Equation 2:

$$RSD = \frac{70 \text{ kg } (10^{-5})}{q_1^*(H)} \quad (2)$$

This risk-specific intake rate is easily converted into a media concentration by dividing by the appropriate consumption assumptions for the exposure medium. For example, assuming a daily intake of 2 l of contaminated water and 0.0065 kg of contaminated fish, the risk specific water concentration (C, in mg/l) is calculated in Equation 3:

$$C = \frac{\text{RSD (mg/day)}}{2 \text{ l/day} + 0.0065 \text{ kg/day X BCF (l/kg)}} \quad (3)$$

where BCF is the bioconcentration factor of the chemical in fish flesh when compared with water in l/kg. The EPA has established an ambient water quality criterion for the pesticide dieldrin, which is a suspected carcinogen and priority pollutant (16). In the criteria document for dieldrin, the EPA estimated the human potency using animal data from a study by Walker et. al (17). The upper 95% confidence limit of the linear coefficient was 2.29 (mg/kg/day)$^{-1}$.

Using Equation 1:

$$q_1^*(H) = \frac{2.29 \text{ (mg/kg/day)}^{-1} \text{ (70kg/0.030 kg)}^{1/3}}{\text{(924d/924d) X (924d/924d)}^3}$$

$$q_1^*(H) = 30.37 \text{ (mg/kg/d)}^{-1}$$

The risk-specific intake is calculated using Equation 2:

$$RSD = \frac{70kg\ (10^{-5})}{30.37\ (mg/kg/d)^{-1}}$$

$$RSD = 2.30 \times 10^{-5}\ mg/d$$

A criterion was derived (from Equation 3) assuming 2 liters of water/day, 0.0065 kg of fish/day and a bioconcentration factor of 4670 l/kg:

$$C = \frac{2.30 \times 10^{-5}\ mg/d}{2\ l/d + 0.0065\ kg/d \times 4670\ l/kg}$$

$$C = 0.71\ ng/l$$

The prediction of cancer risk at a given exposure level uses the same basic approach outlined above and involves similar assumptions. When human data are adequate, the observed human potency is used directly to predict the upper bound of risk. When animal data must be used, and particularly when higher exposure levels are involved, the potency alone is not sufficient, and the complete multistage model needs to be used. The risk (r) at exposure (d) using the multistage model is shown in Equation 4 (15):

$$r(d)=1-\exp(-q_0-q_1d-q_2d^2-...) \qquad (4)$$

where the q_is are parameters in the model to be estimated by curve-fitting procedures. The incremental risk (or "excess risk") is then calculated in in Equation 5:

$$R= \frac{r(d)-r(0)}{1-r(0)} \qquad (5)$$

An estimated upper confidence limit on the excess risk (R) is used as the lifetime risk projection at exposure level (d), suitably modified as above for species differences and for duration if the animal study was for only partial lifetime.

Systemic Toxicants (Threshold Effects). This area is discussed in more detail elsewhere (1,2). To derive criteria based on noncarcinogenic responses, five types of response levels are considered:

° NOEL - No-Observed-Effect Level
° NOAEL - No-Observed-Adverse-Effect Level
° LOEL - Lowest-Observed-Effect Level
° LOAEL - Lowest-Observed-Adverse-Effect Level
° FEL - Frank-Effect Level

Adverse effects are defined as any effects that result in functional impairment or pathological lesions that may affect the performance of the whole organism or that reduce an organism's ability to respond to an additional challenge. Frank effects are defined as overtly or grossly adverse (e.g., severe convulsions, lethality).

These concepts represent landmarks that help to define the threshold region in specific experiments. Thus, if an experiment yields a NOEL, a NOAEL, a LOAEL, and a clearly defined FEL in relatively closely spaced doses, the threshold region has been relatively well-defined. Such data are very useful in deriving ADIs. To derive an ADI in water, the highest NOEL or NOAEL, or the lowest LOAEL (depending on the data available) is divided by one or more uncertainty factors (1) as illustrated in Equations 6 and 7:

$$ADI = \frac{\text{Highest NOAEL or NOEL}}{\text{Uncertainty Factor(s)}} \qquad (6)$$

To derive a criterion, the ADI is divided by exposure assumptions:

$$\text{Criterion} = \frac{\text{ADI}}{\text{Exposure Assumptions}} \qquad (7)$$

These exposure assumptions may include daily water consumption, daily fish consumption, bioconcentration factors, etc.

Some general guidelines have been established for deriving criteria from toxicity data (1). A clearly defined FEL is of little use in establishing criteria when it stands alone, because such a level gives no indication of how far removed it is from the threshold region. Similarly, a freestanding NOEL has little utility because there is no indication of its proximity to the threshold region. If multiple NOELs are available without additional data on LOELs, NOAELs, or LOAELs, the highest NOEL is used to derive a criterion.

NOAELs, LOELs, and LOAELs are most suitable for criteria derivation. A well-defined NOAEL from a chronic (at least 90-day) animal study can be used directly, divided by the appropriate uncertainty factor. A LOEL often corresponds to a NOAEL or a LOAEL. In the case of a LOAEL, an additional uncertainty factor is applied; the magnitude of the additional uncertainty factor is judgmental and should lie in the range of 1 to 10 based on review of all supportive evidence. Caution must be exercised not to substitute other toxic effects for the LOAELs. If, in reasonable closely spaced doses, only a NOEL and a LOAEL of equal quality are available, the appropriate uncertainty factor is applied to the NOEL (1).

The selection and justification of uncertainty factors are critical in using this approach. The National Academy of Science has provided guidelines for using uncertainty factors (13). "Safety factor" or "uncertainty factor" is defined as a number that reflects the degree or amount of uncertainty that must be considered when ADIs are estimated from variable toxicity data bases. It includes extrapolation based on intraspecies (human population) as well as interspecies (from animal to human) variability. When the quality and quantity of experimental data are satisfactory, a low uncertainty factor is used; when data are judged to be inadequate or equivocal, a larger uncertainty factor is needed. In those cases where the data do not completely fulfill the conditions for one category, or appear to be intermediate between two categories, an intermediate uncertainty factor is used. These issues were reviewed by Dourson and Stara (18).

The following example using aldicarb illustrates how to derive an ADI. Aldicarb is a pesticide that has been detected in ground water in Florida and elsewhere. The EPA is currently in the process of establishing a drinking water criterion for aldicarb. Weil and Carpenter (19) studied the effects of aldicarb sulfoxide on rats and determined a NOEL of 0.125 mg/kg/day. This is supported by another rat study by Mirrow et al. (20), which resulted in a NOEL equal to 0.12 mg/kg/day. An ADI is estimated for aldicarb, using the NOEL from the Weil and Carpenter study and an uncertainty factor of 100:

$$\text{ADI} = \frac{0.125\ \text{mg/kg/day}}{100}$$

$$\simeq 0.012\ \text{mg/kg/day (or 84 ug/day for a 70 kg adult)}$$

This factor of 100 accounts for two 10-fold decreases in dose based on the expected intra- and inter- species variability to the toxicity of aldicarb. Equation 7 is used to establish a criterion for a 70 kg adult:

$$\text{Criterion} = \frac{84\ \text{ug/day}}{2\ \text{l/day}}$$

$$= 42\ \text{ug/l}$$

The calculated criterion assumes that 100% of exposure to aldicarb is from drinking water, with an average consumption of 2 l/day.

Approaches to Risk Assessment for Chemical Mixtures. The above methodologies are used to estimate the human health risk from exposure to single chemicals. These methods were developed for use in deriving criteria and establishing protective standards. However, many contamination situations frequently do not deal with a single environmental pollutant, but instead involve mixtures of chemicals in more than one environmental medium (e.g., air and water). Hazardous waste sites usually involve chemical mixtures for which the total health risk to the surrounding population must be determined in order to decide what corrective action may be necessary. Groundwater contamination from the agricultural use of pesticides can also involve chemical mixtures that must be evaluated in toto.

Because there is rarely actual data on the mixture of concern, an additivity approach (adding all the toxic effects of the individual components) is sometimes used. This approach involves a number of assumptions and has certain limitations due to the lack of consideration for potential synergism or antagonism between individual chemical compounds. Further research is needed on human health effects from chemical mixtures to more accurately predict the risks involved in these common environmental situations.

The U. S. EPA has proposed guidelines for health risk assessment of chemical mixtures (7) along with guidelines for exposure, carcinogens, mutagens, and toxic agents affecting reproduction (49 Federal Register 46294). Guidelines for systemic toxicants are also being developed. For a more complete discussion of methods for the health assessment of chemical mixtures, see 3-7.

Discussion

Risk assessment is a rapidly developing science representing many disciplines including toxicology, epidemiology, biomathematics, chemistry, and engineering. As a result, the methods discussed in this paper (i.e., estimations of ADIs for systemic toxicants and risk-specific intakes for carcinogens) are being constantly improved based on new scientific evidence or conceptualizations. The U.S. EPA has been involved in this development for several years. Some examples include uncertainty factors (18), use of epidemiology data (21), hypersusceptible subgroups (22), and novel approaches to the estimation of the ADI (3,4,23).

Many dose/duration gaps exist in the available data base for systemic toxicants. Weil and McCollister (24), Weil et al. (25), and McNamara (26) have attempted to derive empirical relationships between long-term and short-term exposures and the resulting toxic effects. However, variations among chemical assessments and the limited types of chemicals considered would seem to preclude the development of a single "temporal correction factor" for toxic effects. Two areas that have been under study in an attempt to reduce the uncertainties involved in helath risk assessment are discussed below.

Graphic Display. A graphic display of all relevant data can be used to perform risk assessments (3,4). For each chemical under review, an effect-dose-duration plot is constructed, as shown for Mirex in Figure 1. Each symbol in the figure represents an experimental observation plotted on the graph at the exposure duration and average daily dose of the experiment. The dose rate in the graph has been converted to an estimated human dose rate (in mg/day/70 kg) by a model which assumes that equitoxic doses among species is related to body surface area. The exposure duration has been converted to a fraction of the animal lifespan, and this fraction has been assumed to be equivalent to the potential human exposure. Both assumptions are discussed elsewhere (3,4,23). The larger symbols indicate greater confidence in the data. Liver (L), reproductive (RP), and nervous system (NS) effects are also indicated separately.

Depending on the consistency of the pattern, a statistical or judgmental approach (the latter is presented here) could be used to interpolate or extrapolate to exposure durations for which actual data are not available. For example, in Figure 1, an estimate of the best-fitting, highest NOAEL line was made and then a corresponding ADI line was estimated. (NOTE: This graph is presented as an example only. Mirex is considered to be a carcinogen by the oral route and thus derivation of an ADI may not be appropriate). This approach provides the flexibility to estimate an acceptable intake for any duration once the NOAEL line is established. Alternatively, by dividing

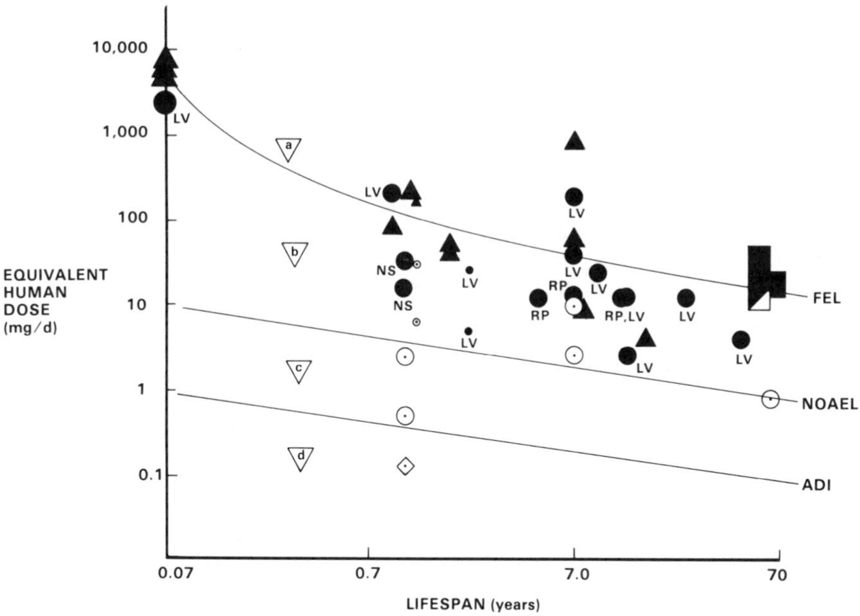

Legend:

■	Cancer Effect Level (statistically or biologically significant increase in tumors observed)
▲	FEL (Frank-Effect Level)
●	AEL (Adverse-Effect Level)
◇	NOAEL (No-Observed-Adverse-Effect Level)
⊙	NOEL (No-Observed-Effect Level)
LV	Liver
NS	Nervous System
RP	Reproductive Effects

Figure 1. Effect-dose-duration plot for all relevant human and animal oral toxicity data for mirex.

the data plot vertically into four duration segments (acute, short-term, subchronic and chronic) one could also estimate an acceptable intake for each duration using the current EPA approach that was described earlier and is used to estimate drinking water health advisories; 1-day health advisories could be estimated from acute data, and 10-day advisories from short-term or subchronic data. (The Office of Drinking Water of the U.S. EPA currently uses this approach but without the dose-duration graph.) Another advantage of the graph is that the dose axis can be divided into areas expected to cause: (a) gross toxicity or death, (b) adverse effects, (c) transitional effects, or (d) no effects (see Figure 1). This allows for the evaluation of various doses and their corresponding effect levels at any exposure duration.

This graphic approach was presented at a workshop sponsored by U.S. EPA in July 1983 and attended by over 50 scientists from academia, industry, environmental groups and government. The consensus opinion was that this approach is generally acceptable and should be used (4). The main reservations were that the approach tends to be somewhat subjective and that a statistical procedure should be designed to account for the overall quality of data in an objective manner. The U.S. EPA is currently engaged in work on these aspects of the new approach.

Verification of Uncertainty Factors. As summarized in several publications, uncertainty factors are currently recommended to estimate acceptable intakes for systemic toxicants (1,13,18). The selection of these factors in general reflects the uncertainty inherent with the use of different human or animal toxicity data (i.e., the weight of evidence plays a major role in the selection of uncertainty factors). For example, an uncertainty factor of less than 10 and perhaps even 1 may be used to estimate an ADI if sufficient data of chronic duration are available on a chemical's critical toxic effect in a known sensitive human population. That is to say that this "ideal" data base is sufficiently predictive of the population threshold dose; therefore, uncertainty factors are not warranted. An overall uncertainty factor of 10 might be used to estimate an acceptable intake based on chronic human toxicity data and would reflect the expected intraspecies variability to the adverse effects of a chemical in the absence of chemical-specific data. An overall uncertainty factor of 100 might be used to estimate ADIs with sufficient chronic animal toxicity data; this would reflect the expected intra- and interspecies variability in lieu of chemical-specific data. However, this overall factor of 100 might be used with subchronic human data; in this case the 100-fold factor would reflect intraspecies variability and a subchronic exposure extrapolation.

An overall uncertainty factor of 1000 is used to estimate ADIs with satisfactory subchronic animal data (if adequate chronic data are unavailable). It incorporates the uncertainty in extrapolating toxicity data from subchronic to chronic exposures as well as the two former uncertainty factors. Of course, additional available evidence, even though scanty, is also considered in these instances. A variable uncertainty factor between 1 and 10 is applied to estimate

ADIs using LOAELs (if NOAELs are unavailable). This uncertainty factor defines an exposure level below the LOAEL expected to be in the range of a NOAEL.

Subsequent publications have focussed on the uncertainty factor that accounts for intra-species variability. For example, Hattis and Ballew (27) investigated the uncertainty associated with human variability and concluded that a 10-fold factor was an appropriate default value but that in special cases this value may be inadequate. Erdreich and Sonich-Mullin (22) identified sensitive subgroups and concluded that such subgroups were not rare when exposures to mixtures of chemicals were considered. Recently, Calabrese (23) published a paper in regard to the variability of human response to the toxicity of chemicals. His conclusion was that large interindividual variation in response to toxic substances exist and exceeding a 10-fold factor is not uncommon.

In addition to the graphic approach for toxicity data and the verification of uncertainty factors, other areas are under study such as route-to-route conversion, high-dose to low-dose extrapolation, approaches to assess the health risk from less-than-lifetime exposures, and refinement of risk assessment approaches for chemical mixtures. All of these areas represent progress in the methods used for risk assessment of single chemicals and chemical mixtures. With the new risk assessment guidelines currently being developed, the U.S. EPA can move forward to better and more consistent health risk assessments.

Literature Cited

1. "Guidelines and Methodology Used in Preparation of Health Effects Assessment Chapters of the Consent Decree Water Criteria Documents," U.S. EPA, 1980. Federal Register 45: 79347, November 28, 1980.
2. Stara, J.F., M.L. Dourson, and C.T. DeRosa. Proc. Environmental Risk Assessment, "How New Regulations Will Affect the Utility Industry," Section 3 - Water Quality Criteria: Methodology and Applications. EPRI Contract No. WS-80-124, 1981.
3. "Approaches to Risk Assessment for Multiple Chemical Exposures." Summary of a Workshop held in Cincinnati, OH, Sept. 29-30, 1982. Contract No. 68-03-3111, Doc. No. ECAO-CIN-400. U.S. EPA, Environmental Criteria and Assessment Office, 1984.
4. "Selected Methods for Risk Assessment of Multiple Chemical Exposures," Summary of a Workshop held in Cincinnati, OH, July 12-13, 1983, Contract No. 68-03-3111, U.S. EPA, Environmental Criteria and Assessment Office, 1984.
5. Stara, J.F., Bruins, R.J.F., Dourson, M.L., Edreich, L.S., and Hertzberg, R.C. "Risk Assessment is a Developing Science: Approaches to Improve Evaluation of Single Chemicals and Chemical Mixtures. 1985. In press.
6. Stara, J.F., Hertzberg, R.C., Bruins, R.J.F., Dourson, M.L., Durkin, P.R, Erdreich, L.S., and Pepelko, W.E. "Approaches to Risk Assessment of Chemical Mixtures." In: "Chemical Safety Regulation and Compliance." Proceedings of a course held in Cambridge, MA, Oct. 24-25, 1983; F. Homburger and J.K. Marquis, Ed., Basel, Switzerland, 1985.

7. "Proposed Guidelines for the Health Risk Assessment of Chemical Mixtures and Request for Comments," U.S. EPA, 50 FR 1170, Jan. 19, 1985.
8. CEQ (Council on Environmental Quality), 14th Annual Report of the Council on Environmental Quality, 1983.
9. CEQ (Council on Environmental Quality), 13th Annual Report of the Council on Environmental Quality, 1982.
10. Pye, V.I., Patrick, R. and Quarles, J. "Groundwater Contamination in the United States," Philadelphia, PA, University of Pennsylvania Press, 1983.
11. Cohen, S.Z., Creeger, S.M., Carsel, R.F. and Enfield, C.G. "Potential for Pesticide Contamination of Ground Water Resulting from Agricultural Uses." In: Treatment and Disposal of Pesticide Wastes, American Chemical Society Symposium Series, 1984.
12. Cotruvo, J.A. and Vogt, C. J. of Am. Water Works Assoc., Nov. 1984; p. 34-38, "Development of Revised Primary Drinking Water Regulations."
13. "Drinking Water and Health," National Academy of Sciences, 1977.
14. Albert, R.E., Train, R.E., and Anderson, E. J. Natl. Cancer Inst. 58: 1537-1541, 1977. "Rationale Developed by the Environmental Protection Agency for the Assessment of Carcinogenic Risks."
15. Crump, K.S. J. Environ. Path. Toxicol. 5: 675-684, 1981. "An Improved Procedure for Low-dose Carcinogenic Risk Assessment from Animal Data."
16. "Ambient Water Quality Criteria for Aldrin/Dieldrin," U.S. EPA, No.440/5-80-019. 1980.
17. Walker et al. Food Cosmet. Toxicol. 11: 415, 1972. "The Toxicology of Dieldrin (HEOD). Long-term Oral Toxicity Studies in Mice."
18. Dourson, M.L. and Stara, J.F. Regul. Toxicol. Pharmacol. 3: 224-238, 1983. "Regulatory History and Experimental Support of Uncertainty (Safety) Factors.
19. Weil, C.S. and Carpenter, C.P. "Temik Sulfoxide. Results of Feeding in the Diet of Rats for Six Months and Dogs for 3 Months." Mellon Inst. Rept. No. 31-141. EPA Pesticide Petition No. 9F0798, 1968.
20. Mirrow, E.J., DePass, L.R. and Frank, F.R. "Aldicarb Sulfone: Aldicarb Sulfoxide Twenty-nine Day Water Inclusion Study on Rats." Mellon Inst. Rept. No. 45-18, 1982.
21. Erdreich, L.S. Envir. Health Persp. 53: 99-104, 1983. "Comparing Epidemiologic Studies of Ingested Asbestos for Use in Risk Assessment."
22. Erdreich, L.S. and Sonich-Mullin, C. "Hypersusceptible Subgroups of the Population in Multichemical Risk Assessment." ECAO-Cin-400. In "Approaches to Risk Assessment for Multiple Chemical Exposures. Summary of a Workshop" held in Cincinnati, OH, Sept. 29-30, 1982, under Contract No. 68-03-3111. U.S. EPA, Environmental Criteria and Assessment Office, Cincinnati, OH, 1983.
23. Dourson, M.D., Hertzberg, R.C., Hartung, R. and Blackburn, K. Toxicology and Industrial Health, 1985. "Novel Methods for the Estimation of Acceptable Daily Intakes."

24. Weil, C.S. and McCollister, D.D. Agric. Food Chem. 11(6): 486-491, 1963. "Relationship Between Short- and Long-term Feeding Studies in Designing an Effective Toxicity Test."
25. Weil, C.S., Woodside, M.D, Bernard, V.R. and Carpenter, C.P. Toxicol. Appl. Pharmacol. 14: 426-431, 1969. "Relationship Between Single Peroral, One-Week and 90-Day Rat Feeding Studies."
26. McNamara, B.P. "Concepts in Health Evaluation of Commercial and Industrial Chemicals." In "Advances in Modern Toxicology," Vol. 1, Part 1, 1976. M.A. Mehlman, R.E. Shapiro and H. Blumenthal, Ed. Hemisphere Publishing Co., Washington, D.C.
27. Hattis, D. and Ballew, M. "Human Variability in Susceptibility to Toxic Chemicals." EPA Contract, L.S. Erdreich, Project Officer, U.S. EPA, Environmental Criteria and Assessment Office, Cincinnati, OH, 1983.
28. Calabrese, Edward J. Regulatory Toxicology and Pharmacology 5: 190-196, 1985. "Uncertainty Factors and Inter-Individual Variation."

RECEIVED November 4, 1985

26

Risk, Uncertainty, and the Legal Process

Sheila Jasanoff

Program on Science, Technology, & Society, Cornell University, Ithaca, NY 14853

Controversies about risk form an important component in the caseload of the federal courts. These highly technical disputes are decided by judges with no special expertise in science and technology. Judicial intervention in risk management sometimes leads to questionable results because the courts do not fully appreciate the complexities of decision-making in this area. However, the courts more than compensate for their technical errors by reinforcing the basic values of fairness and openness in governmental decision-making and by ensuring that technical evidence and arguments are presented in a form that is understandable to the general public.

We live in a world that is increasingly dependent on science and technology, but our ideas about how our governmental institutions should be structured and organized go back hundreds of years before the industrial revolution. This is particularly true of our courts and our legal system. We do not expect our judges to be technical experts. Indeed, the idea of an expert judiciary is alien to our common law tradition. Unlike the countries of Continental Europe, we do not have separate administrative tribunals to review the decisions of government agencies. Writing at the end of the nineteenth century, the great English constitutional scholar Dicey observed that "this idea is utterly unknown to the law of England, and indeed is fundamentally inconsistent with our traditions and customs" (1). While European nations train their judges to become experts in particular fields of law, our legal practices, such as assigning cases by lot, seem designed to prevent any undue concentration of specialized knowledge. When a judge joins the West German Federal Constitutional Court, for example, he or she is assigned an area of competence and deals with all cases arising in that area, such as family law, tax law, religious or cultural matters, and the civil service. By contrast, we expect our Supreme Court justices to deal equally with a whole range of issues from sex discrimination to environmental pollution. We demand no special expertise, not even prior judicial experience.

0097-6156/86/0315-0462$06.00/0
© 1986 American Chemical Society

Specialization, as we know, has its drawbacks. Depth of knowledge can be negated by narrowness or triviality. Our society has jestingly defined a specialist as someone who knows more and more about less and less until he or she knows everything about nothing. Anyone who works in an academic environment knows that there is some substance to this charge. In selecting our judges, however, we often seem to reach for the opposite extreme, seeking the perfect generalist who knows nothing about everything.

Unfortunately, as the century draws on our problems seem to demand decision-makers who know everything about everything. Judges are confronted with issues of increasing technical complexity, and nowhere more so than in the area of chemical risk management. Courts, in our system of government, are the last resort for answering a host of questions about the risks presented by toxic substances to health and the environment. In deciding whether a regulatory decision about a hazardous chemical is legally valid, courts must consider a variety of subordinate issues, many of which are highly technical. If the substance causes cancer in laboratory animals, does it also cause cancer in humans? Has the regulatory agency looked at the best available evidence, used the most reasonable assumptions, and made the most reliable estimate of risk? Will the proposed standard protect only healthy people, or also the old, the sick and the very young? Are the recommended controls both economically and technologically feasible?

Our legal system provides exceptionally liberal opportunities for parties to bring such questions before the courts. It also ensures that arguments about such issues are developed and presented at a high level of technical sophistication. Agencies, for example, are required by law to compile detailed scientific records and to engage in reasoned decision-making. Private firms routinely employ large stores of expertise in preparing their cases against governmental regulation. Congress has even designed laws that try to place environmental and public interest groups on a more equal footing with industry by funding their use of experts in regulatory proceedings. Yet the claims and counterclaims made by all these parties eventually are reviewed by judges who have no formal training in toxicology, engineering, economics, or risk assessment.

Of course, our judges could elect to lighten their responsibilities by deferring to the technical opinions of agencies and other expert authorities. European courts have followed just this course. In the European civil law systems, judges enjoy even greater power than their common law brethren to second-guess the government's technical determinations, but they seldom exercise this power. As American administrative law has evolved in the last fifteen years, the federal courts have interpreted their duty to supervise government agencies much more conscientiously. Under the direction of such noted jurists as Judge Bazelon, Judge Skelly Wright, and the late Judge Leventhal of the D.C. Circuit Court of Appeals, the federal courts have concluded that they should closely scrutinize the way in which agencies collect and analyze technical information. It is a basic rule of administrative law that courts must not substitute their judgment for that of the agencies. Judges, however, can and do make searching inquiries into the records developed by regulatory agencies to ensure that administrators have taken a "hard look" at all the relevant factors, based their factual conclusions on "substantial evidence," and convincingly explained the reasons for their policy choices (2-3).

Many in the scientific community are skeptical about the ability of technically untrained judges to carry out these supervisory tasks. How can judges with no scientific background hope to adjudicate conflicts among experts in highly specialized fields of knowledge? Can courts really understand the criteria that scientists use to judge the relative strengths and weaknesses of scientific arguments? We know that the fine points of risk and probability are lost on most members of the lay public. There is evidence that judges are not likely to fare any better. They too can be easily swayed by precise numbers, failing to understand the multiple uncertainties buried in probabilistic statements about risk (4-5). Indeed, when technical arguments become too complex, judges may be inclined to retreat from the scientific aspects of decision-making and decide cases purely on the basis of policy and politics. These and similar fears account for the continuing popularity of the idea that a "science court" or some other "objective" forum should adjudicate conflicts that seem to lie beyond the grasp of our archaic, generalist judicial system.

In the remainder of this paper, I would like to address some of these concerns in the light of recent judicial decisions involving toxic chemicals. There is no doubt that examples can be found to show that some fears about the lay judiciary are warranted. Judges do get into serious muddles and mistakes when dealing with highly technical issues. Yet I hope to show in the second part of the paper that these problems are counterbalanced by important social and political values that generalist courts preserve in dealing with controversies at the frontiers of scientific knowledge. This is our ultimate justification for retaining the institutions we have inherited from the past.

Courts and Science: A Failed Relationship?

Looking at the recent history of environmental litigation, one can point to numerous cases that raise questions about the judicial capacity to distinguish good science from bad. It is not unusual to find judges agreeing with agency decisions that scientists believe were based on unsound science. The involvement of the courts in EPA's efforts to regulate chemical carcinogens is one example. In the agency's earliest years, the courts were a major force in requiring EPA to take a tougher stand on potentially carcinogenic pollutants. The District of Columbia Circuit Court, which reviewed a series of agency decisions concerning the chlorinated hydrocarbon pesticides, instructed EPA that it would look very suspiciously at decisions not to regulate substances showing positive results in animal tests (6). Critics of EPA's current regulatory priorities charge that such judicial directives were responsible for the agency's disproportionately heavy investment in carcinogen regulation during the past decade.

It is possible to draw up a powerful indictment of the judiciary's role in scientific decision-making from the string of cases involving EPA's regulation of DDT, aldrin/dieldrin, heptachlor/chlordane, and Mirex. First, by attaching great weight to evidence of animal carcinogencity, the courts helped extend the principle of the Delaney Clause into the area of pesticide control. In so doing, they enlarged the applicability of a provision that scientists consider almost synonymous with bad science. Second, the courts arguably deferred too far to EPA on some of its more questionable science policy decisions. For example, the courts accepted EPA's use of "cancer principles" (7) as a basis for making risk determinations, although scientists, then as now, agreed that the principles were conceptually simplistic, unscientific, and dangerously rigid. Third, the judicial decisions reflected and even magnified the public's fear of cancer, affirming the scientifically naive belief that even the smallest contact with a carcinogen is dangerous enough to merit strict regulation. The pesticide cases thus helped undermine reasoned debate about cancer risks. Yet the public's demand for complete protection against chemical carcinogens is unrealistic, and an exclusive focus on identifying and controlling these substances diverts needed resources from regulating more substantial risks to public health.

Another criticism leveled against the courts is that they permit policy considerations to override valid scientific findings. One frequently cited example is the outcome in the Agent Orange case, which involved a massive lawsuit by Vietnam veterans against the manufacturers of the herbicide (8). Judge Weinstein, who presided over the latter stages of this litigation, acknowledged that the proof of causation supplied by the veterans' groups was so weak that it probably would not support a finding of liability against the defendant manufacturers. Why then did the judge encourage and accept a settlement that required the manufacturers to pay millions of dollars to the veterans? Some have argued that the outcome was dictated by the judge's personal policy preferences. His view that some reparation should be made to the Vietnam veterans took precedence over his interest in letting science speak for itself.

Criticism of the courts in the area of risk management is by no means one-sided. Environmental and labor groups are quite prepared to join with industry in complaining about judicial misreadings of scientific information. One decision that has been roundly criticized by public interest groups is the Fifth Circuit Court's overruling of the attempt by the Consumer Product Safety Commission (CPSC) to ban urea-formaldehyde foam insulation (UFFI) (9). In the course of its rule-making on UFFI, CPSC compiled 102 volumes of supporting data and analysis. To facilitate judicial review, the agency identified the parts of the record it relied on, the parts it rejected, and its reasons for doing so. Yet the court concluded that the proposed ban was not supported by substantial evidence.

One of CPSC's compelling reasons for banning UFFI was a study done by the Chemical Industry Institute of Toxicology (CIIT) which showed that formaldehyde causes cancer in rats at two exposure doses (10). CPSC used the data from the CIIT study to estimate the risk of cancer to people living in homes insulated with UFFI, a product that is known to emit gaseous formaldehyde when it is improperly installed. As in any risk assessment, CPSC had to make several assumptions about the probable duration and intensity of human exposure to the toxic agent. Most of these were challenged by the formaldehyde industry. To obtain a numerical estimate of the human risk, CPSC applied to the CIIT data a mathematical model (Global 79) which statistically calculates an upper limit on risk and incorporates some standard assumptions about the mechanism of cancer causation. Formaldehyde manufacturers challenged many of CPSC's assumptions, as well as the legitimacy of applying any mathematical risk extrapolation model to the available toxicological data on formaldehyde. The court agreed with the industry that CPSC's exposure measurements were open to question. More important, the court rejected the agency's attempt to derive a numerically precise estimate of risk from inherently imprecise experimental data.

In tracing the court's reasoning, one can point to apparent gaps in the judges' understanding of the theory and practice of risk assessment. For example, the court took CPSC to task for estimating the cancer risk to humans on the basis of a "single experiment, particularly one involving only 240 subjects." This objection suggests that the court was unfamiliar with the principles of animal testing, which recognize that it is appropriate to test small numbers of animals at high doses in order to detect small increases in risk caused by toxic chemicals. Larger numbers of animals cannot normally be used in bioassays because of the prohibitive costs of running such experiments. The size of the CIIT study was well in accord with accepted standards for cancer testing.

The court was also on questionable ground in suggesting that CPSC should have used more than one study in its risk assessment. Under appropriate circumstances, toxicologists are prepared to accept even one well-designed and well-conducted animal study as a reliable basis for risk extrapolation. Indeed, the court's skepticism about the CIIT data was not shared by a panel of experts subsequently convened by EPA to evaluate the health effects of formaldehyde. This consensus workshop found that the CIIT bioassay results were suitable for use in a risk assessment, whereas the available epidemiological studies on formaldehyde were inappropriate for this purpose (11).

The Fifth Circuit's decision on formaldehyde has won support from scientists who consider CPSC's choice of the Global 79 model ill-advised. But even if one agrees with the case on substantive grounds, one may have to accept the fact that the court reached the right result for the wrong reasons. The formaldehyde opinion creates the impression of a technically naive court demanding far greater certainty than the art of risk assessment is currently able to provide. Moreover, in rejecting CPSC's proposal on apparently scientific grounds, the court overlooked the blending of science and policy that is involved in any attempt to assess carcinogenic risk. The principle of judicial restraint in American administrative law directs that courts should be very careful not to second-guess the regulatory agencies in the area of science policy. Even if an agency decision appears technically unsound, courts should refrain from trying to correct the agency unless there has been a clear abuse of discretion. Not surprisingly therefore, most analysts of the formaldehyde case have concluded that the Fifth Circuit misconstrued its role in reviewing risk assessment decisions and impermissibly substituted its own judgment for that of CPSC (3, 12-13).

The Supreme Court has also spoken on the issue of risk assessment in ways that are troublesome and confusing. The occasion for the High Court's involvement with chemical risks was a lawsuit by the petroleum industry challenging the new workplace standard for benzene promulgated by the Occupational Safety and Health Administration (OSHA) (14). One disturbing aspect of the court's decision was the failure of the nine justices to articulate any clear principles to guide agencies in future cases involving toxic chemicals. Justice Rehnquist exemplified the reluctance of some judges to review the technical aspects of controversies about risk. Sidestepping the complex scientific arguments offered by the parties, Rehnquist resurrected a doctrine considered dead since the early days of the New Deal and declared that the statutory section under which OSHA had issued the benzene standard was unconstitutional. The appropriate solution, in his view, was to return the statute to Congress for clearer legislative guidance.

The remaining eight justices engaged in a detailed examination of the record, but were unable to agree whether or not OSHA had mustered an adequate scientific argument in support of the proposed benzene standard. A plurality of four found that OSHA had not made a crucial factual finding. Specifically, the agency had failed to show that there was a significant risk to the health of workers at the current exposure standard. Yet if we accept the agency's contention that there is no safe threshold of exposure to carcinogens--a position for which there is considerable support in science--then it is arguable that the agency's showing of risk should have been judged sufficient as a matter of law. Benzene is known to cause leukemia in humans, and the no-threshold hypothesis implies that this risk cannot be ruled out at any exposure level. Whether the risk at the 10 ppm exposure standard is "significant" is in large part a policy judgment, and OSHA's determination on this issue was entitled to great deference from the court.

The benzene decision was construed by the federal regulatory agencies as a mandate from the Supreme Court to perform quantitative risk assessments in the course of standard-setting. But in our present woefully incomplete state of knowledge about cancer causation and about the risks of particular chemicals, we can ask whether this was a wise directive. For many toxic substances, the data are too sketchy to permit quantitative analysis, and risk assessment can only create an artificial impression of precision. Indeed, given the uncertainties involved in exposure assessment, high-to-low dose extrapolation, and extrapolations from animal to man, most regulatory decisions based on mathematical models run a serious risk of being dismissed by the courts as "arbitrary and capricious." Yet this is just the opposite of what four members of the Supreme Court concluded in the benzene case.

A Brighter Appraisal

If we concede that courts, at their best, are unable to resolve genuine scientific controversies, and that they often are confused or just plain wrong on technical matters, then why should they continue to play such a major role in risk management? The answer is that, despite their technical shortcomings, courts uphold values that are central to our traditions of public decision-making and that it is crucial to maintain these values in the process of regulating technological risks. I would argue, as well, that courts can successfully assert these values even in cases where they are not in full command of the scientific and technical arguments.

The first of these values is fairness. In the U.S. constitutional tradition, judges are steeped in the notion of due process. This means that they are acutely aware of the individual's right to be heard before being deprived of an important right or liberty through governmental action. Courts are specially sensitive to agency decisions that seem to curtail the right to be heard without providing adequate justification.

The UFFI case I discussed earlier seems much less problematic if we view it as a judicial attempt to ensure fairness in the regulatory process. Congress authorized CPSC to regulate toxic substances under two major statutes, the Consumer Product Safety Act (CPSA) and the Federal Hazardous Substances Act (FHSA). The latter statute calls for more formal rulemaking procedures than the former. Under the FHSA, evidence must be presented according to legal rules of evidence, and parties have a right to cross-examine witnesses. CPSC, however, chose to regulate UFFI under the procedurally less demanding statute. The Fifth Circuit concluded that this choice was not properly motivated, since CPSC had not established to the court's satisfaction that the public interest would be better served by a proceeding under the CPSA. Lawyers may well differ as to whether the court correctly assessed the relationship between the two statutes or the extent of CPSC's discretion to choose between them. But the Fifth Circuit's overall judgment that CPSC paid insufficient heed to the UFFI manufacturers' procedural rights deserves attention. And this part of the court's opinion can be taken as authoritative even if we dismiss the comments on CPSC's risk assessment as technically flawed.

The judicial concern for fairness also explains the outcome in "toxic tort" lawsuits such as the Agent Orange case. In such litigation, the risk is not purely conjectural, since the case starts with someone who has actually been injured. The problem for the plaintiff is to prove that the injury was caused by an identifiable polluting activity or toxic product. But courts have recognized that it is almost impossible for plaintiffs to prove causation in toxic tort cases with as much certainty as in more conventional products liability cases. The long latency and uncertain etiology of diseases induced by toxic substances, as well as the possibility of multiple exposures and synergistic effects, enormously increase the plaintiff's burden of proof (15-16). For courts confronted by complaints about toxic substances, the primary challenge is to find equitable ways of allocating the cost of the uncertainties that prevent straightforward showings of causation. Fairness demands that there be some causal link between the plaintiff's injuries and exposure to the identified toxic agent. Yet fairness also requires that victims be compensated on something less than a watertight factual showing. Otherwise, plaintiffs would rarely recover, and the cost of society's imperfect knowledge would always fall on those who are injured and those who are financially least prepared to bear the cost. The Agent Orange settlement is one dramatic illustration of the fact that judges must undertake a complex weighing of scientific and equitable considerations in toxic tort cases. It underscores the fact that courts are instruments for doing justice as well as for establishing truth. And the outcome in that case teaches us that, when necessary, courts may try to satisfy the demands of justice even if they cannot meet the demands of truth.

A second fundamental value that courts have imposed on government agencies in the area of risk regulation is that state power should not be used in arbitrary fashion. One important corollary is that agencies should not act to restrain commercial activity unless it presents more than a negligible risk to public health and safety. This principle was invoked by the D.C. Court of Appeals in reviewing a decision by the Food and Drug Administration (FDA) to ban plastic beverage containers made of acrylonitrile polymer (17). In supporting the ban, FDA constructed a theoretical argument from the most basic scientific principles. Relying on the second law of thermodynamics, or the diffusion principle, the agency argued that any two substances in contact must diffuse into each other, whether or not the amounts are detectable. Therefore, FDA assumed that free acrylonitrile from the container walls would migrate into the beverage and so become a "food additive." Since acrylonitrile was discovered to be an animal carcinogen, FDA further argued that such migration could not be tolerated under the Delaney clause. The only remedy the agency saw was to prohibit the use of containers made of acrylonitrile.

In his famous essay on two cultures (18), C. P. Snow defined the class of humanists (or non-scientists) as those who do not understand the second law of thermodynamics. If we accept Snow's characterization, then we must conclude that the court in the acrylonitrile case patterned with the humanists, for it remained unpersuaded by FDA's abstract theoretical argument. Writing for the D.C. Circuit, Judge Leventhal held that Congress did not intend FDA to use its regulatory power just because it could cite a scientific principle to show that diffusion might occur. Rather, Leventhal reminded FDA that any statute enacted by Congress implicitly instructs the implementing agency not to regulate insignificant risks to public health and safety. If agencies do not refrain from acting in these de minimis situations, then they are in effect overextending the power delegated to them by the legislature.

The Supreme Court's benzene decision, mentioned earlier, can also be read as an injunction against regulating insignificant risks. By asking OSHA to quantify the risk at the existing exposure standard, the court was seeking reassurance that there was indeed a genuine health hazard confronting workers exposed to benzene. Without such a minimal showing, the court could not countenance OSHA's imposing millions of dollars of additional costs on industry in the form of a stricter exposure limit for the workplace. The agency had a positive obligation to explain why it considered benzene worth regulating, and its explanation should have been couched in terms that the reviewing court could easily understand.

The third and perhaps most important value that courts have incorporated into public decisions about risk is the principle that governmental actions should be explained to the concerned public. It may be hard for us to remember a time when agencies dealt in secret with the interests they were responsible for regulating, and made significant policy decisions without notification or opportunities for the public to comment. Yet the era of open government is not so old even in the United States and its equivalent has yet to materialize in many other Western democracies.

Since Congress enacted the Freedom of Information Act and other laws mandating openness in government, the courts have played a crucial role in ensuring that public officials respect the spirit as well as the letter of this legislation. Under the watchful eye of the courts, openness has gradually become a part of our administrative culture. For example, government agencies, no matter what their ideological or political orientation, are committed to the idea of producing a public record and explaining their decisions in terms of the material contained in that record. It is also accepted by decision-makers both in and out of government that even scientifically complex proceedings should be made accessible to the public. The proliferation of risk assessment guidelines from EPA and other regulatory agencies reflects a desire to make technical decisions as transparent as possible. No doubt these initiatives impose more rigidity on risk assessment than scientists would ideally wish to see. But such methodological guidelines serve two valuable purposes in return. They inform the public how the agency proposes to make risk management decisions, thus opening a potentially arcane process to democratic control. They also introduce regularity and predictability into agency analyses, thus providing safeguards against arbitrary action.

It is a tribute to the success of the courts that sudden, unexplained shifts in policy have relatively little chance of success in the U.S. administrative process. The controversies generated by EPA during the first Reagan term illustrate this point. Public reaction was immediate and highly negative when EPA attempted to change its principles of carcinogenic risk assessment without adequate notice or open discussion (19). Some of the most pointed criticism came from members of the scientific community who recognized the ideological rather than scientific reasons underlying the agency's proposed changes. A similar outcry greeted the decision by EPA's Office of Toxic Substances not to regulate formaldehyde as a priority substance under the Toxic Substances Control Act (20). Again, the major complaint against the agency was its apparent deviation from the scientific principles previously used within EPA for carcinogenic risk assessment. To make matters worse, the departure from prior practice occurred without public explanation and against the recommendations of the agency's own scientific staff. Under pressure from Congress and independent scientists, intensified by the threat of a lawsuit, EPA eventually reopened the formaldehyde case and agreed to reconsider the substance according to its own established guidelines.

The courts have not only insisted that regulatory actions be explained, but have pressed administrators to improve the quality of their explanations. Close scrutiny by the courts was instrumental in getting administrative agencies to sharpen the concept of a record, particularly in the context of informal rulemaking (21). As a result, there is now something approaching a common understanding among federal agencies about what documents and analytical materials should be included in a record and how these should be placed before the public and the reviewing courts. Judicial insistence that explanations be framed in ways that judges can understand has also forced greater clarity in administrative decision-making. Impatience with inconsistencies and omissions in agency reasoning can lead a court to invalidate a regulatory decision and to demand further clarification. In the benzene case, for example, OSHA's somewhat confused presentation of the arguments in support of the new exposure standard clearly troubled the Supreme Court, as did the agency's failure to explain why it had not prepared a formal characterization of the risk at the existing standard (22).

Judges can use the obligation to explain as a powerful weapon against passivity and delay in the regulatory process. Since the early 1970s, federal courts have insisted that it is important for government officials not merely to justify positive decisions to act, but to explain why no action has been taken on a toxic hazard. In an early case involving the regulation of DDT, for example, the Secretary of Agriculture refused to respond to a petition from the Environmental Defense Fund (EDF) that federal registration for the pesticide be suspended (23). The Secretary's apparent intention was to wait for the completion of additional studies before initiating formal regulatory proceedings on DDT. The D.C. Circuit Court, however, held that delay in this case effectively amounted to a refusal of the suspension request. Yet this negative decision was essentially unreviewable because it was reached without explanation and without the creation of a supporting record. Accordingly, the court remanded the case to the Secretary with instructions either to determine the suspension issue afresh or to provide reasons for his silent refusal to act on EDF's petition. Such judicial mandates have helped drive home the point that governmental agencies are accountable to the public whether they choose to regulate or to refrain from action.

The Science Court Reconsidered

I have suggested a number of reasons why courts should retain a major role in shaping risk management decisions. One may well ask, however, whether the burden on the generalist judicial institutions we have today could not be eased by supplementing them with one or more "science courts"--forums that would be specially equipped to deal with technical controversies. Proponents of the concept note that a specialized science court would add to, not replace, the ordinary hierarchy of civil courts (24). The duties of the science court would be restricted to the factual side of technical disputes. Legal issues would still be decided by non-specialized state and federal tribunals, as they are at present. According to advocates of the science court, the two-tier adjudication of technical controversies would have an entirely beneficial effect. Courts staffed by experts would first give a sophisticated hearing to the technical component of the dispute. Subsequent adjudication of the legal issues would take advantage of the understanding of equity and due process built up through centuries of experience in the ordinary courts.

A close look at modern risk controversies suggests why this idea, despite its surface appeal, is not likely to prove workable. To begin with, the concept of the science court is based on the premise that technical controversies can be neatly separated into factual (or technical) and legal components. Yet anyone conversant with risk disputes must quickly conclude that the reverse is more often true. Risk assessment, in particular, is known to be a highly subjective process in which technical determinations are almost invariably bound up with discretionary judgments (25). Experts engaged in risk assessment are guided not merely by their specialized knowledge, but by intuition and even personal values. It is difficult to treat something so central as the selection of a high-to-low dose extrapolation model in carcinogen risk assessment as a purely scientific matter. The choice of a more or less conservative statistical model is likely to be guided by too many extra-scientific considerations, such as the decision-maker's personal appraisal of the overall risk that chemicals present to public health or environmental quality. Under current conditions of uncertainty, the choice of a risk assessment model becomes a mixed scientific and policy issue. Where science does not uniquely dictate the choice, agencies must select the analytical approach most in keeping with their statutory mandate, and non-specialist courts must review the decision to make sure that the law has been correctly understood.

Even if the lines between science and law or facts and values were easier to draw, something significant would be lost by assigning scientific disputes to specialist courts. As I mentioned above, one of the most valuable services the courts can perform in the risk management process is to cement the idea that governmental decisions, however technical, must be explained in ways that judges and private citizens can understand. If this principle were abandoned, then official decision-makers could all too easily protect themselves from public scrutiny, concealing inherently arbitrary decisions beneath a veil of technical discourse. By requiring experts to be answerable to lay standards of rationality and consistency, generalist courts provide an indispensable check against the loss of public control over regulatory decisions. A proliferation of science courts might tilt the balance too far in favor of government by technocrats who are not fully accountable to the public.

It is important to remember, as well, that most of the disputes that come to court in the area of risk management have no clearcut scientific solutions. These controversies arise at the frontiers of science where, almost by definition, there are no established or widely accepted answers to scientific questions. Moreover, disputes seldom come to court until they have matured in forums where the technical issues are thoroughly debated: scientific advisory committees, administrative hearings, internal or external peer review, and pretrial negotiations. If technical conflicts persist after all these processes are exhausted, the reason very probably is that the matter in controversy cannot be settled by scientists according to purely scientific norms.

Finally, let me suggest that we avoid the temptation to overread judicial decisions concerning risk management. It would be silly, for example, to argue as some industry representatives have done that the UFFI decision is a prohibition against worst-case risk analysis or the use of a linearized multistage model for high-to-low dose extrapolation. Similarly, the case should not be read as providing support for any particular techniques of animal testing or exposure assessment. Judges have no authority to issue such highly technical directives to administrative agencies. Courts are at their best in applying concepts that have acquired meaning through generations of prior judicial interpretation, such as due process, reasonableness, or de minimis harm. Their judgments about risk management are most authoritative when explained in terms of these concepts. Let us not fall into the trap of preserving our generalist courts but treating their legal decisions as the pronouncements of remote scientific oracles.

Courts in Context

In criticizing the way courts handle technical disputes, it is easy to lose sight of the wider cultural and historical context within which courts operate. In closing, I would like to call attention to two features of this context, one institutional and the other temporal.

We should recall at the outset that lack of expertise is not uniquely a characteristic of the courts in our political system. Ours is a government of generalists. There are few technical experts in

Congress, and it is debatable how much influence is really exercised by congressional advisory bodies such as the Office of Technology Assessment and the General Accounting Office. Legislation, even in technologically complex areas, emerges through a haphazard process of tradeoffs and compromise. The statutes enacted by Congress are merely frameworks for continuing negotiation between government and private interests. They do not resolve all relevant policy issues and they cannot be regarded as blueprints for scientifically exact decision-making. On the contrary, the federal legislative process often produces technically naive formulations, such as the Delaney Clause, which create severe problems of interpretation and implementation for both courts and administrative agencies.

The preference for generalists rather than technicians carries over to the upper echelons of the executive branch. Law is considered a better training ground for high administrative officials than science. By 1984, for example, seven out of EPA's eight acting and permanent administrators had been lawyers. Unlike Japan and some European countries, we do not seek to build reservoirs of specialized knowledge and experience by centralizing policy-making for science and technology within a single cabinet department. Though our presidents have to display leadership in economic, industrial and defense policy, we do not expect them to approach these tasks with extraordinary technical expertise. Jimmy Carter, who was known for his technical knowledge and grasp of detail, fared much worse in the White House than his successor Ronald Reagan.

Our generalist courts, then, are a natural extension of the much larger network of non-technical institutions that govern the United States. These institutions accurately reflect this country's pluralistic political process. Government's role is not to impose its own policy choices on the people, but to arbitrate among competing interests and to let policy evolve out of interactions among disparate groups following their own self-interests. It would be inconsistent with this vision of government to have any single interest dominate our political or judicial institutions, even so powerful an interest as science and technology. A public mistrustful of elites feels more comfortable placing science and technology on a par with other private interests, competing for resources and power, instead of entrusting them with the reins of government.

The second point to remember about the courts is that they change over time in response to wider social and political processes. Periods of activism, in particular, are succeeded by periods of relative passivity, as, for example, during and after the New Deal. The surge of environmental and health and safety legislation in the 1970s called forth a spirit of active intervention from the courts, who saw themselves as the institutions best qualified to effectuate the apparent will of the legislature. In the mid-1980's, however, we are dealing with an administrative culture that is very different from the one that predominated before the environmental movement began. Agencies now are much more sensitive to the need for openness in technology-related decision-making. I have suggested, too, that the obligation to explain technical decisions in terms understandable to the public is now recognized as an integral part of our administrative process. If this is so, then the courts have accomplished much that they set out to do, and we can expect them to play a less active role in supervising risk decisions over the next decade. Yet the period of activism, despite some errors, flaws, and wrong turns, has played a crucial part in ensuring that our administrative process remains under democratic control.

Conclusion

The relationship between science and the courts in the United States cannot be summed up with a neat formulaic phrase. Recent judicial decisions in the area of risk management provide numerous examples of faulty scientific analysis by courts untrained in the intricacies of risk assessment. There is reason to suspect that courts do not fully understand concepts of probability and uncertainty and that they will avoid looking closely at technical controversies if they can find an alternate basis for resolving disputes. Yet judicial review has imported into the risk management process values that we cherish in public administration: a sense of fairness, an aversion to arbitrary exercises of power, and a preference for openness and understandable reasoning. I would not deny that there are costs associated with our current approach to judicial decision-making. The process is long and untidy and does not always lead to the right or the best results. But in sacrificing efficiency and technical perfection, we may have gained from our courts something far more precious: humanism and an affirmation that science and technology must be put to use in ways consistent with our culture's democratic values.

Literature Cited

1. Dicey, A. V. "Introduction to the Study of the Law of the Constitution"; St. Martin's Press: New York, 1959; p. 203.
2. Rodgers, W. H., Jr. Georgetown Law J. 1979, 67, 699-727.
3. McGarity, T. O. Sci., Tech., and Human Values 1984, 9, 97-106.
4. Tribe, L. H. Harvard Law Rev. 1971, 84, 1329-93, 1810-20.
5. Finkelstein, M. O.; Fairley, W. B. Harvard Law Rev. 1971, 84, 1801-9.
6. EDF v. Ruckelshaus, 439 F.2d 584 (D. C. Cir. 1971).
7. Karch, N. In "Decision Making in the Environmental Protection Agency", IIa; National Academy of Sciences, 1977, pp. 119-206.
8. In Re "Agent Orange" Product Liability Litigation, 597 F.Supp. 740 (E. D. N. Y. 1984).
9. Gulf South Insulation v. CPSC, 701 F.2d 1137 (5th Cir. 1983).
10. Kerns, W. D.; Pavkov, K. L.; Donofrio, D. J.; Gralla, E. J.; Swenberg, J.A. Cancer Res. 1983, 43, 4382-92.
11. "Report on the Consensus Workshop on Formaldehyde"; Little Rock, Ark., 1983, pp. 138-9.
12. Ashford, N. A.; Ryan, C. W.; Caldart, C. C. Science 1983, 222, 894-900.
13. Davis, D. L. Columbia J. Env. Law 1985, 10, 67-109.
14. Industrial Union Dept. AFL-CIO v. Amer. Petroleum Inst. 448, U.S. 607 (1980).
15. Yale Law J. 1981, 90, 840-62.
16. Delgado, R. Calif. Law Rev. 1982, 70, 881-908.
17. Monsanto v. Kennedy 613 F.2d 947 (D. C. Cir. 1979).
18. Snow, C. P. "The Two Cultures and the Scientific Revolution"; Cambridge University Press: New York, 1959.
19. Marshall, E. Science 1983, 220, 36-7.

20. "Hearing on Formaldehyde: Review of Scientific Basis of EPA's Carcinogenic Risk Assessment"; U. S. House of Representatives, 97th Cong., 2nd Sess. 1982.
21. Pedersen, W. F., Jr. Yale Law J. 1975, 85, 38-88.
22. Brickman, R.; Jasanoff, S.; Ilgen, T. "Controlling Chemicals: The Politics of Regulation in Europe and the United States"; Cornell University Press: Ithaca, N. Y., 1985; p. 123.
23. EDF v. Hardin, 428 F.2d 1093 (D. C. Cir. 1970).
24. Kantrowitz, A. Am. Sci. 1975, 63, 505-9.
25. "Risk Assessment in the Federal Government: Managing the Process"; National Academy of Sciences, 1983.

RECEIVED November 4, 1985

REGULATORY ASPECTS

27

Industry Perspective on Pesticide and Ground Water Legislation

Loy C. Newby and Charles G. Rock

Agricultural Division, CIBA-GEIGY Corporation, Greensboro, NC 27419

Groundwater must be protected - society is in agreement. It is the focus and level of regulatory protection which concerns the agricultural chemical industry. Groundwater is not pure. It contains many substances, both natural and man-made, for most of which, minimal safety information is available. With continually improving analytical capability, additional products are certain to be detected in even the most pristine aquifers. A few pesticides have already been detected and it seems likely that others will be found. In most cases, these will be substantially below the levels which could result in adverse effects. Regulatory efforts should be directed to establishing acceptable levels, while industry should act to minimize the movement of all chemicals to groundwaters. Acceptable levels of chemicals in groundwater should be uniform throughout the nation. They should be set at the federal level or by uniform procedures among the state agencies.

A recent national public opinion survey (1) conducted by the National Agricultural Chemicals Association (NACA) revealed that 47% of all Americans believe that pesticides and farm fertilizers are a "major national problem;" 55% include chemicals in drinking water in this category; and 48% of all Americans also consider pollution of underground water as a "major national problem." These results reinforce the timeliness and importance of this three-day symposium.

The viewpoints expressed herein generally reflect those of the agricultural chemical industry as represented by NACA. NACA is a non-profit trade organization located in Washington, DC and represents those companies who manufacture and formulate pest control products employed in agricultural production.

Two and one half years ago, the agricultural chemical industry concluded that groundwater was going to be the issue of the decade. Consequently, NACA established an ad hoc Group on

0097-6156/86/0315-0478$06.00/0
© 1986 American Chemical Society

Groundwater Protection currently chaired by the principal author. The co-author is a member of NACA's State Affairs Committee (SAC) which addresses proposed state legislation and regulations including those dealing with groundwater.

It seems redundant, almost trite, to say that groundwater is a valuable national resource, but it is and it must be protected -- no one disagrees! The degree of protection and the means may not yield the same consensus, however.

Excellent papers have been presented during this symposium on the retrospective and prospective appearance of agricultural chemicals in groundwater. Subject matter has included contributing factors to mobility such as soil and pesticide physical and chemical parameters, how to incorporate these into predictive models and toxicological significance and risk assessment of pesticides in groundwater. It is certainly appropriate that the topic of legislation/regulation follows the subject of toxicological significance because any regulatory action should be dependent on the former. The mere presence of a specific chemical in groundwater does not necessarily mean that there is cause for alarm about health effects.

It is acknowledged that certain agricultural chemicals have been detected in groundwater and others may be found in the future. Detects have been at very low concentrations and the findings have generally been associated with unusual combinations of soil type, high water table and other unique environmental factors. A nationwide generic groundwater problem as a result of pesticide applications for registered uses seems unlikely.

A comprehensive data base to support this or the opposing position is lacking, however, because a nationwide groundwater monitoring program has never been undertaken for agricultural chemicals. It should be noted, though, that a national survey is being planned by EPA at this time. The availability of funding will dictate when the survey will begin. Currently, only localized studies are available and these can be selectively used to support or argue against the extent and seriousness of a pesticide/groundwater problem.

Of the available data, one of the better assessments of overall rural drinking water quality is provided by a Cornell University study completed for EPA in 1982 (2). On a statistical basis, it represents 22 million rural households in the U.S. Basically, representative water samples were analyzed for about 30 separate biological, physical, chemical or radiological properties. The study reported the number of households where the analyses exceeded the Maximum Contaminant Levels (MCLs) as established by the EPA, Office of Drinking Water (ODW). The MCL is basically a standard which defines a safe level.

The Cornell study found that most households had problem levels with at least one of the constituents for which an MCL has been established. The presence of coliform bacteria was the most common problem and was in excess of the standard in 28.9% of the households. Iron, manganese and sodium levels were also found to be high for a large number of the households. Nitrate concentrations above the standard occurred among 2.7% of the households, lead in 16.6%, selenium in 13.7%, mercury in 24.1%, and arsenic in 0.8% of the households.

There are six pesticides which have MCLs and they were also analyzed for in the study. Endrin, Toxaphene, 2,4-D, and 2,4,5-T were not detected and Lindane and methoxychlor were found only at extremely low concentrations at a few sites. These products are no longer used on crops, so significant residues were probably not expected and were not found.

In summary, 64% of all households were found to have excessive concentrations of at least one constituent. Over 30% were too high on two or more constituents. While these findings are somewhat startling and were not expected, widespread water-related health problems were not apparent throughout the rural U.S., according to the Cornell study. Health problems may not be common simply because MCLs generally incorporate substantial safety margins.

Results from a more limited study were recently reported by Kelly (3). Iowa scientists analyzed water samples from 128 wells involving 58 public water supplies between May 1984 and March 1985. One or more synthetic chemicals were found in 57 wells representing 33 water supplies. Trihalomethanes, as a group, were the most commonly occurring substances. The most frequently detected single product was atrazine, found in 24 wells. Overall, only 6, of 34 pesticides analyzed, were detected. None were above established standards or standards that could be set by methodology discussed later in this report. The highest level reported for any pesticide was 16.6 ppb for alachlor.

The most comprehensive information about the presence of pesticides in groundwater seems to be that compiled by the EPA. During this symposium (4), the Agency reported that a total of 16 different pesticides have been found in a total of 23 different states. Findings were attributed to agricultural use.

While the public may have a perception that groundwater should be pure -- perhaps would be pure if not for agro- and other synthetic chemicals -- many studies, in addition to the Cornell study, show the presence of natural substances, some at higher than desired levels. Groundwater is not pure! Drinking water is not pure! Chlorination can form trihalomethanes, the most widely identified toxic organic chemicals in U.S. water supplies (5). Page's review (5) of toxic contaminants in drinking water noted that the EPA's water supply survey of finished drinking water from both surface and ground waters showed practically all systems contained chloroform, one of the trihalomethanes.

Shackleford and Keith (6) reported in a 1977 paper that 1,259 different compounds had been identified in water supplies. A significant paper (7) on this subject was published in 1977 by Dr. Bill Donaldson of the Athens, Georgia EPA Laboratory. Dr. Donaldson says, "More than two million organic chemicals have been identified. The number of these compounds detected in a sample of water is related to the sensitivity of the measurement technique: as the detection level decreases by an order of magnitude, the number of compounds detected increases accordingly. Based on the number of compounds detected by current methods, one would expect to find every known compound at a concentration of 10^{-12} g/l or higher in a sample of treated drinking water." Page (5) in his 1984 article, supports Donaldson's thesis by suggesting that virtually any substance in sufficient quantity and with adequate time has the potential to reach groundwater.

This means synthetic and certainly natural products cannot be kept out of groundwater. In fact, the natural products have probably been there for many lifetimes already and the synthetics much longer than thought.

Many, if not most, people refer to some or most of these substances that have been detected, particularly the synthetics, as contaminants. It seems that any agricultural chemical detected in groundwater is referred to as a contaminant or pollutant. Just what does it mean to contaminate or pollute? Webster's (8) says contaminate means to make unfit for use by introduction of unwholesome or undesireable elements. A recent groundwater publication by the American Chemical Society (9) defines a pollutant as any substance, natural or human-made, that degrades water quality, preventing the use of water for some specific purpose. Mere presence, then, should not connote contamination or pollution. More often than not though, it does, and consequently implies health concerns.

As mentioned at the beginning of this presentation, the industry public opinion survey (1) revealed that about half of the U.S. population believes pollution of underground water is a major national problem. Seven out of 10 believe fertilizers and pesticides used on farms are at least a minor cause of such pollution and 23% think they are a major cause. What factors contribute to these beliefs as expressed in the survey?

Any time a public problem arises, special interests like to blame the media, sometimes with real justification. In this instance, they are only one of several factors contributing to the public perceptions held about pesticides in groundwater. The media do not tell us what to think, but they do influence what we think about. Generally, the data generators, including those in industry, government and academia, must share responsibility.

The investigative focus by scientists, and consequently by media, may be on pesticides simply because there are much more data available on this use category of chemicals than any other class. Contrary to a popular perception, extensive safety and environmental data are required for agricultural chemicals before registration. Sensitive analytical methodology is developed for each product by the registrant and is subsequently published by the registrant and/or the regulatory agency. Basically, pesticides are easier to look for because methodology is available. It is interesting to note that one of the criteria for choosing products to be included in the EPA's National Groundwater Survey is the availability of good, sensitive, analytical methodology.

In conjunction with method availability is the fantastic progress made by the analytical chemists in recent years. Such progress is particularly applicable to the analysis of pesticides. To put this in perspective, chemists could routinely detect a part per million (ppm) in the 1960's, a part per billion (ppb) in the 1970's and low parts per trillion (ppt) here in the 1980's. This represents more than a million fold increase over the past 25 years in our ability to detect agricultural chemicals.

Great pride is taken in such progress, but it has caused a lot of confusion in society. Substances have been discovered which were not known or thought to be present. Terms of parts per million, billion and trillion are used many times without considering what they really mean to the media and the general public. They need to be put into a perspective which can be visualized by everyone. A meaningful example may be: 1 ppm equals 1 minute in two years, 1 ppb equals 1 second in 32 years, and 1 ppt equals 1 second in 32,000 years. Essentially, the analysts have gone from the equivalent of 1 minute in 2 years to almost 1 second in 32,000 years in their ability to routinely detect.

The media and the public more often focus on the analytical findings (a chemical is present) and less often on the association of the levels with any health effect. The pesticide data presentation in the Cornell Groundwater Study and a subsequent media report provide good examples of these observations.

The MCLs for the six pesticides analyzed and the amounts detected are:

Product	MCL (ppb)	Method Sensitivity (ppb)	Detects/No. of Samples	Detects Over MCL	Level in Survey (ppb)
2,4-D	100	*	0/267	0	Not Detected
2,4,5-T	10	*	0/267	0	Not Detected
Toxaphene	5	0.17	0/267	0	<0.17
Lindane	4	0.08	4/267	0	0.08
Endrin	0.2	0.008	0/267	0	<0.008
Methoxychlor	100	0.09	3/267	0	0.09

*Method sensitivity not provided.

An examination of the pesticide data in the report points out that the four Lindane detects were in the southern states and the three methoxychlor detects were in the western states (10). The maximum value found for either product was never more than 2% of the established MCL. Yet, the report, while stating that neither Lindane nor methoxychlor pose health threats at the levels found, goes further and concludes that since these products are poisons, even the presence of these small amounts is undesirable. Hasn't it been known for centuries that all substances can be poisons? More troublesome is an extrapolation reported. From the 1% detects in the subsamples analyzed, the statement is made that 224,000 samples (1% of the 22,000,000 rural homes) had methoxychlor. While perhaps statistically appropriate, this extrapolation seems to ignore the fact that methoxychlor was only found in western state samples.

Equally troublesome is what one publication did with this statement (11). The New Farm report on this particular part of the Cornell study said, "The water in more than 200,000 rural homes contains detectable amounts of methoxychlor...While considered low in toxicity to humans, methoxychlor disrupted nervous and reproductive systems in test animals." No mention was made of an MCL, or the levels the animals were exposed to in the test, only the implication that rural water is unsafe to drink.

From another viewpoint, it can be argued that from a public perspective, all chemicals present in groundwater are in the same category. Pesticides arising from agricultural use are not differentiated from industrial chemicals moving from point sources such as hazardous waste sites. All are looked upon as hazardous chemicals. Since the media focus on hazardous waste sites, the concern about pesticides in groundwater may be heightened unnecessarily.

Is there, then, a groundwater problem arising from the use of agricultural chemicals? One would conclude that there is - but, it is more preceived than real! In either case, the agricultural chemical industry must confront it and must do so both scientifically and from the public relations standpoint. Some critics suggest there is a need for additional legislation and/or regulations. Public perception may demand it.

A review of statutes and regulations currently in place to address the presence of pesticides in groundwater will be useful. Based on a recent review (12) by the Office of Technology Assessment (OTA), there is no comprehensive protection of groundwater offered by any one federal law. Of sixteen environmental statutes reviewed, only three directly or indirectly touch on the subject of pesticides and groundwater.

The Clean Water Act (CWA) of 1977 has as one of its objectives (Section 208) the establishment of a program to control non-point source pollution in rural areas. This is a program aimed at controlling, for example, run-off from agricutural fields. Although minimal implementation has occurred, it could impact both surface and groundwater.

The Federal Insecticide, Fungicide and Rodenticide Act (FIFRA) which largely controls the pesticide industry, does not contain any direct reference to groundwater. It does call for the cancellation or suspension of products having unreasonable adverse effects on the environment, which includes groundwater. The application of this statute to groundwater has recently occurred with the request that "restricted" labeling and groundwater advisory statements be added to certain product labels. Without the establishment of a reference value or standard, this approach will have the effect of banning or severely restricting product use in some areas.

The only other act that addresses groundwater, though not explicitly, is the Safe Drinking Water Act (SDWA) of 1974. It is primarily designed to assure that the drinking water from public water systems is safe. It requires the establishment of standards for certain substances present in public drinking water supplies, the establishment of regulations for underground injection and the protection of sole source aquifers. The statute is administered by the EPA's Office of Drinking Water.

It is under the SDWA that standards have been set for 22 substances in drinking water whether the source is surface or underground. These standards are referred to as Maximum Contaminant Levels (MCLs) and include the six pesticides referred to earlier in this paper. Regulations also provide for the establishment of Recommended Maximum Contaminant Levels (RMCLs). RMCLs are non-enforceable health goals while MCLs are enforceable, but must balance health protection with other factors including the availability and cost of treatment technologies to attain the standard (12). The Office of Drinking Water has plans this year to issue RMCLs for an additional 17 or 18 pesticides.

Should available technology be too costly for a public system to attain an MCL, the EPA may, in the view of some groups (13), employ land use control as a treatment technology option. A watershed or recharge zone could be designated as an area in which certain chemicals could not be used. The Agency has apparently not adopted this interpretation of the statute.

Another regulatory approach to the presence of substances in drinking water has been the establishment of Health Advisories (HAs) by the EPA's Office of Drinking Water (14). These are numbers which simply serve as guidelines to evaluate the health significance of certain chemicals found in groundwater or surface water used for drinking water and for which there are no federal requirements. In general, they are established in response to emergency situations and are based on short term - up to two years - exposure to a particular chemical. Health Advisories were formerly known as SNARLs, i.e., Suggested No Adverse Response Levels. They are not enforceable.

A framework for further addressing chemical presence in groundwater at the federal level is the EPA's groundwater protection strategy (15) issued in August 1984. This strategy deals with groundwater from the aspect of the various laws which the Agency administers. Basically, groundwater is divided into three classes:

1) Special groundwaters - those which are highly vulnerable to contamination because of hydrological characteristics and:

   a. are irreplaceable drinking water sources or
   b. ecologically vital.

2) Current and potential sources of drinking water and waters having other beneficial uses.

3) Groundwaters not considered potential sources of drinking water and are of limited beneficial use.

The Agency will give different levels of protection to these classes.

To prevent contamination of class 1, and presumably class 2, EPA "--policy will be directed toward restricting or banning the use in these areas of those pesticides which are known to leach through soils and are a particular problem in groundwater." The general policy for cleanup of contamination will be to background or accceptable drinking water levels.

Industry does not have a basic argument with the policy, but EPA's interpretation and eventual enforcement remain undefined from a practical standpoint. It is not clear whether the Agency's intent is to allow no pesticide residues in these class 1 and 2 aquifers or whether certain levels are acceptable dependent on the safety of the product. If the former is true, risk will be managed by eliminating it.

Several states have regulatory programs in place for pesticides found in groundwater. These states are, for the most part, represented in this symposium; thus, comments herein will be quite limited.

Legislative initiatives have been put forth by several states to provide necessary regulatory authority for establishing standards for substances that could enter groundwater. This is contrasted with the federal situation which is relying upon existing water quality statutes for standard setting.

Certain specific incidents of substance appearance in groundwater have caused state legislators to move rapidly and decisively on this issue. For example, the states of Wisconsin, New York and Florida enacted enabling legislation that permitted standard setting and large scale monitoring programs to be established.

Sweeping legislation was adopted in Wisconsin (16) that established two precedents, the Acceptable Daily Intake (ADI) methodology and the Preventive Action Limit (PAL) concept. The state of New York, under existing statutory authority (17), proposed by rule, not only the ADI methodology for standard setting, but went even further to allow the adoption of standards by chemical class or similarity in metabolism, structure and toxicological response for organic substances. Wisconsin and New York placed significant weight on the likelihood of appearance or the actual appearance of a substance in water as criteria for standard setting. These programs utilize existing toxicology data from which the ADI and, subsequently, the water quality standard is established.

Florida, on the other hand, maintains a zero degradation philosophy which says, in effect, that if no standard is set, then the mere detection of a substance is a violation of the water quality standards (18). Unlike the two formerly mentioned states, legislative initiatives in Florida did not clearly define a standard setting methodology or specifically mandate rule promulgation in this area.

The State Affairs Committee of the NACA has identified 13 states as being likely to adopt legislation in the area of water quality this year. Should the Federal Government continue to delay implementation of standard setting, more and more pressure will be generated by the public for action by state legislative and regulatory officials. The results of such action have potential for chaos and confusion, particularly when aquifers traverse state lines and water quality standards and methodology vary from state to state.

At the federal level, the Safe Drinking Water Act, the Clean Water Act and FIFRA are all up for reauthorization this year. Groundwater amendments have been proposed for the SDWA and some federal legislators are even suggesting national groundwater legislation.

Obviously, the agricultural chemical industry would prefer to see federal guidance on this issue. Adequate federal authority already exists to establish guideline numbers. For the past two years, the EPA has been urged to establish these numbers for all pesticide products. Although only a few agricultural chemicals have been detected in groundwater to date, it seems likely more will be identified as sampling and analytical capabilities continue to improve. It would be a positive step to have numbers in place prior to detection.

Ironically, while the industry public opinion survey (1) reported 90% of the people nationwide endorse national standards for drinking water, 63% do not believe that water is safe to drink just because the government allows analytically detectable amounts of chemicals. It remains the best way, though, to begin to address the public's concern about the perceived problem of agricultural chemicals in groundwater.

Guideline numbers, advisory levels or standards should be set at the federal level not only to assure uniformity, but because the data base for registered products resides in the EPA's Office of Pesticide Programs. While states would probably have the authority to be more stringent than the federal guideline, such action seems unlikely, particularly if the guidance numbers are based on sound science. Dealing with several different groundwater numbers could present significant problems for registrants from a labeling standpoint.

Methodology for establishing health guidance levels is that presented by Stevens and Sumner (19) in this symposium. Briefly, it incorporates the ADI and a 10 kilogram child consuming one liter of water per day.

$$\text{HGL* (mg/liter)} = \frac{\text{ADI (mg/kg/day)} \times \text{10 kg child}}{\text{1 liter/day}}$$

*Health Guidance Level

This approach is not without precedent. It has been recommended by EPA's OPP (20) and subsequently supported by EPA's Scientific Advisory Panel (SAP) (21). Wisconsin and New York included this concept in their approach to establishing groundwater and drinking water standards. The ODW suggests that the inclusion of a 10 kg child acts as a surrogate for the most sensitive subgroup of the adult population (14).

This guidance number should not reduce the ADI for food intake. The SAP voiced support for this concept because of the safety factors built into the food ADI and because groundwater residues are geographically restricted and in general, transient.

In summary, there is a problem, more perceived than real, associated with the presence of agricultural chemicals in groundwater. To place these findings in perspective requires a guideline number based on health effects which would say to the media and the public that there should not be health concerns below a certain level.

The agricultural chemical industry is very serious about the importance of protecting groundwater and the safe use of agricultural chemicals. Several projects are underway to minimize the presence of these products in groundwater. The industry is also very concerned that efficacious and safe products be maintained on the market and new products registered. Groundwater is a major issue for NACA and its member companies. Initiatives have been taken in working with government agencies - federal, state and local - and other interested parties to help prevent problems involving agricultural chemicals.

Literature Cited

1. "Groundwater Contamination: The Measure of Public Concern," Center for Communication Dynamics: Washington, DC, February 1985.
2. "The Cornell Report to ODW, EPA: A National Statistical Assessment of Rural Drinking Water Conditions," Cornell University, 1983.
3. Kelley, Richard D. "Synthetic Organic Compound Sampling Survey of Public Water Supplies," Iowa Department of Water, Air and Waste Management, 1985.
4. Creeger, S. M. Included in this symposium series, 1985.
5. Page, William G. The Environmentalist, 1984, 4, 131-138.
6. Shackelford, W. M.; Keith, L. H. "Frequency of Organic Compounds Identified in Water," Environmental Research Laboratory, NTIS PB-265470, Athens, GA, 1977.
7. Donaldson, William T. Environ. Sci. & Technol., 1977, 4, 348-51.
8. Webster's Seventh New Collegiate Dictionary, 1224 pp., G. & C. Merriam Company, Publishers, Springfield, MA, 1971.
9. Groundwater Information Pamphlet. American Chemical Society, Washington, DC, 1984.
10. Back, Richard. Personal Communication.
11. Zahradnik, Fred, New Farm, 1984, 6, 30-31.
12. "Protecting the Nation's Groundwater from Contamination"; Volume 1, Washington, DC; U.S. Congress, Office of Technology Assessment, OTA-0-233, October 1984.
13. Gordon, Wendy. In "Citizens Handbook on Groundwater Protection"; Natural Resources Defense Council, Inc.: New York, 1984, p. 90.
14. Lappenbusch, William L.; Moskowitz, Susan B. Proc. AMA/EPA Symp. on Drinking Water and Human Health, Washington, DC, 1984.
15. "Ground-Water Protection Strategy," U.S. EPA-Office of Groundwater Protection, Washington, DC, 1984.
16. State of Wisconsin Act 410, 1983.
17. State of Florida Statutes, Chapter 487, 1983.
18. State of New York, Proposed Changes to Subdivision 701, 1985.
19. Stevens, James T.; Sumner, Darrell D. Included in this symposium series, 1985.
20. Severn, David J.; Offutt, Carolyn K.; Cohen, Stuart Z.; Burnam, William L.; Burin, Gary J. "Assessment of Groundwater Contamination by Pesticides," prepared for FIFRA Scientific Advisory Panel Meeting, June 21-23, 1983.
21. Review of EPA Strategy Paper on: "Assessment of Groundwater Contamination by Pesticides," Minutes, FIFRA SAP Meeting, June 21-23, 1983.

RECEIVED March 25, 1986

28

Ground Water Regulations

Impact, Public Acceptance, and Enforcement

Orlo R. Ehart[1], Gordon Chesters[2], and Kari J. Sherman[2]

[1]Trade and Consumer Protection, Wisconsin Department of Agriculture, Madison, WI 53708

[2]Water Resources Center, University of Wisconsin-Madison, Madison, WI 53706

Sanctioning pesticides for agricultural use was based on equating environmental and public health risks to crop production benefits. More recently, the introduction of groundwater protection into pesticide regulations has modified the equation by adding other elements of protection. During the past decade pesticide issues have been emotionally charged with the extremists' views most notably reported. The public believes that groundwater deserves greater protection than surface water because improving the quality of contaminated groundwater is more difficult and costly. This coupled with the users'--particularly rural residents--concerns that their drinking water is being polluted, has lead to acceptance of groundwater regulations by pesticide users often opposed to regulation. Enforcement of regulations, based on acceptable tolerance levels of pesticides in groundwater, must rely on quality-assured data. Regulatory mechanisms include changes in use practices, moratoria and cancellation of registrations. Mathematical models--used as an information source--could be helpful in evaluating the consequences of pesticide use. Enforcement sanctions follow traditional civil and criminal remedies but innovative cost incentives might be needed to accelerate protection programs. Mechanisms of implementation of land management controls vary from state to state.

The Federal Insecticide, Fungicide and Rodenticide Act, as amended, establishes a policy for determining the acceptability of a pesticide use, or the continuation of that use, according to a risk/benefit assessment. This policy, administered by the U.S. Environmental Protection Agency (EPA), recognizes that pesticides are intentionally placed in the environment for purposes of pest control. Contrary to some public opinion, the risk/benefit assessment does not determine that the use is totally "safe" nor

0097-6156/86/0315-0488$06.00/0
© 1986 American Chemical Society

suggest that all uses of a pesticide will not adversely affect the environment, including effects on human health. As long as benefits outweigh adverse effects uses of a pesticide can be registered, i.e., officially sanctioned by the EPA.

Despite this permissiveness in the law, it has become increasingly difficult to justify the use, or continued use, of many pesticides. The ability to recognize potential risks has improved with advances in technology. The ability to detect increasingly lower concentrations of pesticides, increased knowledge of and concern about potential chronic effects, and the finding of low levels of pesticides in drinking water in several areas of the United States have added more factors to the risk side of the risk/benefit ledger while the benefit side has remained relatively static.

Public Perceptions toward Groundwater Policy and Regulation

A study conducted for the National Academy of Sciences (1) has revealed that the average Californian is becoming increasingly aware of and concerned about toxic chemicals as a threat to groundwater. This important perception is without regard to the contaminant's source, relative toxicity or pervasiveness. The public generally does not understand nor accept the consequences of federal policy, which allows the purposeful, judicious introduction of pesticides into the environment, even if they may have a minimal yet measurable undesired effect. Thus, if only minute quantities of a pesticide reach groundwater, the federal policy of a risk/benefit assessment is rejected by much of the public regardless of any assessment of the benefit of the pesticide in improving crop production and quality.

When easily accessible facts on technical issues are unavailable, the public frequently favors simple policies. In the case of groundwater, this is the policy of nondegradation which allows for no change in water quality. The establishment of numerical groundwater standards is an important policy alternative for assessing the significance of manmade or naturally occurring contamination. The federal government must establish those levels if the public is ever to perceive the difference between the mere presence of a pesticide in groundwater and an acute or chronic health significance associated with exposure to some level of contamination.

The impacted community must accept regulation and cooperate with the regulatory authority if a management approach to solving the problem of contaminants in groundwater is to succeed. Regulation requires access to technological information, the cooperation and dedication of researchers and adequate surveillance to assure violators are deterred. Regulations have costs associated with them, but they encourage use of those pesticides which create minimum risks. However, new regulations may substantially modify the present uses of pesticides. Pesticides can adversely affect humans or the environment if not handled properly. Because of these detrimental consequences, alert surveillance and speedy and just enforcement become essential components of the regulatory program. Outright bans can result but they are the product of a decision-making process rather than the first line of defense.

Most Americans are concerned about the quality of groundwater and believe that it is more difficult and costly to protect and renovate than surface water. When their concerns are coupled with inadequate information and an unwillingness to accept regulatory standards, it becomes difficult if not impossible to establish balanced public policy.

Groundwater is one of the least tangible components of the total environment. It is the decision to use groundwater--as a drinking water source for humans and domestic animals, for irrigation and for industrial uses--that has made its contamination a real and immediate problem. Since groundwater is not visible and its properties and flow patterns are poorly understood by laymen, the public is unwilling to support a policy permitting the use of a pesticide that could be found in groundwater, even if the levels have no significant effect on human health and the environment. Since water has different uses and qualities, variable standards matching safeguards needed for particular uses now and in the future would be appropriate. However, this type of standard setting would be complex and it may be difficult to convince the public of the need for this policy. Nonetheless, such a system would provide greater flexibility in devising best management practices to protect the groundwater resource.

Survey data. The following survey statistics (2) provide some knowledge of the public's perception of groundwater contamination.

- 75% of the U.S. population (75% in Wisconsin) is more worried about pollution now than 5 years ago.
- 48% of Americans (58% in Wisconsin) consider underground water pollution to be a major problem.
- 91% (91% in Wisconsin) endorsed national standards for safe drinking water but only 23% (20% in Wisconsin) are willing to accept "small amounts of chemicals" in water even if such contamination meets government standards.
- 63% of those asked (69% in Wisconsin) disagreed with the idea that if the government allows "small amounts of chemicals in water," the water is safe to drink.

When the public was questioned on agricultural sources of groundwater pollution--particularly from pesticides--the following opinions were obtained.

- Asked if farmers use too many pesticides, nationally 59% of those surveyed (57% in Wisconsin) answered yes.
- 23% of all Americans surveyed (29% in Wisconsin) believe that fertilizers and pesticides used on farms are a major national problem causing water pollution; 76% (82% in Wisconsin) consider them to be at least a minor problem.
- 35% (24% in Wisconsin) believe fertilizers and pesticides are not a problem in their local drinking water while only 6% (3% in Wisconsin) believe they are not a national problem. Although some respondents thought that their local drinking water source was safe, they perceived that a problem exists in some other parts of the country.

A survey (3) showed that 56% of respondents indicated a need to protect farmers when asked: "Do you think that the solutions for water pollution in Wisconsin should be planned so that they do not increase costs for farmers, or are these problems so serious that we have to have solutions even if they mean higher costs for farmers?" Of those surveyed, 50% indicated that solutions (remedial measures) for water pollution should be planned not to interfere with the ability to create new jobs; 61% favor state financial assistance to farmers to control pollution from agricultural activities. In a national Harris poll (4), 63% indicated that clean water was obtainable without affecting jobs; however, if it were an either/or proposition, the population surveyed believed that health is more important than jobs. These surveys and polls provide a snapshot of public opinion but do not necessarily reflect the true beliefs of society. However, they do indicate the paradoxical elements in dealing with the public's perception of complex policy issues.

A statistical survey study (5) conducted for the EPA by Cornell University researchers reported that 63% of rural homes have contaminated drinking water; nearly 30% have excessive bacterial counts but few have detectable pesticide residues. Samples from wells at 267 sites were selected for analysis. The analyses were limited to only those six pesticides for which maximum contaminant levels (MCLs) are nationally established. Results of the study showed:

- Four out of 121 sites in 16 southern states contained lindane residues. The highest level was 0.08 parts per billion (ppb or µg/l). The MCL for lindane is 4 ppb.
- Three out of 32 sites in the west contained methoxychlor with a maximum level of 0.09 ppb. The MCL for methoxychlor is 100 ppb (6).

The executive summary (7) of the study states: "Endrin, lindane, methoxychlor, toxaphene, 2,4-D and 2,4,5-TP were virtually never detected among rural household water supplies." The study draws no conclusions on the extent of contamination of rural water supplies by pesticides beyond this limited sampling. However, this small data base was translated and extrapolated to provide a public information article (8) which states that "lindane is present in 347,000 rural household water supplies" and "water in more than 200,000 rural homes contains detectable amounts of methoxychlor." This type of projection misleads and inflames the emotions of the public regarding the extent and seriousness of contamination. Editors should respond very forcefully to scare tactics and decision-makers must determine the reliability of such statements.

Some states have conducted surveys of water quality. Currently the information available on the status of the quality of the national groundwater resource is inadequate to draw any conclusions. Generally, assessments of water quality are available only for localized areas. The EPA has proposed a national groundwater survey for 1986-88. The preliminary draft calls for sampling 1500-2000 wells for approximately 90 pesticide analytes (9). Even though this information will be valuable, a sample size of 30-50 wells per state is still inadequate to assess the quality of the nation's groundwater for contamination by pesticides.

Need for Scientific Evaluation. Zaki et al. (10) concluded that:

> "Our society seems to be willing to accept tangible and measurable risks in our daily activities as a result of cigarette smoking, excessive food and alcohol intake, and the use of the automobile.
>
> The same society, however, is unable to tolerate potential, intangible and unmeasurable risks from food additives, pesticides, air pollutants, and water contaminants. As public health administrators, we have a responsibility not only to monitor and control these substances but also to help the public become fully aware of their risks and benefits without resort to rhetoric on either side of the issue."

It is difficult to prevent some personal biases from entering professional attitudes. However, professionals serving the public interest must render an unbiased presentation of the facts if the role of public opinion in the decision-making process is to be valid. Professionals must adopt an unbiased attitude, even in those circumstances where they have personal beliefs which may differ from the facts.

Holden (1), in commenting on an, as yet, unpublished report to the National Academy of Sciences, stated that there is also considerable concern that public apprehension about groundwater contamination will grow to the point where statewide or national bans will become politically expedient, even in cases where pesticide contamination is a controllable localized phenomenon. The role of scientists in assuring the use of reliable data in the decision-making process has become an integral part of their professional activities. It is imperative that unbiased, accurate information be presented by trained professionals to secure public acceptance of sound, reasonable public policy regulating the presence of pesticides in groundwater.

The public is not well informed on technical issues and frequently provides a reaction rather than an informed opinion. Although most people can distinguish voluntary from involuntary risk, Orloff (11) points out people do not focus on a particular set of risks simply to safeguard health and safety; the choice often reflects a moral view. The government operates on the premise that public policy should protect against major risks; the public should accept minor risks. By viewing an issue from the moral perspective it is unlikely that public consensus can be reached. It has not been shown that vocal participants represent the spectrum of public opinion. Unless the effectiveness of public participation is enhanced, the advisability of direct public involvement in administrative decisions is debatable (12). Agencies administering state policy must render unbiased decisions based on the information available, use its best judgement to obtain balanced policy, and be accountable for their decisions.

Development and Administration of the Aldicarb Rule in Wisconsin

Wisconsin has had a number of experiences in dealing with traces of pesticide residues in groundwater; the most notable is the presence of aldicarb residues in shallow aquifers in the Central Sands region of the State (Table I).

Table I. Major Events in the Development of the Wisconsin Policy on Pesticides in Groundwater

Date	Event
July 1981	Aldicarb confirmed in Wisconsin groundwater
Nov. 1981	WDATCP proposed emergency rule
Jan. 1982	Legislative Council Groundwater Management Study Committee formed
Mar. 1982	Emergency rule adopted
Mar. 1982	WDATCP was sued and courts upheld rule
Aug. 1982	WDATCP proposed permanent aldicarb rule
Apr. 1983	Permanent aldicarb rule adopted
May 1984	1983 Wisconsin Act 410, Groundwater Law, signed by Governor A. Earl
Nov. 1984	WDATCP proposed permanent fertilizers or pesticides in groundwater rule
Sept. 1985	Permanent groundwater rule adopted

Union Carbide Agricultural Products Company, Inc. sells aldicarb--under the trade name Temik--in the State for use in controlling insect pests on potatoes. In July 1981, the presence of aldicarb residues in Wisconsin groundwater was confirmed in areas where the pesticide had been used. The Wisconsin Department of Agriculture, Trade and Consumer Protection (WDATCP) promulgated an emergency rule which became effective prior to the 1982 growing season. The manufacturer and a growers' organization subsequently sued the State based on whether it had authority to impose such a rule and whether the facts supported the decisions made by WDATCP. The suit--only tried on the authority question--was decided in favor of the State. A trial, based on the merits, would not have been concluded until after the emergency rule had terminated which led the manufacturer and growers to drop their suit.

Because aldicarb was detected in drinking water samples in Wisconsin during 1981, the State with the concurrence and assistance of the manufacturer amended the use directions. This resulted in a federal label amendment which changed the time that aldicarb could be legally applied to potatoes from time of planting to 4 to 6 weeks after planting, reduced application rates from 3 to 2 pounds of active ingredient per acre, limited use to alternate years on any piece of land, placed an environmental warning statement on the label, and made the product a restricted-use pesticide. These requirements were placed in the State's emergency rule along with provisions requiring users to file a notice of intent to use the insecticide and dealers to report sales of the Temik 15G aldicarb formulation; these restrictions also became a part of a permanent aldicarb rule.

The emergency rule also established moratorium areas prohibiting aldicarb use wherever detectable residues were found in drinking water wells. Moratorium areas under the emergency rule were land parcels on which no aldicarb could be used. The moratorium areas were established based on known contamination of drinking water and on available data on direction of groundwater flow, depth to the water table, water use, and aldicarb use. Approximately 70,000 acres of land, some of which was not in agricultural production, were included under the moratorium. Although the emergency rule expired in July 1982, a modified moratorium area concept was extended under the permanent rule. The moratorium on use remains in effect until aldicarb residues in well samples remain below the enforcement standard for 1 year.

In 1983, WDATCP adopted a permanent rule (13) establishing 10 ppb as the "action level" which triggers designation of a moratorium area. Legal concerns over the public's perception of how moratorium boundaries were established under the emergency rule led WDATCP to establish the boundaries as a circle of 1-mile radius (approximately 2800 acres for each moratorium area established from a single contamination site) around locations where aldicarb residues exceeded the enforcement standard. The center of the moratorium area is the center of the quarter/quarter section of land on which the well is located.

There are 50 wells located in 25 quarter/quarter sections of land which were contaminated with aldicarb residues above the 10 ppb level and were used to establish the 1985 aldicarb moratorium areas; 33 wells currently are over the guideline level. Because of overlapping, these sites establish 11 moratorium areas encompassing 36,500 acres. Well owners have been informed of levels of contamination and advised to seek alternate water supplies until the contamination levels drop below the guideline level, or new wells can be installed. Under separate rules of the State, some owners qualify for a compensation fund as a result of the presence of aldicarb in their wells. Only recently has the authority been granted to close private wells; the authority has not been exercised. Since 1982, 12 of the 75 owners whose wells exceeded the 10 ppb level have taken advantage of the aldicarb manufacturer's offer to pay for bottled water. About 40% of the 51 private well owners who have been offered carbon filters have accepted (6).

Despite the WDATCP rule, aldicarb residues are still present in groundwater although contamination levels of most wells within moratorium areas are decreasing with time. However, since aldicarb residues are still detectable, the State's public intervenor--a Wisconsin Assistant Attorney General with responsibility for environmental protection--and some State legislators, environmentalists and concerned citizens believe the rule is ineffective. In their view, the remedial action taken by WDATCP should also rectify an already existing condition, while WDATCP regards the remedial action as protection for the future, providing cleansing time for the past conditions but not a cleanup program *per se*. Staff of WDATCP recognize that protection and cleanup efforts require decades rather than months. Some members of the public are demanding immediate decontamination of an inaccessible resource, and retribution in the form of strict liability for practices performed before the existing conditions were recognized.

Meanwhile, growers have become more supportive of proper regulation of pesticide products which have a potential to reach groundwater. They are concerned about water quality as it affects their families, livestock and economic future. They have adopted irrigation practices to reduce excessive water use which previously they considered unnecessary even though economically advantageous. They have, with few exceptions, filed notices of intent to apply aldicarb. They have cooperated with field investigations and sampling of soil, groundwater and produce for aldicarb residues.

Development, Administration and Enforcement of the Wisconsin Groundwater Law

For a 2-year period, concurrent with the development and administration of the aldicarb rule by WDATCP, the Wisconsin Legislature studied and refined the legal framework for a State groundwater protection program. Although described by many to be a comprehensive groundwater protection program, the groundwater law (14) signed in May 1984 added little authority to existing State programs; it merely codified many existing and emerging efforts of State agencies to protect groundwater. The legislative action is important because it establishes the mechanism to use "health advisory" levels as State "enforcement standards" not to be attained or exceeded in groundwater. It also defines a "preventive action limit" as a fraction of the enforcement standard. This is a contaminant concentration for substances found in groundwater where some agency action is required. The law effects a change in policy from the strict nondegradation philosophy of the past--where action was not mandated as a statutory requirement--to a new numerical standard policy, where an agency decision, even if it chooses to do nothing, must be justified and articulated. This is a major change in public policy. Its importance is that it assures better, although not necessarily absolute, protection of the groundwater resource. However, this assurance is not yet recognized or appreciated by the general public.

The legislation was needed to define the institutional responsibilities of the Wisconsin State agencies with a mandate to protect groundwater. However, its timing had detrimental effects on WDATCP's efforts to conduct an effective aldicarb control program. The WDATCP needed answers to several questions to address the problem of groundwater contamination by aldicarb residues in an informed, responsible fashion. For example, many of the reasons that aldicarb residues move through the vadose zone to groundwater were not understood and are still inadequately researched. The State did not provide adequate research funding because the Legislature was grappling with what the State's general groundwater policy should be. Ironically, the State modeled its groundwater policy on the aldicarb rule; no State dollars, however, were made available directly for research, even though frequent legislative inquiries continue to demand proof that the aldicarb rule is working to minimize or eliminate further intrusion of aldicarb residues into groundwater.

The current State program requires each agency to prioritize potential groundwater pollutants under its jurisdiction. The Wisconsin Department of Natural Resources (WDNR) is required to monitor groundwater to assess problems caused by these priority substances. If pesticide residues are found in groundwater, WDNR will forward the data to WDATCP. When WDATCP determines that a practice regulated by WDATCP is responsible for the pesticide residues and when the level of any pesticide attains or exceeds a preventive action limit or enforcement standard, certain site-specific actions must be taken and/or rules must be promulgated to modify management practices. Changes in practices must be monitored for effectiveness to minimize impacts on groundwater quality. The WDATCP will perform this function to the extent that financial resources are available.

The WDATCP rule (15)--establishing a regulatory program for fertilizers or pesticides in groundwater--became effective in September 1985. Contained in the rule are provisions for WDATCP action following notification of confirmed pesticide residue contamination of groundwater, a description of the agency's investigational procedures, methods for identification of well locations where numerical standards will be applied, and a selection of site-specific responses to be expected when contamination exists. The agency is also committed to promulgate specific rules for substances for which an enforcement standard has been established.

Memorandum of Understanding. Another important provision in the rule is the establishment of a memorandum of understanding between WDATCP and WDNR regarding sample handling and analysis. In the past, preliminary, unconfirmed, analytical data intended for program planning was released to the public. In addition the lack of a specific protocol for sample handling raised questions regarding the comparability and reliability of some data. The WDATCP and WDNR staff recognized that laboratory protocols under pesticide regulatory programs differ from those conducted under mandatory federal water programs. Agreeable, uniform practices to deal with these differing approaches to quality assurance needed to be established and are part of the memorandum of understanding.

The memorandum of understanding contains a description of acceptable collection equipment and containers, collection procedures, record keeping requirements, use of control samples, sample size, storage (i.e., temperature and preservation) and custody requirements. It establishes a requirement for the maintenance of a quality assurance manual and an approval process for analytical methods detailing limits of detection (qualitative) and limits of quantification. The memorandum of understanding also defines when confirmatory analysis by another method or chromatographic column is necessary, and requires development of reliable information on well construction and depth in the aquifer and allows for the development of alternative protocols for sampling and handling when specific substances require it.

This process may seem unduly bureaucratic and unnecessary. However, if governmental programs are to be credible, exacting protocols must be established and adhered to. Education of the public on the need for these programs and general education on groundwater are needed to increase factual awareness. The inexpensive method of cutting corners and not assuring the quality of program efforts has cost far more in credibility than the cost to run programs properly. Without such expenditures to secure quality assurance, the credibility of scientific data in public decision-making will not be maintained.

Mathematical Models. Using models to aid in evaluating the potential for groundwater contamination has been emphasized. Many believe that models should be used as part of a regulatory approach to prevent future contamination. Models are a useful tool for determining potential priorities and possible sensitive areas and crop uses and in developing monitoring plans. However, models are not sufficiently reliable and predictive to allow regulatory actions to be taken on the basis of model findings alone. Therefore, although useful in research and planning, models are not yet sufficiently sophisticated to be used as a basis for regulatory programs.

Enforcement. No groundwater protection program will be effective without proper enforcement. Surveillance of pesticide uses will increase as more management criteria are added to label directions. Compliance will require improved education of users and changes in social attitudes. Calibration of application equipment, always a difficult procedure and subject to individual error, will need more critical attention. The increased potential for legal challenge of decisions will cause enforcement agencies to spend more resources on each investigation. This will entail more extensive record keeping and more detailed reporting.

The reliability of currently available technology for establishing and administering these policies, must be assured or the system is useless. There must be accurate and readily available information to establish enforcement standards for the pesticide products identified as having the potential to reach groundwater. Enforcement or regulatory actions cannot be fair without accurate analytical data. Knowledge of pesticide uses, geology, hydrogeology and environmental fate are needed to predict potential problems, and more importantly, to establish a reasonable, cost effective monitoring program (16). Programs which provide incentives and proper enforcement are needed to make the system effective. A system which assures that agencies tend to the business of administering and academic institutions tend to the business of research must be developed; researchers must be responsive to regulatory issues. Wisconsin has made some strides at developing a system which is technically sound and practically workable but more needs to be done.

Violations of state laws and rules which result in groundwater contamination must be prosecuted by traditional methods. However, to accelerate protection programs, some innovative cost incentives are needed. When the informed public and affected industries are active participants in the decision-making process, the law is more strictly adhered to.

The examples detailed in this paper are from Wisconsin. Geographical, climatological, legal and political variabilities exist in other states which may require modification of the Wisconsin procedures before these methods of land management controls can be adopted by other states.

Literature Cited

1. Holden, P., personal communication.
2. "Groundwater Contamination, The Measure of Public Concern"; Center for Communication Dynamics, Washington, D.C., 1985.
3. Wood, D. "Wisconsin Environmental Priorities Survey Results." Environ. Priorities Conf., Madison, Feb. 14, 1983.
4. Business Week, Jan. 24, 1983. p. 87.
5. Francis, J., personal communication.
6. Back, R., personal communications.
7. Francis, J.; Larson, O. III. National Statistical Assessment of Rural Water Conditions. Executive Summary. Cornell University, Ithaca, 1983.
8. Zahradnik, F. The New Farm Jan. 1984, 30-31.
9. Cohen, S., personal communication.
10. Zaki, M.; Moran, D.; Harris, D. Am. J. Public Health 1982, 72, 1391-1395.
11. Orloff, N. "We Scoff at Big Risks and Scotch Small Ones." Wall Street J., Dec. 3, 1984, p. 34.
12. Schierow, L.-J.; Chesters, G. Water Resources Bull. 1983, 19, 107-114.
13. WDATCP. Aldicarb Use Restrictions; Reporting Requirements. Chapter Ag 29.17, Wis. Admin. Code.
14. 83 Wis. Act 410. Laws of Wis., 1983.
15. WDATCP. Fertilizer or Pesticide Substances in Groundwater: Regulatory Program. Chapter Ag 161, Wis. Admin. Code.
16. Ehart, O. R. Unpublished presentation. Entomol. Soc. Am., Detroit, Nov. 29, 1983.

RECEIVED March 25, 1986

29

Ground Water Contamination by Toxic Substances

A California Assessment

David B. Cohen

Pollutant Investigations Branch, State Water Resources Control Board, Sacramento, CA 95801

Overoptimistic risk perception, coupled with inadequate risk assessment, has led to underprotective risk management of toxic chemicals in ground water. In California, soil nematicides like DBCP are a particular concern. DBCP use was banned in 1977, yet it is still found in over 2500 wells. An estimated 700,000 people in 32 towns are either drinking water with measurable levels of DBCP or have had to seek alternative water supplies. Over 50 other pesticides including 1,2-D, EDB, and aldicarb have been found in California ground water from both point and nonpoint sources. Future risk management strategies must be based on a more conservative risk perception approach. California Water Resources Control Board's ground water "hot spots" program is an example of such an approach.

In semi-arid California, ground water is a precious resource. During the past decade, the State Water Resources Control Board, California's environmental protection agency for water quality, has learned that we have not done enough to prevent the growing problem of toxic chemical contamination of our ground water.

While both industrial and agricultural chemicals have been found in California ground waters, pesticides injected beneath the soil surface to control nematodes are a particular concern. The nematicide 1,2-dibromo-3-chloropropane (DBCP), which was banned in California in 1977, is the most serious example of this problem in California (1-2). Figure 1 shows the increase between 1979 and 1984 in the number of wells found to contain measurable quantities of DBCP. As of April 1984, 2522 wells contained DBCP, with over half these wells (1455) having concentrations in excess of 1.0 part per billion, the current California "action level".

Table I lists over 50 other pesticides (from both point and nonpoint sources) that have been identified in ground water from 28 of California's 58 counties. Figure 2 shows the geographic

0097-6156/86/0315-0499$08.75/0
© 1986 American Chemical Society

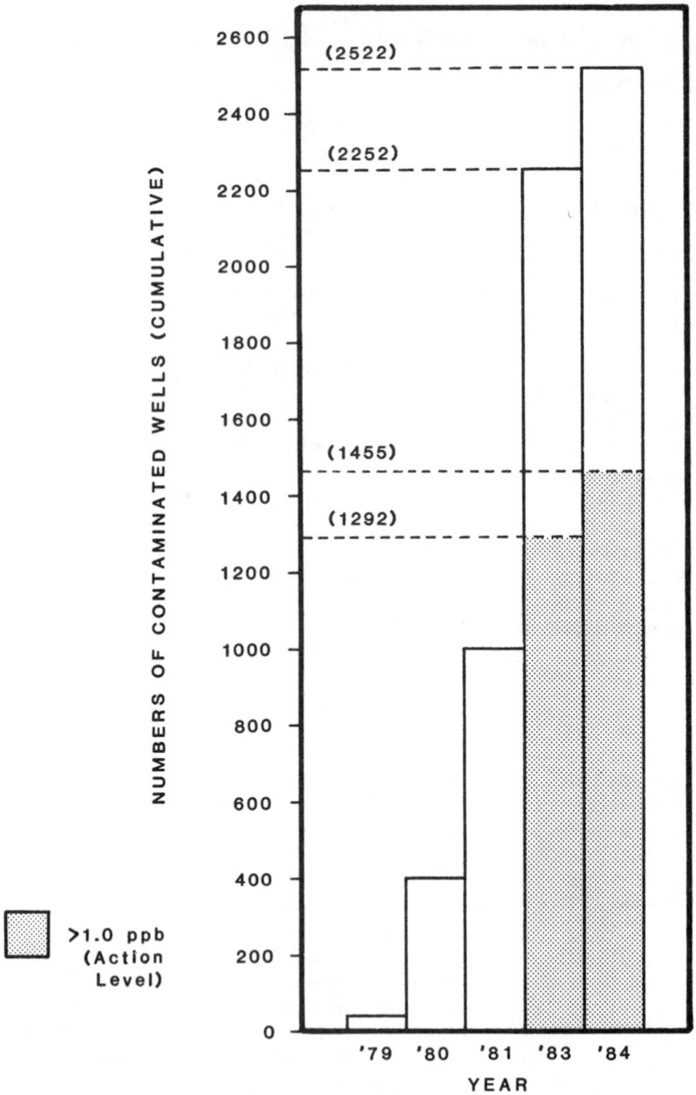

Figure 1. DBCP in California Ground Water - Cumulative Summary (May 1979 - April 1984).

Table I. Pesticides Detected in California Ground Water
(Large Water Systems vs. Small Domestic Systems)

Pesticide	No. Verified Incidents	Pesticide	No. Verified Incidents	Pesticide	No. Verified Incidents
Aldicarb	27	Dieldrin	4	Omite	2
Aldrin	22	Dimethoate	24	Ordram	3
Atrazine	2	Diphenamid	1	Paraoxon	1
Bentazon	1	Disulfoton	6	Parathion, ethyl	4
Benzaldehyde	1	DNBP	11	Parathion, methyl	1
Chlordane	4	DNOC	2	PCNB	1
Chlorpropham	1	Dursban	3	PCP	38
Dacthal	4	EDB	32	Phorate	2
DBCP	2522	Endosulfan	23	Phthalates	4
DDD	4	Endrin	1	Sevin	3
DDE	15	Ethion	5	Simazine	9
DDT	10	Ethylene thiourea	1	TCP	6
DEF	1	Furadan (carbofuran)	2	Toxaphene	5
Delnav	4	Heptachlor	4	Treflen	1
Diazinon	12	Kelthane (dicofol)	3	Zytron	4
Dichlone	1	Lindane	18	2,4-D	10
1,2-Dichloropropane	72	Malathion	5	2,4,5-T	4
1,3-Dichloropropene(cis)	2	Methylene chloride	4	2,4,5-TP	3
1,3-Dichloropropene(trans)	1	Naled	7	TOTAL	2963

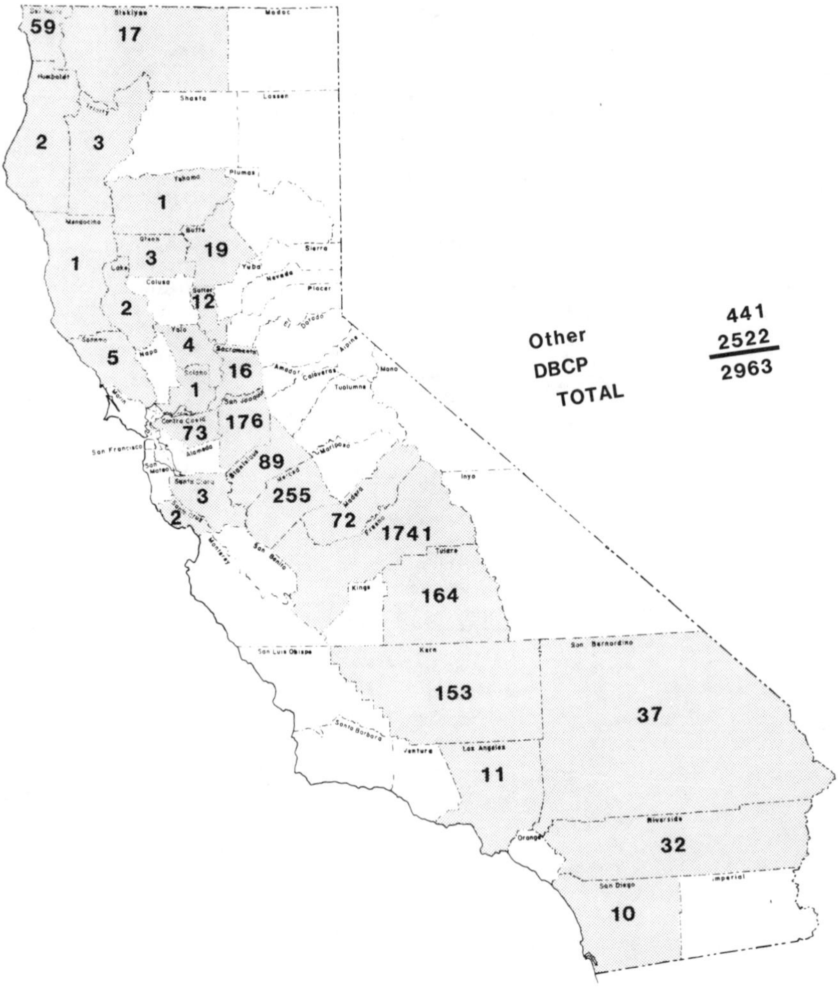

Figure 2. Verified Incidents of Ground Water Contamination with Pesticides (Including DBC).

distribution of all verified incidents of ground water contamination by pesticides (including DBCP) as of November 1984. This paper focuses on California's experiences with pesticides in ground water, potential causes, and possible solutions.

California's Water Quality Protection Program

In 1969, California adopted the Porter-Cologne Act giving the State Water Resources Control Board (State Board) the authority to (1) assure that all water diversions be put to a beneficial use, and (2) maintain and enhance the quality of all waters of the State. One major difference between federal and state water quality protection programs involves protection of ground water. The 1972 Federal Clean Water Act addressed surface water quality ("fishable and swimmable" by 1985), specifically excluding nonpoint source agricultural return flows from regulation. California law protects all waters of the State, including ground water and gives the State Board authority to regulate all sources of contamination including pesticides in agricultural runoff. Historically, such nonpoint sources have rarely been regulated by the State and Regional Boards because of the difficulty in tracing problem events to an unknown number of diffuse sources.

The State is subdivided along drainage basin boundaries into nine regions, each with its own Regional Water Quality Control Board (Figure 3). The regional boards have the authority to (1) issue waste discharge requirements, (2) monitor water quality, (3) take enforcement actions, and (4) adopt narrative or numerical water quality objectives into basin plans. A typical basin plan narrative objective states that "the discharge of pesticides to waters of the region or at locations where the waste may subsequently reach waters of the region is prohibited". In order to protect ground water, the State Board in 1972 adopted regulations governing hazardous waste disposal to land. These regulations (Subchapter 15 of the California Water Code) classified both wastes and sites. Wastes were categorized on the basis of the relative hazard to health and the environment, while sites were prioritized according to potential for leaching and ground water contamination.

As threats from biological pollutants in surface waters lessen, the awareness of threats from toxic chemicals increases. Hardly a week goes by without a new report of toxic organic chemicals in a community water supply or other water resource.

DBCP (1,2-Dibromo-3-chloropropane)

The most widespread pesticide contamination of ground water in California (and, possibly, the nation) has occurred through use of DBCP. The chemical was first discovered in California ground water in 1979 by the Central Valley Regional Board. Eight years after its use was banned, more DBCP-contaminated wells are still being discovered. DBCP in some well waters was at higher levels in 1984 than when first discovered in 1979 (Figure 4). Many of these wells have been shut down permanently, while others are only used on an emergency basis or when DBCP concentrations decrease below 1 ppb.

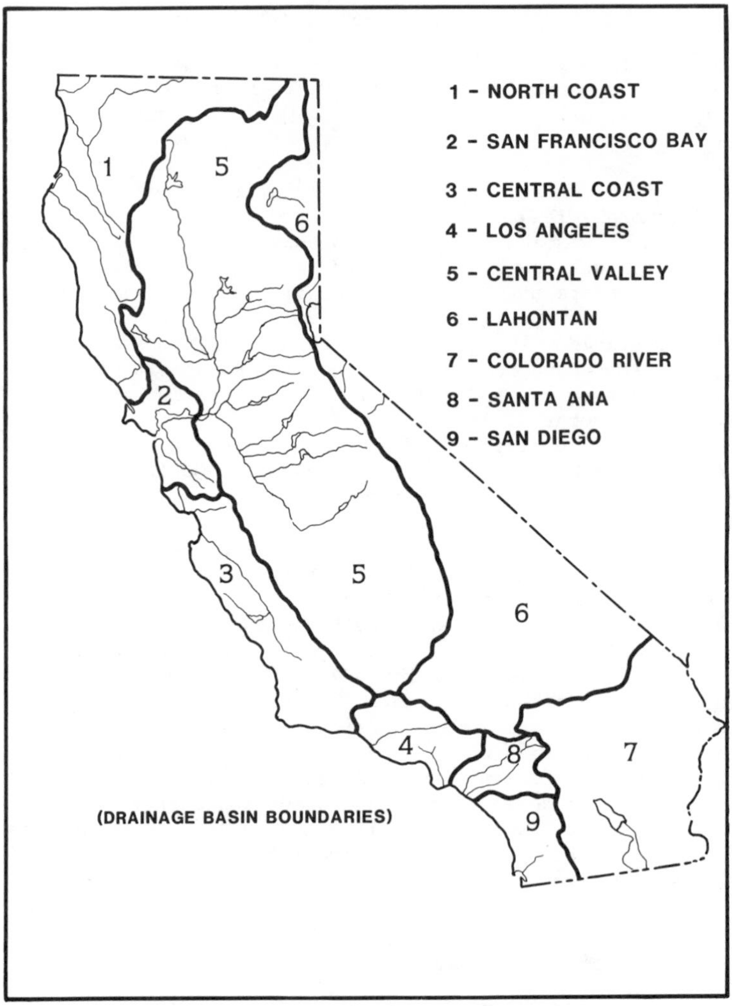

Figure 3. Regional Water Quality Control Boards.

Of 8,190 California wells monitored by November 1984, 30.8 percent were found to contain DBCP. Fresno County had the highest percentage (41.3 percent) and greatest number (1,696) of DBCP contaminated wells. In some areas of Fresno County, over half the private wells sampled contained DBCP above the DHS action level of 1 ppb. California has spent over $10 million in Fresno County alone to rehabilitate DBCP-contaminated ground water supplies.

Many of the worst pollution problems were found in small or private rural systems where a well serves one or more homes. Table II compares the degree of DBCP-contamination for both large public water systems (>200 connections) and smaller systems (5-200 connections) in 11 California counties. Large water systems had 14.6 percent of their wells contaminated with DBCP compared to 33.6 percent for small and domestic water systems which frequently pump from shallower, more vulnerable aquifers. By comparison, a recently inaugurated statewide monitoring program for over 40 toxic chemicals in large community wells (AB 1803) found toxic organic chemicals in 82 of 320 systems (25.6 percent).(3)

DBCP concentrations in some Fresno County domestic wells sampled between 1979-1984 have fluctuated seasonally by an order of magnitude. Figure 5 shows DBCP concentrations in one well fluctuated between 3 to 33 parts per billion within six months during 1981. Sample collection and analytical methods were unchanged during this period. Although the reasons for this variability are not yet clearly understood, these examples demonstrate the need to continue monitoring even after concentrations fall "below detection limits". Plumes of DBCP contaminated ground water can move unpredictably through aquifers to reach previously uncontaminated monitoring wells.

These retrospective DBCP monitoring data are presently being transferred to computer files. When this work is completed, analysis of spatial and temporal trends and possible correlations with soil and water predictive factors will be attempted.

California is fortunate in having a mandatory use reporting system for restricted pesticides.(4) Computer tapes containing the raw data from this program are transferred to the University of California at Davis, which has, with State Board support, developed a computer program for mapping pesticide use information geographically (by county, region, or the State) and temporally (by season, year, or cumulative years). Figure 6, which shows reported California use of DBCP for 1972-1977, is an example of that capability. Unfortunately, not all pesticide use was required to be reported during this period. DBCP was only required to be reported shortly before its use was banned in 1977. The Department of Food and Agriculture has therefore estimated actual use to be approximately three times reported use. The best estimate of actual cumulative DBCP use between 1957 and 1977 is in excess of 50 million pounds. Despite the incompleteness of past pesticide use reports, this pesticide mapping capability is an important tool in focusing State Board field monitoring efforts on areas of greatest potential risk.

A prospective field study of DBCP movement (both horizontal and vertical) through soil to ground water has been proposed by researchers at the University of California, Davis. A particularly

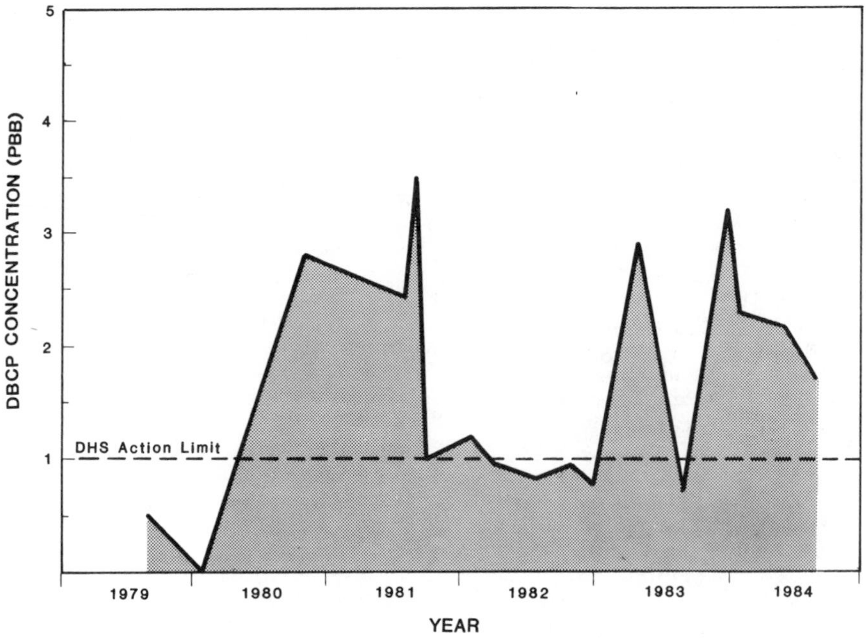

Figure 4. DBCP Concentrations in Escalon Municipal Well No. 3.

Table II. Percent of Wells Contaminated with DBCP
(Point and Nonpoint Source-1884)

County	Large Systems	Small Systems	TOTAL
FRESNO	19.1	43.4	41.3
KERN	5.9	23.3	15.9
MADERA	2.7	22.9	20.7
MERCED	22.7	23.5	23.4
TULARE	10.4	29.4	23.8
SAN JOAQUIN	19.9	19.4	19.6
STANISLAUS	16.2	19.9	18.5
RIVERSIDE	2.7	9.1	5.0
SAN BERNADINO	35.8	-	35.8
SUTTER	100.0	44.4	52.4
BUTTE	-	33.3	33.3
TOTAL	14.6	33.6	30.8

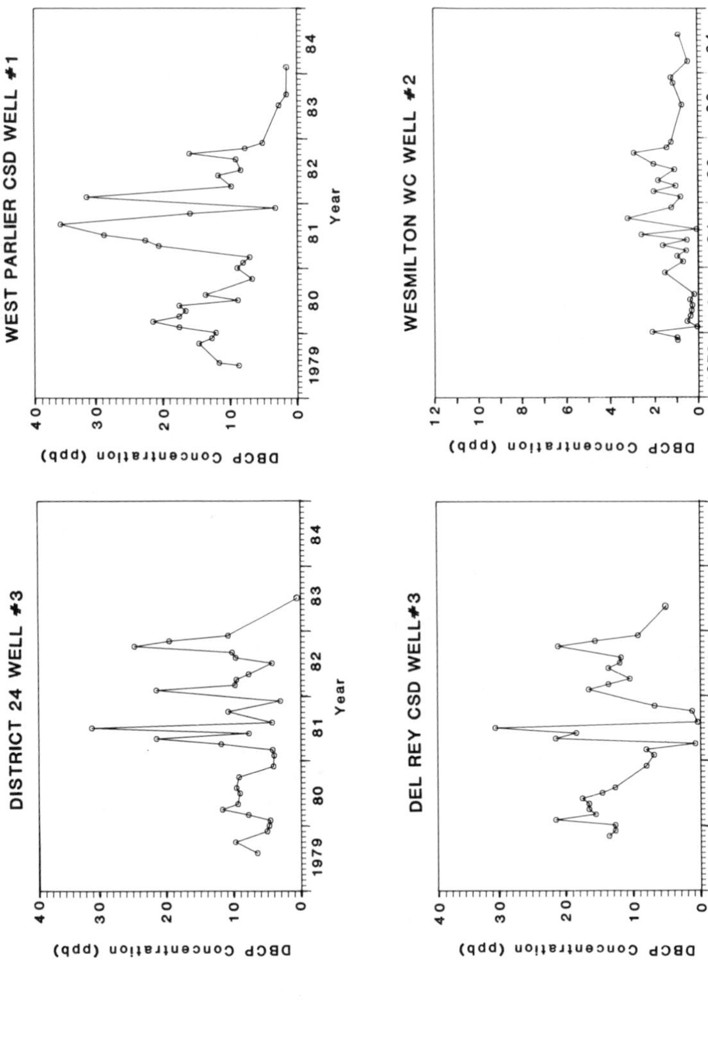

Figure 5. DBCP Concentration Fluctuations in Four Fresno County Wells (1979-84).

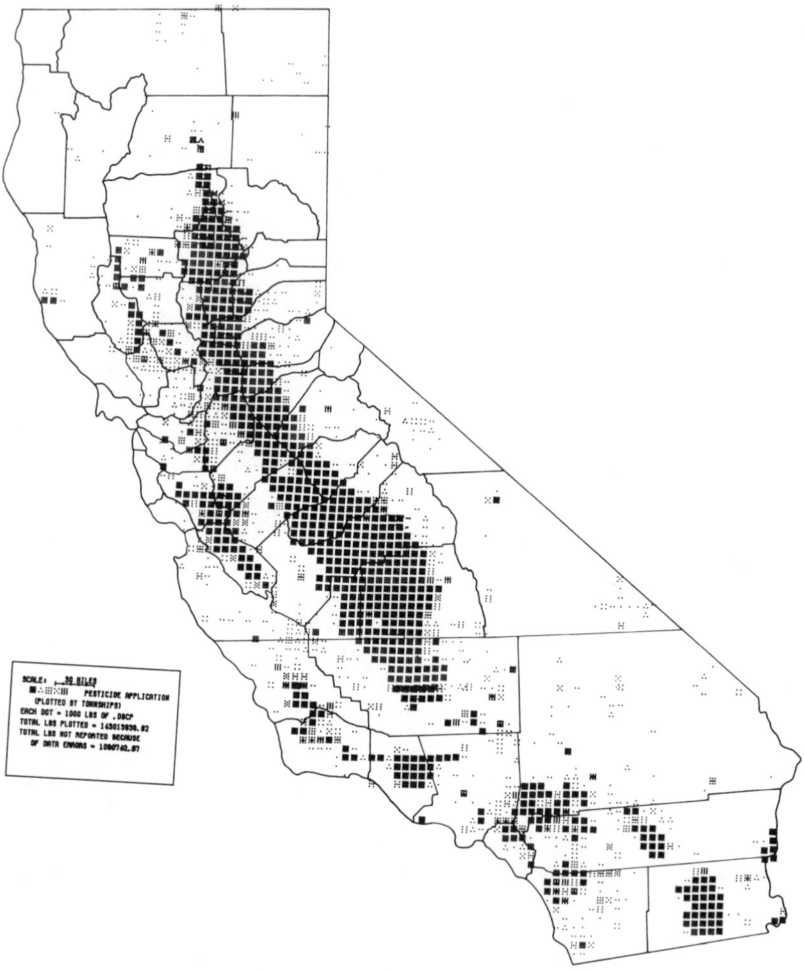

Figure 6. DBCP California Reported Use (1972-1977).

intriguing site is the central valley town of Sanger, near Fresno. DBCP use for 25 years in that area had caused a shallow aquifer used for domestic purposes to become contaminated. A well 600 feet deep was dug to obtain better quality water from the presumably uncontaminated deep aquifer. Special care was taken to prevent transfer of contaminated surface soil particles to the depth sample. Traces of DBCP were found at 420 feet, the greatest depth sampled, indicating an average vertical transport rate of approximately 17 feet per year (5). In this case, digging deeper was not the solution to pollution.

DBCP appears to persist in soils for decades with a half-life of over a century (6). At one California site, DBCP was found to persist in top soil for nearly a decade after a single application (7). This study conducted in 1980 in the Parlier area showed DBCP contamination to a depth of 30 feet throughout the soil profile. The highest concentrations of DBCP were found in the clay and silt layers. DBCP findings in soil and ground water coincided with areas of high DBCP use. This has led State Board staff to conclude that past agricultural use is the most likely source of this widespread contamination.

DBCP Drinking Water Standard

EPA has not yet developed a national drinking water standard for DBCP (8). The 1978 DHS drinking water action level of 1.0 ppb was adopted before current information on carcinogenicity was available. EPA has since estimated a lifetime cancer risk from consuming 2 liters per day of water containing 1 part per billion DBCP as 180 excess cancer incidents per million population exposed (9). Figure 7 depicts the EPA estimated cancer risk at various DBCP water concentrations. The State of Hawaii in 1984 adopted a water quality limit of 0.02 ppb (20 ppt) (10). In view of DBCP's potential cancer impacts at very low concentrations, SWRCB has recommended that DHS lower the California DBCP action level to the current detection limit of 0.02 parts per billion (11).

Population Impacts

Department of Health Services staff has estimated that over 700,000 people in 32 towns are either drinking DBCP-contaminated water or have had to seek alternative water supplies (12). The estimated breakdown by county, town, number of wells contaminated, and population served is shown in Table III.

Large water districts can dig deeper, switch to nonpolluted wells, or blend their supplies to meet current public health advisories. Nevertheless, according to testimony of State Board Chairperson C. Onorato before the California Senate Agriculture and Water Committee on March 5, 1985, "we are approaching the point where toxic pollution is beginning to affect the overall availability of water in California. Instead of moving forward to anticipate future water shortages, we are moving backward by losing some of what we already have". Home buyers in the central valley wishing to obtain a Federal Farm Home Mortgage loan must first provide the lender with evidence that the well supplying their prospective home with domestic water is free of DBCP (12).

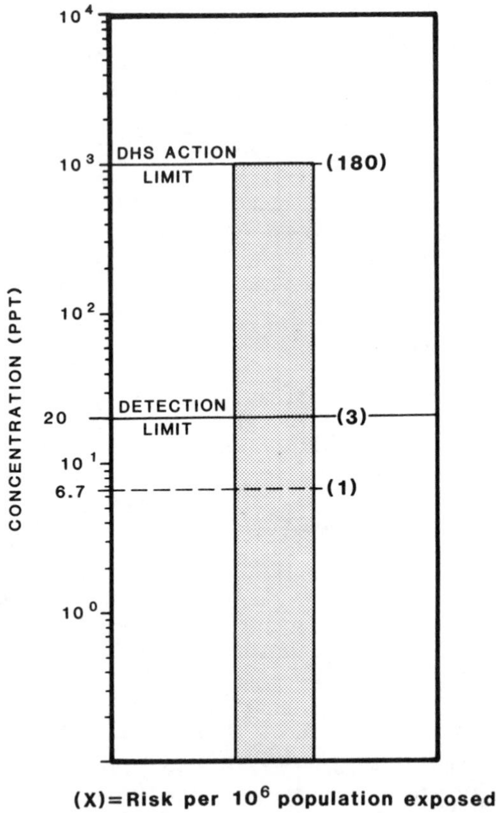

Figure 7. DBCP Water Concentrations Versus Cancer Risk Levels.

Table III. Estimated Population Affected by DBCP (>1 ppb) in Large Water Supplies (April 1983)

County	Towns	Wells Contaminated	Population Served
Fresno	12	21	290,000
Riverside	3	13	195,000
Stanislaus	3	4	90,000
San Bernardino	2	4	50,000
San Joaquin	5	17	50,000
Merced	3	6	15,000
Tulare	2	6	10,000
Sutter	1	2	2,000
Kern	1	2	1,000
Total	32	75	703,000

Central Valley Ground Water--How Much is Contaminated?

Over half (55.6 percent) of all pesticides reported used in California were used in the central valley. (13) The eastern San Joaquin Valley in particular is vulnerable to chemical contamination of ground water because of a unique combination of high pesticide use, shallow water tables, relatively porous soils, and seasonal irrigation. Schmidt has estimated the volume of DBCP polluted ground water in the Fresno urban area in excess of 4 million acre feet (M.A.F.). (14) Similar calculations for other high DBCP use areas in the San Joaquin Valley (8 million acres) yield an estimated volume of DBCP polluted ground water approaching 30 M.A.F. or over one quarter of the usable ground water in the entire valley. The costs for pumping and treating such large volumes of polluted water would be astronomical. Other alternatives, such as carbon filtration, bottled water, or drilling deeper, each have their own technical or economic drawbacks. We are relearning the hard way that an ounce of prevention is better than any cure.

State Board Priority Chemical Investigations

In order to answer the question: are there any more DBCP-type problems out there awaiting discovery, the State Board in May 1980 established a Toxic Substances Control Program. A major focus of this program was development of an "early warning" priority chemicals project to (1) assess risks, (2) recommend water quality objectives, and (3) propose appropriate mitigation measures for agricultural and industrial chemicals most likely to adversely impact surface or ground water quality.

Table IV lists the 13 criteria used for ranking those compounds (from among hundreds of potential candidate chemicals) that posed the highest risk to water quality. The 100 highest reported use pesticides were screened. A semi-quantitative ranking scale was developed and applied in selecting 12 priority chemicals (Figure 8). Field monitoring in potential "hot spot" areas was designed to document presence and extent of priority chemical residues. The various steps in the priority chemical process are shown in Figure 9.

Table IV. SWRCB Criteria for Selection of Priority Chemicals

1. Potential for surface water contamination
2. Potential for ground water contamination
3. Detection in California
4. Use in California
5. Fish/wildlife kills
6. Public concern
7. Actions by other agencies/countries
8. Detection worldwide
9. Bioaccumulation potential
10. Persistence
11. Carcinogenicity, mutagenicity, teratogenicity
12. Aquatic toxicity, acute/chronic
13. Human toxicity, acute/chronic

1,2-D/1,3-D (1,2-dichloropropane/1,3-dichloropropene)

The nematicides "D-D" and "Telone", which contain 1,2-D/1,3-D, were selected for priority chemical investigation because of their relatively high mobility in soil, solubility in water, and chemical similarity to DBCP. (15) Preliminary information also indicated that these chemicals might be mutagenic and carcinogenic. Of particular significance was the increased use of D-D and Telone in California from 4 million pounds in 1977 (the year DBCP was banned) to over 16 million pounds in 1981 (Figure 10).

Prior to 1982 when SWRCB selected these nematicides for priority pesticide investigation, sampling efforts by other state or local agencies had not found 1,2-D in ground water. The State Board's 1,2-D/1,3-D investigation documented over 60 wells contaminated with 1,2-D (25 percent of samples collected). (15) No 1,3-D was found in any of the samples containing 1,2-D. SWRCB subsequently recommended that the Department of Food and Agriculture (DFA) reevaluate all soil fumigants containing 1,2-D. DFA accepted this State Board recommendation.

Use of D-D (approximately 35 percent 1,2-D) was subsequently suspended by the Del Norte County agricultural commissioner. Shell Chemical Company has since withdrawn D-D from the California market.

The State Board recommended that DHS adopt an action level of 10 ppb 1,2-D in domestic water supplies. County health officials now advise private domestic well owners to seek alternative supplies for water containing more than 10 ppb 1,2-D.

All of the chemical constituents of soil fumigants, both active and inert, should be tested for their potential to leach to ground water. The 1,2-D component of 1,2-D/1,3-D formulations was a non-essential contaminant of the major active ingredient 1,3-D.

Most of the documented 1,2-D contamination incidents were in private rural wells. A comprehensive ground water monitoring

AGRICULTURAL CHEMICALS	Human Toxicity Acute/Chronic	Aquatic Toxicity Acute/Chronic	Carcinogen Mutagen Teratogen	Bioaccumulation Potential	Persistence	Public Concern	California Use	Detection in California	Detection Worldwide	Fish/Wildlife Kills	Potential for Ground Water Contamination	Potential for Surface Water Contamination	Action by Other Agencies, States or Nations
TOXAPHENE	M	H	M	H	H	M	H	M	M	L	L	M	M
1,2-DICHLOROPROPANE 1,3-DICHLOROPROPENE	M	L	M	L	M	L	H	L	L	L	H	L	L
ETHYLENE DIBROMIDE (EDB)	H	M	H	L	M	L	M	M	L	L	H	L	L
ENDOSULFAN	H	H	L	L	L	L	M	M	L	H	L	H	L
ARSENICALS	H	L	H	H	H	L	M	M	M	L	L	H	L
RICE HERBICIDES (MOLINATE & THIOBENCARB)	M	H	L	M	L	M	H	H	L	H	L	H	L
INDUSTRIAL CHEMICALS													
POLYCHLORINATED BIPHENYLS (PCB)	H	H	H	H	H	H	H	H	H	L	L	H	H
PENTACHLOROPHENOL (PCP)	M	H	L	M	M	M	H	M	M	M	M	M	O
TRICHLOROETHYLENE (TCE)	H	H	H	M	M	H	H	H	O	O	H	M	O
CHLORINATED ETHANES	H	H	H	M	M	H	H	H	O	O	H	M	O
CYANIDE	H	H	M	M	H	H	H	H	O	O	H	H	O
CHROMIUM (hexavalent)	H	H	H	M	H	M	H	H	H	O	H	M	O

L = LOW RATING
M = MODERATE RATING
H = HIGH RATING

Figure 8. Top Twelve SWRCB Priority Chemicals Selected in 1981.

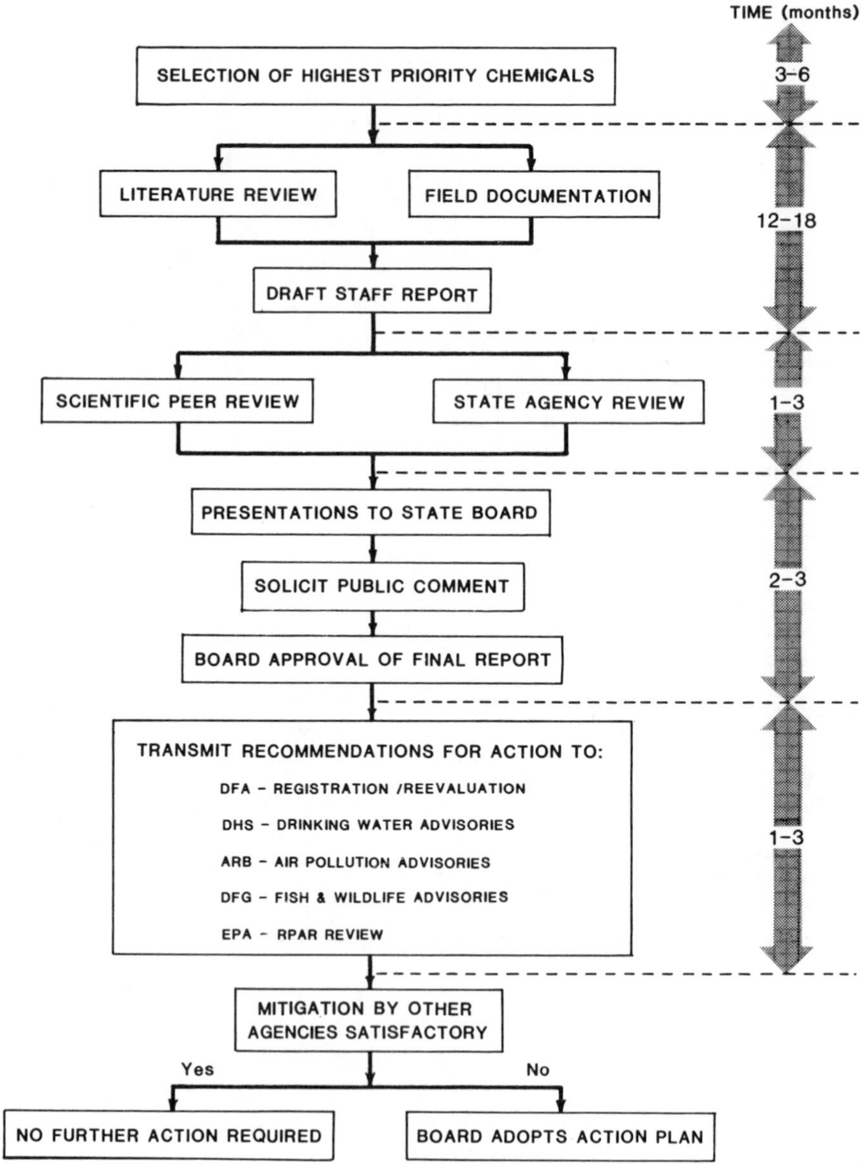

Figure 9. SWRCB Priority Chemical Assessment Process.

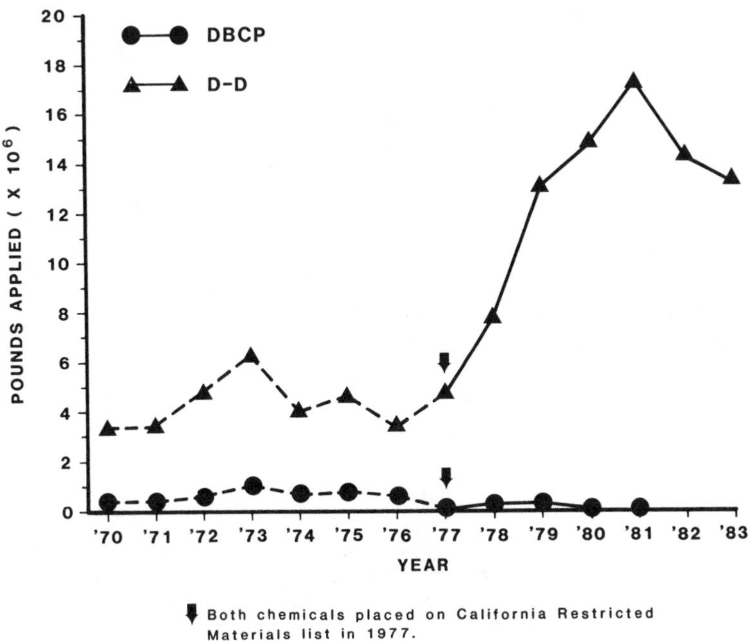

Figure 10. DBCP and 1,2-D/1,30-D (D-D) Reported Use (1970 to 1983).

program should include such wells to provide "early warning" of potential contaminant movement from rural areas to large urban population centers.

The 1,2-D/1,3-D study discovered several cases where chloroform, carbon tetrachloride, and TCE were detected together with DBCP or 1,2-D in a single sample. Should a regulatory action level for a single toxic chemical be revised when other chemicals, particularly carcinogens, are also present? More stringent safety factors may be required for deriving adequately protective water quality standards when multiple exposure to a mixture of two or more chemicals is involved.

Implications of SWRCB Recommendation for 1,2-D

The SWRCB recommendation to reduce 1,2-D in soil nematicides to the lowest practical level (<2 percent) was technically feasible. DFA is presently developing a regulation to limit 1,2-D in nematicides to less than 0.5 percent. Of the two U. S. manufacturers of these nematicides in 1982, one was producing a relatively clean material while the other was producing a "high" 1,2-D product. The manufacturer of D-D subsequently decided to cease production leaving only one manufacturer of 1,2-D/1,3-D type products in the U. S. Consequently, the cost to agriculture for these types of soil fumigants has reportedly increased. This experience highlights the urgent need to accelerate the search for nonchemical nematode control methods which would not contaminate ground water.

Ethelene Dibromide (EDB)

The State Board's EDB priority pesticide study was initiated in 1982.(16) In 1983, DFA announced finding EDB contamination in two domestic wells near Fresno. Because agricultural use was suspected as the cause of this contamination, DFA suspended permits for EDB use as a soil fumigant in Fresno County. Subsequent findings of EDB-contaminated wells in Kern, Merced, San Joaquin, and Stanislaus Counties led to suspension of EDB use in these counties. The California Department of Health Services in 1983 detected EDB in several private wells in Bakersfield. The source of this contamination is still under investigation. On the basis of these and other findings nationwide, EPA, on October 11, 1983, published an emergency order suspending registration of EDB soil fumigant.

EPA's emergency suspension forced states with EDB-contaminated wells to scramble for an EDB drinking water action level. EPA advised Hawaii that EDB residues in drinking water above the detection level of 20 ppt should be considered unsafe. EPA's Scientific Advisory Panel (SAP) has concurred with this decision.

The use of animal carcinogen data to derive safe limits for drinking water continues to be highly controversial. Advocates of less regulatory stringency argue that EPA has over-regulated EDB on the basis of flimsy evidence concerning EDB dietary effects on humans. Others feel that the burden of proof should be on the registrants of products containing animal carcinogens to show why animal test data are inappropriate to derive human health risks.

The SWRCB selection in 1982 of EDB for priority pesticide investigation heightened the awareness of all California regulatory agencies with responsibility in this area to the potential for EDB contamination of ground water. It also led to the recognition of a generic problem with all soil fumigants. The persistence of these nematicides, as well as the unique way they are applied to soils, increases the likelihood that they will cause similar water quality problems. EPA has since decided that in order to avoid the mistakes of the past, all currently registered soil fumigants be reviewed for toxicology, environmental fate, and product chemistry data.

SWRCB in 1984 recommended and DHS established an action level for EDB in drinking water based on an NCI cancer risk assessment that EDB above the then current detection limit (50 ppt) would pose an unacceptable risk to human health. In other words, any confirmed finding of EDB in California drinking water is considered grounds for recommending the consumer seek an alternate source.

A draft cancer risk policy for California (17) expresses as a goal that the risk from exposure to a chemical carcinogen in drinking water should not exceed one person affected per million population exposed. Analytical techniques for known or suspected carcinogens must be refined to enable detection at levels that are relevant to this risk assessment policy.

Current drinking water action levels for many toxic chemicals are based on noncarcinogenic health effects data. As soon as possible after new cancer risk and other chronic effects information becomes available, these action levels should be revised.

Aldicarb

Agricultural use of aldicarb, a post-plant nematicide, has resulted in ground water contamination in six states east of the Mississippi. The Department of Food and Agriculture in 1979 conducted an aldicarb ground water monitoring survey in several California counties. All survey results were negative. In 1983 the North Coast Regional Water Quality Control Board discovered the first evidence of aldicarb in California ground water (up to 47 parts per billion) in Del Norte County. EPA has recommended a limit for total aldicarb in drinking water of 10 ppb.

In response to the regional board findings, DFA and the county agricultural commissioner suspended the use of aldicarb in Del Norte County. Although the aldicarb contamination in Del Norte County was correlated with somewhat unique site-specific conditions, particularly low soil pH, permeable soil, and shallow ground water, other areas with similar conditions might also be vulnerable. Field testing is being conducted in the central valley by the registrant Union Carbide Company, to determine the potential for ground water contamination in this high use area.

Union Carbide on October 31, 1984 announced a significant, unprecedented program to provide activated carbon filters for homeowners whose wells are contaminated with aldicarb above the recommended action level. This is the first time in California that a pesticide manufacturer has provided water treatment systems to clean up private domestic well water contaminated by their product. It should provide added incentive for other manufacturers to thoroughly field test their products before they are marketed for potential to contaminate ground water. Although this testing is costly, the costs of correction after contamination would be far greater.

Problem Causes

A review of California information on ground water contamination indicates that while the causes are site specific, most incidents occurred because of three interrelated problems:

overoptimistic risk perception (RP)
inadequate risk assessment (RA)
underprotective risk management (RM)

The following factors are discussed as relevant to these problems.

Risk Perception (RP).

"Filter Fantasy"
Inadequate testing of "leachers"

Risk Assessment (RA).

Analytical constraints
Monitoring constraints
Toxicological controversy

Risk Management (RM).

Illegal use
Improper discharge
Wells as pollutant pathways
High-risk irrigation/infiltration practices
Nonexistent early warning systems

Society is increasingly concerned that ground water, a scarce resource, deserves much greater efforts toward "foolproof" risk management (RM approaching 100 percent). In order to accomplish this goal, both (RA) and (RP) must also approach 100 percent. The optimal risk management strategy to meet society's expectations would be (1) to err on the conservative side in perceiving potential risks while (2) improving risk assessment by mobilizing increased resources to adequately fill risk and exposure assessment data gaps.

In some cases, economic and technical constraints may require society to settle for protecting less than 100 percent of a particular ground water resource. Even under such circumstances, the conservative approach to risk perception and risk assessment should not change. A simple way of illustrating the interconnectedness of these problems is to assume a hypothetical quantitative relationship where RM = (RP) x (RA). Achievement of only 80 percent risk management (RM = 0.8) would still require near-perfect risk perception and assessment (e.g., RA (0.9) x RP (0.9)).

Filter Fantasy

Filter fantasy is a term used to describe unwarranted trust in the "magical" properties of soil to filter out and prevent toxic chemicals from reaching ground water. This overoptimistic risk perception is based on the belief that most agricultural chemicals are unlikely to reach ground water because of their "immobility and rapid degradation in the environment"(18). Mounting evidence that many soil conditions cannot be relied upon to prevent chemical movement to ground water from either point or nonpoint sources is leading to a changed perception of this risk. Even so-called "impermeable" California hazardous waste disposal sites have been found to leak. Thinking of the soil as a "sieve" rather than a "filter" might help to counter this overoptimistic risk perception.

Inadequate Testing of "Leachers"

Information on pesticide soil mobility can be derived from lab bench scale tests, use of predictive models, or field tests. The failure to predict the potential for ground water contamination of a number of pesticides indicates the need for a change in the risk perception toward multi-year field dissipation testing of cumulative applications. Experience with certain soil applied pesticides such as Oxamyl exemplifies the need for such testing. Field studies with Oxamyl were conducted at a single California site for two years with no evidence of a threat to ground water. Nevertheless, New York recently reported findings of Oxamyl in Suffolk County ground water. Obviously testing is needed at a wide variety of environmental sites under different soil and climatic conditions to predict long-term potential for ground water contamination. The length of time needed to conduct field tests for soil injected herbicides and nematicides may, in some cases, have to exceed two years. One can only speculate whether multi-year field testing of DBCP's soil mobility before it was registered for use would have prevented a ground water contamination problem of unprecedented proportions.

Analytical Constraints

Most of the pesticides detected in California ground water have neither federal nor state drinking water standards. Laboratory analytical protocols are not available for detecting many of them at or below one part per billion. Inability to routinely detect many of these chemicals at levels suspected of having long-term chronic effects is a significant weakness in our ability to adequately assess their risks. Commercial laboratories have been reluctant to develop new pesticide-specific, analytical protocols without assurance that the anticipated volume of lab work would repay their investment.

Our ability to fully assess these risks depends in large part on current analytical methods geared towards detecting the more volatile organic compounds. New, less volatile organic contaminants will likely be discovered with greater frequency in ground water as trace organic analytic methods are further developed. Some of these less volatile compounds may turn out to be of greater toxicological significance than the currently detected volatile organic compounds.

Monitoring Constraints

An ideal ground water monitoring network would continuously provide data for all constituents of concern in all areas of concern. In reality, most ground water monitoring programs have, until recently, looked for very few toxic organic chemicals in very few places, very infrequently.

Monitoring existing wells for toxic chemical contamination has been compared to searching blindfolded for a needle in a haystack. Monitoring wells may be too deep, too shallow, or in the wrong place. Because of hydraulic and pumping variations in the immediate vicinity of a contaminated well, pollutant concentrations vary significantly within very short time periods. Nevertheless, some agencies have stopped monitoring for DBCP after a single "not detected" result. It is important to interpret "not detected" results with caution. Monitoring for persistent chemicals like DBCP in once contaminated areas should continue for several years after the last "not detected" result before concluding that "all is clean".

Special precautions may be needed to prevent loss of certain volatile organic compounds during sample collection and storage. North Coast Regional Board staff(19) conducted a study of 1,2-D loss from a ground water sample as a function of sample storage time. The EPA-recommended method (up to 10 days on ice) was used. One sample was found to contain 250 ppb when analyzed within 24 hours, but only 77 ppb when another aliquot of this sample was reanalyzed after 10 days of storage.

Toxicological Controversy

Our current ability to detect and quantify a growing list of organic chemicals in ground water outstrips our ability to interpret the toxicological significance of these findings. The limits of present knowledge concerning potential long-term damage leads to controversy. Is it more prudent to err on the side of overprotection or underprotection?

Furrer(18) states that "the detectable levels of trace quantities of some pesticide residues in some geographical areas "are generally far below those that may cause any adverse health effects". McCarty(20) states: "Difficulty in demonstrating a cause-and-effect relationship between drinking water and cancer is not sufficient cause to conclude that one does not exist. In the face of such uncertainties, prudence dictates that error should be on the side of caution."

Figure 11 shows that four different mathematical models have a five order of magnitude difference in estimated lifetime cancer risk when drinking water contains 50 ppb TCE. These models were designed to extrapolate animal study results from high-to-low doses to correspond to human exposure. The validity of any of these models for low-dose risk assessment is questionable. Current knowledge does not allow the selection of one particular model, and experimental data are not sufficient to discriminate among competing models. In light of this uncertainty, should regulatory agencies base their policies on the most conservative or the most "reasonable" estimate?

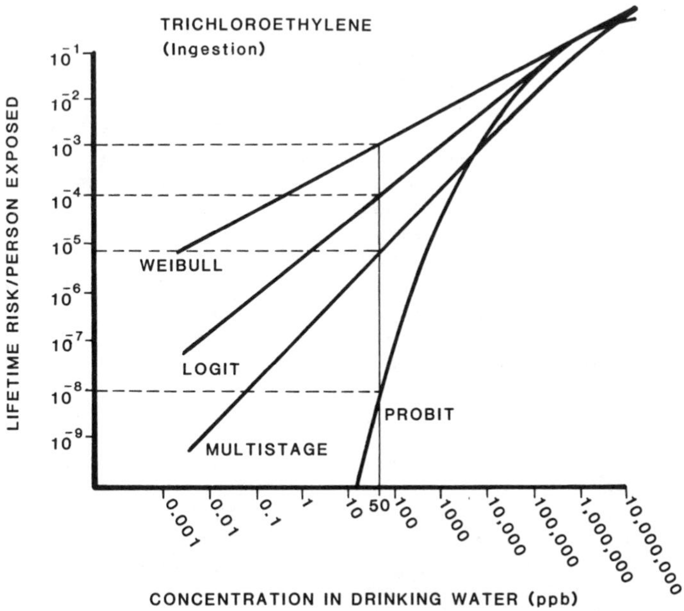

Figure 11. Extrapolation of Dose-Response Models to Low Doses of TCE. Adapted with permission from Ref. 21. Copyright 1984, Regul. Toxicol. Pharmacol.

The U.S. Supreme Court decision on benzene of July 2, 1980, dealt with the issue--what are prudent boundaries for assessing significant risk. The court concluded that if the odds for exposure are 1 in 1,000 (10^{-3}), a reasonable person might consider the risk significant. If the odds are 1-in-1 billion (10^{-9}) that a person will die from cancer by taking a drink of chlorinated water, the risk could clearly be considered not significant. Thus, the Supreme Court has defined the bounds of significance between 10^{-3} and 10^{-9}. Cothern and Marcus (21) have suggested a scheme for using the results of risk estimates in federal and state regulatory decision making. By assuming that the entire population of California was exposed to a chemical at the individual risk rate, three different regulatory levels of population risk can be derived for California: (1) 2,000 or more potential cases per lifetime at risk should trigger a national regulatory decision; (2) between 20 and 2,000 cases per lifetime, state regulatory decision making should be triggered; and (3) less than 20 per lifetime risk would be considered a local problem not requiring state or federal action. The current EPA risk model for DBCP (9) indicates the excess cancer risk is 180 per million, or 4,500 Californians at risk. If this scheme were to be adopted as national policy, California's DBCP in ground water problem would be defined as an unacceptable risk requiring immediate federal action.

Illegal Use

EPA's past decisions to ban certain pesticides often coincided with a decline in their demand because of decreasing efficacy (increasing pest resistance). DBCP, however, is still considered to be a most efficacious nematicide. Thus, illegal use, if proven, could explain some of the rise in DBCP contaminated wells since its use was banned in 1977. The estimated half-life of DBCP in water is 3-1/2 years (8). If illegal use were not occurring, one would expect a decline in DBCP well water concentrations after eight years of nonuse.

On March 11, 1985, the Department of Food and Agriculture announced evidence of illegal DBCP use in a Lodi area, San Joaquin County vineyard. The investigation was started after a farm worker cleaning out equipment became ill, and a fellow worker recognized the distinctive odor of DBCP. Equipment, soil, and well water from the suspected field were analyzed and showed positive DBCP content. CDFA (22) has concluded that this appears to be an isolated incident. If illegal use is still occurring in California, it could not approach the earlier million pounds-per-year usage. Pest control operators found using a banned pesticide would be jeopardizing their license and livelihood. For these reasons illegal use is not considered to be a major reason for the spread of DBCP contaminated ground water throughout the San Joaquin Valley. The most likely source is the large amount of residual DBCP, stored and continuously moving down the soil profile.

Improper Discharge

Formulation, storage, and poor disposal practices have resulted in a number of pesticide ground water contamination incidents in California (2).

One of the most serious and costly of these incidents at the Occidental Chemical Company Plant, Lathrop, involved ground water contamination by DBCP, EDB, and other pesticides in concentrations ranging up to 1,240 ppb. The Central Valley Regional Board determined that the contamination was the result of willful discharge of hazardous wastes and fraudulent self monitoring. (The company intentionally withheld information that they were processing and discharging pesticides into unlined leach ponds.) This "point source" discovery prompted the regional board to investigate the much more widespread nonpoint source DBCP problems.

Discovery of pesticide contamination from any source should be considered evidence that it can occur from agricultural use as well. The discovery in California ground water of DBCP, EDB, and 1,2-D from point source discharges preceeded subsequent discovery of these same pesticides in ground water from nonpoint sources. Thus, point source incidents should be regarded by regulatory agencies as an early warning of the potential for cumulative agricultural use of these chemicals to also contaminate ground water. Ground water monitoring programs should be periodically augmented to include newly discovered pesticides in ground water from any source.

Wells as Pollutant Pathways

"It is ironic that one way ground water quality can decline is through the well. This occurs when, because of inadequate construction, wells provide a physical connection between sources of pollution and usable water. The geologic environment has some natural defenses against pollutants, but each time we penetrate that environment, we may carelessly establish avenues for their uncontrolled introduction."(23)

There are over 750,000 wells in California, most of which were drilled for agricultural irrigation purposes. There is no inventory of an estimated one hundred thousand wells that have been either deliberately or indiscriminately abandoned. The Department of Water Resources has established standards of water well construction and abandonment. These standards are voluntary, however, and various counties apply them in different ways.

During the drought year, 1977, over 20,000 new wells were drilled in California with nearly half of these in the Central Valley. The rate of well construction and abandonment is increasing by approximately 5 percent per year. Current state standards do not require sanitary seals on most agricultural wells. Thus, unless action is taken to reverse this situation, the problem of poorly sealed or repaired water wells which allow surface pollutants to reach ground water is likely to increase.

High Risk Irrigation/Infiltration Practices

The practice of metering pesticides into irrigation systems (Chemigation) can lead to soil and ground water contamination if irrigation pumps fail and chemical injection continues causing chemical backflow into the well. Backflow prevention devices can mitigate this problem. In 1982, Kern County was the first to adopt an ordinance requiring backflow prevention devices for chemigation systems (2).

Capturing and diverting storm water that may contain pesticides or other chemical constituents into an aquifer through "dry wells" may also pose risks to ground water quality. Tens of thousands of such recharge wells have been constructed in the San Joaquin Valley. Several thousand new wells are being constructed each year to replace older wells clogged with sediment and debris. Although some county health departments require a 10-foot separation distance between the dry-well bottom and the underlying water table, provisions for assuring compliance with this requirement are vague, and some drainage wells do penetrate the water table to increase infiltration drainage rates. These practices exemplify inadequate risk perception which may lead to underprotective risk management.

Nonexistent Early Warning System

Almost every significant instance of ground water contamination in California has been discovered only after a drinking water source has been affected. We need, but do not have a reliable early warning "canary in the coal mine" to warn us at an early stage of incipient danger from pesticides in soil moving toward ground water.

Improved ground water sampling and analytical techniques is a necessary step towards the "early warning" goal. One promising technique currently being investigated in California involves automated large volume sampling, concentrating and quantitatively extracting trace amounts of synthetic organic chemicals in ground water using XAD-type resins.

Monitoring of the vadose (unsaturated) zone may, in certain areas, be another promising step towards an "early warning" system. Ground water contamination if detected early enough from this zone, could allow for remedial measures to be implemented sooner and aquifer restoration costs significantly reduced. In areas with a very shallow vadose zone (i.e., Smith River with less than 10 feet depth), there may be insufficient lead time for preventive regulatory action.

Possible Sollutions

The least costly and, ultimately, most effective approach toward solving this problem is preventative. At the federal level, EPA must develop more rigorous pesticide registration requirements. Pesticides with a high potential for leaching should not be registered for nematode control. Industry can play a lead role in preventing future chemical contamination of ground waters by developing alternative biotechnological techniques (e.g., modify root bacteria genetic codes specifically to induce resistance in plants).

Robert Kaufman, Monsanto's director of plant sciences research, is quoted as saying, "It is conceivable that if we're successful with this technique, chemical insecticides as we know them could be phased out in the next 25 years" (24).

Agricultural engineers can help minimize deep percolation of water soluble pesticides below the root zone by conducting research to improve irrigation efficiency (drip method). Local and regional government agencies can promote adoption of better management practices (BMPs) for applying pesticides and water to soils, as well as stepping up enforcement actions against illegal use and improper discharge.

The challenge facing federal, state, and local agencies is to devise a coordinated system that effectively safeguards water quality while recognizing that this complex problem cuts across jurisdictional boundaries of all levels of government.

EPA's national ground water protection strategy is based on promoting state ground water protection programs. California's Secretary of Environmental Affairs on March 16, 1984, designated SWRCB as the lead state agency for developing a coordinated ground water protection strategy. The State Board has submitted to EPA a proposed ground water strategy workplan whose elements include: (1) Inventory of all current ground water quality protection activities by all levels of government including assessment and prioritizing of needs to correct and prevent problems; (2) assessment of EPA ground water basin classification systems and comparison with California current ground water basin designation and water quality objectives; (3) development of a comprehensive computerized ground water quality data management system; (4) refinement of a ground water "hot spots" risk assessment and monitoring methodology.

Ground Water "Hot Spots" Program

There is an urgent need for a systematic methodology to identify emerging ground water contamination problems before they become environmental crises. The State Board's ground water "hot spots" project, which began in July 1984, is an attempt to develop such a predictive approach. It is based on screening, ranking, and assessing chemicals and sites likely to contain contaminated soil and ground water before an actual problem is documented. The success of this approach to predict trouble spots before they assume basinwide proportions will be assessed by comparing "predicted" vs. "observed" data in three regions, the Central Coast, Central Valley, and Los Angeles Basin.

Selection criteria include toxicology, physical/chemical characteristics, and environmental fate information. These data are used to evaluate individual chemicals and estimate the degree of hazard associated with each site. For nonpoint source contamination from pesticide application, rating values include the amount of chemical applied. Final ranking is based on well logs, hydrogeology, and other data derived from on-site reconnaissance.

Criteria may be weighted according to an estimated degree of hazard. If the water table at a site approaches ground surface, ground water contamination is highly likely. The criteria "Depth to Ground Water" would thus receive major emphasis over certain other criteria.

Transparent overlays of site-specific data (e.g., depth to ground water) and chemical use data (e.g., DBCP, 1,2-D, and aldicarb cumulative reported use) allow for visual corroboration and fine-tuning of preliminary selected hot spot sites.

Other objectives of the ground water "hot spots" program include mapping of geologically sensitive areas to provide data for regulatory action by county agricultural commissioners and regional boards. Such maps can be used to allow regulatory agencies at the local level to determine where to allow or not allow application of those chemicals most likely to leach to ground water (Figure 12).

Is the Cup Half Empty or Half Full?

Jay H. Lehr (25) writing about the state of the nation's ground water states that, "We have polluted less than 1 percent of our ground water which has impacted less than 5 percent of our population. In the next 10 years, these numbers will increase less than 10 percent, and in the following decade, they will decline.... ground water's glacially slow movement...affords adequate time to achieve the most reasonable approach to managing these residual problems of our past."

The perception of glacially slow movement is not shared by USDA soil chemist Robert Bowman (26). His research in Arizona of pesticide percolation rates indicates that pesticides applied to both wet and dry soils move through these soils from three to five times faster than previous studies indicated.

Professor Paul Roberts of Stanford University (27) feels that although "less than 1 percent of the nation's ground water is thought to be contaminated by anthropogenic organic pollutants.... this estimate is only a rough approximation; our ground water contamination may turn out to be greater than is now realized...In the context of the rapid expansion in manufacture and use of synthetic organic chemicals since 1940, long-time lags mean that incidence of ground water contamination may continue to be discovered at an increasing rate in the years to come, even if disposal practices are modified immediately to avoid new sources. In that sense, contamination by synthetic organic chemicals is like a water quality time bomb, ticking away for years under many communities before erupting in the form of a contaminated ground water supply".

Perhaps the last word on this subject should be left to Professor Abel Wolman, Professor Emeritus John Hopkins Unitersity (28). Professor Wolman, who has been called "the conscience of the American water supply community", reviewed over 3,000 studies on health effects of trace organic chemicals in drinking water for the National Academy of Science. He concludes that:

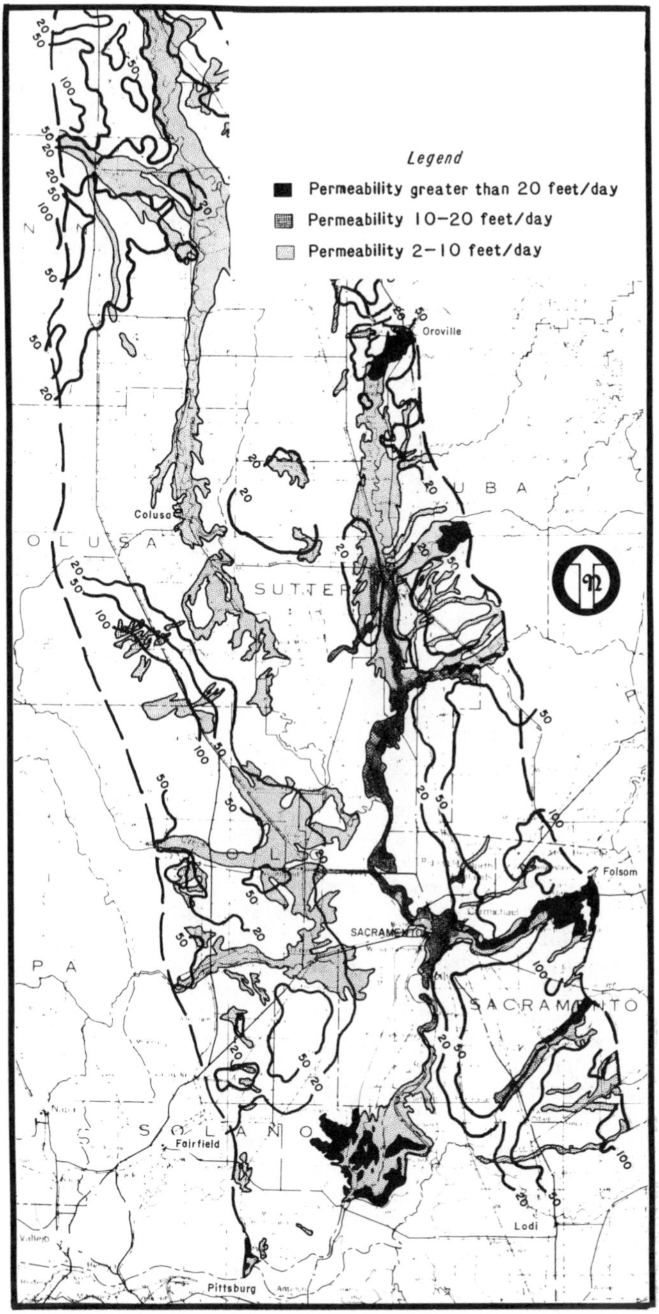

Figure 12. High Permeability Soils Without Known Shallow Underlying Restrictive Layers.

- o We are dealing with an extremely complex universe of ill-defined impacts of organic chemicals upon man. Disconcerting findings, indicative of long-term genetic effects, appear from day to day.
- o We can remove parts per quadrillion without great leaps of technology. Economical techniques are in use or are being developed for removing organic chemicals to nondetectable levels.
- o It remains the part of wisdom that we reduce as much as possible the deleterious effects of organics on our water resources. When in doubt, take it out.
- o Those of us who now suggest these future responsibilities and actions are sometimes accused of surrendering to public concern and fear rather than insisting on waiting upon scientific validation of the impact of toxic chemicals upon man.
- o I am fully aware of what the costs will be. I am fully aware that what I have said is disturbing, but I don't believe we can sit around and wait until all the evidence is in. It never will be.
- o Let's stop demanding proof. We can't wait.

If we, as stewards of our nation's ground water resource adopt Professor Wolman's perception of this risk, future generations looking back on the problem we are grappling with today will be able to say (paraphrasing Will Shakespeare's Hamlet searching for his father's ghost), "T'was here, t'was there, t'is gone!".

Literature Cited

1. "Water Quality and Pesticides: A California Risk Assessment Program (Volume 1)", California State Water Resources Control Board, 1984.
2. "Groundwater Contamination by Pesticides--A California Assessment", California State Water Resources Control Board, 1983.
3. Spath, D., personal communication.
4. "Pesticide Use Report: Annual 1983", California Department of Food and Agriculture, 1984.
5. Vonder Hass, S., personal communication.
6. Burlinson, N.E., Lee, L.A.; Rosenblatt, D.H. Environ. Sci. Technol. 1982, 16, 627.
7. "Investigation of Groundwater Contamination by Dibromochloropropane", California Department of Health Services, 1980.
8. Kim, V., personal communication.
9. "Dibromochloropropane; Intent to Cancel Registrations of Pesticide Products Containing Dibromochloropropane (DBCP)", Federal Register, January 9, 1985 (50 FR1122).
10. Koizumi, M.K., personal communication.
11. Onorato, C.A., personal communication.
12. Redlin, G., personal communication.
13. "Report on Environmental Assessment of Pesticide Regulatory Programs: State Component (Volume 1)", California Department of Food and Agriculture, 1978.
14. Schmidt, K.D. Proc. NWWA: Western Regional Conference on Groundwater Management, 1983.
15. "1,2-Dichloropropane (1,2-D) 1,3-Dichloropropene (1,3-D)", California State Water Resources Control Board, 1983.

16. "Ethylene Dibromide (EDB): A Water Quality Assessment", California State Water Resources Control Board, 1984.
17. "Carcinogen Policy. Section 3: A Policy for Reducing the Risk of Cancer", California Department of Health Services, 1982.
18. "Value of Good Agricultural Practices for Avoidance of Detectable Pesticide Residues in Groundwater", National Agricultural Chemical Association, 1984.
19. Warner, S., personal communication.
20. McCarty, P.L. Proc. National Conf. on Environ. Engineering, 1979.
21. Cothran, C.R.; Marcus, W.L. Regul. Toxicol. Pharmacol. 1984, 4,265.
22. Wells, J., personal communication.
23. "Water Well Standards: State of California", California Department of Water Resources, 1981.
24. "Designer Genes for Field Use: Monsanto Seeking Permit to Test Revolutionary Pesticide", Sacramento Bee, December 23, 1984.
25. Lehr, J.H. Groundwater Monitoring J. 1984.
26. Bowman, R. Groundwater Monitor, 1985, 1,27.
27. Roberts, P.V. JAWWA. 1982, 74, 161.
28. Wolman, A. Engineering News Record, 1983.

RECEIVED March 25, 1986

30

Two Ground Water Contamination Problems

Case Studies

Nancy K. Kim, Anthony J. Grey, Ronald Tramontano, Charles Hudson, and Geoffrey Laccetti

Bureau of Toxic Substance Assessment, New York State Department of Health, Rockefeller Empire State Plaza, Albany, NY 12237

In New York State, three types of chemicals are found to cause most groundwater contamination: solvents, petroleum products, and pesticides. Two different contamination scenarios involving sandy soil and pesticides are described. In Fort Edward, an investigation of drinking water contamination near an industrial plant was conducted. During sampling, chemicals not associated with the industrial plant were detected. This identification resulted in another sampling program which showed drinking water wells and springs contaminated with atrazine and alachlor; the highest levels were 260 ug/l and 660 ug/l respectively. In Suffolk County following the removal of aldicarb from the Long Island market, oxamyl was used as an alternative pesticide under an experimental use permit. After several years of use, monitoring indicated increasing levels of oxamyl contamination in groundwater, and by January 1984, oxamyl was removed from the Long Island market.

Fort Edward, N.Y.

The pesticide contamination in this small, upstate New York town was discovered in March, 1984, but the events leading to these findings began to unfold in 1983. In 1983, the state received taste and odor complaints about drinking water obtained from private wells in the vicinity of a local industrial plant (See Figure 1). The plant, located in a residential area not serviced by public water, used large quantities of solvents, particularly trichloroethylene and PCBs. Although water samples taken several years earlier from the residences near the plant had not contained PCBs, because of the complaints, water samples were taken again from houses on Park Avenue (see map) and analyzed for

0097–6156/86/0315–0530$06.00/0
© 1986 American Chemical Society

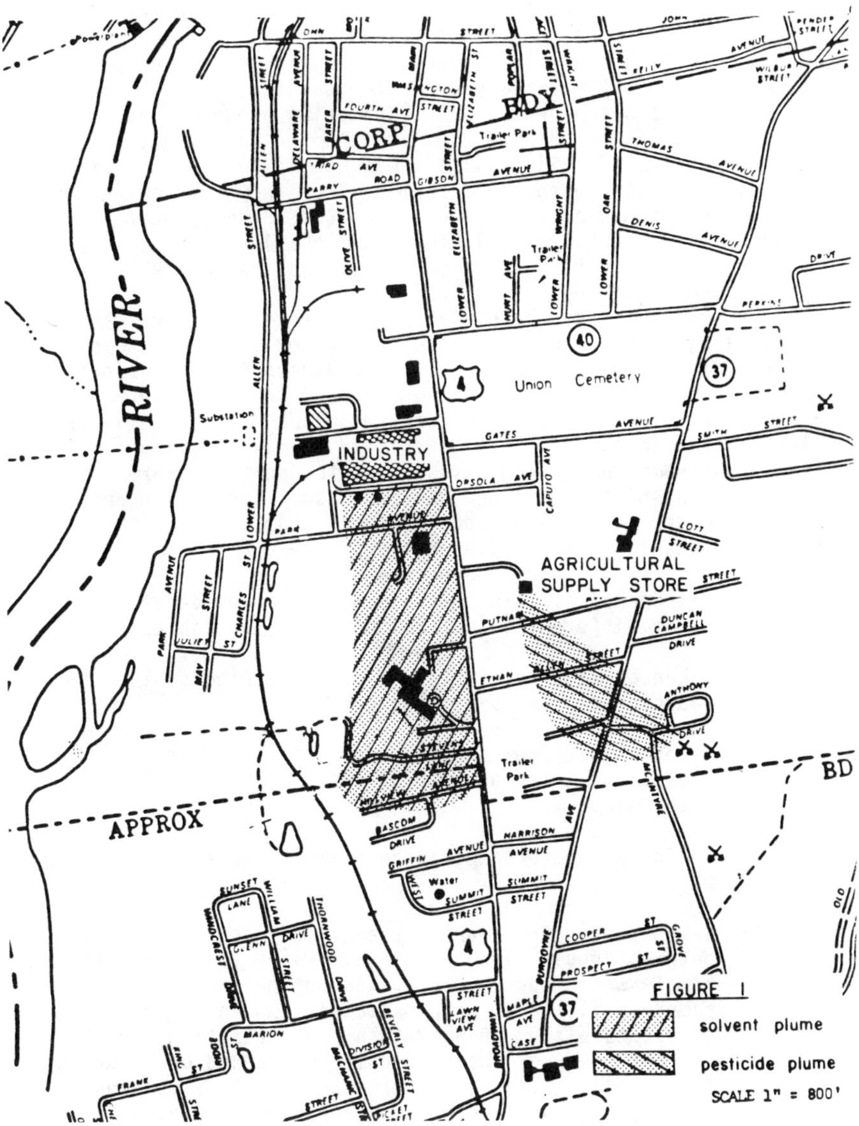

Figure 1. Map of study area in Fort Edward, NY.

PCBs and volatile chlorinated hydrocarbons. The samples were found to contain PCBs, trichloroethylene plus other related solvents.

Because of these findings, groundwater at the plant site was sampled and found to contain trichloroethylene. Apparently, improper handling and spill procedures led to the build-up of a pool of highly contaminated groundwater on-site.

The State began an off-site monitoring program. As contamination was found, the areal extent for sampling private wells was increased. Since little was known about groundwater flow, water samples were taken from most of the surrounding homes and analyzed for volatile solvents and PCBs. The groundwater and well depths were not available for most locations. Generally, the groundwater is at 15 to 20 feet in a shallow aquifer. Well depths ranged from about 10 to 200 feet.

After several rounds of sampling, with most of the sampling concentrated in the area west, southwest, and south of the plant (see Figure 1 - the area closest to the Hudson River), drinking water from homes on Stevens Lane were found to be highly contaminated with trichloroethylene, with the most contaminated wells having concentrations greater than 50 milligrams per liter. The homeowners were advised not to use the water. Because of the high levels and concerns about inhalation and dermal exposure, a limited experiment was undertaken in an attempt to quantify inhalation exposure.

Five air samples (1) were collected at a residence on Stevens Lane at different times or locations (see Table I).

(1) Taken on kitchen counter prior to turning on the faucets; doors and windows closed.

(2) Taken on bathroom sink counter with door and window closed; no water use during test. Taken concurrently with number 1.

(3) Taken on uncovered patio in rear of house concurrently with number 1.

(4) Taken on kitchen counter (same location as number 1) between double sink and stove. Water taken from the home's water supply was boiled on the stove during sample collection. Approximately two inches of tap water was placed in a 12 inch sauce-roasting pan which was not covered. Make-up water was added once during the sampling period. Staff attempted to simulate dishwashing by continously scrubbing and rinsing a pot in the double sink. This sample ran for 57 minutes. Towards the end of this time period, one member of the staff commented several times that he "felt funny."

(5) Taken in bathroom/shower and run concurrently with number 4. Sample head was hung inside shower

enclosure. The shower was run on a "hot" setting. The initial flow rate was three liters/25 seconds at 42oC and final flow rate was three liters/27 seconds at 39oC. The shower ran continuously and the door remained closed during sampling. This sample ran for 22 minutes. When re-entering bathroom to remove the sample, a very slight sweet odor was detected. This sample would simulate conditions in a bathroom with showers being taken sequentially by two residents.

The following water samples were collected (see Table I):

(1) Taken at kitchen sink before air sampling. Cold water.

(2) Taken at bathroom tub spigot before air sampling. Water at "hot" shower temperature.

(3) Taken by sampling water in tub that remained after running shower for air sample. Taken after air sampling.

A hearing was held to obtain public water for the people on Stevens Lane. This street is just north of the nearest public water so an extension was fairly easy to obtain. The industry agreed to pay to have these families hooked up to the public water supply and to undertake a sampling program to determine if any other localized "hot spots" existed in the area; the State analyzed a small number of duplicate samples.

Table I. Trichloroethylene Concentrations in Micrograms Per Liter (ug/l) or Micrograms Per Cubic Meter (ug/m3)

	Kitchen Cold	Bathroom Hot	Bathroom Tub
Water	34,000 ug/l	29,000 ug/l	13,000 ug/l
Air			
	Kitchen	Bathroom	Ambient
ambient background	35 ug/m3	230 ug/m3	35 ug/m3
after water use	2000 ug/m3	38,000 ug/m3*	

*breakthrough on cartridge; minimum level

During the sampling program, several wells from homes on Ethan Allen and Putnam Streets were found to be contaminated with PCBs. Along with the PCBs, the gas chromotograph picked up several unidentified peaks. On March 2, 1984, one well was found to be contaminated with 35 ug/l of PCBs, an unusually high level. Follow-up samples were taken from the same residence and nearby residences. By March 8, 1984, the original PCB result was confirmed and the samples with unknown peaks had been analyzed by a mass spectrophotometer. The compounds present at high concentrations were identified as alachlor and atrazine, chemicals not associated with the industry. The potential source for these compounds was a local agricultural supply store.

Laboratory standards were obtained for these new compoundsand the sampling area was expanded. By March 21, 1984, six private wells were found to be contaminated, two with PCBs and four with pesticides. By the middle of May, 15 wells had been found to be contaminated with pesticides, six with PCBs and/or chlorinated hydrocarbon solvents and one with both pesticides and PCBs. In addition to drinking water wells, springs in the areas were found to be contaminated. The highest levels found for the pesticides were 260 ug/l for atrazine and 660 ug/l for alachlor.

Atrazine

Atrazine is a widely used pre- and post-emergency herbicide for selectively controlling broadleaf and grassy weeds in corn, sorghum, asparagus, millet, pineapple, fruit trees, sugar cane, and summer fallow; it is also used to control vegetation in land not used for crops. Sugar beets, tobacco, oats and many other vegetable crops are very sensitive to atrazine.

Triazine herbicides are moderately persistent in soils and water (10-18 months) with atrazine being one of the more persistent of the group. Laboratory and field data have shown that the disappearance of atrazine from soil is influenced by temperature, pH, moisture content, and organic matter content; the effect of temperature and pH on the half-life of atrazine in soil (silt loam - laboratory study) is as follows (2):

Temperature	pH	Half-life (days)
25°C	4.9	35
25°C	7.0	72
5°C	4.9	250
5°C	7.0	439

Respiratory difficulties, loss of muscular coordination, convulsions, and thyroid dysfunction have been observed in laboratory animals given high levels of atrazine. The oral LD_{50} is 2,090 mg/kg for rats, 1,750 mg/kg for mice, and 750 mg/kg for rabbits. In one acute experiment (3) with sheep and cattle, symptoms of atrazine poisoning included muscular spasms, stiff gait, and increased respiration rates; at necropsy, compound degeneration and discoloration of the adrenal glands, and congestion of the lungs, liver and kidneys were observed. No cases of poisoning in humans were found in the literature.

Atrazine appears to have a relatively low chronic toxicity via the oral route, but the data base is limited. The National Academy of Sciences (4) calculated an acceptable daily intake (ADI) using a no-observed-effect level (NOEL) from an 80 week mouse (dietary) tumorigenicity study (5) and an uncertainty factor of 1,000; a suggested guideline of

150 ppb for atrazine in drinking water was calculated, assuming a human weight of 70 kg, consumption of two liters of water per day and 20% of the ADI allotted to drinking water. Review of the experimental protocol for the mouse study indicates, however, that the average daily dose of atrazine was calculated incorrectly; the revised drinking waterguideline would be 60 ppb. Because only one dose level was used and the study was designed as a screening study for carcinogenicity, it is inadequate to derive a NOEL for non-carcinogenic effects.

In 1981, the U.S. EPA (6) evaluated unpublished toxicological data submitted in support of a proposed tolerance for atrazine in or on orchardgrass and orchardgrass hay, which included:

(1) 2-year rat chronic feeding/oncogenic study
NOEL = 100 ppm (5 mg/kg/day)

(2) 2-year dog chronic feeding study
NOEL = 150 ppm (3.75 mg/kg/day)

(3) Three-generation rat reproduction study
NOEL = 100 ppm (5 mg/kg/day)

The chronic rat study is considered inadequate (some effects observed at 100 ppm) and is being repeated by the manufacturer. Data lacking for this pesticide included teratology studies in two species (preferably rat and rabbit) and an oncogenicity study in mice.

Most of the toxicological studies on atrazine were conducted in the 1960's and the requirements for an adequate study have become much greater over the last 10 years, particularly with regard to pathological, histopathological, and biochemical parameters. Some of the studies were conducted by Industrial Biotest Laboratories and are considered inadequate or at best supplemental for regulatory purposes. In addition to concerns regarding the chronic toxicity data base, another concern involves the possible adverse effects of chronic ingestion on carbohydrate metabolism and DNA, RNA and protein synthesis.

Until the results of the repeat chronic rat feeding study and rat teratology study are available, the 70 ppm dietary level (rat feeding study, in progress) or the 2-year dog chronic feeding study and an uncertainty factor of 1,000 can be used to calculate the ADI. A drinking water guideline of 25 ug/l is calculated assuming a human weight of 70 kg, consumption of two liters of water per day and that 20% of the ADI can come from drinking water.

Alachlor

Alachlor is a herbicide used to control annual grasses and certain broadleaf weeds in soybeans, corn, peanuts, dry beans, sunflowers, milo, and potatoes. The pure ingredient is a solid which melts at about 40° centigrade (C) and is soluble in water to about 240 milligrams per liter (mg/l) at 25°C.

The body of toxicological data on alachlor consists of unpublished manufacturer-sponsored studies (Monsanto Agricultural Products). No long-term animal studies were available in the published literature which would contribute information on reproductive effects or carcinogenic potential of alachlor. The information used by the New York State Department of Health was obtained from the U.S. EPA summary (7-8) and Monsanto (9).

Alachlor does not appear to be highly toxic on an acute basis. The oral LD_{50} value for rats was reported to be 930 mg/kg and the dermal LD_{50} was 13,300 mg/kg in rabbits. Alachlor was reported to be only slightly irritating to rabbit skin.

The U.S. EPA Office of Pesticide Programs (7) reviewed and evaluated the toxicological data base submitted by Monsanto in support of a petition for tolerance levels for alachlor in or on raw agricultural commodities*. Data considered in support of the petition and summarized in the Federal Register, included:

(1) Rat teratology study
No observed teratogenicity at 400 mg/kg [highest dose tested]
NOEL = 150 mg/kg for maternal toxicity and fetotoxicity

(2) 3-generation rat reproduction study
NOEL = 10 mg/kg/day

(3) 18-month mouse chronic feeding/oncogenicity study
Positive oncogenic response in females (bronchiolar alveolar
tumors) at 260 mg/kg/day

(4) 2-year rat chronic feeding/oncogenicity study
Positive oncogenic response at 42 mg/kg/day (both sexes - nasal turbinate tumors) and 126 mg/kg/day (both sexes - nasal and stomach tumors; males - thyroid follicular tumors); no carcinogenic effects were observed at 14 mg/kg/day.

* Recently the U.S. EPA (10) initiated a special review of all pesticide products containing alachlor based on the oncogenic findings.

According to EPA, data lacking included a 2-year chronic feeding study (rats), a 1-year feeding study (dogs), a teratology study (rabbits), a metabolism study (rats), a skin sensitization study (guinea pigs), and mutagenicity studies. The Agency also stated that alachlor had been determined to be an oncogen in mice and rats. In addition to multi-site oncogenicity, concern has also been raised regarding effects on the eyes of rats. An acceptable daily intake has not been established for alachlor. In New York State, notification is requested if drinking water levels exceed 1 ug/l.

In Summary

As a result of these groundwater contamination findings, actions were taken.

(1) Homes with private wells contaminated with PCBs and/or solvents were hooked up to public water supplies by the company.

(2) Wells contaminated with pesticides are on alternate water sources.

Suffolk County, Long Island, N.Y.

By 1978, aldicarb had been applied to potato fields in the north and south forks of Long Island for several years. Approximately 200 farms consisting of 22,000 acres are located in the area. Five to ten thousand water supply wells are estimated to be near these potato farms.

In 1978, at the urging of a technical advisory committee of the Nassau/Suffolk planning board, Union Carbide analyzed samples from skimming wells before and after the application of aldicarb to the potato fields on four different farms. Samples were collected in the spring and aldicarb was detected in 20 of 31 samples. Additional samples were collected in April of 1979 and similar results were found. In August, 1979, 14 water samples were taken from private wells at farm homes near the potato fields, other shallow water wells and somewhat deeper municipal water supply wells. Six of nine farm wells showed aldicarb levels between 4 and 140 ug/l.

These findings led to an extensive survey for aldicarb contamination (11). Once extensive contamination was found, aldicarb was removed from the Long Island market.

With aldicarb being withdrawn from the market, Long Island potato farmers needed an alternate in-ground use pesticide. Several thousand acres of farm land are quarantined for growing potatoes because of the Golden Nematode. A suggested alternative was oxamyl which was being used for foliar applications. The in-ground use of this pesticide occurred under an experimental use permit and with a monitoring program to determine if this compound would contaminate groundwater.

A soil and water monitoring program was carried out throughout the summer of 1980. Initial results looked promising. The compound was not detected in groundwater. The soil sampling results also indicated that oxamyl was decomposing rapidly in the top soil layers and only one site showed a low level of contamination in the 20 inch cores.

For the 1981 season, oxamyl was allowed to be used on the potato fields again with a monitoring program. The program was to address several concerns raised the previous year.

(1) Soil samples were taken further into the fall because some core levels increased during the latter half of the summer.

(2) Rainfall was 6" less than normal in 1980 which may have decreased the migration compared to normal years.

(3) Deeper soil cores were needed because oxamyl was detected in the deepest core sampled.

The 1981 report (12) concluded, "That under normal use conditions oxamyl does not contaminate well water on Long Island."

Oxamyl was used again during the 1982 growing season. However, in 1982, the Suffolk County Health Department took over the Union Carbide (manufacturer of aldicarb) sampling program for aldicarb in drinking water and surveyed areas not previously covered. The County also began monitoring for oxamyl. By November, 1982, 27 homeowner wells had been found to contain oxamyl at concentrations less than the New York State Health Department guideline of 50 ug/l.

In 1983, the decision was again to allow oxamyl to be used in potato fields and for above ground applications because the guideline had not been exceeded. By November, 1983, the County Health Department (13) had detected oxamyl (detection limit - 1 ug/l) at 75 of 3,000 sampling sites. Calculating the percentage of positive sites underestimates the potential contamination because the sampling program was originally designed for aldicarb and the sampling sites were generally not close to farm areas. Oxamyl contamination was probably widespread close to use areas as exemplified by the monitoring results from the Town of Laurel. In 1983 for this town, positive results (16 sites, 23 samples) ranged from 1 ug/l to 106 ug/l with two sites (6 samples) equal to or exceeding the 50 ug/l guideline. Oxamyl's continued use was producing increasing levels in groundwater. Most of the contaminated wells were shallow, but oxamyl was found in wells deeper than 100 feet. In addition, at Laurel, oxamyl had been found to have migrated several hundred feet from the site of application. By January 1984, three wells had been found to contain oxamyl above the guidelines and the manufacturer removed it from the market on Long Island.

Oxamyl

Oxamyl is a systemic or contact insecticide, miticide and nematocide. The half-life for oxamyl in soil reportedly ranges from 1-6 weeks depending on soil type, pH, moisture content, and temperature (14). Some chemical and physical properities are:

Physical State	crystalline solid
Vapor Pressure	2.3 x 10^{-4} mm Hg at 25°C
	7.6 x 10^{-3} mm Hg at 70°C
Specific Gravity	0.097
Solubility in Water ..	280 g/l at 25°C
Melting Point	100°-102°C changing to a different crystalline from which from which melts at 108°-110°C

Oxamyl has a high acute toxicity displaying the typical effects of carbamate insecticides, e.g. rapid onset of cholinesterase inhibition followed by rapid recovery. Symptoms of acute intoxication in animals included tremors, salivation, lacrimation, bulging eyes, and muscular twitching. The oral LD_{50} for oxamyl in the rat is the range of 5 to 15 mg/kg.

The World Health Organization (WHO) (15), in 1981, set a temporary ADI for oxamyl based on a no-effect level from a 2-year dog study. The Committee expressed concern, however, that the dog is somewhat less sensitive to oxamyl than the rat and that a clear no-effect level had not been established in the rat. Additional concern was expressed over the lack of identification of 50% of the tissue residues of oxamyl. The WHO was requiring that by 1983 tissue residues be identified and that the no-effect level in the rat be clarified. Based on the WHO temporary ADI, a drinking water guideline of 70 ug/l can be calculated.

The U.S. EPA has used an ADI based on a 2-year chronic rat dietary study. A drinking water guideline of 175 ug/l was calculated using this ADI.

The New York State Department of Health established a preliminary guideline of 50 ug/l because of uncertainities in the no-effect level in rats and lack of data on cholinesterase inhibition in humans. Additional concerns centered on the results of a rabbit teratology study which indicated NOELs for embryotoxicity and for maternal body weight changes which are lower than the estimated NOEL from chronic rat data.

Literature Cited

(1) "Air Sample collection technique for aromatic hydrocarbons, and related volatile organic compounds," Wadsworth Center for Laboratories and Research, New York State Department of Health, 1981.

(2) Nearpass, D.; Edwards, W.M.; Taylor, A.W. Agronomy Journal 1978, 70, 937-40.

(3) Palmer, J.; Radeleff, R. Ann. N.Y. Acad. Sci. 1964, 111(2), 729-36.

(4) "Drinking Water and Health," National Academy of Sciences, 1977, Volume 1.

(5) Innes, J.; Ulland, B.; Valerio, M.; Petrucelli, L.; Fishkein, L.; Hart, E.; Pallotta, A.; Bates, R.; Falk, H.; Gart, J.; Klein, M.; Mitchell, I.; Peters, J. J. Natl. Canc. Inst. 1969, 42, 1101-14.
(6) "Atrazine: Proposed Tolerance," U.S., Food and Agriculture Organization of the United Nations, Rome, Report Number 56, 1984.
(7) "Tolerances and Exemptions from Tolerances for Pesticide Chemicals in or on Raw Agricultural Commodities: Alachlor," U.S. Environmental Protection Agency, Federal Register, 48(62):13173-4, 1983.
(8) U.S. Environmental Protection Agency, Office of Drinking Water, personal communication, April, 1984.
(9) Monsanto Agricultural Products, Alachlor, Information sheets submitted to the New York State Department of Health, April, 1984.
(10) "Alachlor: Special Review of Certain Pesticide Products," U.S. Environmental Protection Agency, Federal Register, 50(6):115-19, 1985.
(11) Zaki, M.H.; Moran, D.; Harris, D. Am. J. Publ. Hlth. 1982, 72, 1291-95.
(12) E.I. duPont de Nemours and Co., Inc., Biochemicals Department, January 1982, Data supporting the use of Vydate L insecticide/nematicide on potatoes. Groundwater analyses and soil residue determination of oxamyl, Long Island, New York, 1981.
(13) Mr. Baire, November 15, 1983, Oral presentation to Oxamyl Task Force. Suffolk County Department of Health Services, Hauppauge, NY 11788.
(14) Harvey, J.; Han, J. J. Agric. Food Chem. 1978, 26, 536-41.
(15) "Pesticide Residues in Food," FAO/WHO, Food and Agriculture Organization of the United Nations, Rome, Report Number 56, 1984.

Two recent reviews of the toxicological effects of trichloroethylene and PCBs have been published:

Safe, S. CRC Critical Rev. Toxicol. 1984, 13, 319-395.
Kimbrough, R.; Mitchell, F.; Hunk, V. J. Toxic. and Environ. Health 1985, 15, 369-383.

RECEIVED March 25, 1986

31

The Emerging Role of Pesticide Regulation in Florida Due to Ground Water Contamination

Howard L. Rhodes

Florida Department of Environmental Regulation, Tallahassee, FL 32301

This paper describes the sequence of events that led to a stronger environmental pesticide program in the State of Florida. Prior to 1983, the State of Florida did not have an active pesticide program. There was very little impetus to protect the groundwater from the effects of pesticide contamination. The technical findings, the news media publications and the resultant political actions changed the climate in the state to one of serious concern over groundwater contamination by pesticides. In 1983, the State Legislature made broad, major inroads into legislation and funding of programs that provided the state with a comprehensive pesticide program that looked at the environmental impacts of a wide range of pesticide activities. The discoveries of aldicarb and EDB in the groundwater were the two pesticides that made these programs possible.

The process chosen for presenting this paper is one of a story-telling nature as opposed to a rigorous scientific treatise. It tries to show a trail of science, risk-assessment and public policy. It is not meant to show or justify the actions taken nor to detract from them, but rather to portray a series of events leading to a regulatory conclusion.

The nature of pesticides is one of a poison deliberately being applied to plants and animals for the benefit of man. Much work has been done on the methods to do this most efficiently and effectively. Finding pesticides to accomplish this task has been rewarding and has increased both the quality and quantity of mankind's food supply. This is clearly the beneficial and good side of the coin and is a fairly well-known fact in the agricultural community, but much less well-known and appreciated by the consumers in our society.

The reverse side of the coin is what pesticides do to our environment and more specifically to man. While much study in some

0097–6156/86/0315–0541$06.00/0
© 1986 American Chemical Society

aspects of this issue has been done, it is beginning to appear that we will never fully understand all the complex impacts of the introduction of a foreign material into our environment or its resultant impacts on man. Because of a changing political climate and improved scientific technology for detection of pesticides, we can do a better job than has previously been done. Knowledge of people and our environment has expanded our understanding of the interrelationship among various elements within our environment. The role of the media is also extremely effective at setting public opinion that sometimes is otherwise shaped by the status quo. These forces came together in Florida in the last two or three years and forced the state to look more closely at the "other side of the coin."

In mid-1982 on the program "60 Minutes," CBS presented a somewhat frightening story of how water wells on Long Island, New York had been contaminated by the pesticide Temik, and that its use was proliferating, especially in Florida. The story parenthetically added that no one in Florida was monitoring for pesticides in groundwater, nor was anyone really aware of the pesticide Temik's contamination potential.

You can be assured that the following Monday, the state news media and political community were calling to ask what the Florida Department of Environmental Regulation (FDER) proposed to do to test groundwater. Needless to say, the state was not prepared for the on-slaught. We were not testing groundwater to any significant degree at the time and had only just passed rules to regulate the discharge of pollutants to groundwater. In addition, our laboratory chemist did not even know how to analyze for the chemical.

What CBS did not mention was that their "investigative" reporter had called this speaker some two months earlier inquiring about the state's pesticide program. When queried about any potential problem we should be aware of, he merely stated that there was no problem--he just wanted to find out about the pesticide program in the state. This appears to have been the major contact with the state when it was found little was being done in the area of groundwater contamination caused by pesticides. This deception was responsible for several months delay in finding the pesticides in Florida's groundwater.

Within days, staff was detailed to study at least two or three sites in the state. During the interim, a contingent of the manufacturer's agricultural chemicals division met with the staffs of the Florida Department of Agriculture and Consumer Services (FDACS) and the FDER. Assurances were made that there was no way for the material to be found in groundwater in Florida because of its warm climate and more alkaline waters. In retrospect, we believe these highly technically oriented people really believed these statements, but their credibility suffered when our laboratory reported positive findings a few weeks later.

News reporters called my staff daily for reports on findings and literally started getting results and wanting my comments before I had seen the results. This led to a requirement that results not be released before I read them, which meant I would be the first contact with the media, but not the last. To have imposed a restraint requiring staff not to talk to the media in a state like Florida that has an open records policy would have led to accusations of censorship, which was not the intended purpose. This did lead to various "quotes" of the day by various staff members interpreting the positive finds over the next two or three months. It is quite likely that no one really knew the significance of the positive findings except that a pesticide was present in groundwater.

In early December 1982, the Commissioner of Agriculture, under some pressure from the media, held a hearing for several state agencies. The Commissioner stated that he would like to have some recommendations on how to address this contamination event. The FDER and the Florida Department of Health and Rehabilitative Services (FDHRS) suggested several constraints, including that the material not be applied within 1,000 feet of drinking water wells. It was felt that with the unknowns surrounding the pesticide Temik a response was called for, but not a complete ban. Four agencies were represented.

When a vote was called for on the health and environmental recommendations, the agricultural community had disproportionate voting power and got five of the seven votes on the committee. Each subsequent substantive environmental recommendation was voted down 5-2. The following day there was much news coverage on this meeting. Most of the coverage was not very complimentary to the agricultural portion of the committee.

Several weeks later when an actual drinking water well was found positive, not the groundwater positives found at the test site, the Commissioner of Agriculture banned the use of the pesticide in the state for one year.

Some people felt that the environmental and health agencies had pressured the Commissioner to take such an action. They had not. It was more likely that the press coverage, as well as a desire by the Commissioner that no citizen in the state be allowed to drink water contaminated with pesticides, prompted the action.

Such is the case in the political arena when nothing seems to be clearly black and white. Such action clearly alienated some in the agricultural community for depriving them of an effective pesticide. Some environmental groups were upset that action had not taken place sooner. The press got good coverage of these actions.

There were a couple of exceptions to the ban: 1) totally self-contained plants in nurserys, and 2) a small potato growing area of North Florida. This potato area was initially in the ban, but after extensive testing by the health and agricultural agencies found no pesticide in any well in the area, the exemption was granted.

Concurrent with the announcement of the ban, the Commissioner requested the manufacturer, the University of Florida Agricultural Research Branch and the FDER to research the effects of the pesticide on groundwater in Florida and to make recommendations to him on what should be done.

As a result, the manufacturer set up two virgin research sites in conjunction with the University and the FDER during the next seven to eight months to study the effects of the pesticide on groundwater. These sites had never had previous applications of aldicarb. Thus, there could not be a source of unknown application time and rate to interfere with the study. The manufacturer also volunteered to monitor the other three state sites where contamination had been found. Test applications of aldicarb were conducted at these virgin sites to determine the length of time for contamination to occur and also the rate of movement in the groundwater and degradation rates.

Recommendations resulting from the study were nearly unanimous to allow the pesticide's use again in September 1983, with the only discussion being how far from drinking water wells the material should be applied. This resulted from the research showing the material did get in the shallow groundwater and except for the central ridge area of the state, the material hydrolized quickly and no trace could be found after a few weeks. In the central ridge, the degradation was much slower. Models developed by the manufacturer, and concurred with by the University, suggested a 300 foot buffer. The FDER felt a safety factor of 100 percent should be added based on data collected from two of the sites. The Commissioner chose to go with the 300 foot buffer.

Sometime later it became clear that one of the virgin sites which had an application of aldicarb in the ridge area showed leaching to groundwater took almost a year and then spread slowly and erratically. After the model was shown to be incorrect, the manufacturer requested the Commissioner to change the buffer to 1,000 feet.

During the whole process, the state was visited by many high ranking officials of the company, including the president of the agricultural chemicals division on two or three occasions. Even today, the manufacturer is continuing to monitor the site with the contamination to learn more of what happens in groundwater in the Highland ridge area.

In late 1982, the new Speaker of the Florida House of Representatives had assumed duties, and quickly appointed a blue ribbon task force on water quality. Many of these previously reported events happened during the time the task force was holding hearings. Other chemicals were also being found in drinking water during this time. This led to a report that became a partial blueprint for the 1983 Florida Legislature on a bill later known as the Water Quality Assurance Act.

These and other events relating to water contamination became the major bill during the session, putting many legislators in the position of supporting legislation that would have been impossible to support without the attendant publicity and without the strong desires of the House leadership. Included as one provision was a section setting up a data product evaluation bureau in the FDACS, a pesticide section in the FDER and a toxicologist in the FDHRS. In addition, a Pesticide Review Council was abolished and then reconstituted to include more environmental and health representatives.

The Legislature went into extended session to pass the bill, by which time the Senate and House had become fully committed. The legislation was the most comprehensive environmental legislation in twelve years and set the course for many new initiatives in not only pesticides and groundwater but also hazardous waste and a state grant program for sewage plants.

The story could end at this point except that EDB was discovered in Hawaii's and California's groundwater in early 1983. The FDACS by now was fully sensitized to the problems of groundwater contamination by pesticides. The agency knew that it had an ongoing program for applying large quantities of EDB to the soil in and around citrus groves and had done so for years to control nematodes. It tested thirty or forty wells in early July 1983 and found EDB contamination. The FDER and FDHRS were promptly notified of the results. The news media was not far behind. The publicity was widespread, and the results predictable.

The Chairman of the Florida House Community Affairs Committee, whose district previously had a groundwater contamination event caused by a Florida Department of Transportation testing laboratory disposal area, was very interested in the whole affair. He assigned staff to interview the affected well owners and ask questions such as "Are you satisfied with the way the state has treated you?" "What do you think the state should do to make you happy?" "Are you satisfied with the answers you are getting from the state?" Needless to say, most answers were negative.

The Governor, the Commissioner of Agriculture and the House Community Affairs Chairman met to discuss the problem. The result was the appointment of an EDB Task Force composed of the health agency, the agricultural agency, the community affairs committee and the environmental agency with charges to provide tests for all the wells that were contaminated, an information newsletter and an "800" hot-line, and then to find a solution and provide it for inclusion in the budget in January 1984. The task force met and worked on the logistics.

The logistical problems were legion. The state has between 500-700,000 private wells in addition to public water supplies. With a number of exceptions, these difficulties were worked out. The most difficult exceptions were finding and plotting locations where the material was applied and getting enough people to sample and analyze the samples. No extra provisions were made for new staff in any of the agencies to accomplish this and as a result, a number of activities suffered.

We could find nothing in the literature that offered effective ways to remove EDB. Research was done under contract with a private firm in cooperation with the U.S. Environmental Protection Agency to determine the best way to remove the material from the water. This research indicated that activated carbon was the most effective technique.

Through the fall of 1983, the media continued to give wide coverage to the constant number of findings of EDB in the wells. This translated into continued political awareness of groundwater contamination right to the top of the political structure in Florida. At the same time, the FDER had elected not to use its laboratory for EDB work, but to concentrate on toxic dumps and the analysis of groundwater at these sites. Each one of these sites then became a media event when positives were found.

The task force requested funds to correct contaminated wells in areas where the EDB had been state applied. The Governor's Office accepted the request in January 1984, only weeks before the state budget was to be presented to the state Legislature.

The request for funds passed through the House easily, but

faulted in a very conservative Senate until the President of the Senate found his own district had some of the most contaminated wells. Some 3.1 million dollars was ultimately made available, but only after a proviso was attached that the well owners had to release the state from property liability in exchange for the relief. This provision has caused about half the well owners to refuse to sign the release, and thus no filter can be provided.

The EDB situation is still present in Florida today with the logistics of sampling, analysis, notification of well owners and retrofit of wells continuing. New contamination is found every week with a total of 7,729 wells having been tested and 817 positives having been found as of February 27, 1985.

While the day-to-day issues of EDB have dominated many hours, weeks and months of my staff and other agencies' staff time, a greater good has evolved. For a year-and-a half, weekly meetings of all concerned agencies have resulted in decisions being made on how things are done in the state. The close working relationships developed among the five agencies mean resources are used to their fullest and the agencies' dirty laundry is aired weekly. Also, the decisions of one agency are subject to intense scrutiny by another agency. This has led to a tremendous amount of informed decision making. The environmental and health agencies have learned a lot about agriculture and pesticide use. The agricultural community has learned significant facts about chemical toxicology, groundwater movement and means of contamination and about drinking water supply sources in the state. This open discussion in the various agencies has led to a better appreciation of the concerns of affected interests.

The result has been a significant concern in the agricultural agency for groundwater and surface water contamination potential of pesticides. This is manifested by a bureau being created to deal with pesticide registration in the state and the hiring of hydrologists, biologists and toxicologists in the agricultural agency to question the potential impacts of new pesticide registrations and renewals of pesticides on the environment in the state.

There has been a commensurate impact at the environmental agency, which had not given much priority to pesticide contamination until the Temik and EDB situations. The agency now has a pesticide section that is systematically looking at groundwater in areas of high pesticide use and in areas with soil likely to allow pesticide migration. The purpose is to determine if any other pesticides are leaching into Florida's fragile groundwater.

Contrary to the media's implications, EDB is a fairly localized phenomena in certain areas of the state. The whole water supply is not contaminated. The media's interest, and therefore much of the political interest, has waned; but due to the sensitive nature of this issue, the least amount of a toxic substance find can create considerable consternation among the public as a whole.

The state level agencies are cooperating in a style not experienced in recent Florida history. Many agricultural meetings that had previously been closed are being opened to environmental agency personnel. While there is still a degree of mistrust, these groups now recognize that in addition to food production, there is a growing awareness of the stewardship role that the agricultural community must play in protecting the earth's resources.

In the state Legislature, discussions on pesticides are no longer the solely vested interest of the agricultural committees. Discussions now occur in natural resource committees where most environmental legislation is discussed. This activity causes all points of an issue about pesticides to be analyzed.

Recently, we were advised that the main university research institute in Florida has decided it will no longer do efficacy testing for new pesticides unless environmental fate studies unique to Florida are done concurrently or have already been completed. This will have significant impacts on the registration process, but will go a long way toward ensuring that new pesticides will not have a detrimental impact on Florida's environment.

While regulation can never fully be the answer to problems, in Florida it has begun to assist with the basic issue all of us must grapple with--the protection of this planet earth and its inhabitants. Food, fiber, housing, clean air and water are vital to our existence.

RECEIVED March 25, 1986

32

Considering Pesticide Potential for Reaching Ground Water in the Registration of Pesticides

Samuel M. Creeger

Office of Pesticide Programs (TS-769C), U.S. Environmental Protection Agency, Washington, DC 20460

Ground water was once thought immume from chemicals applied on the soil's surface, but current evidence shows the presence of at least 1 of 16 different pesticides in the ground water of 23 different states. This paper describes procedures the U.S. Environmental Protection Agency is using in considering the potential of pesticides to reach ground water in the pesticide registration process.

About 379,500 metric tons of pesticide active ingredients were applied agriculturally in the United States in 1984 (1). In the case of insecticides, it has been estimated that <<1% of a typical, agricultural insecticide application is responsible for the insecticidal effect on the insect (2). The remaining 99+% of the application does not directly perform its intended pesticidal function since it misses the target insect, degrades, or moves from the treatment area due to drift and other transport mechanisms. One can deduce that significant portions of typical herbicide and fungicide applications are similarly not directly responsible in controlling their target pest(s). It is a portion of this 99+% of pesticide applications that is available for leaching through soil and reaching ground water.

The complexity involved in detecting pesticide residues in ground water and in monitoring and cleaning up ground water containing agriculturally applied pesticides is a factor that contributes to the public's fear for the loss of purity of the nation's ground water supplies; water that is used for drinking, bathing, and irrigating/watering food and feed commodities. Pesticides are intentionally applied to crops for the purpose of pest control and a tolerance is set allowing an acceptable level of residues to remain on the treated crop at the time of use/consumption of the raw agricultural commodity (3). If the tolerance is exceeded or illegal pesticide residues are found in a food or feed commodity, the residue-containing food or feed can be removed from the market. In contrast to this, pesticide residues in ground water are not put there intentionally, serve no useful purpose there and may not dissipate from the ground water system for decades.

This chapter not subject to U.S. copyright.
Published 1986, American Chemical Society

Table Ia. SUMMARY OF ENVIRONMENTAL FATE DATA REQUIREMENTS FOR TERRESTRIAL USE PATTERNS

	TERRESTRIAL USES					
DATA REQUIREMENTS*	DOMESTIC OUTDOOR	GREEN-HOUSE	NON-CROP	ORCHARD CROP	FIELD AND VEGE. CROP	FORES-TRY
DEGRADATION						
HYDROLYSIS	R	R	R	R	R	R
PHOTOLYSIS						
-WATER			R	R	R	R
-SOIL				CR	CR	CR
METABOLISM						
AEROBIC SOIL	R	R	R	R	R	R
ANAEROBIC SOIL					R	
MOBILITY						
LEACHING	R	R	R	R	R	R
AGED LEACHING	R	R	R	R	R	R
FIELD DISSIPATION						
SOIL	R		R	R	R	
FOREST						R
ACCUMULATION						
ROTATIONAL CROP					CR	
FISH			CR	CR	CR	CR
AQUATIC NONTARGET						CR

* STUDIES REQUIRED ONLY UNDER CERTAIN CONDITIONS (SUCH AS PHOTOLYSIS IN AIR, VOLATILITY STUDIES, TANK MIX STUDIES AND IRRIGATED CROP STUDIES) ARE NOT INCLUDED IN THIS TABLE.

R = REQUIRED; CR = CONDITIONALLY REQUIRED; UNDERLINE (R, CR) INDICATES DATA REQUIREMENTS WHEN AN EXPERIMENTAL USE PERMIT IS SOUGHT.

Table Ib. SUMMARY OF ENVIRONMENTAL FATE DATA REQUIREMENTS FOR AQUATIC AND AQUATIC IMPACT USE PATTERNS

	AQUATIC USES	
DATA REQUIREMENTS*	FOOD CROP	NON-CROP
DEGRADATION		
HYDROLYSIS	R (underlined)	R (underlined)
PHOTOLYSIS		
-WATER	R	R
METABOLISM		
AEROBIC AQUATIC	R (underlined)	R (underlined)
ANAEROBIC AQUATIC	R	R
MOBILITY		
LEACHING**	R (underlined)	R (underlined)
FIELD DISSIPATION		
SOIL (SEDIMENT)	R	R
WATER	R	R
ACCUMULATION		
ROTATIONAL CROP	CR (underlined)	
IRRIGATED CROP	CR	CR
FISH	CR (underlined)	CR (underlined)
AQUATIC NONTARGET		CR

* STUDIES REQUIRED ONLY UNDER CERTAIN CONDITIONS (SUCH AS PHOTOLYSIS IN AIR, VOLATILITY STUDIES AND TANK MIX STUDIES) ARE NOT INCLUDED IN THIS TABLE.

** A BATCH EQUILIBRIUM (ADSORPTION/DESORPTION) STUDY.

R = REQUIRED; CR = CONDITIONALLY REQUIRED; UNDERLINE (R, CR) INDICATES DATA REQUIREMENTS WHEN AN EXPERIMENTAL USE PERMIT IS SOUGHT.

The nation's ground water supply was once considered to be isolated and immune from activities occurring on the Earth's surface. The belief that organic chemicals would completely degrade in soil and that ground water cannot be directly observed supported that position (4). However, ground water monitoring done in the past 10 years does show ground water in many areas to be vulnerable to organic chemicals, such as pesticides, applied at the soil surface. Evidence (5) published in 1984 recorded the ground water of 18 states to contain at least 1 of 12 agriculturally applied pesticides; information received since that 1984 publication now shows the ground water of 23 different states to contain at least 1 of 16 agriculturally applied pesticides (6).

How does the Environmental Protection Agency now determine potential for pesticides to reach ground water and how is that knowledge used in the pesticide registration process? In the registration process, there are three mechanisms that bring pesticide chemicals under consideration for potential to reach ground water. They are: (1) New chemical registration, (2) Reregistration (also known as the Registration Standards Process), and (3) An Amendment to an Existing Registration.

1. A new chemical registration involves submission of a data package by the registrant to EPA. That package contains, in part, studies on the environmental fate of the active ingredient. The number of studies to be submitted depends on the proposed use pattern as shown by information presented in Tables Ia and Ib taken from the Subdivision N Guidelines (7). 2. These tables can also be used to determine the data needed under the Registration Standards process, which is the second of the three registration processes mentioned in the previous paragraph.

Of the studies required in support of the registration or re-registration of a pesticide product, the results from the following studies have the greatest impact on determining a pesticide's potential to reach ground water (Table II). Data on the topics in Table II are also required as part of the Ground Water Data Call In discussed at the end of this paper.

Table II. ENVIRONMENTAL FATE STUDIES USED IN DETERMINING A PESTICIDE'S POTENTIAL TO REACH GROUND WATER

Hydrolysis
Photolysis in Water
Photolysis on Soil
Aerobic Soil Metabolism
Anaerobic Soil Metabolism
Anaerobic Aquatic Metabolism
Leaching
Field Dissipation (Terrestrial, Aquatic or Forestry)

3. The other registration mechanism that brings pesticides under consideration for potential to reach ground water is an amendment to an existing registration. The following amendments to a pesticide product may change the pesticide's potential to reach ground water (Table III):

Table III. TYPES OF AMENDMENTS TO EXISTING REGISTRATIONS

Increased Application Rates
Change in Application Timing
Change in Formulation
New Use Site

When a registrant wishes to amend an existing registration by adding a new crop use, then data needed to support that new use must be submitted. For example, if a fungicide product is used on greenhouse-grown eggplant seedlings and the registrant wants to extend use of that product to field grown tomatoes, green peppers and eggplants, then the differences in the data requirements (as given in Table Ia) between greenhouse use and field vegetable crop use will have to be satisfied (Table IV):

Table IV. ADDITIONAL DATA TO SUPPORT CHANGE IN USE FROM GREENHOUSE TO TERRESTRIAL FIELD CROP

Photolysis in Water
Photolysis on Soil
Anaerobic Soil Metabolism
Field Dissipation
Rotational Crop
Fish Accumulation

(The underlined studies are used in determining potential to reach ground water).

In this example, the results of the underlined studies in Table IV will be considered with the previously submitted hydrolysis, aerobic soil metabolism and leaching studies which supported the greenhouse-grown eggplant seedling use, to determine the potential for the pesticide to reach ground water when used in the field as proposed. The results of the studies, along with water solubility, vapor pressure and octanol-water partitioning data are used to determine potential for the pesticide to reach ground water. If the review of the studies shows the pesticide to meet at least one of the criteria (5) in Table V when used as proposed, then the pesticide is categorized as having potential to reach ground water.

Table V. CRITERIA FOR POTENTIAL TO REACH GROUND WATER

Water Solubility - Greater than about 30 ppm
K_d - Less than 5
Hydrolysis half-life - Greater than about 25 weeks
Soil half-life (field) - Greater than about 2-3 weeks

Designating pesticides as potential leachers based on only one criteria may appear to be an overly strict approach. For example, would not one expect a pesticide with a water solubility of 50 ppm but a soil half-life of less than 2 weeks to degrade in soil before leaching deep enough to reach ground water? However, the Agency's concern about such chemicals is that they may be subjected to a heavy rainfall or irrigation so soon after application that they will leach through the top portion (2-3 feet) of the soil with the highest populations of microbial pesticide degraders, reach the deeper soil layers where they will persist and be available for further leaching and eventual reaching of the ground water.

If the registrant chooses to pursue registration of a pesticide designated as having potential to reach ground water, then additional information and sometimes extensive field studies under representative but worst case use conditions will have to be performed to determine whether or not leaching will actually occur. The results of these field studies will have to describe the level of pesticide residues in the ground water, if any, that can be expected when used as proposed. Alternatively, the registrant may propose label restrictions or label changes (such as a change in application rates and/or timing, geographical restrictions against use in sandy soils or areas with shallow or unconfined aquifers) that will remove the potential for ground water contamination. This information will be subjected to a risk-benefit analysis to determine if registration should be granted.

Below are some descriptions of pesticides with potential for reaching ground water or that have already been found in ground water and how their leaching characteristics have (in part) affected their registration status. The first two examples involve suspension and/or cancellation and the last five describe requests that are in "object status". Object status means that the particular request has not been granted but the registrant may resubmit the request with new information and the Agency will reconsider its position. The registrants involved in the last 5 examples have all elected to do additional work which may alleviate the Agency's ground water concerns.

(1) DBCP (1,2-dibromo-3-chloropropane), a soil nematicide/fumigant, was found in ground water and found to pose significant health risks. In 1979, all uses of DBCP in the continental U.S. were suspended (8) and the remaining use in pineapple fields in Hawaii was cancelled in 1985 (9).

(2) EDB (1,2-dibromoethane), a soil nematicide/fumigant, was found in ground water and found to pose significant health risks. Its uses were suspended in 1983 (10).

(3) Carbofuran is a soil nematicide/insecticide and has been found in the ground water of NY, WI and MD due to agricultural use (6). The registrant submitted requests at different times for a change in application timing on cotton to "at planting", for use on non-bearing citrus and raspberries (new crops), and for use on sorghum at increased rates. These amendments have not been granted based on ground water concerns (11).

(4) Carbosulfan is a new soil nematicide/insecticide. The registrant has requested use on non-bearing fruit, nut and citrus and on alfalfa, sorghum and corn. Based on ground water concerns, none of these new uses have been granted and the first product containing carbosulfan as an active ingredient has yet to receive a registration (11).

(5) Oxamyl is a soil nematicide/insecticide and has been found in the ground water of NY and RI due to agricultural use (6). The registrant has proposed amending its product label to include uses on lettuce, endive, escarole, cabbage, mint and beans. Based on ground water concerns, these amendments have not been granted (11).

(6) Aldicarb is a soil nematicide/insecticide and has been found in the ground water of 15 states due to agricultural use (6). The registrant has proposed amending its product labels to include uses on corn, tomatoes and tobacco. Based on ground water concerns, these amendments have not been granted (11). Use in New Jersey on eggplant seedlings in greenhouses was granted under Section 18 of the Federal Insecticide, Fungicide, and Rodenticide Act (FIFRA) (12) since data showed the aldicarb residues to rapidly degrade in the soil during the two week post-application period in the greenhouse before being transplanted to the field, and the Section 18 was to last only 1 year (11).

(7) Aldoxycarb is a soil nematicide/insecticide. Registration of a product for use on tobacco, peanuts, sweet potatoes and cole crops was pursued by the registrant but registration was not granted based on ground water concerns. The registrant was granted a registration for a plant fertilizer spike product containing the pesticide based on the Agency's conclusion that such a low pound per acre use would not result in significant residues reaching ground water (11).

This ends a brief description of how the Agency considers potential for pesticides to reach ground water in the pesticide registration process. This paper will close with a description of the Ground Water Data Call In initiated by the Agency in 1984 for the purpose of determining which of the 600 most widely used active ingredients have potential to reach ground water when applied under use conditions. A Data Call In is the means by which EPA acquires additional data relevant to existing registrations and is authorized by Section 3(c)(2)(B) of FIFRA (12). Criteria for selecting the candidates from the 600 most highly used active ingredients are given in Table V. If one or more of the criteria were met, then the pesticide was placed on the Ground Water Data Call In list. The resultant list contained 141 chemicals (see Appendix). However, 51 of those chemicals were on a different data call-in list called the Registration Standards development list and the data needed to assess ground water contamination potential had already been or is to be submitted shortly with that data call in. Therefore, 90 chemicals remained for which the Agency asked for the data in Table II plus water solubility, vapor pressure and octanol-water partitioning data. Those pesticides with use patterns not requiring a study(ies) in support of their registration (consult Tables Ia and Ib) will not normally need that study(ies) in assessing its leaching potential. In some cases, this would be the first time environmental fate data assessing the leaching potential of the pesticide would be submitted to EPA since the registration of some pesticide products pre-date the environmental fate data requirements instituted in 1970 (13).

A pesticide reviewed under the ground water data call-in and designated as a leacher will be subject to the following responses as provided by FIFRA (11) (Table VI):

Table VI. REGULATORY OPTIONS FOR PESTICIDES FOUND TO HAVE POTENTIAL TO REACH GROUND WATER

(1) Suspension

(2) Cancellation

(3) Restricted Use(s) including geographic restrictions

(4) Advisory Labels

(5) Field Studies and further study

APPENDIX

The 141 Pesticides to be Evaluated Under the Ground Water Data Call-in

GW - Data to be submitted under the Ground Water Data Call-in
RS - Data to be submitted under the Registration Standards Data Call-in

Acephate - RS
Acifluorfen - GW
Alachlor - RS
Aldicarb - RS
Ametryn - GW
Aminocarb - GW
Amitrole - RS
Ammonium sulfamate - RS
Amobam - GW
Anilazine - RS
Aspon - RS
Asulam - GW
Atrazine - RS
Bendiocarb - GW
Bentazon - RS
Bromacil - RS
Bromoxynil - GW
Bufencarb - GW

(continued)

APPENDIX (continued)

Butylate - RS
Cacodylic acid - GW
Calcium arsenate - GW
Carbofuran - RS
Carbophenothion - RS
Carboxin - RS
CDAA - GW
Chloramben - RS
4-Chloropyridine N-oxide - GW
Chlordimeform - RS
Chlordimeform HCl - RS
Chlormequat - GW
Chlorpropham - GW
Chlorpyrifos - RS
Chloropicrin - RS
Chlorothalonil - RS
Crotoxyphos - GW
Cryolite - RS
Cyanazine - RS
Cycloate - GW
Dalapon - GW
Dacthal - GW
Daminozide - RS
Dazomet - GW
DCNA - RS
Demeton - GW
Desmedipham - GW
Dialifor - RS
Diallate - RS
Diazinon - GW
Dicamba - RS
Dichlobenil - GW
Dichlone - RS
1,3-Dichloropropene - GW
Diclofop methyl - GW
Dicrotophos - RS
Difenzoquat - GW
Diflubenzuron - GW
Dimethoate - RS
Dinoseb - GW
Diphenamid - GW
Dipropetryn - GW
Diquat - GW
Disulfoton - RS
2,2-Dithiobisbenzthiazole - GW
Diuron - RS
DSMA - GW
Dyfonate - RS
Endothall - GW
EPTC - RS
Ethiofencarb - GW
Ethoprop - RS
Fenaminosulf - RS
Fenamiphos - GW
Fensulfothion - RS
Ferbam - GW
Fluometuron - GW
Formetanate - RS
Fosthietan - GW
Guthion - GW
Hexazinone - RS
Imidan - GW
Karathane - GW
Krenite - GW
Lead arsenate - GW
Linuron - RS
Maleic hydrazide - GW
Mancozeb - GW
Maneb - GW
MCPA - RS
Mecoprop - GW
Mefluidide - GW
Mesurol - GW
Metam-sodium - GW
Methidathion - RS
Methomyl - RS
Methyl bromide - GW
Methyl isothiocyanate - GW
Metolachlor - RS
Metribuzin - GW
Mobam - GW
Molinate - GW
Monocrotophos - GW
Monuron - RS
MSMA - GW
Nabam - GW
Napropamide - GW
Naptalam - GW
Neburon - GW
Oftanol - GW
Oxamyl - GW
Oxydemeton-methyl - GW
Paraquat - GW
Pebulate - GW
Phenmedipham - GW
Phorate - RS
Phosalone - RS
Phosdrin - GW
Phosphamidon - GW
Picloram - RS
Polyram - GW
Prometon - GW
Prometryn - GW
Pronamide - GW
Propachlor - GW
Propazine - GW
Propham - GW
Propanil - GW
Siduron - GW
Simazine - RS

Tebuthiuron - GW
Terbacil - RS
Terbutol - GW
Terbutryn - GW
Thiabendazole - GW
Thidiazuron - GW
Triallate - GW
Triadimefon - GW
Trichlorobenzoic acid - GW
Trichlorfon - RS
Vernolate - GW
Zineb - GW
Ziram - GW

Literature Cited

1. "Pesticide Industry Sales and Usage - 1984 Market Estimates," U.S. Environmental Protection Agency, 1985.
2. von Rumker, R.; Lawless, E.W.; Meiners, G. In "A Study of the Efficiency of the Use of Pesticides in Agriculture," von Rumker, R.; Kelso, G.L. U.S. Environmental Protection Agency, Contract # 68-01-2608, 1975, p. 10.
3. Federal Food, Drug and Cosmetic Act, approved June 25, 1936, 52 Stat. 1040 et seq., as amended.
4. Parfit, M. "Ground Water: Out of Sight, Out of Mind", Smithsonian 1983, 13, p. 51-61.
5. Cohen, S.Z.; Creeger, S.M.; Carsel, R.F.; Enfield, C.G. "Potential Pesticide Contamination of Groundwater Resulting from Agricultural Uses"; In Treatment and Diposal of Pesticide Wastes, Eds. Krueger, R.F.; Seiber, J.N., ACS Symposium Series 259, Washington, DC, 1984. p. 297-325.
6. Cohen, S.Z.; Eiden, C.; Lorber, M.N. "Monitoring Ground Water for Pesticides in the U.S.A." (in press, this volume).
7. "Pesticide Assessment Guidelines Subdivision N, Chemistry: Environmental Fate," EPA 540/9-82-021, U.S. Environmental Protection Agency, 1982.
8. Federal Register, October 29, 1979, p. 65161.
9. Federal Register, January 1985, p. 1122-1130.
10. Federal Register, October 11, 1983, p. 46228.
11. Registration files maintained by the U.S. Environmental Protection Agency.
12. Federal Insecticide, Fungicide and Rodenticide Act, as amended, 1978; U.S. Environmental Protection Agency.
13. PR Notice 70-15, Notice to Manufacturers, Formulators, Distributors, and Registrants of Economic Poisons. United States Department of Agriculture, June 23, 1970.

RECEIVED March 25, 1986

Author Index

Subject Index

R

Production by Keith B. Belton and Joan C. Cook
Indexing by Keith B. Belton
Jacket design by Pamela Lewis

Elements typeset by Hot Type Ltd., Washington, DC
Printed and bound by Maple Press Co., York, PA